Birkhäuser

Progress in Mathematics

Volume 338

More information about this series at http://www.springer.com/series/4848

Pierre Charollois • Gerard Freixas i Montplet
Vincent Maillot
Editors

Arithmetic L-Functions and Differential Geometric Methods

Regulators IV, May 2016, Paris

Birkhäuser

Editors
Pierre Charollois
Institut de Mathématiques de Jussieu
Sorbonne Université
Paris, France

Gerard Freixas i Montplet
Institut de Mathématiques de Jussieu
CNRS
Paris, France

Vincent Maillot
Institut de Mathématiques de Jussieu
CNRS
Paris, France

ISSN 0743-1643 ISSN 2296-505X (electronic)
Progress in Mathematics
ISBN 978-3-030-65202-9 ISBN 978-3-030-65203-6 (eBook)
https://doi.org/10.1007/978-3-030-65203-6

Mathematics Subject Classification (2020): 11G55, 11G30, 11S80, 14C40, 14F30, 19F27, 58J52

This book is published under the imprint Birkhäuser, www.birkhauser-science.com by the registered company Springer Nature Switzerland AG
The registered company address is: Gewerbestrasse 11, 6330 Cham, Switzerland

Contents

Preface

From early 2012 to late 2016, a string of activities on regulator maps, supported by the French public funding agency ANR (grant ANR-12-BS01-0002), was organised jointly at the Mathematics Institute of Jussieu (IMJ-PRG) in Paris and at the ENS in Lyon. This program culminated with the international conference *Arithmetic L-functions and Differential Geometric Methods* (*Regulators IV*), which was held from May 23 to May 28, 2016, at the Sophie Germain building of the Université de Paris and was organised by Pierre Charollois (ed.), Frédéric Déglise, Gerard Freixas i Montplet (ed.), Xionan Ma and Vincent Maillot (ed.).

The series of *Regulators* conferences itself was initiated in May 1998 at the Mathematisches Forschungsinstitut Oberwolfach and had been pursued by a second occurence in December 2005 at the Banff International Research Station and by a third one in July 2010 at the Universitat de Barcelona. For this forth iteration of the series, members of the scientific committee strongly encouraged the organisers to extend the initial scope of the conference in order to account for the most exciting and recent developments in the field. This choice (incidentally made visible in the long title of the conference) brought at the same place and for one week people coming from very different horizons and resulted into some intense and innovative discussions.

After the meeting, it was felt that the community of mathematicians working in the field of Regulators could benefit from a volume collecting revised versions of some research talks given on this occasion as well as some additional contributions (including original works and specialised surveys) written afterwards. This book is the result of this line of thought. Each paper was refereed and is in final form. We warmly thank all the authors and referees who have generously and patiently contributed.

This conference would not have been possible without the involvement of the IMJ-PRG and its staff. In particular, Élodie Destrebecq went above and beyond the call of duty in taking in charge the concrete organisation of the event. We wish to express our gratitude to everyone who contributed to the success of the conference and the production of these proceedings.

At the time of writing this preface, there have been some conversations regarding a fifth episode of the series of *Regulators* conferences, but these plans haven't come to fruition yet. We hope that this volume will bring additional motivation in making this possible.

The editors

The group picture of some of the participants, from left to right:

Jennifer BALAKRISHNAN
Gerard FREIXAS i MONTPLET
Not identified
Wayne RASKIND
Ramesh SREEKANTAN
José Ignacio BURGOS
Deepam PATEL
Not identified
James LEWIS
Aurélien RODRIGUEZ
Amnon BESSER
Andreas LANGER
Wieslawa NIZIOL
Christopher DENINGER
Giuseppe ANCONA
Masanori ASAKURA
Vincent MAILLOT
Michael NEURURER
Jean-Michel BISMUT
Ishai DAN-COHEN
Javier FRESÁN
Rob DE JEU
Spencer BLOCH
Matt KERR
Xiaonan MA (hidden)
Marie-José BERTIN
Masha VLASENKO
Odile LECACHEUX
Annette HUBER
Madhav NORI

Progress in Mathematics, Vol. 338, 1–30

Regulators of K_1 of Hypergeometric Fibrations

Masanori Asakura and Noriyuki Otsubo

Abstract. We study a deformation of what we call hypergeometric fibrations. Its periods and K_1-regulators are described in terms of hypergeometric functions ${}_3F_2$ in a variable given by the deformation parameter.

Mathematics Subject Classification (2010). 14D07, 19F27, 33C20 (primary), 11G15, 14K22 (secondary).

Keywords. Periods, regulators, hypergeometric functions.

1. Introduction

In [1] and [2] we studied the periods and regulators for a certain class of fibrations, which we call *hypergeometric fibrations* (see §3 for the definition). The purpose of this paper is to extend the main results in [2].

Let $f : X \to \mathbb{P}^1$ be a hypergeometric fibration in the sense of §3.1. Let $\pi : \mathbb{P}^1 \to \mathbb{P}^1$ be a map given by $t \mapsto t^l$ with $l \geq 1$ an integer. Let

$$\begin{array}{ccccc} X^{(l)} & \xrightarrow{i} & X \times_{\pi,\mathbb{P}^1} \mathbb{P}^1 & \longrightarrow & X \\ & \searrow^{f^{(l)}} & \downarrow & & \downarrow f \\ & & \mathbb{P}^1 & \xrightarrow{\pi} & \mathbb{P}^1 \end{array}$$

be a Cartesian diagram with i a desingularization. One of the main results in [2] is the period formula which describes the periods of $X^{(l)}$, and the other is the regulator formula which describes Beilinson's regualtor map on the motivic cohomology group $H^3_{\mathscr{M}}(X^{(l)}, \mathbb{Q}(2))$, especially on elements supported on certain singular fibers of $f^{(l)}$. In particular we described the regulator in terms of the special values of the generalized hypergeometric functions ${}_3F_2\left({\alpha_1,\alpha_2,\alpha_3, \atop \beta_1,\beta_2}; z\right)$ at $z = 1$.

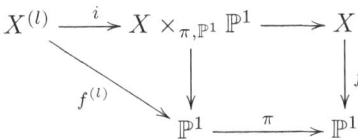

We extend those results in the following way. Our idea is simple, just replacing π with a map π_λ given by $t \mapsto \lambda - t^l$ for $\lambda \in \mathbb{C} \setminus \{0, 1\}$. We then obtain fibrations $f^{(l)}_\lambda : X^{(l)}_\lambda \to \mathbb{P}^1$ in the same way as above, and they are parametrized by λ. We discuss the periods and regulators for $X_\lambda^{(l)}$. Since the fibrations are parametrized by λ, the periods and regulators are no longer complex numbers but analytic functions. The main results of this paper are to describe them in terms of hypergeometric functions (Theorems 4.1, 5.1).

Taking the limits $\lambda \to 0$ of mixed Hodge structures ($\Leftrightarrow$ the nearby cycle cohomology functor $\psi_{\lambda=0}$), one can derive the main results of [2] from our main results. However we make a somewhat strong assumption "$\alpha_1^X \in \mathbb{Z}$" throughout this paper, so that they do *not* cover all of [2].

Our another motivation is the *logarithmic formula* in [4] where we gave a sufficient condition for that the special value of ${}_3F_2$ at $z = 1$ is written by a linear combination of log of algebraic numbers. Theorem 5.9 (=a precise version of Theorem 5.1) enables us to obtain its functional version, namely we can give a sufficient condition for that ${}_3F_2(z)$ is written in terms of the logarithmic functions. This will be discussed in a paper [3].

At the conference "Regulator IV" in Paris (May 2016), S. Bloch asked the first author whether results in the author's talk gave examples to the following question of V. Golyshev.

Question: Let

$$P_{\mathrm{HG}} = D_z \prod_{i=1}^{p-1}(D_z + \beta_i - 1) - z\prod_{i=1}^{p}(D_z + \alpha_i), \quad D_z := z\frac{d}{dz}$$

be the hypergeometric differential operator and let $M = D_S/D_S P_{\mathrm{HG}}$ the D_S-module on $S = \mathbb{P}^1 \setminus \{0, 1, \infty\}$ where D_S denotes the sheaf of differential operators. Suppose that M is reducible, equivalently $\exists \alpha_i \in \mathbb{Z}$ or $\alpha_j - \beta_k \in \mathbb{Z}$ for some j, k, so that there is an exact sequence

$$0 \longrightarrow N \longrightarrow M \longrightarrow Q \longrightarrow 0$$

of D_S-modules. Then does it underly a variation of mixed Hodge structures of geometric origin? If so, does the extension data arise from Beilinson's regulator map on a motivic cohomology group? Moreover, is the regulator described in terms of hypergeometric functions which are solutions of P_{HG}?

See Theorem 5.8. Our regulator formula (Theorem 5.1) gives an affirmative answer in case $p = 3$ and $\alpha_1 = \alpha_2 = 1$. However we do not have a general solution to his question.

Acknowledgements

We would like to thank Spencer Bloch for asking Golyshev's question. This work is supported by JSPS Grant-in-Aid for Scientific Research, 24540001 and 25400007.

Notations

For $\alpha \in \mathbb{C}$ and an integer $n \geq 0$, $(\alpha)_n = \prod_{i=0}^{n-1}(\alpha + i)$ is the Pochhammer symbol and the generalized hypergeometric function is defined by

$$ {}_pF_{p-1}\left(\begin{matrix}\alpha_1,\dots,\alpha_p\\ \beta_1,\dots,\beta_{p-1}\end{matrix};x\right) = \sum_{n=0}^{\infty} \frac{\prod_{i=1}^{p}(\alpha_i)_n}{\prod_{j=1}^{p-1}(\beta_j)_n}\frac{x^n}{n!}. $$

When $p = 2$, this is called the Gauss hypergeometric function. We use the standard notation for the product of values of the gamma function $\Gamma(s)$

$$ \Gamma\left(\begin{matrix}\alpha_1,\dots,\alpha_p\\ \beta_1,\dots,\beta_q\end{matrix}\right) = \frac{\prod_{i=1}^{p}\Gamma(\alpha_i)}{\prod_{j=1}^{q}\Gamma(\beta_j)}. $$

Throughout this paper, we fix an embedding $\overline{\mathbb{Q}} \hookrightarrow \mathbb{C}$, and think $\overline{\mathbb{Q}}$ of being a subfield. For a variety X over $\overline{\mathbb{Q}}$, $H^n_{\mathrm{dR}}(X) = H^n_{\mathrm{dR}}(X/\overline{\mathbb{Q}})$ denotes the algebraic de Rham cohomology and $H^n(X, \mathbb{Q})$ denotes the Betti cohomology of the analytic manifold $X^{an} = (X \times_{\overline{\mathbb{Q}}} \mathbb{C})^{an}$.

2. Betti–de Rham structures, Hodge–de Rham structures and periods

2.1. Betti–de Rham structures and Hodge–de Rham structures

Let k_B, k_{dR} be fields with fixed embeddings $k_B \hookrightarrow \mathbb{C}$ and $k_{\mathrm{dR}} \hookrightarrow \mathbb{C}$. A *Betti–de Rham structure* over (k_B, k_{dR}) (abbreviated BdR) is a datum $(H_B, H_{\mathrm{dR}}, \iota)$ consisting of

- a finite-dimensional vector space H_B (resp. H_{dR}) over k_B (resp. k_{dR}),
- a comparison isomorphism $\iota\colon \mathbb{C} \otimes_{k_{\mathrm{dR}}} H_{\mathrm{dR}} \xrightarrow{\sim} \mathbb{C} \otimes_{k_B} H_B$.

A *Hodge–de Rham structure* over k_{dR} (abbreviated HdR) is a datum $(H_B, H_{\mathrm{dR}}, F^\bullet, \iota)$ consisting of

- a finite-dimensional vector space H_B (resp. H_{dR}) over $\mathbb{Q}$ (resp. k_{dR}),
- a finite decreasing filtration $F^\bullet$ on H_{dR}
- a comparison isomorphism $\iota\colon \mathbb{C} \otimes_{k_{\mathrm{dR}}} H_{\mathrm{dR}} \xrightarrow{\sim} \mathbb{C} \otimes_{\mathbb{Q}} H_B$

such that $(H_B, \mathbb{C}\otimes_{k_{\mathrm{dR}}} H_{\mathrm{dR}}, \mathbb{C}\otimes_{k_{\mathrm{dR}}} F^\bullet, \iota)$ is a Hodge structure in the usual sense. A *mixed Hodge–de Rham structure* $(H_B, W_B, H_{\mathrm{dR}}, F^\bullet, W_{\mathrm{dR}}, \iota)$ over k_{dR} (abbreviated MHdR) is defined in the similar way where W_B (resp. W_{dR}) is a finite increasing filtration on H_B (resp. H_{dR}). The Tate twists $\mathbb{Q}(r) = (\mathbb{Q}, k_{\mathrm{dR}}, F^\bullet, \iota)$ is defined as $F^{-r}k_{\mathrm{dR}} = k_{\mathrm{dR}}$, $F^{-r+1}k_{\mathrm{dR}} = 0$ and the comparison $\iota : k_{\mathrm{dR}} \to \mathbb{C}$ given by $1 \mapsto (2\pi i)^{-r}$. The *dual* and *tensor products* of BdR, HdR and MHdR are defined in the customary way.

In this paper we usually consider the case $k_{\mathrm{dR}} = \overline{\mathbb{Q}} \hookrightarrow \mathbb{C}$ with the fixed embedding.

A *filtered Betti–de Rham structure* over (k_B, k_{dR}) is a datum $(H_B, H_{\mathrm{dR}}, F^\bullet, \iota)$ consisting of a Betti–de Rham structure $(H_B, H_{\mathrm{dR}}, \iota)$ and a finite decreasing filtration $F^\bullet$ on H_{dR}. The category Filt-BdR = Filt-BdR$_{k_B,k_{\mathrm{dR}}}$ of filtered Betti–de

Rham structures over (k_B, k_{dR}) is *not* abelian but exact category. The Yoneda extension groups

$$\mathrm{Ext}^{\bullet}_{\text{Filt-BdR}}(H, H')$$

are defined in the canonical way (cf. [5] 1.1). The following isomorphism is well known (cf. [2] Proposition 2.1).

Proposition 2.1 (Carlson's isomorphism). *Let $H = (H_B, H_{\mathrm{dR}}, F^\bullet, \iota)$ be a filtered Betti–de Rham structure. Then there is a natural isomorphism*

$$\mathrm{Ext}^1_{\text{Filt-BdR}}(k, H) \cong (\mathbb{C} \otimes_{k_{\mathrm{dR}}} H_{\mathrm{dR}})/(F^0 H_{\mathrm{dR}} + \iota^{-1} H_B)$$

where $k = (k_B, k_{\mathrm{dR}}, F^\bullet, \mathrm{id})$ denotes the unit object which is defined as $F^0 k_{\mathrm{dR}} = k_{\mathrm{dR}}$, $F^1 k_{\mathrm{dR}} = 0$ and the comparison is the identity.

2.2. Periods

For a Betti–de Rham structure $H = (H_B, H_{\mathrm{dR}}, \iota)$, the *period matrix* of H is defined to be the representation matrix of ι with respect to the k_B, k_{dR}-lattices H_B, H_{dR}, and we denote by

$$\mathrm{Per}(H) \in \mathrm{GL}_r(k_B)\backslash \mathrm{GL}_r(\mathbb{C})/\mathrm{GL}_r(k_{\mathrm{dR}})$$

where r is the rank of H.

2.3. Multiplication

A *multiplication* on a Betti–de Rham structure H by a commutative $\mathbb{Q}$-algebra R is defined as a ring homomorphism $R \to \mathrm{End}_{\mathrm{BdR}}(H)$ to the endomorphism ring of Betti–de Rham structures. The *tensor product* $H_1 \otimes_R H_2$ over R is defined to be

$$H_1 \otimes_R H_2 = (H_{1,B} \otimes_{k_B \otimes R} H_{2,B}, H_{1,\mathrm{dR}} \otimes_{k_{\mathrm{dR}} \otimes R} H_{2,\mathrm{dR}}, \iota_1 \otimes \iota_2)$$

endowed with multiplication by R. The multiplication on the dual Betti–de Rham structure

$$H^* = (\mathrm{Hom}_{k_B}(H_B, k_B), \mathrm{Hom}_{k_{\mathrm{dR}}}(H_{\mathrm{dR}}, k_{\mathrm{dR}}), \iota)$$

is defined in such a way that $r\phi := \phi \circ r$ for $\phi \in \mathrm{Hom}(H_B, k_B)$ and $r \in R$.

A multiplication on a filtered BdR, HdR, MHdR and its χ-parts are defined in the same way as above.

Assume $\mathrm{Im}(k_B \hookrightarrow \mathbb{C}) \subset \overline{\mathbb{Q}}$ and $\mathrm{Im}(k_{\mathrm{dR}} \hookrightarrow \mathbb{C}) \subset \overline{\mathbb{Q}}$ (note that $\overline{\mathbb{Q}} \hookrightarrow \mathbb{C}$ is fixed throughout the paper). For a homomorphism $\chi : R \to \overline{\mathbb{Q}}$, we define the *$\chi$-part* of a BdR structure H as

$$H(\chi) := (H_B(\chi), H_{\mathrm{dR}}(\chi), \iota)$$

$$H_B(\chi) := \overline{\mathbb{Q}} \otimes_{k_B \otimes R} H_B, \quad H_{\mathrm{dR}}(\chi) := \overline{\mathbb{Q}} \otimes_{k_{\mathrm{dR}} \otimes R} H_{\mathrm{dR}},$$

where $k_B \otimes R \to \overline{\mathbb{Q}}$ and $k_{\mathrm{dR}} \otimes R \to \overline{\mathbb{Q}}$ are induced from χ and the embeddings $k_B \hookrightarrow \overline{\mathbb{Q}}$ and $k_{\mathrm{dR}} \hookrightarrow \overline{\mathbb{Q}}$. Then $H(\chi)$ is a BdR over $(\overline{\mathbb{Q}}, \overline{\mathbb{Q}})$. We call its period matrix

$$\mathrm{Per}(H(\chi)) \in \mathrm{GL}_r(\overline{\mathbb{Q}})\backslash \mathrm{GL}_r(\mathbb{C})/\mathrm{GL}_r(\overline{\mathbb{Q}})$$

the *χ-part of the period matrix* of H. The χ-part of filtered BdR, HdR and MHdR are defined in the same way.

Suppose that R is a semisimple and finite-dimensional $\mathbb{Q}$-algebra. Then the functor Filt-BdR$_{k_B,k_{\mathrm{dR}}} \to$ Filt-BdR$_{\overline{\mathbb{Q}},\overline{\mathbb{Q}}}$ given by $H \mapsto H(\chi)$ is exact. Composing with the forgetting functor $\mathrm{MHdR}_{k_{\mathrm{dR}}} \to$ Filt-BdR$_{\mathbb{Q},k_{\mathrm{dR}}}$ one has a map

$$\mathrm{Ext}^1_{\mathrm{MHdR}_{k_{\mathrm{dR}}}}(\mathbb{Q}, H) \longrightarrow \mathrm{Ext}^1_{\text{Filt-BdR}_{\overline{\mathbb{Q}},\overline{\mathbb{Q}}}}(\overline{\mathbb{Q}}, H(\chi)), \quad M \longmapsto M(\chi) \tag{2.1}$$

and we call $M(\chi)$ the *χ-part of extension class M*.

Let X be a smooth projective variety over k_{dR}. If $i \neq 2j$, one can construct the Beilinson regulator map

$$\mathrm{reg} : H^i_{\mathscr{M}}(X, \mathbb{Q}(j)) \longrightarrow \mathrm{Ext}^1_{\mathrm{MHdR}}(\mathbb{Q}, H^{i-1}(X, \mathbb{Q}(j))) \tag{2.2}$$

to the group of 1-extensions of MHdR's in the following way. Firstly there is the (usual) Beilinson regulator map

$$H^i_{\mathscr{M}}(X, \mathbb{Q}(j)) \longrightarrow \mathrm{Ext}^1_{\mathrm{MHM}(X_{\mathbb{C}})}(\mathbb{Q}, \mathbb{Q}(j)) \cong \mathrm{Ext}^1_{\mathrm{MHS}}(\mathbb{Q}, H^{i-1}(X_{\mathbb{C}}, \mathbb{Q}(j))) \tag{2.3}$$

to the extension group of mixed Hodge modules on $X_{\mathbb{C}} := X \times_{k_{\mathrm{dR}}} \mathbb{C}$ where the isomorphism follows from the assumption $i \neq 2j$ ([7]). Note that $\mathrm{MHS} = \mathrm{MHdR}_{\mathbb{Q},\mathbb{C}}$ in our notation. On the other hand, let $\mathrm{MFW}(D_X)$ be the category of the triplet $(M, F^\bullet, W_\bullet)$ where M is a regular holonomic D_X-module, $F^\bullet$ is a good filtration and $W_\bullet$ is a finite increasing filtration on M (see [8] 1.1. or [9] 1.8 (ii) for details). Then one also has the natural map

$$\begin{aligned} H^i_{\mathscr{M}}(X, \mathbb{Q}(j)) &\longrightarrow \mathrm{Ext}^i_{\mathrm{MFW}(D_X)}(\mathscr{O}_X, \mathscr{O}_X(j)) \\ &\cong \mathrm{Ext}^1_{\text{Filt-BdR}_{\mathbb{Q},k_{\mathrm{dR}}}}(\mathbb{Q}, H^{i-1}(X, \mathbb{Q}(j))). \end{aligned} \tag{2.4}$$

Since the construction of (2.4) is compatible with that of (2.3), one can associate a 1-extension of MHdR to each element of $H^i_{\mathscr{M}}(X, \mathbb{Q}(j))$. Thus the regulator map (2.2) is defined.

Suppose that the Hodge–de Rham structure $H := H^{i-1}(X, \mathbb{Q}(j))$ over k_{dR} has a multiplication by R. Then we call the composition

$$\mathrm{reg}(\chi) : H^i_{\mathscr{M}}(X, \mathbb{Q}(j)) \longrightarrow \mathrm{Ext}^1_{\mathrm{MHdR}}(\mathbb{Q}, H) \overset{(2.1)}{\longrightarrow} \mathrm{Ext}^1_{\text{Filt-BdR}_{\overline{\mathbb{Q}},\overline{\mathbb{Q}}}}(\overline{\mathbb{Q}}, H(\chi))$$

the *χ-part of regulator map*.

2.4. Variations of Hodge–de Rham structures

Let S be a smooth variety over k_{dR}. A *filtered Betti–de Rham structure on S* consists of a datum $(H_B, H_{\mathrm{dR}}, F^\bullet, \nabla, \iota)$ where

- H_B is a local system of finite-dimensional k_B-modules on S^{an},
- H_{dR} is a locally free $\mathscr{O}_S$-module of finite rank, and $F^\bullet$ is a finite decreasing filtration which is locally a direct summand,
- $(H_{\mathrm{dR}}, \nabla)$ is a connection with regular singularities on S such that $\nabla(F^p) \subset \Omega^1_S \otimes F^{p-1}$,
- ι: $\mathscr{O}^{an}_S \otimes_{a^{-1}\mathscr{O}_S} a^{-1}H_{\mathrm{dR}} \overset{\sim}{\to} \mathscr{O}^{an}_S \otimes_{k_B} H_B$ is a comparison isomorphism such that ∇ annihilates the lattice H_B, where $a : S^{an} \to S^{\mathrm{zar}}$ is the canonical map from the analytic site to the Zariski site and $\mathscr{O}^{an}_S$ denotes the sheaf of analytic functions on S^{an}.

A filtered BdR on S is called a *variation of Hodge–de Rham structure* (abbreviated VHdR) if $k_B = \mathbb{Q}$ and $(H_B, \mathscr{O}_S^{an} \otimes_{\mathscr{O}_S} H_{\mathrm{dR}}, \mathscr{O}_S^{an} \otimes_{\mathscr{O}_S} F^\bullet, \iota, \nabla)$ is a variation of Hodge structure in the usual sense. We also define, in a customary way, a *variation of mixed Hodge–de Rham structure* (abbreviated VMHdR) on S which consists of a datum $(H_B, W_\bullet^B, H_{\mathrm{dR}}, F^\bullet, W_\bullet^{\mathrm{dR}}, \iota, \nabla)$.

3. Hypergeometric fibrations

In what follows we work over the base field $k_{\mathrm{dR}} = \overline{\mathbb{Q}}$.

3.1. Definition

Let R be a finite-dimensional semisimple $\mathbb{Q}$-algebra. Let $e : R \to E$ be a surjection onto a number field E. Let X be a smooth projective variety over k_{dR}, and $f : X \to \mathbb{P}^1$ a surjective map. We say f is a *hypergeometric fibration with multiplication by* (R, e) if it is endowed with a multiplication on $R^1 f_*\mathbb{Q}|_U$ by R where $U \subset \mathbb{P}^1$ is the maximal Zariski open set such that f is smooth over U and the following conditions hold. We fix an inhomogeneous coordinate $t \in \mathbb{P}^1$.

- f is smooth over $\mathbb{P}^1 \setminus \{0, 1, \infty\}$.
- $\dim_E(R^1 f_*\mathbb{Q})(e) = 2$ where we write $V(e) := E \otimes_{e,R} V$ the e-part.
- Let $\mathrm{Pic}_f^0 \to \mathbb{P}^1 \setminus \{0, 1, \infty\}$ be the Picard fibration whose general fiber is the Picard variety $\mathrm{Pic}^0(f^{-1}(t))$, and let $\mathrm{Pic}_f^0(e)$ be the component associated to the e-part $(R^1 f_*\mathbb{Q})(e)$ (this is well defined up to isogeny). Then $\mathrm{Pic}_f^0(e) \to \mathbb{P}^1 \setminus \{0, 1, \infty\}$ has a totally degenerate semistable reduction at $t = 1$.

The last condition is equivalent to say that the local monodromy T on $(R^1 f_*\mathbb{Q})(e)$ at $t = 1$ is unipotent and the rank of log monodromy $N := \log(T)$ is maximal, namely $\mathrm{rank}(N) = \frac{1}{2}\dim_{\mathbb{Q}}(R^1 f_*\mathbb{Q})(e)$ $(= [E : \mathbb{Q}]$ by the second condition).

Example 3.1 (Gauss type). Let $f : X \to \mathbb{P}^1$ be the fibration over $\overline{\mathbb{Q}}$ whose general fiber is defined by an affine equation

$$y^N = x^a(1-x)^b(t-x)^{N-b}$$

with $0 < a, b < N$ and $\gcd(N, a, b) = 1$. Let $\mu_N \subset \overline{\mathbb{Q}}$ be the group of Nth roots of unity. It gives automorphisms $(x, y, t) \mapsto (x, \zeta_N y, t)$ for $\zeta_N \in \mu_N$, and then it defines a multiplication by $R = \mathbb{Q}[\mu_N]$ (=group ring) on $R^1 f_*\mathbb{Q}$. Then one can show that f is a hypergeometric fibration with multiplication by (R, e) if and only $d := \sharp\mathrm{Ker}[e : \mu_N \to E^\times]$ satisfies $ad/N,\ bd/N \notin \mathbb{Z}$.

Example 3.2 (Fermat type). Let $f : X \to \mathbb{P}^1$ be the fibration over $\overline{\mathbb{Q}}$ defined by an affine equation

$$(x^n - 1)(y^m - 1) = 1 - t.$$

The group ring $R = \mathbb{Q}[\mu_n \times \mu_m]$ acts on $R^1 f_*\mathbb{Q}$ in a natural way. Then f is a hypergeometric fibration with multiplication by (R, e) if and only if $e : \mathbb{Q}[\mu_n \times \mu_m] \to E$ does not factor through the projections $\mu_n \times \mu_m \to \mu_n$ nor $\mu_n \times \mu_m \to \mu_m$.

The reason why we call this "Fermat type" is the following. Letting $u = x^{-1}$, $v = y^{-1}$ and $s = tx^{-n}y^{-m}$,

$$(x^n - 1)(y^m - 1) = 1 - t \iff u^n + v^m = 1 + s.$$

3.2. Basic properties

We sum up some properties on our hypergeometric fibrations, which will be used in later sections.

Proposition 3.3 ([2], Prop. 3.7).

$$\dim_{\overline{\mathbb{Q}}} F^1 H^1_{\mathrm{dR}}(X_t)(\chi) = 1 \quad \textit{and} \quad \dim_{\overline{\mathbb{Q}}} \mathrm{Gr}^0_F H^1_{\mathrm{dR}}(X_t)(\chi) = 1$$

where X_t is a general fiber.

Proposition 3.4 (loc. cit. Lemmas 3.6, 5.1). *$(R^1 f_*\mathbb{Q})(\chi)$ is an irreducible $\overline{\mathbb{Q}}[\pi_1(\mathbb{P}^1 \setminus \{0, 1, \infty\})]$-module. Moreover let $\alpha_0^\chi, \beta_0^\chi$ (resp. α^χ, β^χ) be rational numbers such that $e^{2\pi i \alpha_0^\chi}$, $e^{2\pi i \beta_0^\chi}$ (resp. $e^{2\pi i \alpha^\chi}$, $e^{2\pi i \beta^\chi}$) are eigenvalues of the local monodromy on $(R^1 f_*\mathbb{Q})(\chi)$ at $t = 0$ (resp. $t = \infty$). Then, none of $\alpha_0^\chi + \alpha^\chi$, $\alpha_0^\chi + \beta^\chi$, $\beta_0^\chi + \alpha^\chi$, $\beta_0^\chi + \beta^\chi$ is an integer.*

Proposition 3.5 (loc. cit. (3.3), (3.4)). *Let $\psi_{t=1}$ denote the nearby cohomology functor at $t = 1$ and let $W_\bullet$ be the weight monodromy filtration induced by the log monodromy $N_1 = \log(T_1)$. Then there are isomorphisms*

$$\begin{aligned}
\mathrm{Gr}^W_2 \psi_{t=1} R^1 f_*\mathbb{Q}(e) &= \mathrm{Coker}(N_1) \cong E \otimes \mathbb{Q}(-2),\\
\mathrm{Gr}^W_0 \psi_{t=1} R^1 f_*\mathbb{Q}(e) &= \mathrm{Ker}(N_1) \cong E,\\
\mathrm{Gr}^W_j \psi_{t=1} R^1 f_*\mathbb{Q}(e) &= 0, \quad j \neq 0, 2
\end{aligned}$$

of Hodge–de Rham structure with compatible E-action, where E is endowed with a trivial Hodge–de Rham structure of type $(0, 0)$.

4. Period formula

4.1. Setting

Let R_0 be a finite-dimensional semisimple $\mathbb{Q}$-algebra and $e_0 : R_0 \to E_0$ be a surjection onto a number field E_0. Let $f : X \to \mathbb{P}^1$ be a hypergeometric fibration over $\overline{\mathbb{Q}}$ with multiplication by (R_0, e_0) in the sense of §3.1.

Let $S := \mathbb{A}^1 \setminus \{0, 1\}$ be defined over $\overline{\mathbb{Q}}$ with coordinate λ. Write $\mathbb{P}^1_S := \mathbb{P}^1 \times S$. Put $\mathscr{U} := (\mathbb{A}^1 \setminus \{0, 1\} \times S) \setminus \Delta$ where Δ is the diagonal subscheme. Let $l \geq 1$ be an integer. Let $\pi : \mathbb{P}^1_S \to \mathbb{P}^1_S$ be a morphism over S given by $(s, \lambda) \mapsto (\lambda - s^l, \lambda)$. Then we consider a variation of Hodge–de Rham structures

$$\mathscr{M} := \pi_*\mathbb{Q} \otimes \mathrm{pr}_1^* R^1 f_*\mathbb{Q}|_{\mathscr{U}}, \quad \mathrm{pr}_1 : \mathbb{P}^1_S = \mathbb{P}^1 \times S \to \mathbb{P}^1$$

on $\mathscr{U}$ and a variation of mixed Hodge–de Rham structures

$$\mathscr{H} := R^1 \mathrm{pr}_{2*} \mathscr{M} = (\mathscr{H}_B, W^B_\bullet, \mathscr{H}_{\mathrm{dR}}, F^\bullet, W^{\mathrm{dR}}_\bullet, \nabla, \iota), \quad \mathrm{pr}_2 : \mathscr{U} \to S$$

on S. Since $\mathscr{M}$ is a variation of Hodge–de Rham structures of pure weight 1, the weights of $\mathscr{H}$ are at most 2, 3 and 4. We have an exact sequence

$$0 \longrightarrow W_2\mathscr{H} \longrightarrow \mathscr{H} \longrightarrow \mathscr{H}/W_2\mathscr{H} \longrightarrow 0 \tag{4.1}$$

with $W_2\mathscr{H}$ a variation of Hodge–de Rham structures of pure weight 2.

The group $\mu_l \subset \overline{\mathbb{Q}}^\times$ of lth roots of unity acts on the cyclic covering π. Thus the group ring $R := R_0[\mu_l]$ acts on the sheaf $\mathscr{M}$ and hence on $\mathscr{H}$. In what follows we fix $e : R \to E$ a surjection onto a number field E such that $\mathrm{Ker}(e) \supset \mathrm{Ker}(e_0)$. There is a unique embedding $E_0 \hookrightarrow E$ making the diagram

$$\begin{array}{ccc} R_0 & \xrightarrow{e_0} & E_0 \\ \downarrow & & \downarrow \\ R & \xrightarrow{e} & E \end{array}$$

commutative. Then the e-part

$$\mathscr{M}(e) := E \otimes_{e,R} \mathscr{M} \cong E \otimes_{E_0[\mu_l]} (\pi_*\mathbb{Q} \otimes R^1 f_*\mathbb{Q}(e_0))$$

is of rank 2 over E.

Let $\chi : R \to \overline{\mathbb{Q}}$ be a homomorphism factoring through e. This also induces $R_0 \to \overline{\mathbb{Q}}$ which we also write χ by abuse of notation. Define $k \in \{0, 1, \dots, l-1\}$ by $\chi(\zeta_l) = \zeta_l^k$ for all $\zeta_l \in \mu_l$. Put $q^\chi := k/l$. Moreover, let α_0^χ, β_0^χ (resp. α^χ, β^χ) be rational numbers in the interval $[0, 1)$ such that $e^{2\pi i \alpha_0^\chi}$ and $e^{2\pi i \beta_0^\chi}$ (resp. $e^{2\pi i \alpha^\chi}$ and $e^{2\pi i \beta^\chi}$) are eigenvalues of the local monodromy T_0 at $t = 0$ (resp. T_∞ at $t = \infty$) on the χ-part $(R^1 f_*\mathbb{Q})(\chi) = \overline{\mathbb{Q}} \otimes_{\chi, R_0} R^1 f_*\mathbb{Q}$. Equivalently, these are congruent mod $\mathbb{Z}$ to the eigenvalues of the residue $\mathrm{Res}_{t=0}(\nabla)$ (resp. $\mathrm{Res}_{t=\infty}(\nabla)$) of the connection on the χ-part of the bundle $R^1 f_*\Omega^\bullet_{X/\mathbb{P}^1}|_{\mathbb{P}^1\setminus\{0,1,\infty\}}$. Note that the local monodromy T_1 at $t = 1$ is unipotent. Since $T_0 T_1 T_\infty$ is the identity, we have

$$\alpha_0^\chi + \beta_0^\chi + \alpha^\chi + \beta^\chi \in \mathbb{Z}.$$

4.2. Theorem on periods

Let $\mathscr{O}^{an}$ be the sheaf of analytic functions on S^{an}, $\mathscr{O}^{\mathrm{zar}}$ the Zariski sheaf of rational functions (with coefficients in $\overline{\mathbb{Q}}$) on S with coordinate λ. Let a be the canonical morphism from the analytic site to the Zariski site. We put

$$W_2\mathscr{H}_{\mathrm{dR}}^{an} := \mathscr{O}^{an} \otimes_{a^{-1}\mathscr{O}^{\mathrm{zar}}} a^{-1} W_2\mathscr{H}_{\mathrm{dR}}, \quad W_2\mathscr{H}_B^{an} := \mathscr{O}^{an} \otimes_{\mathbb{Q}} W_2\mathscr{H}_B$$

sheaves on the analytic site. The comparison isomorphism of $W_2\mathscr{H}$ gives an analytic section

$$\iota \in \Gamma(S^{an}, \mathit{Isom}(W_2\mathscr{H}_{\mathrm{dR}}^{an}, W_2\mathscr{H}_B^{an})).$$

There is a canonical map ($d := \mathrm{rank} W_2\mathscr{H}$)

$$\mathit{Isom}(W_2\mathscr{H}_{\mathrm{dR}}^{an}, W_2\mathscr{H}_B^{an}) \longrightarrow \mathrm{GL}_d(\mathbb{Q})\backslash \mathrm{GL}_d(\mathscr{O}^{an})/\mathrm{GL}_d(a^{-1}\mathscr{O}^{\mathrm{zar}})$$

of sheaves by associating the representation matrices with respect to the lattices $W_2\mathscr{H}_B$ and $W_2\mathscr{H}_{\mathrm{dR}}$.

We call the image of ι the *period matrix* of $W_2\mathscr{H}$:

$$\mathrm{Per}(W_2\mathscr{H}) \in \Gamma(S^{an}, \mathrm{GL}_d(\mathbb{Q})\backslash \mathrm{GL}_d(\mathscr{O}^{an})/\mathrm{GL}_d(a^{-1}\mathscr{O}^{\mathrm{zar}})).$$

The χ-part of $W_2\mathscr{H}$ defines the χ-part of the period matrix ($r := \mathrm{rank} W_2\mathscr{H}(\chi)$)

$$\mathrm{Per}(W_2\mathscr{H}(\chi)) \in \Gamma(S^{an}, \mathrm{GL}_r(\overline{\mathbb{Q}})\backslash \mathrm{GL}_r(\mathscr{O}^{an})/\mathrm{GL}_r(a^{-1}\mathscr{O}^{\mathrm{zar}})).$$

Theorem 4.1 (Period formula). *Assume $\alpha_0^\chi = 0$ ($\Leftrightarrow \beta_0^\chi + \alpha^\chi + \beta^\chi \in \mathbb{Z}$) and that $q^\chi \not\equiv 0$, α^χ, β^χ, $\alpha^\chi + \beta^\chi$ (mod $\mathbb{Z}$). Then the rank of $W_2\mathscr{H}(\chi)$ is 2. For some $\mu > 1$, $\mu \equiv q^\chi$ (mod $\mathbb{Z}$), the period matrix is locally given by*

$$\mathrm{Per}(W_2\mathscr{H}(\chi)) = 2\pi i \begin{pmatrix} \Theta F_\mu(\lambda) & \Theta G_\mu(\lambda) \\ \partial_\lambda \Theta F_\mu(\lambda) & \partial_\lambda \Theta G_\mu(\lambda) \end{pmatrix},$$

where we put

$$F_\mu(\lambda) := \frac{1}{\mu}(\lambda - 1)^\mu {}_2F_1\left({\alpha^\chi, \beta^\chi \atop \mu + 1} ; 1 - \lambda \right),$$

$$G_\mu(\lambda) := (-1)^\mu \Gamma\left({\mu, \mu + 1 - \alpha^\chi - \beta^\chi \atop \mu + 1 - \alpha^\chi, \mu + 1 - \beta^\chi} \right) {}_2F_1\left({\alpha^\chi - \mu, \beta^\chi - \mu \atop \alpha^\chi + \beta^\chi - \mu} ; \lambda \right),$$

and Θ is a differential operator of the form

$$\Theta = q(\lambda) + r(\lambda)\partial_\lambda, \quad q(\lambda), r(\lambda) \in \overline{\mathbb{Q}}[\lambda, 1/\lambda(\lambda - 1)].$$

4.3. Proof of period formula: Part 1

We first show $\dim_{\overline{\mathbb{Q}}} W_2\mathscr{H}(\chi) = 2$. We write $U_a := \mathrm{pr}_2^{-1}(a) \cong \mathbb{P}^1 \setminus \{0, 1, a, \infty\}$ for $a \in S^{an} = \mathbb{C} \setminus \{0, 1\}$ where $\mathrm{pr}_2 : \mathscr{U} \to S$. Moreover we put the fibers

$$\mathscr{M}_a := \mathscr{M}|_{\mathrm{pr}_2^{-1}(a)} \cong \pi_{a*}\mathbb{Q} \otimes R^1 f_* \mathbb{Q}, \quad H_a := \mathscr{H}|_{\{a\}} \cong H^1(\mathrm{pr}_2^{-1}(a), \mathscr{M}_a),$$

where $\pi_a : \mathbb{P}^1 \to \mathbb{P}^1$ is the map given by $s \mapsto a - s^l$. When $a \in \overline{\mathbb{Q}} \setminus \{0, 1\}$, we endow $\mathscr{M}_a$ and $\mathscr{H}_a$ with HdR structure over $\overline{\mathbb{Q}}$ induced from the $\overline{\mathbb{Q}}$-frames on $\mathscr{M}_{\mathrm{dR}}$ and $\mathscr{H}_{\mathrm{dR}}$ respectively. We then want to show $\dim_{\overline{\mathbb{Q}}} W_2 H_a(\chi) = 2$ for $a \in \overline{\mathbb{Q}} \setminus \{0, 1\}$. The weight filtration induces an exact sequence

$$0 \longrightarrow W_2 H_a(\chi) \longrightarrow H_a(\chi) \longrightarrow H_a(\chi)/W_2 H_a(\chi) \longrightarrow 0.$$

There are canonical isomorphisms

$$W_2 H_a \cong H^1(\mathbb{P}^1, j_* \mathscr{M}_a), \quad j : \mathbb{P}^1 \setminus \{0, 1, a, \infty\} \hookrightarrow \mathbb{P}^1, \tag{4.2}$$

$$H_a / W_2 H_a \cong H^0(\mathbb{P}^1, R^1 j_* \mathscr{M}_a) \tag{4.3}$$

of mixed Hodge–de Rham structures. Let $\varepsilon_k : \mathbb{Q}[\mu_l] \to \overline{\mathbb{Q}}$ be given by $\varepsilon(\zeta_l) = \zeta_l^k$ and $\overline{\mathbb{Q}}(\varepsilon_k) := \overline{\mathbb{Q}} \otimes_{\varepsilon_k, \mathbb{Q}[\mu_l]} \pi_* \mathbb{Q}$ a one-dimensional local system on $\mathbb{P}^1 \setminus \{a, \infty\}$. Then there is a natural isomorphism

$$\mathscr{M}_a(\chi) \cong \overline{\mathbb{Q}}(\varepsilon_k) \otimes_{\overline{\mathbb{Q}}} (R^1 f_* \mathbb{Q})(\chi)$$

of $\pi_1(\mathbb{P}^1\setminus\{0,1,a,\infty\})$-modules. Since $(R^1f_*\mathbb{Q})(\chi)$ is an irreducible $\pi_1(\mathbb{P}^1\setminus\{0,1,\infty\})$-module (Proposition 3.4), so is $\mathscr{M}_a(\chi)$ as a $\pi_1(\mathbb{P}^1\setminus\{0,1,a,\infty\})$-module. In particular $H^0(\mathrm{pr}_2^{-1}(a),\mathscr{M}_a(\chi))=0$. Hence

$$\begin{aligned}\dim_{\overline{\mathbb{Q}}} H_a(\chi) &= \dim_{\overline{\mathbb{Q}}} H^1(\mathrm{pr}_2^{-1}(a),\mathscr{M}_a(\chi)) = -\chi(\mathrm{pr}_2^{-1}(a),\mathscr{M}_a(\chi))\\ &= -\chi^{\mathrm{top}}(\mathrm{pr}_2^{-1}(a))\cdot\dim_{\overline{\mathbb{Q}}}\mathscr{M}_a(\chi) = -(-2)\cdot 2 = 4.\end{aligned}$$

Thus $\dim_{\overline{\mathbb{Q}}} W_2H_a(\chi)=2 \Leftrightarrow \dim_{\overline{\mathbb{Q}}} H_a(\chi)/W_2=2$.

Let T_0,T_1,T_a,T_∞ be the local monodromy on $\mathscr{M}_a(\chi)$ at $t=0,1,a,\infty$, respectively. T_1 is unipotent with trivial action on $\overline{\mathbb{Q}}(\varepsilon_k)$, and T_a is multiplication by $e^{2\pi i q^\chi}$ with trivial action on $(R^1f_*\mathbb{Q})(\chi)$. The eigenvalues of T_0 (resp. T_∞) are

$$e^{2\pi i\alpha_0^\chi},\ e^{2\pi i\beta_0^\chi}\quad(\text{resp. } e^{2\pi i(-q^\chi+\alpha^\chi)},\ e^{2\pi i(-q^\chi+\beta^\chi)}).$$

Recall (4.3). We then have

$$\begin{aligned}H_a/W_2 &\cong H^0(\mathbb{P}^1,R^1j_*\mathscr{M}_a)\\ &\cong \bigoplus_{p=0,1,a,\infty}\mathrm{Coker}[T_p-1:\psi_{t=p}\mathscr{M}_a\to\psi_{t=p}\mathscr{M}_a\otimes\mathbb{Q}(-1)]\end{aligned}$$

where $\psi_{t=p}$ denotes the nearby cohomology at $t=p$. By the assumption $q^\chi\not\equiv 0$, α^χ, β^χ (mod $\mathbb{Z}$), T_a and T_∞ have no eigenvalue 1 on the χ-part of $\psi\mathscr{M}_a$. Hence $\mathrm{Coker}[T_a-1]=\mathrm{Coker}[T_\infty-1]=0$. Note

$$\begin{aligned}&\mathrm{Coker}[T_p-1:\psi_{t=p}\mathscr{M}_a(\chi)\to\psi_{t=p}\mathscr{M}_a(\chi)\otimes\mathbb{Q}(-1)]\\ &\cong\overline{\mathbb{Q}}(\varepsilon_k)\otimes_{\overline{\mathbb{Q}}}\mathrm{Coker}[T_p-1:(R^1f_*\mathbb{Q})(\chi)\to(R^1f_*\mathbb{Q})(\chi)],\quad p=0,1.\end{aligned}$$

It follows from Proposition 3.5 that we have $\dim_{\overline{\mathbb{Q}}}\mathrm{Coker}(T_1-1)=1$. There remains to show $\dim_{\overline{\mathbb{Q}}}\mathrm{Coker}(T_0-1)=1$. By the assumption $\alpha_0^\chi=0$, one has $\dim_{\overline{\mathbb{Q}}}\mathrm{Coker}(T_0-1)\geq 1$. If $\dim_{\overline{\mathbb{Q}}}\mathrm{Coker}(T_0-1)=2$ this means that T_0 is trivial on $(R^1f_*\mathbb{Q})(\chi)$. Then $(R^1f_*\mathbb{Q})(\chi)$ cannot be irreducible as $\overline{\mathbb{Q}}[\pi_1(\mathbb{P}^1\setminus\{0,1,\infty\})]$-module. This contradicts with Proposition 3.4. We thus have $\dim_{\overline{\mathbb{Q}}}\mathrm{Coker}(T_0-1)=1$, and hence $\dim_{\overline{\mathbb{Q}}} H_a(\chi)/W_2=2$. This completes the proof of $\dim_{\overline{\mathbb{Q}}} W_2H_a(\chi)=2$.

In the discussion above, we obtained the following.

Proposition 4.2. *Assume* $\alpha_0^\chi\in\mathbb{Z}$ *and* $q^\chi\not\equiv 0$, α^χ, β^χ (mod $\mathbb{Z}$). *Then there is an isomorphism*

$$\mathscr{H}(e)/W_2\cong\bigoplus_{p=0,1}\mathrm{Coker}[T_p-1:\psi_{t=p}\mathscr{M}(e)\to\psi_{t=p}\mathscr{M}(e)\otimes\mathbb{Q}(-1)]$$

of variations of mixed Hodge–de Rham structures on $\mathbb{P}^1\setminus\{0,1,\infty\}$.

Each $\mathrm{Coker}(T_p-1)$ *is one-dimensional over* E. *Moreover,* $\mathrm{Coker}(T_1-1)$ *is endowed with a Hodge–de Rham structure of type* $(2,2)$ (*Proposition 3.5*).

4.4. Relative 1-form ω^χ

The χ-part $(f_*\Omega^1_{X/\mathbb{P}^1})(\chi)|_{\mathbb{P}^1\setminus\{0,1,\infty\}}$ of the Hodge bundle has rank one (Proposition 3.3). Hence it is a trivial line bundle on $\mathbb{P}^1 \setminus \{0,1,\infty\}$. In what follows we fix a relative 1-form

$$\omega^\chi \in \Gamma(\mathbb{P}^1 \setminus \{0,1,\infty\}, (f_*\Omega^1_{X/\mathbb{P}^1})(\chi))$$

with coefficients in $\overline{\mathbb{Q}}$ which is everywhere nonzero (until the end of the paper). Let $X_t = f^{-1}(t)$ be the general fiber. We fix (nonzero) homology cycles

$$\gamma = \gamma_t \in H_1(X_t, \overline{\mathbb{Q}})(\chi) \cap \mathrm{Ker}(T_0 - 1), \quad \delta = \delta_t \in H_1(X_t, \overline{\mathbb{Q}})(\chi) \cap \mathrm{Ker}(T_1 - 1).$$

Note that each $H_1(X_t, \overline{\mathbb{Q}})(\chi) \cap \mathrm{Ker}(T_p - 1)$ is one-dimensional over $\overline{\mathbb{Q}}$ (Proposition 4.2).

Lemma 4.3 (Key Lemma). *There is a differential operator $\theta = p_0(t) + p_1(t)\frac{d}{dt}$ with $p_i(t) \in \overline{\mathbb{Q}}[t]$ such that*

$$\int_\gamma \omega^\chi = B(\alpha^\chi, \beta^\chi) \cdot \theta\, {}_2F_1\left(\begin{matrix}\alpha^\chi, \beta^\chi \\ \alpha^\chi + \beta^\chi\end{matrix}; t\right), \quad \int_\delta \omega^\chi = 2\pi i \cdot \theta\, {}_2F_1\left(\begin{matrix}\alpha^\chi, \beta^\chi \\ 1\end{matrix}; 1-t\right). \tag{4.4}$$

Here $B(\alpha, \beta) := \Gamma(\alpha)\Gamma(\beta)/\Gamma(\alpha+\beta)$ is the beta function.

Proof. Since δ is T_1-invariant, $\int_\delta \omega^\chi$ is a single-valued meromorphic function at $t = 1$. So is $\int_\gamma \omega^\chi$ at $t = 0$ as γ is T_0-invariant. Therefore (4.4) follows from [2], Lemmas 5.2, 5.3 and 5.4. □

The differential equation

$$(t(D+\alpha)(D+\beta) - (D+\alpha+\beta-1)D)f = 0, \quad D = t\frac{d}{dt}$$

has the Riemann scheme

$$\left\{\begin{matrix} t=0 & t=1 & t=\infty \\ 0 & 0 & \alpha \\ 1-\alpha-\beta & 0 & \beta \end{matrix}\right\}.$$

Among Kummer's 24 solutions, we used in the preceding lemma

$$f_1(t) := {}_2F_1\left(\begin{matrix}\alpha, \beta \\ \alpha+\beta\end{matrix}; t\right), \quad f_2(t) := {}_2F_1\left(\begin{matrix}\alpha, \beta \\ 1\end{matrix}; 1-t\right). \tag{4.5}$$

Later in Section 5.4, we will use the solution

$$f_3(t) := t^{1-\alpha-\beta}{}_2F_1\left(\begin{matrix}1-\alpha, 1-\beta \\ 2-\alpha-\beta\end{matrix}; t\right), \tag{4.6}$$

which has the characteristic exponent $1-\alpha-\beta$ at $t = 0$. These satisfy the linear relation (cf. [6, §2.9])

$$\Gamma\left(\begin{matrix}1-\alpha-\beta \\ 1-\alpha, 1-\beta\end{matrix}\right) f_1(t) - f_2(t) + \Gamma\left(\begin{matrix}\alpha+\beta-1 \\ \alpha, \beta\end{matrix}\right) f_3(t) = 0.$$

This can be written, using functional equations of the gamma function

$$B(\alpha,\beta)=\Gamma\begin{pmatrix}\alpha,\beta\\ \alpha+\beta\end{pmatrix},\quad \Gamma(s+1)=s\Gamma(s),$$

as

$$B(\alpha,\beta)f_1(t)+2\pi i\frac{1-e^{2\pi i(\alpha+\beta)}}{(1-e^{2\pi i\alpha})(1-e^{2\pi i\beta})}f_2(t)-B(1-\alpha,1-\beta)f_3(t)=0. \quad (4.7)$$

4.5. Rational 2-forms $s^{m-1}ds\wedge\omega^\chi$.

By taking an embedded resolution, we may assume that the reduced divisor

$$D:=(f^{-1}(0)+f^{-1}(1)+f^{-1}(\infty))_{\mathrm{red}}$$

of the singular fibers is a NCD. Recall from Lemma 4.3 (Key Lemma) the differential operator $\theta=p_0(t)+p_1(t)\frac{d}{dt}$. By replacing ω^χ with $t^n(1-t)^m\omega^\chi$ for some $n,m\geq 0$, we may assume without loss of generality

P1: $p_i(t)$ are polynomials and $t(1-t)|p_1(t)$,

P2: $\omega^\chi\in\Gamma(\mathbb{P}^1\backslash\{\infty\},f_*\Omega^1_{X/\mathbb{P}^1}(\log D))$, where the locally free sheaf $\Omega^1_{X/\mathbb{P}^1}(\log D)$ is defined by the exact sequence

$$0\longrightarrow f^*\Omega^1_{\mathbb{P}^1}(\log(0+1+\infty))\longrightarrow\Omega^1_X(\log D)\longrightarrow\Omega^1_{X/\mathbb{P}^1}(\log D)\longrightarrow 0.$$

Let $a\in\mathbb{C}\setminus\{0,1\}$ and $\pi_a:\mathbb{P}^1\to\mathbb{P}^1$ a morphism given by $s\mapsto a-s^l$ as in §4.3. To distinguish the source and target of π_a, we denote by $\mathbb{P}^1$ the target with inhomogeneous coordinate t, and by $\mathbb{P}^1_a$ the source with inhomogeneous coordinate s. Let

$$\begin{array}{ccccc} X_a & \xrightarrow{\ i\ } & \mathbb{P}^1_a\times_{\mathbb{P}^1}X & \longrightarrow & X\\ & \searrow^{f_a} & \downarrow & & \downarrow f\\ & & \mathbb{P}^1_a & \xrightarrow{\ \pi_a\ } & \mathbb{P}^1\end{array}$$

where i is the desingularization such that the inverse image $D_a\subset X_a$ of D is a NCD. Let $U_a\subset X_a$ (resp. $\overline{U}_a\subset X_a$) be the inverse image of $\mathbb{P}^1\setminus\{0,1,a,\infty\}$ (resp. $\mathbb{P}^1\setminus\{\infty\}$) under $\pi_a\circ f_a$. By the projection formula,

$$\mathscr{M}_a=\pi_{a*}\mathbb{Q}\otimes R^1f_*\mathbb{Q}\cong\pi_{a*}(\pi_a^*R^1f_*\mathbb{Q})=\pi_{a*}(R^1f_{a*}\mathbb{Q}).$$

This implies

$$H_a=H^1(\mathbb{P}^1\setminus\{0,1,a,\infty\},\mathscr{M}_a)\cong H^1(\mathbb{P}^1_a\setminus\{s^l=0,a,a-1,\infty\},R^1f_{a*}\mathbb{Q})$$

and hence we have an exact sequence

$$0\longrightarrow H_a\longrightarrow H^2(U_a,\mathbb{Q})\longrightarrow H^2(f_a^{-1}(s),\mathbb{Q}).$$

In particular, by taking the weight 2 piece, we have a canonical isomorphism

$$\begin{aligned}W_2H_a&\xrightarrow{\cong}\mathrm{Ker}[W_2H^2(U_a,\mathbb{Q})\to H^2(f_a^{-1}(s),\mathbb{Q})]\\ &=\mathrm{Ker}[H^2(X_a,\mathbb{Q})/H^2_{D_a}(X_a)\to H^2(f_a^{-1}(s),\mathbb{Q})].\end{aligned}\quad (4.8)$$

Consider the rational 2-form

$$s^{m-1}ds\,\omega^\chi = s^{m-1}ds\wedge\omega^\chi \in \Gamma(\mathscr{U},\Omega^2_{\mathscr{U}/\overline{\mathbb{Q}}})$$

for $m \geq 1$. By the assumption **P2**,

$$\begin{aligned} s^{m-1}ds\wedge\omega^\chi|_{\lambda=a} &\in \mathrm{Im}[\Omega^1_{\mathbb{P}^1_a\setminus\{\infty\}}\wedge\Omega^1_{\overline{U}/\mathbb{P}^1}(\log D) \longrightarrow \Omega^2_{\overline{U}_a}(\log D_a)], \\ &\subset \mathrm{Im}[t(1-t)\Omega^1_{\mathbb{P}^1_a}(\log(\pi_a^{-1}(0+1)))\wedge\Omega^1_{\overline{U}/\mathbb{P}^1}(\log D)\to\Omega^2_{\overline{U}_a}(\log D_a)] \\ &\subset t(1-t)\Omega^2_{\overline{U}_a}(\log D_a) = \Omega^2_{\overline{U}_a}, \end{aligned}$$

so that we have

$$s^{m-1}ds\wedge\omega^\chi|_{\lambda=a} \in \Gamma(\overline{U}_a,\Omega^2_{\overline{U}_a}).$$

Let $[s^{m-1}ds\wedge\omega^\chi]|_{\lambda=a} \in H^2_{\mathrm{dR}}(\overline{U}_a/\mathbb{C})$ denote the de Rham cohomology class. Obviously, its restriction to the general fiber $f_a^{-1}(s)$ vanishes. Thus

$$[s^{m-1}ds\wedge\omega^\chi]|_{\lambda=a} \in H^1(\mathbb{P}^1\setminus\{0,1,a,\infty\},\mathscr{M}_a)\cap\mathrm{Im}H^2_{\mathrm{dR}}(\overline{U}_a/\mathbb{C}).$$

If $m \equiv k$ modulo l, then $[s^{m-1}ds\wedge\omega^\chi]|_{\lambda=a}$ belongs to the χ-part. By Proposition 4.2 together with the commutative diagram

$$\begin{array}{ccc} H_a & \longrightarrow & H^2(U_a,\mathbb{Q}) \\ \downarrow & & \downarrow \\ H_a/W_2 & \xrightarrow{\ i\ } & H^3_{D_a}(X_a,\mathbb{Q}) \end{array}$$

with injective i, we have

$$[s^{m-1}ds\wedge\omega^\chi]|_{\lambda=a} \in W_2H_{a,\mathrm{dR}}(\chi) = W_2H^1(\mathbb{P}^1\setminus\{0,1,a,\infty\},\mathscr{M}_a)(\chi).$$

This means

$$[s^{m-1}ds\wedge\omega^\chi] \in \Gamma(S,W_2\mathscr{H}(\chi)).$$

Summarizing, we have:

Lemma 4.4. *Let ω^χ be a relative 1-form satisfying the condition* **P2**. *Then for any integer $m \geq 1$ such that $m \equiv k \pmod{l}$, the rational 2-form $s^{m-1}ds\wedge\omega^\chi \in \Gamma(\mathscr{U},\Omega^2_{\mathscr{U}/\overline{\mathbb{Q}}})$ defines a de Rham cohomology class*

$$[s^{m-1}ds\wedge\omega^\chi] \in \Gamma(S,W_2\mathscr{H}(\chi)).$$

As is shown in §4.3, $W_2H_{a,\mathrm{dR}}(\chi)$ is two-dimensional. We shall show in below that it is spanned by $[s^{m-1}ds\wedge\omega^\chi]|_{\lambda=a}$ and $[s^{m-l-1}ds\wedge\omega^\chi]|_{\lambda=a}$ for some m. We note that it is never obvious to show even the non-vanishing.

4.6. Proof of period formula: Part 2

We compute the period matrix $\mathrm{Per}(W_2\mathscr{H}(\chi))$.

Let $A_{\mathrm{dR}} \subset H^2_{\mathrm{dR}}(\overline{U}_a)$ be the $\overline{\mathbb{Q}}$-subspace spanned by $[s^{m-1}ds \wedge \omega^\chi]|_{\lambda=a}$ with $m > 0$ such that $m \equiv k \bmod l$ (cf. Lemma 4.4). Consider the commutative diagram

$$\begin{array}{ccccc} A_{\mathrm{dR}} & \longrightarrow & \mathrm{Hom}(H_2^B(\overline{U}_a), \mathbb{C}) & = & H^2_B(\overline{U}_a, \mathbb{C}) \\ \downarrow & & \downarrow & & \downarrow \\ W_2H_{a,\mathrm{dR}}(\chi) & \longrightarrow & \mathrm{Hom}(H_2^B(U_a), \mathbb{C}) & = & H^2_B(U_a, \mathbb{C}). \end{array}$$

As is easily shown,

$$H^2(\overline{U}_a)_{\mathrm{fib}} := \mathrm{Ker}[H^2(\overline{U}_a) \to H^2(D_a \cap \overline{U}_a)] \longrightarrow H^2(U_a)$$

is injective (cf. [2] §6.1), and A_{dR} is obviously contained in $H^2_{\mathrm{dR}}(\overline{U}_a)_{\mathrm{fib}}$. Our goal is to find m_i and $Z_i \in H_2^B(\overline{U}_a, \overline{\mathbb{Q}})$ $(i = 1, 2)$ such that

$$\begin{vmatrix} \int_{Z_1} s^{m_1-1}ds \wedge \omega^\chi|_{\lambda=a} & \int_{Z_2} s^{m_1-1}ds \wedge \omega^\chi|_{\lambda=a} \\ \int_{Z_1} s^{m_2-1}ds \wedge \omega^\chi|_{\lambda=a} & \int_{Z_2} s^{m_2-1}ds \wedge \omega^\chi|_{\lambda=a} \end{vmatrix} \neq 0 \tag{4.9}$$

and show that the entries (regarded as analytic functions of variable a) are as in Theorem 4.1. Then this gives the period matrix $\mathrm{Per}(W_2\mathscr{H}(\chi))$.

Recall from §4.4 the homology cycle $\delta \in H_1(f^{-1}(t), \overline{\mathbb{Q}})(\chi)$. We think it being a homology cycle in a fiber $f_a^{-1}(s)$. We take the Lefschetz thimble $\Delta \subset \overline{U}_a$ over a segment from $s = 0$ to $s = \sqrt[l]{a-1}$ (a fixed lth root). Let $\zeta \in \mu_l$ be a primitive lth root of unity and $\sigma_\zeta \in \mathrm{Aut}(\pi_a)$ be the corresponding automorphism. We denote the automorphism of X_a induced from $\sigma_\zeta \times \mathrm{id}_X$ by the same symbol σ_ζ. Δ has no boundary over $s = \sqrt[l]{a-1}$, but may have boundary over $s = 0$. Since σ_ζ acts on the fiber over $s = 0$ as identity, $(1 - \sigma_\zeta)\Delta$ has no boundary:

$$(1 - \sigma_\zeta)\Delta \in H_2(\overline{U}_a, \overline{\mathbb{Q}}).$$

Let $T_{a=0}$ denote the local monodromy at $a = 0$ on $H_2(\overline{U}_a, \overline{\mathbb{Q}})$ and we put

$$Z_1 := (1 - \sigma_\zeta)\Delta, \quad Z_2 := T_{a=0}(Z_1) \in H_2(\overline{U}_a, \overline{\mathbb{Q}}). \tag{4.10}$$

Let $p_i(t) \in \overline{\mathbb{Q}}[t]$ be polynomials which satisfy **P1** in the beginning of §4.5. Put

$$a_i(\lambda) := \frac{(-1)^i}{i!}\partial_\lambda^i p_0(\lambda), \quad b_i(\lambda) := \frac{(-1)^i}{i!}\partial_\lambda^i p_1(\lambda), \tag{4.11}$$

so that

$$p_0(t) = \sum_{i=0}^N a_i(\lambda)(\lambda - t)^i, \quad p_1(t) = \sum_{i=0}^N b_i(\lambda)(\lambda - t)^i, \tag{4.12}$$

for a sufficiently large N.

Lemma 4.5. *Let $F_\mu(\lambda)$ and $G_m(\lambda)$ be as defined in Theorem 4.1. Then we have, for $\mu > 1$,*

$$\partial_\lambda F_\mu(\lambda) = (\mu - 1)F_{\mu-1}(\lambda), \tag{4.13}$$
$$\partial_\lambda G_\mu(\lambda) = (\mu - 1)G_{\mu-1}(\lambda). \tag{4.14}$$

Proof. Write $\alpha = \alpha^\chi$, $\beta = \beta^\chi$. Since

$$\partial_\lambda {}_2F_1\left({\alpha, \beta \atop \gamma}; \lambda\right) = \frac{\alpha\beta}{\gamma} {}_2F_1\left({\alpha+1, \beta+1 \atop \gamma+1}; \lambda\right) \tag{4.15}$$

in general, we have

$$\partial_\lambda F_\mu(\lambda) = (\lambda - 1)^{\mu-1}\Bigg({}_2F_1\left({\alpha, \beta \atop \mu+1}; 1-\lambda\right) + \frac{\alpha\beta}{\mu(\mu+1)}(1-\lambda) {}_2F_1\left({\alpha+1, \beta+1 \atop \mu+2}; 1-\lambda\right)\Bigg).$$

Hence (4.13) is equivalent to

$$\frac{(\alpha)_n(\beta)_n}{(\mu+1)_n(1)_n} + \frac{\alpha\beta}{\mu(\mu+1)} \frac{(\alpha+1)_{n-1}(\beta+1)_{n-1}}{(\mu+2)_{n-1}(1)_{n-1}} = \frac{(\alpha)_n(\beta)_n}{(\mu)_n(1)_n} \quad (n \geq 1),$$

and this is easily verified. One proves (4.14) similarly, using (4.15) and the functional equation $\Gamma(s+1) = s\Gamma(s)$. □

Proposition 4.6. *For $a \in \mathbb{C} \setminus \{0, 1\}$ and any positive integer $m \equiv k \pmod{l}$, put*

$$P_m(a) := \int_\Delta s^{m-1} ds \wedge \omega^\chi|_{\lambda=a}, \quad \mu = \frac{m}{l}$$

and regard it as an analytic function $P_m(\lambda)$ of variable λ. Suppose $\mu > 1$. Then we have

$$P_m(\lambda) = \frac{2\pi i}{l} \sum_{i=0}^{N} \left(a_i(\lambda) + b_i(\lambda)\partial_\lambda\right) F_{\mu+i}(\lambda). \tag{4.16}$$

Moreover we have

$$\partial_\lambda P_m(\lambda) = (\mu - 1)P_{m-l}(\lambda). \tag{4.17}$$

Proof. Letting $t = \lambda - s^l$ we have by (4.4)

$$\begin{aligned} P_m(\lambda) &= \frac{2\pi i}{l} \int_1^\lambda (\lambda - t)^{\mu-1}(p_0(t) + p_1(t)\partial_t) {}_2F_1\left({\alpha^\chi, \beta^\chi \atop 1}; 1-t\right) dt \\ &= \frac{2\pi i}{l} \int_1^\lambda \left((\lambda - t)^{\mu-1}p_0(t) - \partial_t\left((\lambda-t)^{\mu-1}p_1(t)\right)\right) {}_2F_1\left({\alpha^\chi, \beta^\chi \atop 1}; 1-t\right) dt. \end{aligned}$$

Here the second equality follows from integration by parts and the assumption $1-t \mid p_1(t)$ in **P1**. By (4.12) and letting $1-t=(1-\lambda)u$, we have

$$P_m(\lambda) = \frac{2\pi i}{l} \sum_{i\geq -1} (a_i(\lambda)+(\mu+i)b_{i+1}(\lambda)) \int_1^\lambda (\lambda-t)^{\mu+i-1} {}_2F_1\left({\alpha^\chi, \beta^\chi \atop 1}; 1-t\right) dt$$
$$= \frac{2\pi i}{l} \sum_{i\geq -1} (a_i(\lambda)+(\mu+i)b_{i+1}(\lambda))$$
$$\times (\lambda-1)^{\mu+i} \int_0^1 (1-u)^{\mu+i-1} {}_2F_1\left({\alpha^\chi, \beta^\chi \atop 1}; (1-\lambda)u\right) du.$$

By the integral representation of ${}_3F_2$ (cf. [10], (4.1.2))

$$\int_0^1 {}_2F_1\left({a, b \atop d}; xt\right) t^{c-1}(1-t)^{e-c-1}\, dt = B(c, e-c)\, {}_3F_2\left({a, b, c \atop d, e}; x\right), \tag{4.18}$$

we have

$$\int_0^1 (1-u)^{\mu+i-1} {}_2F_1\left({\alpha^\chi, \beta^\chi \atop 1}; (1-\lambda)u\right) du$$
$$= B(1, \mu+i)\, {}_3F_2\left({\alpha^\chi, \beta^\chi, 1 \atop 1, \mu+i+1}; 1-\lambda\right) = \frac{1}{\mu+i} {}_2F_1\left({\alpha^\chi, \beta^\chi \atop \mu+i+1}; 1-\lambda\right).$$

Hence we obtain

$$P_m(\lambda) = \frac{2\pi i}{l} \sum_{i\geq -1} (a_i(\lambda)+(\mu+i)b_{i+1}(\lambda)) F_{\mu+i}(\lambda).$$

Now (4.16) follows using (4.13), from which (4.17) follows using

$$\partial_\lambda a_i(\lambda) = -(i+1)a_{i+1}(\lambda), \quad \partial_\lambda b_i(\lambda) = -(i+1)b_{i+1}(\lambda),$$

and (4.13). □

Proposition 4.7. *Let the notations be as in Proposition 4.6. There is a differential operator $\Theta = q(\lambda) + r(\lambda)\partial_\lambda$ with $q(\lambda), r(\lambda) \in \overline{\mathbb{Q}}[\lambda, 1/\lambda(\lambda-1)]$, such that*

$$P_m(\lambda) = 2\pi i \cdot \Theta F_\mu(\lambda).$$

Proof. By (4.13), we have $F_{\mu+i}(\lambda) = \frac{(\mu)_i}{(\mu)_N} \partial_\lambda^{N-i} F_{\mu+N}(\lambda)$ $(i = 0, 1, \ldots, N)$. Hence by Proposition 4.6, we have $P_m(\lambda) = 2\pi i \cdot \Theta_1 F_{\mu+N}(\lambda)$ with

$$\Theta_1 = \frac{1}{l} \sum_{i=0}^{N} \left(\frac{(\mu)_i}{(\mu)_N} (a_i(\lambda)\partial_\lambda^{N-i} + b_i(\lambda)\partial_\lambda^{N+1-i}) \right).$$

Recall that $F_\mu(\lambda)$ is a solution of the differential equation satisfied by

$${}_2F_1\left({\alpha^\chi - \mu, \beta^\chi - \mu \atop \alpha^\chi + \beta^\chi - \mu}; \lambda\right),$$

i.e., $\mathscr{D} F_\mu(\lambda) = 0$ with

$$\mathscr{D} := \lambda(1-\lambda)\partial_\lambda^2 + \{\alpha^\chi + \beta^\chi - \mu - (\alpha^\chi + \beta^\chi - 2\mu + 1)\lambda\}\partial_\lambda - (\alpha^\chi - \mu)(\beta^\chi - \mu). \tag{4.19}$$

Hence we have

$(\alpha^\chi-\mu)(\beta^\chi-\mu)F_\mu=(\{\alpha^\chi+\beta^\chi-\mu-(\alpha^\chi+\beta^\chi-2\mu+1)\lambda\}-\lambda(1-\lambda)\partial_\lambda)(\mu-1)F_{\mu-1}.$

Applying this iteratively, we obtain a differential operator Θ_2 of degree N such that

$$F_{\mu+N}(\lambda)=\Theta_2F_\mu(\lambda).$$

By reducing the degree of $\Theta_1\Theta_2$ using (4.19), we obtain the proposition. □

To compute the period along Z_2 (see (4.10)), we prepare the following.

Proposition 4.8. *Assume $m>l$, $m\equiv k \pmod l$, and that $\mu:=m/l$ satisfies $\mu\not\equiv 0,\ \alpha^\chi,\ \beta^\chi,\ \alpha^\chi+\beta^\chi \pmod{\mathbb{Z}}$. Let Θ be as in Proposition 4.7. Then we have*

$$\begin{pmatrix}\int_{Z_1}s^{m-1}ds\ \omega^\chi & \int_{Z_2}s^{m-1}ds\ \omega^\chi\\ \int_{Z_1}s^{m-l-1}ds\ \omega^\chi & \int_{Z_2}s^{m-l-1}ds\ \omega^\chi\end{pmatrix}$$
$$=2\pi i(1-\zeta^m)\begin{pmatrix}1&0\\0&\frac{1}{\mu-1}\end{pmatrix}\begin{pmatrix}\Theta F_\mu(\lambda)&\Theta G_\mu(\lambda)\\ \partial_\lambda\Theta F_\mu(\lambda)&\partial_\lambda\Theta G_\mu(\lambda)\end{pmatrix}\begin{pmatrix}1&\xi\\0&1-\xi\end{pmatrix}.$$

Here $\int_{Z_i}s^{m-l-1}ds\ \omega^\chi$ is also regarded as an analytic function of variable λ as in Proposition 4.6.

Proof. Firstly, since σ_ζ acts on $s^{m-1}ds$ as multiplication by ζ^m, we have, by Proposition 4.7,

$$\int_{Z_1}s^{m-1}ds\ \omega^\chi=(1-\zeta^m)P_m(\lambda)=2\pi i(1-\zeta^m)\Theta F_\mu(\lambda).$$

Secondly, $G_\mu(\lambda)$ is a solution of the differential equation (4.19), and its monodromy at $\lambda=0$ is given by

$$T_{\lambda=0}(F_\mu,G_\mu)=(F_\mu,G_\mu)\begin{pmatrix}\xi&0\\1-\xi&1\end{pmatrix},\quad \xi:=e^{2\pi i(\mu-\alpha^\chi-\beta^\chi)}$$

(see [6, §2.9 (43)]). Note that ξ depends only on μ mod $\mathbb{Z}$ and $\xi\neq 1$ by the assumption. Hence we have

$$\int_{Z_2}s^{m-1}ds\ \omega^\chi=2\pi i(1-\zeta^m)\Theta T_{\lambda=0}F_\mu(\lambda)=2\pi i(1-\zeta^m)\Theta(\xi F_\mu(\lambda)+(1-\xi)G_\mu(\lambda)).$$

Then the computation for $\int_{Z_i}s^{m-l-1}ds\ \omega^\chi$ $(i=1,2)$ follows by Proposition 4.6. □

Now we finish the proof of Theorem 4.1. By Proposition 4.8, it suffices to show

$$\begin{vmatrix}\Theta F_\mu(\lambda)&\Theta G_\mu(\lambda)\\ \partial_\lambda\Theta F_\mu(\lambda)&\partial_\lambda\Theta G_\mu(\lambda)\end{vmatrix}\neq 0$$

for some m. This is equivalent to that $\Theta F_\mu(\lambda)/\Theta G_\mu(\lambda)$ is non-constant. Suppose that $\Theta F_\mu(\lambda)=C\Theta G_\mu(\lambda)$ for some constant C. Then $F_\mu(\lambda)-CG_\mu(\lambda)$ is a solution of both $\Theta f=0$ and (4.19). Since (4.19) is irreducible by the assumption that $\mu\not\equiv \alpha,\ \beta,\ \alpha+\beta \pmod{\mathbb{Z}}$, and the order of Θ is one, we have $\Theta=0$. Suppose that this is the case for any $m > l$ with $m \equiv k \pmod{l}$. Then $P_m(\lambda) = \int_\Delta s^{m-1} ds\, \omega^\chi = 0$ for any such m. Applying the elementary lemma below (replace s with $x^{1/l}$), it follows that $\int_\delta \omega^\chi = 0$, hence a contradiction. □

Lemma 4.9. *Let f be a continuous function on the closed interval $[0,1]$ whose zeros have no accumulation point. If $\int_0^1 f(x)x^{n-1}dx = 0$ for all $n \in \mathbb{Z}_{>0}$, then $f \equiv 0$.*

Proof. By replacing $f(x)$ with $(x-1)f(x)$ if necessary, we can assume $f(1) = 0$. Suppose that $f \not\equiv 0$. By replacing f with $-f$ if necessary, there exists $a \in (0,1)$ such that $f(x) > 0$ for any $x \in (a,1)$. Put $M = \max_{x\in[0,a]} |f(x)|$ and choose $b, c \in (a,1)$ $(b < c)$ and $m > 0$ such that $f(x) \geq m$ for any $x \in [b,c]$. Then

$$\begin{aligned}\int_0^1 f(x)x^{n-1}\,dx &> -M\int_0^a x^{n-1}\,dx + m\int_b^c x^{n-1}\,dx\\ &= \frac{mc^n}{n}\left(1 - \left(\frac{b}{c}\right)^n - \frac{M}{m}\left(\frac{a}{c}\right)^n\right).\end{aligned}$$

Since the right-hand side is positive for sufficiently large n, this contradicts the assumption. □

5. Regulator formula

We keep the setting and the notations in §4.1. Put

$$C := \mathrm{Gr}_2^W \psi_{t=1}\mathscr{M} \cong \pi_*\mathbb{Q}|_{\{1\}\times S} \otimes (\mathrm{Gr}_2^W \psi_{t=1} R^1 f_*\mathbb{Q}) \tag{5.1}$$

a VHdR on S. In this section we discuss the exact sequences

$$\begin{array}{ccccccccc} 0 & \longrightarrow & W_2\mathscr{H}(e) & \longrightarrow & \mathscr{H}(e) & \longrightarrow & (\mathscr{H}/W_2\mathscr{H})(e) & \longrightarrow & 0 \\ & & \| & & \uparrow & & \uparrow & & \\ 0 & \longrightarrow & W_2\mathscr{H}(e) & \longrightarrow & \mathscr{H}'(e) & \longrightarrow & C(e)\otimes\mathbb{Q}(-1) & \longrightarrow & 0 \end{array} \tag{5.2}$$

of mixed Hodge–de Rham structures on $S = \mathbb{P}^1 \setminus \{0,1,\infty\}$ arising from (4.1), where the right vertical inclusion is as in Proposition 4.2. Since $\mathrm{Gr}_2^W \psi_{t=1}(R^1 f_*\mathbb{Q})(e_0) \cong E_0$ is a constant VHdR of type $(1,1)$, $C(e)$ is one-dimensional over E and endowed with Hodge type $(1,1)$ (however the monodromy is non-trivial).

5.1. Setting

Let $Q : R^1 f_*\mathbb{Q} \otimes R^1 f_*\mathbb{Q} \to \mathbb{Q}(-1)$ be a polarization form which also induces a polarization on the e_0-part $(R^1 f_*\mathbb{Q})(e_0)$. It naturally extends to a non-degenerate pairing $Q : \mathscr{M} \otimes \mathscr{M} \to \mathbb{Q}(-1)$ which is compatible with the action of $\mathrm{Aut}(\pi) \cong \mu_l$, namely $Q(\sigma x, \sigma y) = Q(x,y)$ for $\sigma \in \mathrm{Aut}(\pi)$. This also induces a polarization on the e-part $\mathscr{M}(e)$. We have isomorphisms

$$(\mathscr{M})^* \cong \mathscr{M} \otimes \mathbb{Q}(1), \quad (\mathscr{M}(e))^* \cong \mathscr{M}(e) \otimes \mathbb{Q}(1) \tag{5.3}$$

induced from Q where $(-)^*$ denotes the dual sheaf. Let $j : \mathscr{U} \hookrightarrow \mathbb{P}^1_S$ and $\mathrm{pr}_2 : \mathbb{P}^1_S = \mathbb{P}^1 \times S \to S$. Then there are isomorphisms

$$(\mathscr{H})^* = (R^1\mathrm{pr}_{2*}Rj_*\mathscr{M})^* \cong R^1\mathrm{pr}_{2*}j_!\mathscr{M}^* \otimes \mathbb{Q}(1) \cong R^1\mathrm{pr}_{2*}j_!\mathscr{M} \otimes \mathbb{Q}(2) \qquad (5.4)$$

induced from the Verdier duality and (5.3). We show that (5.4) induces an isomorphism

$$(W_2\mathscr{H})^* \cong W_2\mathscr{H} \otimes \mathbb{Q}(2). \qquad (5.5)$$

Let $i : Z = \mathbb{P}^1_S \setminus \mathscr{U} \hookrightarrow \mathbb{P}^1_S$ be the complement. Note that Z is finite etale over S. There is an exact sequence

$$0 \longrightarrow i_*i^*j_*\mathscr{M} \longrightarrow j_!\mathscr{M}[1] \longrightarrow j_*\mathscr{M}[1] \longrightarrow 0$$

of mixed Hodge modules where $j_*\mathscr{M} = R^0j_*\mathscr{M}$. Applying $R\mathrm{pr}_{2*}$, one has

$$\begin{array}{ccccccc} 0 \longrightarrow & i_*i^*j_*\mathscr{M} & \longrightarrow & R^1\mathrm{pr}_{2*}j_!\mathscr{M} & \longrightarrow & R^1\mathrm{pr}_{2*}j_*\mathscr{M} & \longrightarrow 0 \\ & & & & & \Big\| {\scriptstyle (4.2)} & \\ & & & & & W_2\mathscr{H} & \end{array}$$

because $(R^0\mathrm{pr}_{2*}j_*\mathscr{M})_a = H^0(U_a, \mathscr{M}_a) = 0$ for $a \in \mathbb{C} \setminus \{0,1\}$ (see §4.3 for the notation). The mixed Hodge–de Rham structure

$$i_*i^*j_*\mathscr{M}_a = \bigoplus_{p=0,1,a,\infty} \mathrm{Ker}[T_p - 1 : \psi_{t=p}\mathscr{M}_a \to \psi_{t=p}\mathscr{M}_a \otimes \mathbb{Q}(-1)]$$

has weight ≤ 1. This implies $\mathrm{Gr}^W_2 R^1\mathrm{pr}_{2*}j_!\mathscr{M} = R^1\mathrm{pr}_{2*}j_*\mathscr{M} = W_2\mathscr{H}$. We thus have (5.5) by taking the graded piece of (5.4) of weight -2.

The isomorphisms (5.3) and (5.5) are *not* compatible with respect to the multiplication by R. Here the multiplication on the left-hand side of (5.3) or (5.5) is given as in §2.3. For $r \in E$, we denote by ${}^tr$ the multiplication on $\mathscr{M}(e)$ such that

$$Q(rx, y) = Q(x, {}^try), \quad \forall\, x, y.$$

The multiplication by r on the left corresponds to ${}^tr$ in the right of (5.3). Note ${}^t\sigma = \sigma^{-1}$ for $\sigma \in \mathrm{Aut}(\pi)$. For $\chi : R \to \overline{\mathbb{Q}}$, we denote $(-)({}^t\chi)$ the subspace on which ${}^tr$ acts by multiplication by $\chi(r)$ for all $r \in E$. Then (5.3) and (5.5) induce

$$(\mathscr{M}(\chi))^* \cong \mathscr{M}({}^t\chi), \quad (W_2\mathscr{H}(\chi))^* \cong W_2\mathscr{H}({}^t\chi).$$

5.2. Theorem on regulators

Let $\mathscr{O}^{\mathrm{zar}}$ be the Zariski sheaf of polynomial functions (with coefficients in $\overline{\mathbb{Q}}$) on $S = \mathbb{A}^1_{\overline{\mathbb{Q}}} \setminus \{0,1\}$ with coordinate λ. Let $\mathscr{O}^{an}$ be the sheaf of analytic functions on $S^{an} = \mathbb{C}^{an} \setminus \{0,1\}$. Let a be the canonical morphism from the analytic site to the Zariski site. Set

$$\mathscr{J} := \mathrm{Coker}[a^{-1}F^2W_2\mathscr{H}_{\mathrm{dR}} \oplus \iota^{-1}W_2\mathscr{H}_B \to \mathscr{O}^{an} \otimes_{a^{-1}\mathscr{O}^{\mathrm{zar}}} a^{-1}W_2\mathscr{H}_{\mathrm{dR}}],$$

$$\mathscr{J}^* := \mathrm{Coker}[a^{-1}\mathcal{H}om(W_2\mathscr{H}_{\mathrm{dR}}/F^1, \mathscr{O}^{\mathrm{zar}}) \oplus \iota^{-1}W_2\mathscr{H}_B^* \to \mathcal{H}om(a^{-1}W_2\mathscr{H}_{\mathrm{dR}}, \mathscr{O}^{an})]$$

sheaves on the analytic site $\mathbb{C}^{an} \setminus \{0,1\}$. Note $\mathscr{J}^* \cong \mathscr{J}$ by (5.5).

Let C be the VHdR on S as defined in (5.1). Let $h : \widetilde{S} \to S$ be a generically finite and dominant map such that $\sqrt[l]{\lambda - 1} \in \overline{\mathbb{Q}}(\widetilde{S})$. Then h^*C is a constant VHdR of type $(1,1)$. Let

$$\begin{aligned} \delta : h^*C(e) \otimes \mathbb{Q}(1) = \mathrm{Hom}_{\mathrm{VMHdR}}(\mathbb{Q}, h^*C \otimes \mathbb{Q}(1)) \\ \longrightarrow \mathrm{Ext}^1_{\mathrm{VMHdR}}(\mathbb{Q}, h^*W_2\mathscr{H}(e) \otimes \mathbb{Q}(2)). \end{aligned}$$

be the connecting homomorphism arising from the exact sequence of (5.2). Let ρ be the composition of maps

$$\begin{aligned} h^*C(e) \otimes \mathbb{Q}(1) \xrightarrow{\delta} \mathrm{Ext}^1_{\mathrm{VMHdR}}&(\mathbb{Q}, h^*W_2\mathscr{H}(e) \otimes \mathbb{Q}(2)) \\ &\longrightarrow \Gamma(\widetilde{S}^{an}, h^*\mathscr{J}(e)) \\ &\xrightarrow{\sim} \Gamma(\widetilde{S}^{an}, h^*\mathscr{J}^*(e)) \end{aligned}$$

where the second arrow is constructed in a similar way to the proof of Proposition 2.1. Let $\rho({}^t\chi)$ be ${}^t\chi$-part of ρ, namely the composition of maps

$$\begin{aligned} \overline{\mathbb{Q}} \cong (h^*C(e) \otimes \mathbb{Q}(1))({}^t\chi) \longrightarrow \mathrm{Ext}^1_{\text{Filt-BdR}}&(\overline{\mathbb{Q}}, h^*W_2\mathscr{H}({}^t\chi) \otimes \mathbb{Q}(2)) \\ &\xrightarrow{\sim} \mathrm{Ext}^1_{\text{Filt-BdR}}(\overline{\mathbb{Q}}, h^*(W_2\mathscr{H})^*(\chi)) \\ &\longrightarrow \Gamma(\widetilde{S}^{an}, h^*\mathscr{J}^*(\chi)). \end{aligned}$$

Here $\mathscr{J}^*(\chi)$ is defined by replacing $W_2\mathscr{H}_{\mathrm{dR}}$ with $W_2\mathscr{H}_{\mathrm{dR}}(\chi)$, and $W_2\mathscr{H}_B^*$ with $W_2\mathscr{H}_B^*(\chi) = \mathcal{H}om_{\overline{\mathbb{Q}}}(W_2\mathscr{H}_B(\chi), \overline{\mathbb{Q}})$.

Theorem 5.1 (Regulator formula). *Let the assumptions, μ and Θ be as in Theorem 4.1. Let $\rho({}^t\chi)(1) \in (h^*\mathscr{O}^{an})^2 \cong \mathcal{H}om(a^{-1}W_2\mathscr{H}_{\mathrm{dR}}(\chi), h^*\mathscr{O}^{an})$ be a local lifting where the isomorphism is with respect to a $\overline{\mathbb{Q}}$-frame of $W_2\mathscr{H}_{\mathrm{dR}}(\chi)$. Then we have*

$$\rho({}^t\chi)(1) \equiv (\Theta H_\mu(\lambda), (\mu - 1)^{-1}\partial_\lambda \Theta H_\mu(\lambda)) \pmod{\overline{\mathbb{Q}(\lambda)}^2},$$

where we put

$$H_\mu(\lambda) := \frac{1}{(1-\alpha^\chi)(1-\beta^\chi)}(\lambda - 1)^{\mu - 1} {}_3F_2\left({1, 1, 1-\mu \atop 2-\alpha^\chi, 2-\beta^\chi}; \frac{1}{1-\lambda} \right).$$

The map ρ is related to *Beilinson's regulator map* in the following way. Let $\mathbb{P}^1_{\widetilde{S}} := \mathbb{P}^1 \times \widetilde{S}$ and $\widetilde{\pi} : \mathbb{P}^1_{\widetilde{S}} \to \mathbb{P}^1$ given by $(s, \lambda) \mapsto h(\lambda) - s^l$. Let

$$\begin{array}{ccccc} X_{\widetilde{S}} & \xrightarrow{i} & \mathbb{P}^1_{\widetilde{S}} \times_{\mathbb{P}^1} X & \longrightarrow & X \\ & \searrow^{f_{\widetilde{S}}} & \downarrow & & \downarrow f \\ & & \mathbb{P}^1_{\widetilde{S}} & \xrightarrow{\widetilde{\pi}} & \mathbb{P}^1 \end{array}$$

with i desingularization. Let $g := \mathrm{pr}_2 \circ f_{\widetilde{S}} : X_{\widetilde{S}} \to \widetilde{S}$ be a projective smooth map. Let

$$\mathrm{reg} : H^3_{\mathscr{M}}(X_{\widetilde{S}}, \mathbb{Q}(2)) \longrightarrow \mathrm{Ext}^1_{\mathrm{VMHdR}}(\mathbb{Q}, R^2 g_* \mathbb{Q}(2))$$

be the Beilinson regulator map. Let $j : \widetilde{\pi}^{-1}(\{0,1,\infty\}) \hookrightarrow \mathbb{P}^1_{\widetilde{S}}$ be the embedding and put

$$(R^2 g_*\mathbb{Q})_0 := \mathrm{Ker}[R^2 g_*\mathbb{Q} \to \mathrm{pr}_{2*} j_* j^{-1} R^2 f_{\widetilde{S}*}\mathbb{Q}].$$

There is a canonical surjective map $(R^2 g_*\mathbb{Q})_0 \to h^* W_2 \mathscr{H}$ (cf. (4.8)), so that we have

$$\mathrm{reg}_0 : H^3_{\mathscr{M}}(X_{\widetilde{S}}, \mathbb{Q}(2))_0 \longrightarrow \mathrm{Ext}^1_{\mathrm{VMHdR}}(\mathbb{Q}, h^* W_2 \mathscr{H} \otimes \mathbb{Q}(2)),$$

where $H^3_{\mathscr{M}}(X_{\widetilde{S}}, \mathbb{Q}(2))_0 \subset H^3_{\mathscr{M}}(X_{\widetilde{S}}, \mathbb{Q}(2))$ is the inverse image of

$$\mathrm{Ext}^1_{\mathrm{VMHdR}}(\mathbb{Q}, (R^2 g_* \mathbb{Q}(2))_0)$$

by reg. Let $E_{\widetilde{S}} \subset X_{\widetilde{S}}$ be the inverse image of the singular fiber $f^{-1}(1) \subset X$, then there is a canonical map

$$H^3_{\mathscr{M}, E_{\widetilde{S}}}(X_{\widetilde{S}}, \mathbb{Q}(2)) \to H^3_{B, E_{\widetilde{S}}}(X_{\widetilde{S}}, \mathbb{Q}(2)) \cap H^{0,0} \to \mathrm{Hom}_{\mathrm{VMHdR}}(\mathbb{Q}, h^* C \otimes \mathbb{Q}(1))$$

from the motivic cohomology group with support in $E_{\widetilde{S}}$, and the following diagram is commutative

$$\begin{array}{ccc}
H^3_{\mathscr{M}, E_{\widetilde{S}}}(X_{\widetilde{S}}, \mathbb{Q}(2)) & \longrightarrow & \mathrm{Hom}_{\mathrm{VMHdR}}(\mathbb{Q}, h^* C \otimes \mathbb{Q}(1)) \\
\Big\downarrow & & \Big\downarrow \cap \\
 & & \mathrm{Hom}_{\mathrm{VMHdR}}(\mathbb{Q}, h^* \mathscr{H}/W_2 \otimes \mathbb{Q}(2)) \\
 & & \Big\downarrow \\
H^3_{\mathscr{M}}(X_{\widetilde{S}}, \mathbb{Q}(2))_0 & \xrightarrow{\mathrm{reg}_0} & \mathrm{Ext}^1_{\mathrm{VMHdR}}(\mathbb{Q}, h^* W_2 \mathscr{H} \otimes \mathbb{Q}(2)) \\
 & & \Big\downarrow \\
 & & \Gamma(\widetilde{S}^{an}, h^* \mathscr{J}^*).
\end{array}$$

(with a curved arrow ρ from $\mathrm{Hom}_{\mathrm{VMHdR}}(\mathbb{Q}, h^* C \otimes \mathbb{Q}(1))$ to $\Gamma(\widetilde{S}^{an}, h^* \mathscr{J}^*)$)

Taking the e-part, one easily sees that the top horizontal arrow is surjective, and hence the element $\rho({}^t\chi)(1)$ comes from an element of $H^3_{\mathscr{M}, E_{\widetilde{S}}}(X_{\widetilde{S}}, \mathbb{Q}(2))$ (see also [2] Prop. 4.8).

5.3. Proof of regulator formula: Part 1

Let

$$\begin{array}{ccccc}
X_a & \xrightarrow{i} & \mathbb{P}^1_a \times_{\mathbb{P}^1} X & \longrightarrow & X \\
 & \searrow^{f_a} & \Big\downarrow & & \Big\downarrow f \\
 & & \mathbb{P}^1_a & \xrightarrow{\pi_a} & \mathbb{P}^1
\end{array}$$

be as in §4.5. Let $D_a \subset X_a$ be the inverse image of the singular fibers of f. We denote the coordinate of $\mathbb{P}^1$ (resp. $\mathbb{P}^1_a$) by t (resp. s). We also use the notation in §4.3. Note $\mathscr{M}_a \cong \pi_{a*} R^1 f_{a*}\mathbb{Q}$ is endowed with the de Rham structure induced from a $\overline{\mathbb{Q}}$-frame on $\mathscr{M}_{\mathrm{dR}}$. The distinguished triangle

$$0 \longrightarrow j_! \pi_{a*} f_{a*}\mathbb{Q} \longrightarrow j_! \tau_{\le 1} \pi_{a*} R f_{a*}\mathbb{Q} \longrightarrow j_! \pi_{a*} R^1 f_{a*}\mathbb{Q}[-1] \longrightarrow 0$$

and the fact that

$$H^2(\mathbb{P}^1_a, j_!\pi_{a*}\mathbb{Q}) = H^0(\mathbb{P}^1_a, Rj_*\pi_{a*}\mathbb{Q})^* = H^0(\mathbb{P}^1_a \setminus \{0,1,a,\infty\}, \pi_{a*}\mathbb{Q})^* = 0$$

implies

$$H^2(\mathbb{P}^1, j_!\pi_{a*}\tau_{\leq 1}Rf_{a*}\mathbb{Q}) \xrightarrow{\cong} H^1(\mathbb{P}^1, j_!\mathscr{M}_a).$$

Hence we have an injective map

$$H^1(\mathbb{P}^1, j_!\mathscr{M}_a) \longrightarrow H^2(\mathbb{P}^1, j_!\pi_{a*}Rf_{a*}\mathbb{Q}) = H^2(\mathbb{P}^1, \pi_{a*}Rf_{a*}j_!\mathbb{Q}) = H^2(X_a, D_a; \mathbb{Q}).$$

We have a commutative diagram with exact rows

$$\begin{array}{ccccccccc}
0 & \longrightarrow & W_2H_a(2) & \longrightarrow & H_a(2) & \longrightarrow & H_a/W_2(2) & \longrightarrow & 0 \\
 & & \| & & \| & & \| & & \\
0 & \longrightarrow & (W_2H_a)^* & \longrightarrow & H^1(\mathbb{P}^1, j_!\mathscr{M}_a)^* & \longrightarrow & H^1(\mathbb{P}^1, j_!\mathscr{M}_a)^*/W_{-2} & \longrightarrow & 0 \\
 & & \uparrow{\scriptstyle a_1} & & \uparrow{\scriptstyle a_2} & & \uparrow{\scriptstyle a_3} & & \\
0 & \longrightarrow & H_2(X_a, \mathbb{Q})/H_2(D_a) & \longrightarrow & H_2(X_a, D_a; \mathbb{Q}) & \longrightarrow & H_1(D_a) & \longrightarrow & 0 \\
 & & \uparrow{\scriptstyle b_1} & & \uparrow{\scriptstyle b_2} & & \uparrow{\scriptstyle b_2} & & \\
0 & \longrightarrow & H_2(\overline{U}_a, \mathbb{Q})/H_2(D^\circ_a) & \longrightarrow & H_2(\overline{U}_a, D^\circ_a; \mathbb{Q}) & \longrightarrow & H_1(D^\circ_a) & &
\end{array}$$

where $D^\circ_a := D_a \cap \overline{U}_a$. Since a_2 is surjective, so are a_1 and a_3. Moreover a_3 is bijective (local invariant cycle theorem).

Let Z_1, Z_2 be the homology cycles (4.10), and $(m_1, m_2) = (l\mu, l\mu - l)$ where μ is as in Theorem 4.1. Note that $(W_2H_a)^*(\chi)$ is spanned by the images of Z_1 and Z_2. Hence

$$\begin{aligned}
&H_{a,B}({}^t\chi)/W_2 \cap H^{0,0} \cong \overline{\mathbb{Q}} \\
&\xrightarrow{=} H^B_1(D^\circ_a)(\chi) \cap H^{0,0} \\
&\longrightarrow \mathrm{Ext}^1_{\text{Filt-BdR}}(\overline{\mathbb{Q}}, H_2(\overline{U}_a, \mathbb{Q})/H_2(D^\circ_a)(\chi)) \\
&\longrightarrow \mathrm{Ext}^1_{\text{Filt-BdR}}(\overline{\mathbb{Q}}, (W_2H_a)^*(\chi)) \\
&\xrightarrow{\sim} \mathrm{Coker}[(W_2H_{a,B})^*(\chi) \oplus \mathrm{Hom}(W_2H_{a,\mathrm{dR}}/F^1, \overline{\mathbb{Q}}) \to \mathrm{Hom}(W_2H_{a,\mathrm{dR}}(\chi), \mathbb{C})] \\
&\xrightarrow{=} \mathrm{Coker}[\mathrm{Hom}(W_2H_{a,\mathrm{dR}}(\chi)/F^1, \overline{\mathbb{Q}}) \to \mathrm{Hom}(W_2H_{a,\mathrm{dR}}(\chi), \mathbb{C})/\mathrm{Im}H_2(\overline{U}_a, \overline{\mathbb{Q}})]
\end{aligned}$$

and the composition of the above coincides with the restriction $\rho({}^t\chi)|_{\lambda=a}$. Moreover the image of $H_2(\overline{U}_a, \overline{\mathbb{Q}})$ is given by the period matrix

$$\mathrm{Per}(W_2\mathscr{H}(\chi))|_{\lambda=a} = \begin{pmatrix} \int_{Z_1} s^{m_1-1}ds\omega & \int_{Z_2} s^{m_1-1}ds\omega \\ \int_{Z_1} s^{m_2-1}ds\omega & \int_{Z_2} s^{m_2-1}ds\omega \end{pmatrix}\Bigg|_{\lambda=a} : \overline{\mathbb{Q}}^2 \longrightarrow \mathbb{C}^2$$

under the isomorphism

$$\mathrm{Hom}(W_2H_{a,\mathrm{dR}}(\chi), \mathbb{C}) \cong \mathbb{C}^2$$

given by the $\overline{\mathbb{Q}}$-basis $\{s^{m_1-1}ds\omega, s^{m_2-1}ds\omega\}$ of $W_2H_{a,\mathrm{dR}}(\chi)$. Let $D_a^{ss} \subset D_a$ be the inverse image of $f^{-1}(1)$. Note $D_a^{ss} \subset \overline{U}_a$. Then for $1 \in \overline{\mathbb{Q}} \cong H_1(D_a^{ss})(\chi) \subset H_1(D_a^{\circ})(\chi) \cap H^{0,0}$, we want to compute a lifting

$$\rho({}^t\chi)(1) = (\phi_1(a), \phi_2(a)) \in \mathbb{C}^2.$$

Lemma 5.2. *Let* $\Gamma(a) \in H_2^B(\overline{U}_a, (\pi_a f_a)^{-1}(1); \overline{\mathbb{Q}})$ *be a lifting of the homology cycle* $1 \in \overline{\mathbb{Q}} \cong H_1(D_a^{ss})(\chi)$*. Then there is an algebraic function* $R(\lambda) \in \overline{\mathbb{Q}(\lambda)}$ *of variable* λ *such that*

$$\phi_i(a) = \int_{\Gamma(a)} s^{m_i-1}ds\ \omega^{\chi}|_{\lambda=a} + R(a)$$

for a *in a small neighbourhood of* $\mathbb{C}^{an} \setminus \{0,1\}$.

Proof. See [2], Proposition 7.2; the situation there is slightly different but the same discussion works. □

5.4. Proof of regulator formula: Part 2

Recall from §4.4 the homology cycles $\delta, \gamma \in H_1(X_t, \overline{\mathbb{Q}})(\chi)$. By the local invariant cycle theorem, there is an exact sequence

$$H_1(X_t, \mathbb{Q}) \overset{T_0-1}{\longrightarrow} H_1(X_t, \mathbb{Q}) \longrightarrow H_1(f^{-1}(0), \mathbb{Q}) \longrightarrow 0, \tag{5.6}$$

where T_0 is the local monodromy at $t = 0$. We note that $H_1(f^{-1}(0), \mathbb{Q})$ has multiplication by R_0 induced from (5.6) (recall from §4.1 that $R^1f_*\mathbb{Q}$ has multiplication by R_0). An element $\gamma' \in H_1(X_t, \overline{\mathbb{Q}})(\chi)$ vanishes as $t \to 0$ if and only if it belongs to the one-dimensional space

$$\mathrm{Ker}[H_1(X_t, \overline{\mathbb{Q}})(\chi) \to H_1(f^{-1}(0), \overline{\mathbb{Q}})] = \mathrm{Im}[T_0 - 1 : H_1(X_t, \overline{\mathbb{Q}})(\chi) \to H_1(X_t, \overline{\mathbb{Q}})(\chi)].$$

Lemma 5.3. *Put*

$$\gamma' := \gamma + \frac{1 - e^{2\pi i(\alpha^\chi + \beta^\chi)}}{(1 - e^{2\pi i\alpha^\chi})(1 - e^{2\pi i\beta^\chi})}\delta.$$

Then γ' *is a basis of* $\mathrm{Ker}[H_1(X_t, \overline{\mathbb{Q}})(\chi) \to H_1(f^{-1}(0), \overline{\mathbb{Q}})]$*, and we have*

$$\int_{\gamma'} \omega^\chi = B(1 - \alpha^\chi, 1 - \beta^\chi)\theta\left(t^{1-\alpha^\chi-\beta^\chi}{}_2F_1\left(\begin{matrix}1 - \alpha^\chi, 1 - \beta^\chi \\ 2 - \alpha^\chi - \beta^\chi\end{matrix}; t\right)\right).$$

Proof. If $\alpha^\chi + \beta^\chi = 1$, then T_0 is unipotent and $\mathrm{Im}(T_0 - 1) = \mathrm{Ker}(T_0 - 1)$, to which $\gamma' = \gamma$ belongs. The assertion about the period is Lemma 4.3. Suppose $\alpha^\chi + \beta^\chi \neq 1$. Recall the notations (4.5), (4.6) and the relation (4.7). The assertion about the period follows from (4.7). Since $f_1(t)$ and $f_3(t)$ form a basis of the local solutions near $t = 0$, and $T_0 - 1$ annihilates $f_1(t)$ but not $f_3(t)$, the γ' generates $\mathrm{Im}(T_0 - 1)$. □

We regard γ' as a homology cycle in a general fiber of f_a. Fix lth roots $\sqrt[l]{a}$ and $\sqrt[l]{a-1}$. Let $\Gamma_a \subset \overline{U}_a$ be the Lefschetz thimble over a path from $s = \sqrt[l]{a-1}$ to $s = \sqrt[l]{a}$ with fiber γ' (s is the coordinate of $\mathbb{P}_a^1$). Note that δ vanishes as

$t \to 1$ but γ does not. Therefore Γ_a has a non-trivial boundary supported on $f_a^{-1}(\sqrt[l]{a-1}) \cong f^{-1}(1)$:

$$\Gamma_a \in H_2^B(\overline{U}_a, f_a^{-1}(\sqrt[l]{a-1}); \overline{\mathbb{Q}}), \quad \partial\Gamma_a \neq 0 \in H_1^B(f_a^{-1}(\sqrt[l]{a-1})) \cong H_1^B(f^{-1}(1)).$$

Note that $H_1^B(f^{-1}(1), \mathbb{Q})$ has multiplication by R_0 via the local invariant cycle theorem (cf. (5.6)). The $\chi|_{R_0}$-part $H_1^B(f^{-1}(1), \overline{\mathbb{Q}})(\chi|_{R_0})$ is one-dimensional, spanned by $\partial\Gamma_a$. Hence the χ-part

$$H_1^B((\pi_a f_a)^{-1}(1), \overline{\mathbb{Q}})(\chi) \cong H_1^B(f^{-1}(1), \overline{\mathbb{Q}})(\chi|_{R_0}) \otimes H_0^B(\pi_a^{-1}(1))(\chi|_{\mu_l})$$

is one-dimensional, spanned by the sum $\sum_{\sigma \in \mu_l} \chi(\sigma)^{-1} \cdot \sigma(\partial\Gamma_a)$. Therefore, in Lemma 5.2 we may take $\Gamma(a)$ to be the sum $\sum_{\sigma \in \mu_l} \chi(\sigma)^{-1} \cdot \sigma\Gamma_a$. Then we have

$$\int_{\Gamma(a)} s^{m-1} ds\, \omega^\chi|_{\lambda=a} = \sum_{\sigma \in \mu_l} \chi(\sigma)^{-1} \int_{\sigma\Gamma_a} s^{m-1} ds\, \omega^\chi|_{\lambda=a} = l \int_{\Gamma_a} s^{m-1} ds\, \omega^\chi|_{\lambda=a}.$$

Lemma 5.4. *Let $H_\mu(\lambda)$ be as defined in Theorem 5.1. Then we have*

$$H_\mu(\lambda) = B(1-\alpha^\chi, 1-\beta^\chi) \int_0^1 (\lambda - t)^{\mu-1} t^{1-\alpha^\chi-\beta^\chi} {}_2F_1\left(\begin{matrix} 1-\alpha^\chi, 1-\beta^\chi \\ 2-\alpha^\chi-\beta^\chi \end{matrix}; t\right) dt, \tag{5.7}$$

and

$$\partial_\lambda H_\mu(\lambda) = (\mu - 1) H_{\mu-1}(\lambda). \tag{5.8}$$

Proof. Recall the integral representation of ${}_2F_1$ (cf [10], (4.1.2))

$$B(b, c-b) {}_2F_1\left(\begin{matrix} a, b \\ c \end{matrix}; x\right) = \int_0^1 (1-xt)^{-a} t^{b-1} (1-t)^{c-b-1}\, dt. \tag{5.9}$$

Applying this, we have, writing $\alpha = \alpha^\chi$, $\beta = \beta^\chi$,

$$B(1-\alpha, 1-\beta) {}_2F_1\left(\begin{matrix} 1-\alpha, 1-\beta \\ 2-\alpha-\beta \end{matrix}; t\right) = \int_0^1 (1-ts)^{\alpha-1} s^{-\beta} (1-s)^{-\alpha}\, ds.$$

Letting $u = 1-t$, $v = 1-st$, we have

$$\begin{aligned} &\int_0^1 \int_0^1 (\lambda - t)^{\mu-1} t^{1-\alpha-\beta} (1-ts)^{\alpha-1} s^{-\beta} (1-s)^{-\alpha}\, ds\, dt \\ &= (\lambda - 1)^{\mu-1} \int_0^1 \int_0^v \left(1 - \frac{u}{1-\lambda}\right)^{\mu-1} v^{\alpha-1} (v-u)^{-\alpha} (1-v)^{-\beta}\, du\, dv \\ &= (\lambda - 1)^{\mu-1} \int_0^1 \int_0^1 \left(1 - \frac{wv}{1-\lambda}\right)^{\mu-1} (1-w)^{-\alpha} (1-v)^{-\beta}\, dw\, dv. \end{aligned}$$

Then, using (5.9) and (4.18), we obtain (5.7). Since

$$\partial_\lambda {}_3F_2\left(\begin{matrix} 1, 1, 1-\mu \\ 2-\alpha, 2-\beta \end{matrix}; \frac{1}{1-\lambda}\right) = \frac{1-\mu}{(2-\alpha)(2-\beta)} {}_3F_2\left(\begin{matrix} 2, 2, 2-\mu \\ 3-\alpha, 3-\beta \end{matrix}; \frac{1}{1-\lambda}\right)$$

similarly as (4.15), the proof of (5.8) amounts to show

$$\frac{(1)_n(1)_n(1-\mu)_n}{(2-\alpha)_n(2-\beta)_n(1)_n}+\frac{1}{(2-\alpha)(2-\beta)}\cdot\frac{(2)_{n-1}(2)_{n-1}(2-\mu)_{n-1}}{(3-\alpha)_{n-1}(3-\beta)_{n-1}(1)_{n-1}}$$
$$=\frac{(1)_n(1)_n(2-\mu)_n}{(2-\alpha)_n(2-\beta)_n(1)_n}\quad(n>0),$$

and this is elementary. □

Proposition 5.5. *For a positive integer* $m\equiv k \pmod{l}$, *put*

$$Q_m(a):=\int_{\Gamma_a}s^{m-1}ds\,\omega^\chi|_{\lambda=a},\quad \mu=\frac{m}{l}$$

regarded as an analytic function for a. *Then we have*

$$Q_m(\lambda)=\frac{1}{l}\sum_{i=0}^{N}(a_i(\lambda)+b_i(\lambda)\partial_\lambda)H_{\mu+i}(\lambda),\tag{5.10}$$

where $a_i(\lambda)$, $b_i(\lambda)$ *are as defined in* (4.11). *Moreover we have, if* $\mu>1$,

$$\partial_\lambda Q_m(\lambda)=(\mu-1)Q_{m-l}(\lambda).\tag{5.11}$$

Proof. Since $\frac{-dt}{\lambda-t}=l\frac{ds}{s}$, we have by Lemma 5.3

$$Q_m(\lambda)=\frac{1}{l}B(1-\alpha^\chi,1-\beta^\chi)\int_0^1(\lambda-t)^{\mu-1}\theta\left(t^{1-\alpha^\chi-\beta^\chi}{}_2F_1\left(\begin{matrix}1-\alpha^\chi,1-\beta^\chi\\2-\alpha^\chi-\beta^\chi\end{matrix};t\right)\right)dt.$$

By Lemma 5.4, the same argument as in the proof of Proposition 4.6 works to prove the proposition. □

Lemma 5.6. *Let the differential operator* $\mathscr{D}$ *be as defined in* (4.19). *Then we have*

$$\mathscr{D}H_\mu(\lambda)=-(\lambda-1)^{\mu-1}.$$

Proof. Put $x=\frac{1}{1-\lambda}$, $F(x)={}_3F_2\left(\begin{smallmatrix}1,1,1-\mu\\2-\alpha^\chi,2-\beta^\chi\end{smallmatrix};x\right)$ and $D=x\frac{d}{dx}$. Using $Dx^n=nx^n$, one easily verifies

$$(x(D+1)(D+1-\mu)-(D+1-\alpha^\chi)(D+1-\beta^\chi))F=-(1-\alpha^\chi)(1-\beta^\chi).$$

So H_μ satisfies $\mathscr{D}_1H_\mu=-1$ with

$$\begin{aligned}\mathscr{D}_1&=(x(D+1)(D+1-\mu)-(D+1-\alpha^\chi)(D+1-\beta^\chi))(-x)^{\mu-1}\\&=(-x)^{\mu-1}(x(D+\mu)D-(D-\alpha^\chi+\mu)(D-\beta^\chi+\mu)).\end{aligned}$$

Since $D=(1-\lambda)\partial_\lambda$, where $\partial_\lambda=\frac{d}{d\lambda}$, one obtains $\mathscr{D}_1=(\lambda-1)^{1-\mu}\mathscr{D}$. Hence the lemma follows. □

Now we finish the proof of Theorem 5.1. Let $\mu=m/l$ be as in Theorem 4.1. By Lemma 5.2, (5.10) and (5.11), we have

$$\phi_1(\lambda)\equiv Q_m(\lambda)=\frac{1}{l}\sum\nolimits_{i=0}^{N}(a_i(\lambda)+b_i(\lambda)\partial_\lambda)H_{\mu+i}(\lambda),\quad \mod \overline{\mathbb{Q}(\lambda)},\tag{5.12}$$

$$\phi_2(\lambda)\equiv Q_{m-l}(\lambda)=(\mu-1)^{-1}\partial_\lambda Q_m(\lambda),\quad \mod \overline{\mathbb{Q}(\lambda)}.\tag{5.13}$$

By (5.8), we have similarly as in the proof of Proposition 4.7, that $Q_m(\lambda) = \Theta_1 H_{\mu+N}(\lambda)$ where Θ_1 is the same differential operator as in the proof of Proposition 4.7. By Lemma 5.6, we have

$$H_{\mu+N}(\lambda) \equiv \Theta_2 H_\mu(\lambda) \mod \overline{\mathbb{Q}(\lambda)}$$

where Θ_2 is the same differential operator in the proof of Proposition 4.7. Hence $\phi_1(\lambda) \equiv \Theta_1\Theta_2 H_\mu(\lambda) = \Theta H_\mu(\lambda)$ as desired. □

5.5. Question of Golyshev

We give an affirmative answer to the question of Golyshev in a special case.

Lemma 5.7. *Let*

$$P_{\mathrm{HG}} := D_\lambda(D_\lambda - \mu + \alpha^\chi + \beta^\chi - 1) - \lambda(D_\lambda + \alpha^\chi - \mu)(D_\lambda + \beta^\chi - \mu)$$

be the hypergeometric differential operator. Put

$$Q_{\mathrm{HG}} := \theta_\lambda P_{\mathrm{HG}}, \quad \theta_\lambda := (1-\lambda)D_\lambda + (\mu - 1)\lambda,$$

and local systems of $\mathbb{C}$-modules on $S := \mathbb{P}^1 \setminus \{0, 1, \infty\}$

$$V_P := Sol(D_S/D_S P_{\mathrm{HG}}), \quad V_Q := Sol(D_S/D_S Q_{\mathrm{HG}}),$$

where D_S denotes the ring of differential operators on S. Let

$$0 \longrightarrow V_P \longrightarrow V_Q \longrightarrow V_Q/V_P \longrightarrow 0$$

be the exact sequence obtained by applying the solution functor

$$Sol(\bullet) := \mathcal{H}om_{D_S}(\bullet, \mathscr{O}_S)$$

on

$$0 \longrightarrow D_S/D_S\theta_\lambda \longrightarrow D_S/D_S Q_{\mathrm{HG}} \longrightarrow D_S/D_S P_{\mathrm{HG}} \longrightarrow 0.$$

If $\alpha^\chi, \beta^\chi \notin \mathbb{Z}$, then, for any generically finite dominant map $h : T \to S$, the exact sequence

$$0 \longrightarrow h^*V_P \longrightarrow h^*V_Q \longrightarrow h^*(V_Q/V_P) \longrightarrow 0 \tag{5.14}$$

of $\mathbb{C}[\pi_1(T)]$-modules does not split.

Proof. We first note that $P_{\mathrm{HG}} = \lambda\mathscr{D}$ where $\mathscr{D}$ is the differential operator (4.19). Let $F_\mu(\lambda)$, $G_\mu(\lambda)$ be as in Theorem 4.1, and $H_\mu(\lambda)$ as in in Theorem 5.1. Then the solutions of P_{HG} are $F_\mu(\lambda), G_\mu(\lambda)$ (cf. the proof of Propositions 4.7 and 4.8), and the solutions of Q_{HG} are $F_\mu(\lambda), G_\mu(\lambda), H_\mu(\lambda)$ (cf. Lemma 5.6):

$$V_P = \langle F_\mu(\lambda),\, G_\mu(\lambda)\rangle_{\mathbb{C}}, \quad V_Q = \langle F_\mu(\lambda),\, G_\mu(\lambda),\, H_\mu(\lambda)\rangle_{\mathbb{C}}.$$

Since $\mathrm{Ext}_{\pi_1(S)}(V_Q/V_P, V_P) \to \mathrm{Ext}_{\pi_1(T)}(h^*(V_Q/V_P), h^*V_P)$ is injective, we may assume $T = S$. Assume that the sequence (5.14) splits. This means that there are $c_1, c_2 \in \mathbb{C}$ such that $\mathbb{C}(H_\mu(\lambda) + c_1 F_\mu(\lambda) + c_2 G_\mu(\lambda))$ is stable under the action of $\pi_1(S, \lambda)$. The eigenvalues of the local monodromy T_∞ at $\lambda = \infty$ on V_P are $e^{2\pi i(\alpha^\chi - \mu)}$, $e^{2\pi i(\beta^\chi - \mu)}$. On the other hand $H_\mu(\lambda)$ is the eigenvector with eigenvalue $e^{-2\pi i \mu}$. Since $\alpha^\chi, \beta^\chi \notin \mathbb{Z}$, this implies $c_1 = c_2 = 0$, namely $H_\mu(\lambda)$ is stable under the action of $\pi_1(S, \lambda)$. The eigenvalues of the local monodromy T_0 at $\lambda = 0$ on V_Q (resp. V_P) are $1, 1, e^{2\pi i(\mu - \alpha^\chi - \beta^\chi)}$ (resp. $1, e^{2\pi i(\mu - \alpha^\chi - \beta^\chi)}$). Therefore the eigenvalue of T_0 on $V_Q/V_P \cong \mathbb{C}$ is 1, namely the trivial action. Therefore $T_1 = T_\infty^{-1}$ acts on $H_\mu(\lambda)$ by multiplication by $e^{2\pi i \mu}$. Thus the function

$$(1-\alpha^\chi)(1-\beta^\chi)(\lambda-1)^{1-\mu} H_\mu(\lambda) = {}_3F_2\left(\begin{matrix}1,1,1-\mu\\ 2-\alpha^\chi, 2-\beta^\chi\end{matrix}; (1-\lambda)^{-1}\right)$$

has the trivial monodromy, and this means that this is a rational function. This is impossible. Indeed let $\sum_n a_n z^n$ be the Laurent expansion of the above with respect to variable $z = 1 - \lambda$. Then this satisfies a differential equation

$$Q = (D_z - 1)(D_z - 1)(D_z - 1 + \mu) - z D_z (D_z - 1 + \alpha^\chi)(D_z - 1 + \beta^\chi).$$

Hence

$$(n-1)^2(n-1+\mu)a_n - (n-1)(n-2+\alpha^\chi)(n-2+\beta^\chi)a_{n-1} = 0, \quad \forall n.$$

Then $a_n = 0$ for all $n \leq 0$ as $a_n = 0$ for $n \ll 0$. Moreover

$$a_n = \frac{(n-2+\alpha^\chi)(n-2+\beta^\chi)}{(n-1)(n-1+\mu)} a_{n-1} = \frac{(\alpha^\chi)_{n-1}(\beta^\chi)_{n-1}}{(n-1)!(1+\mu)_{n-1}} a_1, \quad n \geq 1.$$

We thus have

$$\sum_n a_n z^n = a_1 z \cdot {}_2F_1\left(\begin{matrix}\alpha^\chi, \beta^\chi\\ 1+\mu\end{matrix}; z\right).$$

The right-hand side has non-trivial monodromy unless $a_1 = 0$. □

Theorem 5.8. *Let the notation and assumption be as in Theorem* 5.1. *Suppose further that* $\alpha^\chi, \beta^\chi \notin \mathbb{Z}$. *Then the dual of the exact sequence*

$$0 \longrightarrow W_2 \mathscr{H}_{\mathrm{dR}}({}^t\chi) \longrightarrow \mathscr{H}'_{\mathrm{dR}}({}^t\chi) \longrightarrow C({}^t\chi) \longrightarrow 0 \tag{5.15}$$

of connections which underlies (5.2) *is isomorphic to*

$$0 \longrightarrow D_S/\theta_\lambda D_S \longrightarrow D_S/D_S Q_{\mathrm{HG}} \longrightarrow D_S/D_S P_{\mathrm{HG}} \longrightarrow 0. \tag{5.16}$$

In particular, the extension (5.15) *is non-trivial by Lemma* 5.7. *In other words, the regulator* $\rho({}^t\chi)$ *in Theorem* 5.1 *does not vanish.*

Proof. By the Riemann–Hilbert correspondence, it is enough to show that there is an isomorphism

$$\begin{array}{ccccccccc}
0 & \longrightarrow & W_2\mathscr{H}_B({}^t\chi) & \longrightarrow & \mathscr{H}'_B({}^t\chi) & \longrightarrow & C({}^t\chi) & \longrightarrow & 0\\
 & & \downarrow{\cong} & & \downarrow{\cong} & & \downarrow{\cong} & & \\
0 & \longrightarrow & V_P & \longrightarrow & V_P & \longrightarrow & V_Q/V_P & \longrightarrow & 0
\end{array} \tag{5.17}$$

where the top sequence is the underlying local systems of $\mathbb{C}$-modules.

We know that the local system $W_2\mathscr{H}_B({}^t\chi) \cong (W_2\mathscr{H}_B(\chi))^*$ are spanned by Z_1, Z_2, and $C({}^t\chi)$ by the boundary of $\Gamma(\lambda)$ (see (4.10) and Lemma 5.2 for the notation). It is not hard to see that the monodromy representation of $C({}^t\chi)$ is isomorphic to that of V_Q/V_P (cf. (5.1)). The monodromy representation of $\langle Z_1, Z_2\rangle_{\mathbb{C}}$ is isomorphic to that of

$$\left\langle \int_{Z_1} s^{m_1-1}ds\omega,\ \int_{Z_2} s^{m_1-1}ds\omega \right\rangle = \langle \Theta F_\mu(\lambda),\ \Theta G_\mu(\lambda)\rangle$$
$$\overset{\cong}{\leftarrow} \langle F_\mu(\lambda),\ G_\mu(\lambda)\rangle = V_P$$

by Theorem 4.1 (Period formula).

We show that the monodromy representation of $\mathscr{H}'_B({}^t\chi) = \langle Z_1, Z_2, \Gamma(\lambda)\rangle$ is isomorphic to that of

$$\int_{Z_1} s^{m_1-1}ds\omega, \quad \int_{Z_2} s^{m_1-1}ds\omega, \quad \int_{\Gamma_x(\lambda)} s^{m_1-1}ds\omega. \tag{5.18}$$

To do this we need to check that the above integrals are linearly independent over $\mathbb{C}$. The first and second integrals are spanned by $\Theta F_\mu(\lambda)$, $\Theta G_\mu(1-\lambda)$ and they are linearly independent by Theorem 4.1 (Period formula). The third integral is equal to $\Theta H_\mu(\lambda)$ modulo an algebraic function by Theorem 5.1 (Regulator formula). Suppose that the third integral is a linear combination of the first and second integrals. Then

$$\Theta H_\mu(\lambda) = c_1\Theta F_\mu(\lambda) + c_2\Theta G_\mu(\lambda) + (\text{an algebraic function}), \quad (\exists c_i \in \mathbb{C}).$$

Let T_∞ be the local monodromy at $\lambda = \infty$. Then the eigenvalues on $V_P = \langle \Theta F_\mu(\lambda), \Theta G_\mu(\lambda)\rangle$ are $e^{2\pi i(\alpha^\times-\mu)}$, $e^{2\pi i(\beta^\times-\mu)}$ and $\Theta H_\mu(\lambda)$ is the eigenvector with eigenvalue $e^{-2\pi i\mu}$. Applying $(T_\infty - e^{2\pi i(\alpha^\times-\mu)})(T_\infty - e^{2\pi i(\beta^\times-\mu)})$ to the above, we have that $\Theta H_\mu(\lambda)$ is an algebraic function. However since (5.14) is a non-trivial extension (Lemma 5.7), there is some $g \in \mathbb{C}[\pi_1(S,\lambda)]$ such that $gH_\mu(\lambda) = c'_1F_\mu(\lambda) + c'_2G_\mu(\lambda) \neq 0$. Applying Θ, we have that $f(\lambda) := c'_1\Theta F_\mu(\lambda) + c'_2\Theta G_\mu(\lambda) = g\Theta H_\mu(\lambda)$ is also an algebraic function. Since $\Theta F_\mu(\lambda)$, $\Theta G_\mu(\lambda)$ are linearly independent over $\mathbb{C}$, $f(\lambda) \neq 0$. It generates the 2-dimensional space $\langle \Theta F_\mu(\lambda), \Theta G_\mu(\lambda)\rangle \cong V_P$ as $\mathbb{C}[\pi_1(S)]$-module since V_P is irreducible. Hence the monodromy representation of V_P factors through a finite quotient. This is a contradiction. Hence the three integrals (5.18) are linearly independent over $\mathbb{C}$.

We now have that the monodromy representation of $\mathscr{H}'_B({}^t\chi)$ is isomorphic to that of

$$\Theta F_\mu(\lambda),\ \Theta G_\mu(\lambda),\ \Theta H_\mu(\lambda) + (\text{an algebraic function}).$$

Therefore letting $h : T \to S$ be a finite covering which trivializes the monodromy of the algebraic function, we have an isomorphism $h^*\mathscr{H}'_B({}^t\chi)) \cong h^*V_Q$ in a canonical way. Thus the extension data $[h^*V_Q] \in \mathrm{Ext}^1_{\pi_1(T)}(h^*(V_Q/V_P), h^*V_P)$ coincides with $[h^*\mathscr{H}'_B({}^t\chi)] \in \mathrm{Ext}^1_{\pi_1(T)}(h^*\mathrm{Coker}(T_1-1)({}^t\chi), h^*W_2\mathscr{H}'_B({}^t\chi))$ under the natural isomorphisms $V_P \cong W_2\mathscr{H}'_B({}^t\chi)$ and $V_Q/V_P \cong \mathrm{Coker}(T_1-1)({}^t\chi)$. Now the assertion follows from the injectivity of $\mathrm{Ext}^1_{\pi_1(S)}(V_Q/V_P,V_P)\to\mathrm{Ext}^1_{\pi_1(T)}(h^*(V_Q/V_P),h^*V_P)$. This completes the proof. □

5.6. Complement: Precise formula of regulators

Applying the 3-term relation on ${}_3F_2$ (e.g., [2] Lemma 7.5) to (5.12) and (5.13), one can obtain a more explicit description of $\phi_i(\lambda)$ as $\overline{\mathbb{Q}}(\lambda)$-linear combinations of $H_\mu(\lambda)$ and $H_{\mu-1}(\lambda)$:

$$H_\mu(\lambda) := (1-\alpha^\chi)^{-1}(1-\beta^\chi)^{-1}(\lambda-1)^{\mu-1}{}_3F_2\left({1,1,1-\mu \atop 2-\alpha^\chi,2-\beta^\chi};(1-\lambda)^{-1}\right), \quad (5.19)$$

$$H_{\mu-1}(\lambda) := (1-\alpha^\chi)^{-1}(1-\beta^\chi)^{-1}(\lambda-1)^{\mu-2}{}_3F_2\left({1,1,2-\mu \atop 2-\alpha^\chi,2-\beta^\chi};(1-\lambda)^{-1}\right) \quad (5.20)$$

where $\mu := m/l$ (cf. Prop. 4.8). The following theorem is used in [3].

Theorem 5.9 (Regulator formula – precise version). *Let the notation and assumption be as in Theorem 5.1. Let $\mu = m/l$ be as in Theorem 4.1. Let $a_i(\lambda), b_i(\lambda)$ be as in (4.11). Put $a := 2-\alpha^\chi$, $b := 2-\beta^\chi$, and*

$$e_i(s) := (-1)^i(a_i(\lambda)+(s+i)b_{i+1}(\lambda))(1-\lambda)^i$$
$$= \begin{cases} \left(\frac{d^i p_0(\lambda)}{d\lambda^i}+(s+i)\frac{d^{i+1}p_1(\lambda)}{d\lambda^{i+1}}\right)\frac{(1-\lambda)^i}{i!} & i \geq 0, \\ -(s-1)p_1(\lambda)/(1-\lambda) & i=-1, \end{cases}$$

$$A(s) := \frac{s(a+b+2s-3-s(1-\lambda)^{-1})}{(a+s-1)(b+s-1)}, \quad B(s) := \frac{s(1-s)\lambda}{(a+s-1)(b+s-1)}$$

with indeterminate s. Define $C_i(s)$ and $D_i(s)$ by

$$\begin{pmatrix} C_{i+1}(s) \\ D_{i+1}(s) \end{pmatrix} = \begin{pmatrix} A(s) & (\lambda-1)^{-1} \\ B(s) & 0 \end{pmatrix}\begin{pmatrix} C_i(s+1) \\ D_i(s+1) \end{pmatrix}, \quad \begin{pmatrix} C_{-1}(s) \\ D_{-1}(s) \end{pmatrix} := \begin{pmatrix} 0 \\ 1 \end{pmatrix}.$$

Put

$$E_1^{(r)}(s) := \sum_{i\geq -1} e_i(s+r)C_{r+i}(s), \quad E_2^{(r)}(s) := \sum_{i\geq -1} e_i(s+r)D_{r+i}(s).$$

Then

$$\phi_1(\lambda) \equiv C_1(1-\lambda)^n[E_1^{(n)}(\mu)H_\mu(\lambda)+E_2^{(n)}(\mu)H_{\mu-1}(\lambda)],$$
$$\phi_2(\lambda) \equiv C_2(1-\lambda)^{n-1}[E_1^{(n-1)}(\mu)H_\mu(\lambda)+E_2^{(n-1)}(\mu)H_{\mu-1}(\lambda)]$$

modulo $\overline{\mathbb{Q}(\lambda)}$ with some $C_1, C_2 \in \overline{\mathbb{Q}}^\times$. Here we note that $E_i^{(r)}(\mu) \in \overline{\mathbb{Q}}(\lambda)$ are rational functions of variable λ.

Proof. The 3-term relation on ${}_3F_2$ implies that C_i and D_i satisfy

$${}_3F_2\left({1,1,1-s-i \atop a,b};x\right) \equiv C_i(s,x){}_3F_2\left({1,1,1-s \atop a,b};x\right)+D_i(s,x){}_3F_2\left({1,1,2-s \atop a,b};x\right)$$

modulo $\mathbb{Q}(s,x)$. Hence

$$(1-\lambda)^{m/l+i-1}{}_3F_2\left({1,1,1-m/l-i \atop 2-\alpha^\chi,\ 2-\beta^\chi};(1-\lambda)^{-1}\right)$$
$$\equiv(1-\alpha^\chi)(1-\beta^\chi)(1-\lambda)^{r+i}(C_{i+r}(\mu,x)H_\mu(\lambda)+D_{i+r}(q^\chi,x)H_{\mu-1}(\lambda))$$

for $m=k+lr$, $r\in\mathbb{Z}$. Apply this to (5.12) and (5.13). The rest is a direct computation (left to the reader). □

References

[1] M. Asakura and N. Otsubo, *CM periods, CM regulators, and hypergeometric functions, I*, Canad. J. Math. **70** (2018), 481–514.

[2] M. Asakura and N. Otsubo, *CM periods, CM regulators and hypergeometric functions, II*, Math. Z. **289** (2018), no. 3-4, 1325–1355.

[3] M. Asakura and N. Otsubo, *A functional logarithmic formula for hypergeometric function* ${}_3F_2$, Nagoya Math. J. **236** (2019), 29–46.

[4] M. Asakura, N. Otsubo and T. Terasoma, *An algebro-geometric study of special values of hypergeometric functions* ${}_3F_2$, Nagoya Math. J. **236** (2019), 47–62.

[5] A. Beilinson, J. Bernstein and P. Deligne, *Faisceaux pervers* In: Analyse et topologie sur les espaces singuliers I, Astérisque **100** (1982).

[6] A. Erdélyi et al. ed., *Higher transcendental functions*, Vol. 1, California Inst. Tech, 1981.

[7] M. Saito, Mixed Hodge modules. Publ. RIMS, Kyoto Univ. **26**, 221–333 (1990)

[8] M. Saito, *Arithmetic mixed sheaves.* Invent. Math. **144** (2001), no. 3, 533–569.

[9] M. Saito, *On the formalism of mixed sheaves.* RIMS-preprint **784**, Aug. 1991.

[10] L.J. Slater, *Generalized hypergeometric functions*, Cambridge Univ. Press, Cambridge 1966.

Masanori Asakura
Department of Mathematics
Hokkaido University
Sapporo, 060-0810 Japan
e-mail: `asakura@math.sci.hokudai.ac.jp`

Noriyuki Otsubo
Department of Mathematics and Informatics
Chiba University
Chiba, 263-8522 Japan
e-mail: `otsubo@math.s.chiba-u.ac.jp`

Progress in Mathematics, Vol. 338, 31–74

Two Recent p-adic Approaches Towards the (Effective) Mordell Conjecture

J.S. Balakrishnan, A.J. Best, F. Bianchi, B. Lawrence, J.S. Müller, N. Triantafillou, J. Vonk

Abstract. We give an introductory account of two recent approaches towards an effective proof of the Mordell conjecture, due to Lawrence–Venkatesh and Kim. The latter method, which is usually called the method of Chabauty–Kim or non-Abelian Chabauty in the literature, has the advantage that in some cases it has been turned into an effective method to determine the set of rational points on a curve, and we illustrate this by presenting three new examples of modular curves where this set can be determined.

Mathematics Subject Classification (2010). 11G18, 11G50, 11Y50, 14G05.

Keywords. Effective Mordell, Lawrence–Venkatesh method, Chabauty–Kim method, rational points on curves.

1. Introduction

1.1. The Mordell conjecture

Many important developments in arithmetic geometry were motivated by the Mordell conjecture, stated nearly a century ago. Let X be a smooth projective curve, defined over the field of rational numbers $\mathbf{Q}$. Its set of rational points $X(\mathbf{Q})$, which consists of all the projective solutions with rational coordinates to a finite set of equations defining X in some projective space, is an interesting arithmetic quantity. In 1922, Mordell [Mor22] made the following conjecture:

Conjecture 1.1 (Mordell). *Suppose that X is of genus at least two. Then $X(\mathbf{Q})$ is finite.*

In a monumental paper, Faltings [Fal83] proved this conjecture. The method of Faltings is ingenious, and merits a thorough treatment on its own. Indeed, many such are available in the literature, see for instance [CS86] for an early account. In this paper, we wish to give an introductory account of two recent alternative approaches towards this conjecture, due to Lawrence–Venkatesh [LV20] and Kim

[Kim05, Kim09, Kim10]. The latter method, which is usually called the method of *Chabauty–Kim* or *non-abelian Chabauty* in the literature, has the advantage that in some cases it has been turned into an effective method to *determine* the set of rational points $X(\mathbf{Q})$, and we illustrate this by presenting three new examples of modular curves where this set can be determined, due to Best, Bianchi, and Triantafillou.

Remark 1.2. The Mordell conjecture, as well as many of the results discussed below, admit analogues where X is replaced by a smooth *hyperbolic curve*, including also the cases of a punctured elliptic curve and $\mathbf{P}^1 \setminus \{0, 1, \infty\}$, when the set of rational points $X(\mathbf{Q})$ is replaced by the set of S-integral points, where S is a finite set of primes. In this setting, the finiteness of S-integral points is known as Siegel's theorem. Both methods presented here are expected to apply to non-proper hyperbolic curves as well. We discuss the S-unit equation in the context of [LV20] below. Kim [Kim05] proved the finiteness of integral points on $\mathbf{P}^1 \setminus \{0, 1, \infty\}$, and explicit Chabauty–Kim methods for S-unit equations are due to Dan–Cohen and Wewers [DCW15, DCW16, DC20]. Chabauty–Kim theory for integral points on punctured elliptic curves of rank 0 and 1 is discussed in [Kim10] and [BDCKW18].

Remark 1.3. For the purpose of exposition, we only consider the base field $\mathbf{Q}$. It should be noted that many results admit appropriate generalizations to number fields [Sik13, Dog19, BBBM19]. The only exception is our discussion of the method of Lawrence–Venkatesh, where field extensions play an essential role.

1.2. Two recent approaches

After Faltings' proof, two notable new methods for proving finiteness of $X(\mathbf{Q})$ for X of genus $g \geq 2$ have emerged. In broad strokes, they follow a similar strategy: We start by choosing a prime p at which the curve X has good reduction, and we study the set of rational points through the inclusion

$$X(\mathbf{Q}) \subset X(\mathbf{Q}_p). \tag{1}$$

For any field K, we write $G_K = \mathrm{Gal}(\overline{K}/K)$ for its absolute Galois group. The starting point of both the methods of Chabauty–Kim and Lawrence–Venkatesh is the association of a certain finite-dimensional Galois representation over $\mathbf{Q}_p$ to every point on the curve, giving maps

$$\rho : X(K) \longrightarrow \mathrm{Rep}(G_K), \tag{2}$$

for K equal to $\mathbf{Q}$ or $\mathbf{Q}_p$. In both the approaches of Lawrence–Venkatesh and Chabauty–Kim, finiteness of the set $X(\mathbf{Q})$ is obtained from the consideration of a commutative diagram of the following shape:

$$\begin{array}{ccccc} X(\mathbf{Q}) & \longrightarrow & X(\mathbf{Q}_p) & \overset{\mathrm{per}_p}{\dashrightarrow} & \\ \downarrow{\scriptstyle\rho} & & \downarrow{\scriptstyle\rho} & & \\ \mathrm{Rep}(G_{\mathbf{Q}}) & \xrightarrow{\mathrm{res}_p} & \mathrm{Rep}(G_{\mathbf{Q}_p}) & \xrightarrow{\mathrm{D_{cris}}} & \mathrm{MF}^{\phi}/\simeq. \end{array} \tag{3}$$

While the nature of ρ is very different in the two approaches, the horizontal maps are the same. First of all, the map from $X(\mathbf{Q})$ to $X(\mathbf{Q}_p)$ is simply the natural inclusion, and res_p is the restriction of Galois representations, making the diagram commutative in both approaches. The map $\mathrm{D}_{\mathrm{cris}}$ is defined using p-adic Hodge theory. More precisely, it is Fontaine's crystalline Dieudonné functor from p-adic Galois representations to filtered ϕ-modules. Finally, per_p is defined to be the composite of this map with ρ, and will be referred to as the (p-adic) *period map.*

As mentioned above, the maps ρ which feature in the methods of Lawrence–Venkatesh and Chabauty–Kim are of a very different nature, and are responsible for the drastic differences between the two approaches. They may roughly be described as follows:

- The method of Lawrence–Venkatesh starts by considering a family of curves $\mathcal{C} \longrightarrow X$. This is a so-called *Parshin family*, where the fibre $\mathcal{C}_x$ of a point x in $X(K)$ is itself a disjoint union of finite coverings of X, unramified away from the point x. The association ρ is then simply
$$\rho\ :\ x \longmapsto \mathrm{H}^1_{\text{ét}}(\overline{\mathcal{C}}_x, \mathbf{Q}_p).$$
A lemma of Faltings can be used to show that the number of global representations in $\rho(X(\mathbf{Q}_p))$ is finite. The main part of the argument of Lawrence–Venkatesh is to establish that the map per_p is finite-to-one. The argument starts by realizing per_p as the quotient of the Hodge filtration map $\Phi_p : X(\mathbf{Q}_p) \longrightarrow \mathrm{Gr}(g, 2g)$ by the Frobenius centralizer, and showing that on every residue disk
    1. every orbit of the Frobenius centralizer has positive codimension in $\mathrm{Gr}(g, 2g)$, and
    2. the image of Φ_p is Zariski dense.

  The former is established via carefully extending the base field and exploiting the semi-linearity of the Frobenius operator, whereas the latter is established using a monodromy calculation for the family $\mathcal{C}$. The finiteness of $X(\mathbf{Q})$ follows easily from the above commutative diagram.

- In the method of Chabauty–Kim, one chooses a rational base point $b \in X(\mathbf{Q})$ and obtains the association ρ by considering certain well-chosen *unipotent* quotients $\mathrm{U}(b)$ of the algebraic fundamental group $\pi_1^{\text{ét}}(\overline{X}; b)$. This choice of quotient typically depends on the specifics of the curve X under consideration. The association ρ in the method of Chabauty–Kim is then of the form
$$\rho\ :\ x \longmapsto \mathrm{U}(b, x)$$
where $\mathrm{U}(b, x)$ is obtained by twisting the unipotent quotient $\mathrm{U}(b)$ by the path torsor $\pi_1^{\text{ét}}(\overline{X}; b, x)$. This carries the structure of a $\mathbf{Q}_p$-representation of G_K whenever x is in $X(K)$. All these Galois representations are twists of $\mathrm{U}(b)$, whose unipotence provides a certain rigidity that is crucial for arithmetic applications. More precisely, Kim shows that the image of ρ is contained in a set that naturally carries the structure of an algebraic variety, which is

usually referred to as a *Selmer variety*, such that the map res_p between the global and local Selmer varieties is algebraic.

This rigidity provides us with a clear strategy to prove finiteness, in the style of the classical method of Chabauty (see below). Indeed, if we can establish that

1. the global Selmer variety has positive codimension in the local Selmer variety, and
2. the image of $X(\mathbf{Q}_p)$ is Zariski dense,

then the intersection of the two sets (which contains the set of rational points) must be finite. Property (2) is true in great generality, whereas (1) typically requires additional information. Note the amusing similarity with the two steps in the proof of Lawrence–Venkatesh discussed above.

In spite of the apparent similarity of the two strategies, the different nature of the maps ρ already lays bare a crucial difference: In contrast with the unconditional proof of Lawrence–Venkatesh, an additional piece of information is needed to deduce finiteness from the method of Chabauty–Kim. Typically this either takes the form of a *geometric* assumption, such as having a large Néron–Severi rank [BD18, BD20], or the assumption of a geometric conjecture, such as the Bloch–Kato conjecture, see [Kim09].

1.3. Finding rational points explicitly

At first glance, it may seem from the above comments that the conditional nature of the proof of finiteness obtained from the method of Chabauty–Kim puts the method at a significant disadvantage, especially when compared to the unconditional proof of Lawrence–Venkatesh. However, recent developments [BD18, BD20, BDM$^+$19] have shown that in certain examples where additional geometric information is known, the method for proving finiteness can in fact be turned into a method to *explicitly determine* the finite set $X(\mathbf{Q})$.

To explain the ideas, we briefly remind the reader of the method of Chabauty–Coleman [Cha41, Col85], of which an excellent exposition may be found in McCallum–Poonen [MP12]. In this method, one chooses a rational base point b in $X(\mathbf{Q})$ and attaches to every other point a torsor of the p-adic Tate module V of the Jacobian J. More precisely, if K is either $\mathbf{Q}$ or $\mathbf{Q}_p$, this torsor is obtained by the composition

$$\rho \,:\, X(K) \longrightarrow J(K) \longrightarrow \mathrm{H}^1_f(G_K, V) \tag{4}$$

where the first map is the Abel–Jacobi embedding attached to the choice of base point b, and the second map attaches to a point x in $J(K)$ the torsor of V obtained from the inverse limit of the preimages of x under the multiplication-by-p^n map on the Jacobian, i.e.,

$$\mathbf{Q}_p \otimes_{\mathbf{Z}_p} \left(\varprojlim_n \, [p^n]^{-1}(x) \right). \tag{5}$$

Such torsors are classified by the cohomology group $\mathrm{H}^1(G_K, V)$ and satisfy certain *Selmer conditions*[1] which are denoted by the subscript f. This association ρ is familiar from the context of the classical method of descent, used to compute the Mordell–Weil group of the Jacobian.

We now obtain the commutative diagram:

$$\begin{array}{ccccc} X(\mathbf{Q}) & \longrightarrow & X(\mathbf{Q}_p) & \overset{\mathrm{per}_p}{\dashrightarrow} & \\ \downarrow{\scriptstyle\rho} & & \downarrow{\scriptstyle\rho} & & \searrow \\ \mathrm{H}^1_f(G_{\mathbf{Q}}, V) & \xrightarrow{\mathrm{res}_p} & \mathrm{H}^1_f(G_{\mathbf{Q}_p}, V) & \xrightarrow{\sim} & \mathrm{H}^0(X, \Omega^1_X)^\vee \end{array} \tag{6}$$

representing perhaps the simplest instance of the Chabauty–Kim strategy towards the Mordell conjecture discussed above, where U is taken to be the *abelianization* V of the fundamental group. In this situation, the relevant filtered ϕ-modules are classified by the dual to the space of holomorphic differentials on X, which is of dimension g, and the isomorphism is provided by the Bloch–Kato logarithm. With suitable finiteness conditions f, the dimension of $\mathrm{H}^1_f(G_{\mathbf{Q}}, V)$ can be bounded above by the rank r of the $\mathbf{Q}$-rational points of the Jacobian of X. The discussion of how to prove finiteness of $X(\mathbf{Q})$ using the method of Chabauty–Kim then specializes to the classical argument of Chabauty, who deduces finiteness under the assumption that $r < g$.

Going one step further, we note that the p-adic period map per_p has the following concrete description:

$$\mathrm{per}_p(x) = \left(\omega \longmapsto \int_b^x \omega\right) \tag{7}$$

where the integration is taken in the sense of Coleman [Col85]. Our ability to compute Coleman integrals [BBK10, Bal15, BT20] often results in an explicit determination of the set $X(\mathbf{Q})$. Since there already exist several excellent expositions of this method [MP12], we will simply explain the method by showing it in action for a single example.

Example. Let X be the genus 3 hyperelliptic curve with minimal model[2]

$$w^2 + (z^4 + z^2 + z + 1)w = -z^5 - z^2.$$

A search for points with small coordinates gives that

$$\{\infty^\pm, (-1,-2), (-1,0), (0,-1), (0,0)\} \subseteq X(\mathbf{Q}), \tag{8}$$

where $\infty^+ = (1:0:0)$ and $\infty^- = (1:-1:0)$ are the points at infinity. In order to determine the full set of rational points $X(\mathbf{Q})$, we apply the Chabauty–Coleman

[1] We are deliberately vague about these finiteness conditions here, but mention that the discussion below can be made unconditional on the finiteness of the Tate–Shafarevich group of the Jacobian.
[2] Here X is the curve of absolute discriminant and conductor both equal to $60329 = 23 \cdot 43 \cdot 61$ from the database [BPSS].

method with $p = 3$; for convenience, we work with the following model for X:

$$X' \colon w^2 = z^8 + 2z^6 - 2z^5 + 3z^4 + 2z^3 - z^2 + 2z + 1.$$

We embed X' into its Jacobian J via the Abel–Jacobi map corresponding to the base point $b = (0, 1)$ in $X'(\mathbf{Q})$. A computation in `Magma` [BCP97] shows that the Mordell–Weil rank of J is equal to 1, and the above discussion then implies that the codimension of the image of res_3 is at least 2. In fact, it is precisely equal to 2: the set $\{\omega_i = z^i \frac{dz}{w} : 0 \le i \le 2\}$ is a basis for $\mathrm{H}^0(X', \Omega^1_{X'})$ and we have

$$\begin{aligned} \mathrm{per}_3(0,-1)(\omega_0) &\equiv 3(3 + 3^3 + 2\cdot 3^4) && \mod 3^6\\ \mathrm{per}_3(0,-1)(\omega_1) &\equiv 3(1 + 3^3 + 3^4) && \mod 3^6\\ \mathrm{per}_3(0,-1)(\omega_2) &\equiv 3(1 + 3^2 + 2\cdot 3^3 + 2\cdot 3^4) && \mod 3^6. \end{aligned}$$

Thus, we may choose generators $\alpha = a_0\omega_0 - a_1\omega_1$ and $\beta = b_0\omega_0 - b_2\omega_2$ for the $\mathbf{Q}_3$-vector space

$$\{\omega \in \mathrm{H}^0(X', \Omega^1_{X'}) : \mathrm{res}_3(c)(\omega) = 0 \text{ for all } c \in \mathrm{H}^1_f(G_{\mathbf{Q}}, V)\}$$

such that

$$\begin{aligned} a_0 &\equiv 1 + 3^3 + 3^4 \mod 3^5 & a_1 &\equiv 3 + 3^3 + 2\cdot 3^4 \mod 3^5\\ b_0 &\equiv 1 + 3^2 + 2\cdot 3^3 + 2\cdot 3^4 \mod 3^5 & b_2 &\equiv 3 + 3^3 + 2\cdot 3^4 \mod 3^5. \end{aligned} \tag{9}$$

By construction, we have

$$X'(\mathbf{Q}) \subseteq \{x \in X'(\mathbf{Q}_3) : \mathrm{per}_3(x)(\alpha) = 0 \text{ and } \mathrm{per}_3(x)(\beta) = 0\} =: \mathcal{T}; \tag{10}$$

a computation shows that $\mathcal{T}$ contains precisely 6 points and hence that the inclusion in (8) is in fact an equality. Explicitly, suppose for instance that we want to compute all $x \in \mathcal{T}$ which reduce to the point $(1:1:0)$ in $X'(\mathbf{F}_3)$. For $\gamma \in \{\alpha, \beta\}$ we have

$$\mathrm{per}_3(x)(\gamma) = \mathrm{per}_3(1:1:0)(\gamma) + \int_{(1:1:0)}^x \gamma = \int_{(1:1:0)}^x \gamma;$$

expanding in terms of the local parameter $t = z(x)^{-1}$ and formally integrating yields

$$\begin{aligned} \mathrm{per}_3(x)(\alpha) &= (2\cdot 3 + 3^2)\cdot t^2 + (2\cdot 3^{-1} + 2 + 2\cdot 3 + 3^2)\cdot t^3 && \mod (3^3, t^4)\\ \mathrm{per}_3(x)(\beta) &= 3\cdot t + (2\cdot 3^{-1} + 1 + 3 + 2\cdot 3^2)\cdot t^3 && \mod (3^3, t^4). \end{aligned}$$

The ith coefficient of the local expansion of $\mathrm{per}_3(x)(\alpha)$ or $\mathrm{per}_3(x)(\beta)$ has valuation bounded from below by $-\mathrm{ord}_3(i)$; from Newton polygon considerations, we deduce that

- $\mathrm{per}_3(x)(\alpha)$ has a double zero at $t = 0$, a simple zero at some $t \in \mathbf{Z}_3$ which satisfies $t \equiv 2\cdot 3^2 \mod 3^3$, and no other zero in $3\,\mathbf{Z}_3$;
- $\mathrm{per}_3(x)(\beta)$ has a simple zero at $t = 0$, two simple zeros congruent modulo 3^2 to $2\cdot 3$ and 3, respectively, and no other zero in $3\,\mathbf{Z}_3$.

Therefore, the intersection of the zero sets of $\mathrm{per}_3(x)(\alpha)$ and $\mathrm{per}_3(x)(\beta)$ in the residue disk of the point $(1 : 1 : 0)$ in $X'(\mathbf{F}_3)$ is precisely $[(1 : 1 : 0)] \subset X'(\mathbf{Q})$.

We emphasize that neither α nor β on their own would have sufficed to determine $X(\mathbf{Q})$, as each of $\mathrm{per}_3(x)(\alpha)$ and $\mathrm{per}_3(x)(\beta)$ vanishes at some points $x \in X'(\mathbf{Q}_3) \setminus X'(\mathbf{Q})$ which we can only compute modulo 3^n for a choice of n. More generally, for curves X of genus g with rank $g-1$ Jacobians, the Chabauty–Coleman method typically provides us with only one locally analytic function whose zero set $\mathcal{T}$ contains $X(\mathbf{Q})$. It is then often the case that $\mathcal{T}$ contains some points that we cannot recognize as points in $X(\mathbf{Q})$. In such situations, the Mordell–Weil sieve (see §6.7 for a discussion) can often be used to prove that the p-adic approximations of these points that we have computed cannot be approximations of points in $X(\mathbf{Q})$.

1.4. Integral points on higher-dimensional varieties

Both methods (Lawrence–Venkatesh and Chabauty–Kim) could be applied to the problem of finding integral points on higher-dimensional varieties as well. To fix ideas, let X be a smooth variety over $\mathbf{Q}$. If X has large nonabelian fundamental group, one can hope to construct a non-trivial p-adic local system on X. This will attach a Galois representation to every point $x \in X(\mathbf{Q})$, giving rise to a period map ρ as above. In the higher-dimensional setting, one can no longer conclude finiteness of integral points; rather, these methods give the weaker result that $X(\mathbf{Z})$ is not dense for the p-adic analytic topology. See [Had11, Section 9] for a result of this form. (In dimension one, non-density for the analytic topology is equivalent to finiteness.)

It is sometimes possible to strengthen p-adic non-density to Zariski using tools from transcendence theory. In [LV20, Section 9] it is shown that in certain moduli spaces of hypersurfaces, the integral points are not Zariski dense. The key input is a recent transcendence result for period mappings, due to Bakker and Tsimerman [BT19]. This opens the possibility that one might prove finiteness of integral points by an inductive approach: taking X' to be the Zariski closure of the integral points in X, one would hope to use the method of [LV20] to prove that, if $\dim X' \geq 1$, the integral points cannot be dense in X'. To make this work one would need uniform control on the monodromy of the given family, restricted to all subvarieties $X' \subseteq X$.

2. The method of Lawrence–Venkatesh: Finiteness

In this section, we will discuss the main ideas of the approach towards the Mordell conjecture due to Lawrence and Venkatesh. For simplicity, our main focus will be to explain the method in the case of $X = \mathbf{P}^1 \setminus \{0, 1, \infty\}$ where the proof is especially simple. Finally, we make some comments about the obstacles one faces in making this approach effective, in the example of the 2-unit equation.

Recall from the introduction that we start by constructing a map ρ which attaches a Galois representation to any point on X. In the method of Lawrence–

Venkatesh, the map ρ arises from the cohomology of the fibres of a certain Parshin family $\mathcal{C} \longrightarrow X$, see §2.2. In the case of $X = \mathbf{P}^1 \setminus \{0, 1, \infty\}$, which we discuss first, this family is a simple modification of the classical *Legendre family* of elliptic curves.

2.1. The S-unit equation

To explain some of the ideas in the proof, we discuss the case of the S-unit equation in more detail. This has the benefit of being substantially simpler, while still containing many of the main ideas that go into the proof of the Mordell conjecture. To illustrate the ideas of the proof, we will start with a version of the Parshin family for which the period map per_p fails to be finite-to-one. Then we will give a correct argument, in which a non-trivial Galois action on H^0 of the fibers supplies the key missing ingredient.

Take $K = \mathbf{Q}$ and S a finite set of primes. We denote the set of S-units by $\mathcal{O}_S^\times$ and will consider the S-*unit equation* given by

$$x + y = 1, \qquad x, y \in \mathcal{O}_S^\times, \tag{11}$$

whose solution set is finite by Siegel's theorem. This statement represents an attractive toy case for the Mordell conjecture; its geometric proof along the lines sketched above takes place on $X = \mathbf{P}^1 \setminus \{0, 1, \infty\}$. Note that we may enlarge S without loss of generality, so that we may as well assume that S contains 2.

The role of the Parshin family is played by the *Legendre family* over $\mathcal{O}_S$. Denoting x for the coordinate on $\mathcal{X} = \mathbf{P}^1_{\mathcal{O}_S} \setminus \{0, 1, \infty\}$, this family $\mathcal{C} \longrightarrow \mathcal{X}$ is given by the equation

$$\mathcal{C} \; : \; w^2 = z(z-1)(z-x). \tag{12}$$

Let p be a prime not below any primes in S and which is unramified in K. The Legendre family gives us a Galois representation $\rho(x)$ on the étale cohomology group $\mathrm{H}^1_{\text{ét}}(\mathcal{C}_{\overline{x}}, \mathbf{Q}_p)$. This gives the following diagram:

$$\begin{array}{ccccc} \mathcal{X}(\mathcal{O}_S) & \longrightarrow & \mathcal{X}(\mathcal{O}_v) & \overset{\mathrm{per}_p}{\dashrightarrow} & \\ \downarrow{\scriptstyle\rho} & & \downarrow{\scriptstyle\rho} & & \\ \{\text{Iso classes } \rho\} & \longrightarrow & \{\text{Iso classes } \rho_v\} & \xrightarrow{\mathrm{D_{cris}}} & \mathrm{MF}^\phi/\simeq \end{array} \tag{13}$$

Let us now make a first attempt to deduce finiteness from the above diagram. There are two major considerations to the strategy, corresponding to *global* and *local* aspects. The local considerations revolve around a careful analysis of the period map, via a monodromy calculation.

a. Global representations. The Mordell conjecture will ultimately be reduced to a finiteness statement about a certain set of global Galois representations, due to Faltings. More precisely, the proof of [Fal83, Satz 5] deduces the following consequence from the classical theorem of Hermite–Minkowski:

Lemma 2.1. *Fix integers $w, d \geqslant 0$, and fix a number field K and a finite set S of primes of $\mathcal{O}_K$. There are, up to conjugation, only finitely many semisimple Galois representations $\rho : G_K \to \mathrm{GL}_d(\mathbf{Q}_p)$ such that*

(a) *ρ is unramified outside S, and*
(b) *ρ is pure of weight w, i.e., for every prime $\mathfrak{p} \notin S$ all the eigenvalues of Frobenius at $\mathfrak{p}$ are algebraic integers, all of whose conjugates have complex absolute value $|\mathcal{O}_K/\mathfrak{p}|^{w/2}$.*

It should be noted that this does not make the approach of Lawrence–Venkatesh depend on the work of Faltings in an essential way, as this lemma is comparatively simple in Faltings' overall argument.

The semisimplicity hypothesis in Faltings' lemma is essential: there can be infinitely many non-trivial extensions between Galois representations[3]. In fact, Faltings shows that all the representations we consider – which arise as subquotients of the étale cohomology of a curve – are semisimple. This fact requires the full weight of Faltings' argument in [Fal83]. In order to give an independent proof of Mordell's conjecture, it is necessary to contemplate the possibility that some of these representations might not be semisimple. In potential algorithmic applications, we know this situation cannot arise, so we will content ourselves here with mentioning that in [LV20] this is addressed by showing that all but finitely many representations in our family must be simple. This is a consequence of results of the following form:

1. If the global representation $\rho(x)$ has a (global) subrepresentation, then the local representation must be of a particularly special form.
2. There are finitely many x in $X(\mathbf{Q}_p)$ where the local representation $\rho(x)$ takes this special form.

b. The period map. The more subtle points of the argument of Lawrence–Venkatesh lie in the study of the period map per_p, where one systematically enlarges the base field to gain control over the Frobenius centralizers. Let us explain the need for this step, by first approaching the problem naively using the unadjusted Legendre family above.

Recall that we want to show finiteness of the set of solutions to the S-unit equation. Since we already established the finiteness of the set of isomorphism classes of global representations $\rho(x)$ that can arise, it is tempting to try and show that the fibres of the period map per_p are finite. However, this is **not** true: The filtered ϕ-modules that arise in the image of per_p necessarily are of the form $\mathrm{H}^1_{\mathrm{dR}}(\mathcal{C}_x, \mathbf{Q}_p)$, and on every good residue disk of $X(\mathbf{Q}_p)$ the Frobenius operator ϕ has a constant characteristic polynomial

$$f = aT^2 + bT + c \quad \in \mathbf{Z}_p[T] \tag{14}$$

[3] As we will see in the next section, the existence of families of non-trivial extensions of a fixed set of Galois representations is precisely what underlies the method of Chabauty–Kim.

which has two roots in $\mathbf{C}_p$ whose valuations sum up to 1. The number of residue disks is finite, and for each of these finitely many polynomials f, the filtered ϕ-module belongs to a finite number of possible isomorphism classes, which is most easily seen with a simple case-by-case analysis:

- If f is irreducible, then we may pick a basis e_1 for Fil^1 and set $e_2 = \phi(e_1)$. Then $\{e_1, e_2\}$ is a basis for $\mathrm{H}^1_{\mathrm{dR}}$. With respect to this basis, we have $\mathrm{Fil}^1 = \langle e_1 \rangle$ and
$$\phi = \begin{pmatrix} 0 & -b \\ 1 & -a \end{pmatrix}.$$
- If f is reducible, then it must have distinct roots of valuations 0 and 1, corresponding to eigenvectors e_1, e_2 which necessarily span $\mathrm{H}^1_{\mathrm{dR}}$. Then we either have $\mathrm{Fil}^1 = \langle e_1 \rangle$ or $\langle e_2 \rangle$, or we can rescale the eigenvectors to obtain $\mathrm{Fil}^1 = \langle e_1 + e_2 \rangle$.

In conclusion, we see that there is only a *finite number* of possible isomorphism classes of filtered ϕ-modules attached to the representations $\rho(x)$, and therefore the period map appearing in (13) cannot possibly have finite fibres! Furthermore, we see from this discussion exactly what the problem is, since we had in each case so much freedom in choosing our basis, so as to move around the Hodge filtration Fil^1 while respecting the Frobenius operator.

We can rephrase the problem as follows. Fix a pair (V, ϕ) of a two-dimensional vector space and linear endomorphism; in our situation, (V, ϕ) will arise as the crystalline cohomology $\mathrm{H}^1_{\mathrm{cris}}(\mathcal{C}_x \,/\, \mathbf{Z}_p)$, which only depends on the reduction of x modulo p. The possible filtrations $\mathrm{Fil}^1 \subseteq V$ are classified by the Grassmannian $\mathrm{Gr}(\mathrm{Fil}^1 \subseteq V)$. The centralizer $Z(\phi)$ acts on $\mathrm{Gr}(\mathrm{Fil}^1 \subseteq V)$, and the orbits of this action are in bijection with isomorphism classes $(V, \phi, \mathrm{Fil}^1)$ of filtered ϕ-module with underlying ϕ-module (V, ϕ). In the setting just described, $Z(\phi)$ has a Zariski-dense orbit on $\mathrm{Gr}(\mathrm{Fil}^1 \subseteq V)$, so most such filtered ϕ-modules belong to a single isomorphism class.

Interlude: Semilinearity. Let us take a short break to recall some crystalline theory. So far we have been applying p-adic Hodge theory, in particular the crystalline comparison theorem, to schemes $\mathcal{C}_x$ over $\mathbf{Q}_p$. In general, suppose L_p is an unramified extension of $\mathbf{Q}_p$, and $\mathcal{C}_x$ is a scheme over L_p, admitting a smooth model over $\mathcal{O}_{L_p}$. Then L_p is Galois over $\mathbf{Q}_p$, with cyclic Galois group generated by a Frobenius element Fr that acts as the pth power map on the residue field. The crystalline de Rham cohomology $\mathrm{H}^1_{\mathrm{dR}}(\mathcal{C}_x \,/ L_p)$ has the structure of a filtered ϕ-module, where ϕ is now a *semilinear* operator: it satisfies

$$\phi(\lambda v) = \mathrm{Fr}(\lambda)\phi(v). \tag{15}$$

This is important because semilinear automorphisms have small centralizers: it's not easy for an automorphism of V to both respect the action of L_p and commute with ϕ. This is made precise in the following lemma, which was proved in Lawrence–Venkatesh [LV20, Lemma 2.1].

Lemma 2.2. *Let L_p be an unramified extension of $\mathbf{Q}_p$ of degree e, and let $\mathrm{Fr} : L_p \to L_p$ be the Frobenius endomorphism that acts as the pth power map on the residue field. Let V be an L_p-vector space of dimension d, and $\phi : V \to V$ a Fr-semilinear automorphism. Define the centralizer $\mathrm{Z}(\phi)$ of ϕ in the ring of L_p-linear endomorphisms of V via*

$$\mathrm{Z}(\phi) = \{f : V \longrightarrow V \text{ an } L_p\text{-linear map}, \quad f\phi = \phi f\};$$

it is a $\mathbf{Q}_p$-vector space. Then

$$\dim_{\mathbf{Q}_p} \mathrm{Z}(\phi) = \dim_{L_p} \mathrm{Z}(\phi^e),$$

where $\phi^e : V \to V$ is now L_p-linear. In particular, $\dim_{\mathbf{Q}_p} \mathrm{Z}(\phi) \leqslant (\dim_{L_p} V)^2$.

c. Finiteness. Armed with this tool, we now return to the failed finiteness argument above, and take advantage of semilinearity to resolve the issues we were having. More precisely, we bound the size of Frobenius centralizers by considering instead the modified Parshin family

$$E : w^2 = z(z-1)(z-t), \qquad t^8 = x. \tag{16}$$

For every field K and x in $X(K)$, the fiber E_x is a geometrically disconnected curve whose H^0 is the algebra $K[t]/(t^8 - x)$. Suppose $K = \mathbf{Q}_p$ and x is a unit in $\mathbf{Q}_p$ which is not a square[4]. Then E_x is a curve defined over $L_p = \mathbf{Q}_p[x^{1/8}]$, the degree-8 unramified extension of $\mathbf{Q}_p$. We want to show that the map

$$\mathrm{per}_p : \mathcal{X}(\mathcal{O}_v) \longrightarrow (\mathrm{MF}^{\phi}/\simeq)$$

is finite-to-one. On each p-adic residue disk $\Omega_v \subseteq \mathcal{X}(\mathcal{O}_v)$, the ϕ-module $(V, \phi) = \mathrm{H}^1_{\mathrm{cris}}(E_x)$ is constant; only the filtration varies. Thus we can regard per_p as a map

$$\mathrm{per}_p : \Omega_v \longrightarrow \mathrm{Gr}(\mathrm{Fil}^1 \subseteq V) \longrightarrow \mathrm{Gr}(\mathrm{Fil}^1 \subseteq V)/Z(\phi)\,.$$

Since per_p is an analytic map from a one-dimensional source, to show it is finite-to-one, we need only show that it is not constant; in other words, that the image of $\Phi_p \colon \Omega_v \to \mathrm{Gr}(\mathrm{Fil}^1 \subseteq V)$ is not contained in a single orbit of $Z(\phi)$. This follows from the following two results:

1. Every orbit of $Z(\phi)$ has positive codimension in $\mathrm{Gr}(\mathrm{Fil}^1 \subseteq V)$.
2. The image of Φ_p is Zariski dense.

The first of these two follows from the bound in Lemma 2.2; the second, from a complex monodromy calculation. It is essential that L_p have large degree over $\mathbf{Q}_p$, which comes from the assumption that x is not a square in $\mathbf{Q}_p$.

The Zariski density of the image of Φ_p is obtained by comparing it with the complex period map $\Phi_{\mathbf{C}}$. Let us recall the construction of $\Phi_{\mathbf{C}}$. The family E of elliptic curves over X gives rise to a variation of Hodge structure on X. Let $\Omega_{\mathbf{C}}$ be a contractible open subset of X^{an}, containing some basepoint x_0 in $X(K)$, for K

[4]It is enough to consider x of this form by an elementary argument based on the fact that, if x is both a square and a solution to the S-unit equation in some number field K, then $\pm\sqrt{x}$ satisfy the S-unit equation as well. However, this does necessitate some care in the choice of p.

a number field. Over $\Omega_{\mathbf{C}}$, the family E splits as the disjoint union of eight families of elliptic curves $E^{(1)}, \ldots, E^{(8)}$. (The monodromy action of $\pi_1(X)$ preserves the splitting but permutes the eight components.) Choose an integral basis $\mathbf{B}$ for the fiberwise Betti cohomology of each elliptic curve $V_{\mathbf{C}}^{(i)} = \mathrm{H}_B^1(E_{x_0}^{(i)})$ over x_0. With respect to this basis, the Hodge filtration is described by a map

$$\Phi_{\mathbf{C}} \colon \Omega_{\mathbf{C}} \longrightarrow \prod_{i=1}^{8} \mathrm{Gr}(\mathrm{Fil}^1 \subseteq V^{(i)}),$$

where the Grassmannian classifies one-dimensional subspaces of the two-dimensional $V^{(i)}$.

The importance of $\Phi_{\mathbf{C}}$ to us comes from the fact that $\Phi_{\mathbf{C}}$ and Φ_p are, in a suitable sense, the same. (See [LV20, Section 3.4] for details.) Both period maps satisfy the same algebraic differential equation, coming from the Gauss–Manin connection. It follows that in suitable local coordinates, the (complex) power series representation of $\Phi_{\mathbf{C}}$ and the (p-adic) power series representation of Φ_p both have all their coefficients in the number field K, and the two power series agree. This means we can compare the images of the two period maps, and Lemmas 3.1 and 3.2 of [LV20] yield the following result:

Lemma 2.3. *The image of* $\Phi_{\mathbf{C}}$ *is Zariski dense if and only if the image of* Φ_p *is Zariski dense.*

The advantage of this result is that establishing the Zariski-density of the map $\Phi_{\mathbf{C}}$ boils down to an explicit monodromy calculation, see [LV20, Eqn. 3.11].

Lemma 2.4. *If the image of the monodromy representation of* E *contains a Zariski-dense subset of* Sp_2^d*, then the image of* $\Phi_{\mathbf{C}}$ *is Zariski dense in* $\mathrm{Gr}(\mathrm{Fil}^1 \subseteq V)$.

Proof. Let $\widetilde{X}$ be the universal cover of X, and extend $\Phi_{\mathbf{C}}$ to a map

$$\Phi_{\mathbf{C}} \colon \widetilde{X} \longrightarrow \mathrm{Gr}(\mathrm{Fil}^1 \subseteq V).$$

This map $\Phi_{\mathbf{C}}$ is $\pi_1(X)$-equivariant, where $\pi_1(X)$ acts on the Grassmannian through the monodromy representation. Since the image of monodromy is Zariski dense, the extended $\Phi_{\mathbf{C}}$ has Zariski-dense image. By analytic continuation, the restriction of $\Phi_{\mathbf{C}}$ to $\Omega_{\mathbf{C}}$ has Zariski-dense image as well. □

Lemma 2.5. *The image of the monodromy representation*

$$\pi_1(X, x_0) \longrightarrow \mathrm{Aut}\left(\prod_{i=1}^{8} \mathrm{H}_B^1(E_{x_0}^{(i)})\right)$$

contains a Zariski-dense subset of $\mathrm{Sp}_2(\mathbf{Z})^8$.

Proof. This is a calculation in classical topology, see [LV20, Lemma 4.3]. □

2.2. The Mordell conjecture over general K

After our discussion of the S-unit equation, we now make a brief foray into the general case, and outline how to adapt this argument to prove Mordell's conjecture. Suppose X is a smooth projective curve of genus at least 2 over K. We will define the *Parshin family* over X, implicitly dependent on a parameter q. It will replace the Legendre family in the S-unit argument.

Let $q \geq 3$ be a prime number, and let $\mathrm{Aff}(q)$ be the non-abelian group of affine-linear transformations $x \mapsto ax + b$ over $\mathbf{F}_q$. The action of $\mathrm{Aff}(q)$ on $\mathbf{F}_q$ realizes $\mathrm{Aff}(q)$ as a subgroup of the symmetric group S_q. Note also that $\mathrm{Aff}(q)$ surjects onto $\mathbf{F}_q^\times$.

Definition 2.6. Let X be a curve over K, and $x \in X(K)$ a point of X. An $\mathrm{Aff}(q)$-cover of X, branched at x, is a curve Z and a map $Z \to X$, satisfying the following properties.

- $Z \to X$ is étale over $X - \{x\}$, but not étale over x.
- $Z \to X$ is of degree q.
- For any choice of basepoint x_0, and for an appropriate identification of the fiber over x_0 with $\mathbf{F}_q$, the monodromy map $\pi_1(X, x_0) \to S_q$ corresponding to the cover Z has image $\mathrm{Aff}(q)$.

For every x in $X(K)$, there are finitely many isomorphism classes of $\mathrm{Aff}(q)$-covers $Z \to X$ branched at x. The Parshin family $Y \to X$ is characterized by the property that the fiber Y_x is geometrically the disjoint union of these finitely many curves.

In the S-unit argument, the key semilinearity bound came from taking an 8th root of x (along with the elementary assumption that x is not a square). Here the corresponding bound comes from the torsion on the Jacobian J of X, which is guaranteed to have a non-trivial Galois structure. Specifically, for any $\mathrm{Aff}(q)$-cover of X, the composed map

$$\pi_1(X - \{x\}) \longrightarrow \mathrm{Aff}(q) \longrightarrow \mathbf{F}_q^\times$$

gives a degree-$(q-1)$ cover of X that turns out to be unramified everywhere, even over x. This cover in turn corresponds to a $(q-1)$-torsion point on the dual of the Jacobian. We choose q and p so that the Frobenius at p acts with sufficiently large orbits on $J[q-1]$; this in turn guarantees that the components of each fiber Y_x are defined over large p-adic fields, so we can leverage the semilinearity Lemma 2.2.

As with the S-unit equation, a calculation in the classical topology is needed to show that the Parshin family has big monodromy. Fix X and x, and let $Z_1, \ldots, Z_N$ be the $\mathrm{Aff}(q)$-covers of X branched at x. We want to determine the image of the monodromy action

$$\mathrm{Mon}\colon \pi_1(X, x) \longrightarrow \mathrm{Aut}\left(\prod_i \mathrm{H}^1_B(Z_i)\right)$$

as an algebraic group. The cohomology of each Z_i contains a copy of $\mathrm{H}^1_B(X)$; define[5]

$$\mathrm{H}^1_{\mathrm{Pr}}(Z_i) = \mathrm{H}^1_B(Z_i)/\,\mathrm{H}^1_B(X).$$

The map Mon descends to an automorphism of $\prod_i \mathrm{H}^1_{\mathrm{Pr}}(Z_i)$. We need the following big monodromy result.

Theorem 2.7. *The Zariski closure of the image of*

$$\mathrm{Mon}\colon \pi_1(X,x) \longrightarrow \mathrm{Aut}\left(\prod_i \mathrm{H}^1_{\mathrm{Pr}}(Z_i)\right)$$

contains the group

$$\prod_i \mathrm{Sp}(\mathrm{H}^1_{\mathrm{Pr}}(Z_i)).$$

This theorem is really saying that the image of monodromy is as big as possible: we know for abstract reasons that the identity component of the Zariski closure of the image is no larger than the product of symplectic groups. We say a few words about the main ideas that go into the proof:

The monodromy action of $\pi_1(X,x)$ on the covers Z_i extends to an action of the full mapping class group[6] $\mathrm{MCG}(X-\{x\})$. By the Birman exact sequence, $\pi_1(X,x)$ is a normal subgroup of $\mathrm{MCG}(X-\{x\})$. Since the symplectic group is simple modulo center, we can deduce Theorem 2.7 if we know that the Zariski closure of the image of

$$\mathrm{Mon}\colon \mathrm{MCG}(X-\{x\}) \longrightarrow \mathrm{Aut}\left(\prod_i \mathrm{H}^1_{\mathrm{Pr}}(Z_i)\right)$$

contains said product of symplectic groups. The benefit to working with the full mapping class group is that we now have access to Dehn twists, a particularly simple class of automorphism that is amenable to explicit calculation. Dehn twists map to unipotent automorphisms via Mon, and the proof concludes by producing a collection of unipotent automorphisms that generates the full symplectic group.

The study of mapping class group representations like Mon is a big subject in geometric topology. Looijenga [Loo97] studied the analogous question for abelian covers. Grunewald, Larsen, Lubotzky, and Malestein [GLLM15] study (unramified) covers of *compact* surfaces, and in a recent paper [ST20] Salter and Tshishiku study covers whose covering group is the Heisenberg group. These results are stronger than those of Lawrence–Venkatch: they all show that the image of the representation has finite index in an appropriate arithmetic group, rather than merely being Zariski dense.

---

[5]The symbol "Pr" stands for "primitive."

[6]The mapping class group is the group of *topological* automorphisms of the topological surface X fixing the point x, up to isotopy fixing x. The book of Farb and Margalit [FM12] is an excellent introduction and reference on mapping class groups.

3. The method of Lawrence–Venkatesh: Effectivity

We now discuss the extent to which we expect the work of Lawrence–Venkatesh to yield a method for *explicitly determining* the set $X(K)$ in examples. Since it is so recent, it is unsurprising that this aspect of the method of Lawrence–Venkatesh does not yet seem to be addressed in the literature. In this section we adopt a more speculative tone, merely making some brief comments about various ingredients that would likely be needed to parlay this method into an algorithm for bounding the number of rational points on a curve over a number field. This would yield, in a weak sense, a form of "algorithmic Mordell."

Roughly speaking, a potential form of such a hoped-for algorithm is as follows.

Algorithm 3.1. Take as input a number field K, a smooth projective curve X over K, and a power v^n of a good[7] prime ideal v of $\mathcal{O}_K$. Return as output a finite list of points[8] in $\mathcal{X}(\mathcal{O}_v)$, to any desired finite precision which is guaranteed to include all the rational points of X.

It should be mentioned that such an algorithm, until an efficient implementation proves the contrary, is at risk of being prohibitively slow so as to be useless from a practical standpoint. The essential difficulty lies in enumerating Galois representations with prescribed ramification; modularity results for the representations in question, if known, could speed up the algorithms significantly. One possible approach to the calculation is proposed in what follows. It has four essential components, each of which we briefly discuss below. It should be noted that whereas many of the separate ingredients have been extensively studied in the literature, the method of Lawrence–Venkatesh has so far not been made effective, and therefore the ideas in this section are tentative. It would be very interesting to explore the effectivity of this method further, and make a serious attempt at a computational version of this method.

Remark 3.2. An algorithm of the above form may return extraneous points, not corresponding to a rational point. This phenomenon arises also in Chabauty's method, though in the example in §1 it was circumvented by exhibiting two independent analytic sets, which was possible since $g-r=2$. Likewise, it is conceivable that one can circumvent it in the method of Lawrence–Venkatesh by varying the choice of q in the covering group $\mathrm{Aff}(q)$. Alternatively, one could attempt to apply the Mordell–Weil sieve, see §6.7.

3.1. Enumerating global Galois representations

Faltings' finiteness lemma for Galois representations (Lemma 2.1) can be made effective; we expect this to be the most computation-intensive part of the algorithm.

[7] The method of LV requires p to satisfy a certain Galois-theoretic condition; here we will simply call primes satisfying that condition "good" primes. The condition is needed to guarantee that a certain extension of K_p is of large degree, and is analogous to the requirement in the S-unit equation above that x not be a square in $\mathbf{Q}_p$. Choosing a good p presents no algorithmic difficulty.

[8] possibly with multiplicities

Recall that we want to enumerate all global Galois representations

$$\rho : G_K \longrightarrow \mathrm{GL}_d(\mathbf{Z}_p)$$

that could arise from our family, in the sense of Lemma 2.1. We know the following about ρ:

- We are given a finite set S of places of K, outside of which ρ is unramified.
- For every prime $\mathfrak{p} \notin S$, all the eigenvalues of Frobenius at $\mathfrak{p}$ are Weil numbers of weight $1/2$.
- The representation ρ is semisimple.

On the one hand, we can list all possible mod-p^n representations for any n. First, one enumerates all possible residual representations

$$\rho_1 : G_K \longrightarrow \mathrm{GL}_d(\mathbf{F}_p).$$

This is a straightforward application of Hermite–Minkowski finiteness. The residual representation has finite image, so it corresponds to an extension L_1 of K of degree at most $|\mathrm{GL}_d(\mathbf{F}_p)|$. The ramification condition translates to a bound on the discriminant of L_1. One can find all possible number fields L_1 by a targeted Hunter search [Coh00, §9.3]. However, the time complexity of such a search (for fixed K and S) is doubly exponential in the degree $[L_1 : K]$, so it may be necessary to further refine the search using more specifics of the situation at hand.

Second, for each residual representation, one lifts successively to mod-p^n representations

$$\rho_n : G_K \longrightarrow \mathrm{GL}_d(\mathbf{Z}/p^n),$$

which correspond to a tower of fields L_n. The successive extensions L_{n+1}/L_n are abelian, so they can be found by class field theory. (Everything we need from class field theory can be done algorithmically; see [Coh00].) To do this, we need to compute ideal class groups and unit groups of number fields whose degrees grow exponentially in n; this is again a computationally expensive task.

On the other hand, given the residual representation ρ_1, the Faltings–Serre method (see for example [Del85]) allows one to compute effectively a finite set of primes $\mathfrak{p}_1, \ldots, \mathfrak{p}_s$ such that for any semisimple ρ lifting ρ_0, the *rational* representation

$$G_K \longrightarrow \mathrm{GL}_d(\mathbf{Q}_p)$$

is determined by the Frobenius traces

$$\mathrm{Tr}(\mathrm{Fr}_{\mathfrak{p}_i} | \rho)$$

at these finitely many primes. (In general, there may be multiple isomorphism classes of *integral* representation, as the rational representation may have more than one stable $\mathbf{Z}_p$-lattice.) The condition on Frobenius eigenvalues guarantees that there are only finitely many possible values for $\mathrm{Tr}(\mathrm{Fr}_{\mathfrak{p}_i} | \rho)$, for each i. We can choose n_0 large enough that, for each fixed $i \in \{1, \ldots, s\}$, no two of these possible values are congruent modulo p^{n_0}. Then any mod-p^{n_0} representation ρ_{n_0} can lift to at most one semisimple p-adic representation.

The strategy, then, is as follows. First, make a list of all (finitely many) possible tuples

$$(\mathrm{Tr}(\mathrm{Fr}_{\mathfrak{p}_i}|\rho))_{i\in\{1,\ldots,s\}};$$

we'll call such a tuple a *candidate*. As described above, we can enumerate all mod-p^n representations for some $n \geq n_0$. We compute their Frobenius traces and match them with candidates, discarding candidates that don't match any representation, and vice-versa. We can repeat this procedure for any desired n; the list of candidates will get shorter, as spurious candidates are deleted.

3.2. Computing the Parshin family

Before we get to the purely local part of the computation, which consists of describing the p-adic period map per_p, we are faced with the problem of finding an explicit set of algebraic equations defining the Parshin family

$$\mathcal{C} \longrightarrow X,$$

whose fibres are finite covers of X branched over the variable point x. This is an instance of the Riemann–Hurwitz problem. Computational work on branched covers of curves is particularly well developed in the case of Belyĭ covers of $\mathbf{P}^1$; see [SV14] for an overview. The covers appearing in our setting are solvable, and we expect that explicit calculations on the Jacobian could provide a fruitful approach.

The solvability of the covering group $\mathrm{Aff}(q)$ has the following geometric interpretation. Suppose $Z \to X$ is an $\mathrm{Aff}(q)$-cover, branched at x. Let Z^{Gal} be the Galois closure of Z; this is a cover of X of degree $q(q-1)$, ramified only above x and having Galois group $\mathrm{Aff}(q)$. The quotient map $\mathrm{Aff}(q) \to \mathbf{F}_q^\times$ gives a curve Z^{ab}, corresponding by the Galois correspondence to $\mathbf{F}_q^\times$. Thus we have the tower of covers

$$Z^{\mathrm{Gal}} \longrightarrow Z^{\mathrm{ab}} \stackrel{\pi}{\longrightarrow} X.$$

In this tower, Z^{ab} is an unramified abelian cover of X of degree $q-1$, and Z^{Gal} is an abelian cover of Z^{ab} of degree q, ramified at exactly the points of $\pi^{-1}(x)$. The curve Z can be recovered as a quotient of Z^{Gal}.

This suggests the following strategy to compute the Parshin family Y, each of whose fibers is a union of $\mathrm{Aff}(q)$-covers Z. First, we attempt to compute abelian covers (both unbranched and branched) of arbitrary curves, by finding torsion points on algebraic generalized Jacobians. To describe one strategy[9], we will restrict attention to unramified covers and the (ungeneralized) Jacobian. In this setting, we want to find a divisor D on the curve X, along with a meromorphic function f on X such that

$$\mathrm{div}(f) = rD,$$

which amounts to looking for r-torsion on the Jacobian of X. The Jacobian has an algebraic incarnation as a variety classifying divisor classes on X and an analytic incarnation as a complex torus. It is of course trivial to identify torsion points on

[9] An alternative approach to computing covers of curves is by Hensel lifting from a finite field, as in [Mas20].

the *analytic* Jacobian; what we need is to describe them as points on the algebraic Jacobian.

Fix a basepoint $b \in X(\mathbf{C})$. By integration we can compute coordinates on the analytic torus $\operatorname{Jac} X$, along with the analytic Abel–Jacobi map

$$X \longrightarrow \operatorname{Jac} X.$$

In the other direction, let g be the genus of X. We want to invert the map

$$\operatorname{Sym}^g X \longrightarrow \operatorname{Jac} X, \tag{17}$$

to realize a point of the analytic Jacobian as a divisor on X. This map is a birational equivalence, but not an isomorphism. On a Zariski-dense subset of $\operatorname{Jac} X$, the map can be inverted, for example, by theta function methods [Mum83, Theorem II.3.1], by Puiseux series methods [CMSV19], or by computations in Grassmannians arising from Riemann–Roch theory [KM04, CMSV19]. A general algorithm appears in [CMSV19, §3.3].

If we can compute arbitrary abelian covers, we could try to determine all the covers Z^{Gal} for any fixed point x; from there one computes Z by Galois theory on the function field. In other words, we can compute the fiber Y_x of the Parshin family over any given point $x \in X$. To compute the Parshin family as an algebraic family, we are faced with the need to interpolate these fibers, perhaps by Puiseux series methods.

3.3. Computing the p-adic period map

We now come to the local part of the computation, where a description of the p-adic period map per_p reduces to a computation with p-adic cohomology in families. There is a vast literature on this subject, and this step is therefore likely to be more accessible and efficient than the others[10]. We give a brief overview of some results in the literature, for more detailed treatments that address also the history of the subject, see Kedlaya [Ked09, Ked07].

The basic problem is the following: Suppose we are given a curve $\mathcal{C}_x$ over a p-adic field K_v and want to compute the filtered ϕ-module structure of $\mathrm{H}^1_{\mathrm{dR}}(\mathcal{C}_x/K_v)$. Representing this space by differentials of the second kind, the Hodge filtration is easily worked out, and it is the Frobenius operator ϕ that forms the essence of this problem. When $\mathcal{C}_x$ is hyperelliptic, Kedlaya [Ked01] introduced an efficient algorithm, a variant of which we will see in action for the examples of the genus 2 curves in §6. There are two main ingredients for the computation:

- An appropriate lift of Frobenius on the functions in a (p-adic analytic) open subset of $\mathcal{C}_x$,
- A reduction algorithm in de Rham cohomology, that writes an arbitrary differential as the sum of an exact differential and a linear combination of our basis differentials.

[10]Indeed, the algorithms mentioned here are crucial ingredients for the effective method of Chabauty–Kim, as we will see in §6.

By applying the reduction in cohomology to the image of a set of basis differentials under this Frobenius lift, we may obtain a matrix of the Frobenius operator ϕ, up to some precision v^n.

This method has seen extensive developments since [Ked01], notably by Lauder [Lau04, Lau06] who introduced the *fibration method.* This method makes use of the Frobenius structure on the sheaf of relative qth de Rham cohomology $\mathcal{H}^q_{\mathrm{dR}}(X/S)$ of a smooth morphism $X \longrightarrow S$ between smooth varieties over K_v. The variation of the de Rham cohomology of the fibres in this family is described by the Gauß–Manin connection

$$\nabla_{\mathrm{GM}} : \mathcal{H}^q_{\mathrm{dR}}(X/S) \longrightarrow \Omega^1_{X/S} \otimes \mathcal{H}^q_{\mathrm{dR}}(X/S),$$

which gives a system of differential equations known as the *Picard–Fuchs equations*, whose study was taken up in the 19th century. Suppose we find a local lift of Frobenius ϕ on S, then the pullback of the relative de Rham cohomology $\mathcal{H}^q_{\mathrm{dR}}(X/S)$ by ϕ is isomorphic *as a vector bundle with connection* to the original one. In concrete terms, let us suppose that S is a curve, then we may express this in matrix form as

$$NFdt + \frac{\partial}{\partial t}F = \left(\frac{\partial}{\partial t}\phi(t)\right) F\phi(N)dt \tag{18}$$

by choosing a local coordinate t on S, and a basis of the relative de Rham cohomology, with respect to which we obtain a matrix $F(t)$ describing the Frobenius operator on the fibres, and $N(t)dt$ describing the Gauß–Manin connection. This equation is very useful. For instance, if $F(t)$ can be computed for a single value of $t = t_0$, then we may solve these p-adic differential equation using $F(t_0)$ as an initial condition. Lauder [Lau04, Lau06] uses this idea to compute the Frobenius action in families. It is surprisingly versatile, applying both to individual curves with a map to $\mathbf{P}^1$ as well as families of curves. It has been developed in many subsequent papers of which we mention the recent algorithms of Tuitman [Tui16, Tui17], and the references contained therein, which vastly extend the range of applicability of these ideas.

3.4. Compare the global Galois representations with the p-adic periods

We suggest two approaches. The first is to use p-adic Hodge theory, along the lines of Fontaine–Laffaille theory [FL82]. We are given a mod-p^n global Galois representation, presented as a polynomial whose splitting field is its kernel. We can determine the corresponding local representation at p, in terms of extensions of $\mathbf{Q}_p$. Fontaine and Laffaille define a functor $\underline{U}_S$ from a certain category of finite-length filtered ϕ-modules to the category of Galois representations [FL82, §0.6]. One expects that Fontaine–Laffaille theory can be made algorithmic: given a mod-p^n Galois representation, we should be able to determine whether it is in the image of this functor, and if so, describe the underlying filtered ϕ-module. We can then compare these ϕ-modules with the ϕ-modules arising from the p-adic period map, to determine a list of candidate points.

Our second approach avoids filtered ϕ-modules entirely, by working directly with Galois representations. It is a consequence of Fontaine–Laffaille theory that the mod-p^n local Galois representation ρ_x depends only on the reduction of x modulo v^{n+1}. Using this, we can compute explicitly all the possibilities for the local Galois representation at p, and match them explicitly with the list of "candidates" from the global Galois calculation. In other words, for each candidate ρ, we obtain a list of mod-v^{n+1} points of X, the local representations at which agree modulo p^n with ρ. For each of these mod-v^{n+1} residue classes, we then use the period map per_p to compute a bound on the number of rational points in the class.

4. The method of Chabauty–Kim: Finiteness

In this section, we discuss an approach to Mordell's conjecture due to Minhyong Kim. It follows the same pattern as the method of Chabauty–Coleman discussed in the introduction, and as such it depends on some geometric input, replacing the condition $r < g$ by something weaker, which may be done at the cost of replacing the p-adic Tate module V by a more sophisticated quotient of the fundamental group. We discuss in some detail the particular case of a quotient arising from a geometric correspondence [BD18, BD20, BDM$^+$19] using the geometric language of Edixhoven–Lido [EL19].

4.1. Quotients of the fundamental group

To motivate an interest in unipotent quotients of the algebraic fundamental group for Diophantine applications, it is instructive to first recall the *section conjecture* of Grothendieck [Gro97], which states that the map

$$\begin{aligned} \rho \; : \; X(\mathbf{Q}) & \longrightarrow \mathrm{H}^1\big(G_{\mathbf{Q}}, \pi_1^{\text{ét}}(\overline{X}, b)\big)\,, \\ x & \longmapsto [\ \pi_1^{\text{ét}}(\overline{X}; b, x)\ ] \end{aligned}$$

which attaches to every rational point the class of the Galois representation defined by the corresponding *path torsor* of the algebraic fundamental group, should be an isomorphism. In other words, every torsor of the fundamental group should necessarily arise from a rational point. This provides us with the tantalizing possibility of studying the set of such torsors in lieu of the set $X(\mathbf{Q})$. Unfortunately, the cohomology set that classifies these torsors does not seem to have much structure with which we can work.

On the other end of the spectrum, we already saw that the twists of the p-adic Tate module V of the Jacobian J of X, which is essentially the abelianization of the fundamental group, are classified by an object which is very closely related to J, and which therefore has a tremendous amount of structure. That said, this association only gives us enough information under the additional assumption that $r < g$.

In summary, we could roughly describe the situation by saying that the association

$$\rho \; : \; X(\mathbf{Q}) \longrightarrow \mathrm{H}^1\big(G_{\mathbf{Q}}, \pi_1^{\text{ét}}(\overline{X}, b)\big) \tag{19}$$

in the section conjecture has a target with *too little* structure, whereas the association

$$\rho \,:\, X(\mathbf{Q}) \longrightarrow \mathrm{H}^1_f(G_{\mathbf{Q}}, V) \tag{20}$$

appearing in the method of Chabauty–Coleman has a target with *too much* structure. The latter statement is meant in the sense that ρ factors through the Jacobian, and in situations where $r \geq g$ this kills some crucial non-abelian information needed to understand $X(\mathbf{Q})$. In the method of Chabauty–Kim, we allay the difficulties inherent to both settings by working with a suitable intermediate quotient, balancing the availability of structure on the sets H^1 against our ability to explicitly describe the target. We consider quotients of the fundamental group that are unipotent[11].

The strategy for proving finiteness follows the same pattern as our discussion of the method of Lawrence–Venkatesh. First, one attempts to gain sufficient control over the set of global representations involved, and second, one studies the local representations via the analytic properties of an associated period map.

a. Global representations. A general theorem of Kim ([Kim05, Proposition 2] and [Kim09, p. 118]) states that if U is a unipotent quotient satisfying certain technical assumptions which we will not state here, the set $\mathrm{H}^1_f(G_K, \mathrm{U})$ carries the structure of an algebraic variety, dubbed *Selmer variety*, such that the localization map

$$\mathrm{H}^1_f(G_{\mathbf{Q}}, \mathrm{U}) \longrightarrow \mathrm{H}^1_f(G_{\mathbf{Q}_p}, \mathrm{U}) \tag{21}$$

between the global and local Selmer varieties is algebraic. The algebraic nature of this map allows us to gain control over the image of the global Selmer variety, typically by showing that the global Selmer variety is of lower dimension than the local Selmer variety, so that the image cannot be Zariski dense.

b. The period map. As was the case in the method of Lawrence–Venkatesh, the control of global representations can be turned into a proof of finiteness by controlling a p-adic period map. In the method of Chabauty–Kim, this means concretely that one establishes that the association

$$\rho \,:\, X(\mathbf{Q}_p) \longrightarrow \mathrm{H}^1_f(G_{\mathbf{Q}_p}, \mathrm{U}), \tag{22}$$

of the path torsor of U attached to a point, has an image which is Zariski dense. Typically, the quotient U is of a "motivic" nature, in which case the association ρ in (22) has a de Rham realisation

$$\mathrm{per}_p \,:\, X(\mathbf{Q}_p) \longrightarrow \mathrm{MF}^{\phi} \tag{23}$$

which can be expressed as a linear combination of iterated Coleman integrals of differentials. A general theorem of Kim [Kim09, Theorem 1] establishes the linear independence of such iterated integrals, which often implies the Zariski density of the image of (22) by p-adic Hodge theory.

[11]Strictly speaking, quotients of the $\mathbf{Q}_p$-unipotent étale fundamental group studied in Deligne [Del89].

c. Finiteness. In conclusion, we are left with the following attractive strategy to study the set of rational points $X(\mathbf{Q})$: Suppose that we can construct a specific finite-dimensional unipotent quotient U satisfying the technical hypotheses required for the representability of the Selmer varieties, such that furthermore

1. we can prove that $\dim \mathrm{H}^1_f(G_{\mathbf{Q}}, \mathrm{U}) < \dim \mathrm{H}^1_f(G_{\mathbf{Q}_p}, \mathrm{U})$,
2. the quotient is "motivic", so that we have a p-adic period map
$$\mathrm{per}_p \;:\; X(\mathbf{Q}_p) \longrightarrow \mathrm{MF}^{\phi}$$
which is a linear combination of iterated integrals of differentials on X, and
3. we can find a *computable* condition on elements of the image of per_p to come from a point in $X(\mathbf{Q})$.

Once we manage to find a quotient U satisfying these conditions, we consider the diagram

$$\begin{array}{ccccc} X(\mathbf{Q}) & \longrightarrow & X(\mathbf{Q}_p) & \overset{\mathrm{per}_p}{\dashrightarrow} & \\ \downarrow{\scriptstyle \rho} & & \downarrow{\scriptstyle \rho} & & \searrow \\ \mathrm{H}^1_f(G_{\mathbf{Q}}, \mathrm{U}) & \longrightarrow & \mathrm{H}^1_f(G_{\mathbf{Q}_p}, \mathrm{U}) & \xrightarrow{\mathbf{D}_{\mathrm{cris}}} & \mathrm{MF}^{\phi}. \end{array} \tag{24}$$

The first two conditions on U are the active ingredients for deducing finiteness. The first condition is the analogue of the condition "$r < g$" appearing in the method of Chabauty–Coleman, and allows us to control the image of the global Selmer variety. When combined with a concrete understanding of the period map per_p provided by the second condition (for instance, enough to show Zariski-density of (22), see [Kim09, Theorem 1]) the above commutative diagram implies that $X(\mathbf{Q}_p)$ intersects the image of the global Selmer variety in a finite set of points. In particular, this shows that $X(\mathbf{Q})$ is finite.

Finding suitable quotients that satisfy the first two conditions is the subject of many works, and is typically done by considering quotients U arising from powers of the augmentation ideal. See for instance Kim [Kim05, Kim09], Coates–Kim [CK10] and Ellenberg–Hast [EH17]. The third condition is relevant for the *explicit determination* of $X(\mathbf{Q})$ and will reappear later.

4.2. Geometric correspondences on X

We now discuss one instance where such a quotient can be constructed, under the additional assumption that the Jacobian J of X has non-trivial Néron–Severi rank, following [BD18, BDM$^+$19]. To offer a different perspective on the constructions in *loc. cit.* we opt for the more geometric reformulation of this theory following the beautiful work of Edixhoven–Lido [EL19]. It should be noted that in [EL19] this geometric viewpoint is retained to find a method for the effective determination of $X(\mathbf{Q})$, but in §5 we instead opt for the cohomological language of [BD18, BDM$^+$19].

Recall that the Néron–Severi group of a smooth proper variety is the group of components of its Picard scheme. In the situation at hand, we have chosen a base point b in $X(\mathbf{Q})$, which gives us an associated Abel–Jacobi map $X \longrightarrow J$. By functoriality, we obtain the following diagram:

$$\begin{array}{ccccccccc} 1 & \longrightarrow & \mathrm{Pic}^0(J) & \longrightarrow & \mathrm{Pic}(J) & \longrightarrow & \mathrm{NS}(J) & \longrightarrow & 0 \\ & & \| & & \downarrow & & \downarrow & & \\ 1 & \longrightarrow & \mathrm{Pic}^0(X) & \longrightarrow & \mathrm{Pic}(X) & \longrightarrow & \mathbf{Z} & \longrightarrow & 0. \end{array} \tag{25}$$

The Néron–Severi group $\mathrm{NS}(J)$ is a finitely generated group, of rank $\mathrm{rk}_{\mathrm{NS}}$ which is called the *Néron–Severi rank* of J. Now suppose (see Remark 4.1) that we have a non-trivial class Z in $\mathrm{NS}(J)$ which maps to zero in $\mathbf{Z} \simeq \mathrm{NS}(X)$ in the above diagram. Then, by the identification of $\mathrm{Pic}^0(J)$ with $\mathrm{Pic}^0(X)$ there is a unique lift of Z to an element of $\mathrm{Pic}(J)$ which is trivial when restricted to X. In other words, Z uniquely determines a (non-trivial) line bundle $\mathscr{L}_Z$ on J which is trivial when restricted to X, and hence we obtain a lift of the Abel–Jacobi map

$$\begin{array}{ccc} & \overset{}{\dashrightarrow} & \mathscr{L}_Z \\ \nearrow & & \downarrow \\ X & \longrightarrow & J. \end{array} \tag{26}$$

This lifting of the Abel–Jacobi map, or equivalently this trivialization of the line bundle $\mathscr{L}_Z$ restricted to X, is a priori uniquely determined up to multiplication by elements of $\mathbf{Q}^\times$. As explained in Edixhoven–Lido [EL19], one can determine it up to $\mathbf{Z}^\times = \{\pm 1\}$, and hence essentially uniquely, at the cost of taking a small[12] tensor power of $\mathscr{L}_Z$ by spreading out the geometry over $\mathbf{Z}$ and working with the Néron model of J. In conclusion, we obtain an essentially unique lift

$$X \longrightarrow \mathscr{L}_Z^\times := \mathbf{Isom}_J(\mathcal{O}, \mathscr{L}_Z). \tag{27}$$

The scheme $\mathscr{L}_Z^\times$ is a $\mathbf{G}_m$-torsor[13] over the Jacobian J. We define U to be the $\mathbf{Q}_p$-étale fundamental group of $\mathscr{L}_Z^\times$. This group is non-abelian, and may be understood geometrically as follows. One can show (see for instance Bertrand–Edixhoven [BE20, §4] for the arguments in the $\mathbf{C}$-analytic setting) that there is a co-final system of étale coverings

$$\pi_n : (\mathscr{L}_Z^\times)^{\otimes n} \longrightarrow \mathscr{L}_Z^\times$$

obtained by composing the pullback of the map $[n]$ on the Jacobian with the nth power map on fibres. The Galois group U_n of this étale cover is a central extension

$$1 \longrightarrow \mu_n \longrightarrow \mathrm{U}_n \longrightarrow J[n] \longrightarrow 0,$$

so that U is a Heisenberg group, and as a Galois representation it is an extension of V by $\mathbf{Q}_p(1)$. Suppose that x is a point in $X(K)$ for K equal to $\mathbf{Q}$ or $\mathbf{Q}_p$. Then

[12] It suffices to take the least common multiple of the exponents of the Néron component groups at all primes of bad reduction.
[13] Since the class of its line bundle in $\mathrm{Pic}(J)$ maps to a non-zero element of $\mathrm{NS}(J)$, this $\mathbf{G}_m$-torsor is not a group.

in analogy with (5) we obtain a torsor of U from the inverse limit of the preimages of x under the maps π_n, i.e.,

$$\mathbf{Q}_p \otimes_{\mathbf{Z}_p} \left(\varprojlim_n \pi_n^{-1}(x) \right), \qquad \pi_n : (\mathscr{L}_Z^\times)^{\otimes n} \longrightarrow \mathscr{L}_Z^\times. \tag{28}$$

Such torsors are classified by the cohomology group $\mathrm{H}^1(G_K, \mathrm{U})$ and satisfy certain local conditions which we will not make explicit here. In conclusion, we obtain an association ρ analogous to (4):

$$\rho : X(K) \longrightarrow \mathscr{L}_Z^\times(K) \longrightarrow \mathrm{H}^1_f(G_K, \mathrm{U}). \tag{29}$$

Remark 4.1. Note that the above discussion hinges on the assumption that we can find a non-trivial class Z in $\mathrm{NS}(J)$ which maps to zero in $\mathrm{NS}(X)$. Such a class always exists when $\mathrm{rk}_{\mathrm{NS}} > 1$, which is true for many examples of interest. Notably, this includes modular curves, which typically have a large supply of such classes induced by Hecke correspondences. See Siksek [Sik17] for more details.

Remark 4.2. The work of Alexander Betts [Bet17] establishes a precise relationship between the association of path torsors to points in $\mathscr{L}_Z^\times$ and the theory of p-adic heights. He proves that a certain quotient

$$\mathscr{L}_Z^\times(\mathbf{Q}_p) \longrightarrow \mathrm{H}^1_f(G_K, \mathrm{U}) \longrightarrow \mathbf{Q}_p$$

defined by purely Galois theoretic conditions coincides with the Néron log-metric on the pointed line bundle $(\mathscr{L}_Z, b)$, in the language of *loc. cit.* In Section §5, the theory of p-adic heights will play a central role.

4.3. Finiteness of $X(\mathbf{Q})$

The quotient U attached to a Néron–Severi class Z as above has dimension $2g+1$ as a $\mathbf{Q}_p$-vector space. More precisely, as a Galois representation, it is an extension of the form

$$0 \longrightarrow \mathbf{Q}_p(1) \longrightarrow \mathrm{U} \longrightarrow V \longrightarrow 0, \tag{30}$$

where $\mathbf{Q}_p(1)$ is the one-dimensional representation given by the cyclotomic character. The simple nature of this one-dimensional graded piece is responsible for the proof that the quotient U satisfies the first condition on our wish list in §4.1. Indeed, this can be deduced from the statements

$$\begin{aligned} \mathrm{H}^1_f(G_{\mathbf{Q}}\ , \mathbf{Q}_p(1)) &= \mathbf{Z}^\times \mathbin{\widehat{\otimes}} \mathbf{Q}_p = 0, \\ \mathrm{H}^1_f(G_{\mathbf{Q}_p}, \mathbf{Q}_p(1)) &= \mathbf{Z}_p^\times \mathbin{\widehat{\otimes}} \mathbf{Q}_p = \mathbf{Q}_p \end{aligned} \tag{31}$$

which result, via the simple argument in [BD18, Lemma 3.1], in the statements

$$\begin{aligned} \dim \mathrm{H}^1_f(G_{\mathbf{Q}}\ , \mathrm{U}) &\leq r, \\ \dim \mathrm{H}^1_f(G_{\mathbf{Q}_p}, \mathrm{U}) &= g+1. \end{aligned} \tag{32}$$

The quotient U is also *motivic* in nature, as its geometric definition via the $\mathbf{G}_m$-torsor $\mathscr{L}_Z^\times$ shows. In particular, besides the Galois representation U, there is also a de Rham realisation U^{dR}, which is a quotient of the de Rham fundamental group of X, see [Kim05, Kim09] for more precise definitions. The theorem of Kim [Kim09, Theorem 1] discussed in §4.1 then implies that the image of $X(\mathbf{Q}_p)$ under ρ is Zariski dense.

This allows us to deduce finiteness of $X(\mathbf{Q})$ for certain curves X, as we now explain. Suppose that $r = g$, so that we are just outside of the range where the method of Chabauty–Coleman applies, and assume furthermore that the Néron–Severi rank $\mathrm{rk}_{\mathrm{NS}}$ of J is at least 2, so that there exists a quotient U as above. The diagram (25) implies, via the two properties we just discussed, that the intersection of $X(\mathbf{Q}_p)$ with the global Selmer variety is finite. Since this intersection contains $X(\mathbf{Q})$, finiteness of $X(\mathbf{Q})$ follows.

In fact, one can refine the above discussion by constructing a quotient which is an extension of V by the direct sum of characters $\mathbf{Q}_p(1)^{\oplus(\mathrm{rk}_{\mathrm{NS}}-1)}$, resulting via the same reasoning in the following finiteness statement, which is a special case of Balakrishnan–Dogra [BD18, Lemma 3.2].

Theorem 4.3. *Suppose that X is a smooth projective curve over $\mathbf{Q}$. Then $X(\mathbf{Q})$ is finite whenever*

$$r \; < \; g + \mathrm{rk}_{\mathrm{NS}} - 1. \tag{33}$$

Many other instances of finiteness are known to follow from the method of Chabauty–Kim. In general finiteness was proved by Kim [Kim09] under the assumption of the Bloch–Kato conjecture. We will not discuss these results here, but rather turn to the question of how to explicitly determine the set $X(\mathbf{Q})$.

5. The method of Chabauty–Kim: Effectivity

In this section, we discuss how to use the method of Chabauty–Kim to compute the rational points on X in the simplest instance of Theorem 4.3: the case $r = g$ and $\mathrm{rk}_{\mathrm{NS}} > 1$. Whereas the method of Chabauty–Coleman relies on detecting global points via *linear* relations in the image of per_p, we will provide a computable condition on filtered ϕ-modules in the image of per_p to come from a point in $X(\mathbf{Q})$ via *bilinear* relations, thereby addressing the third item in our wish list in §4.1.

5.1. Heights on Selmer varieties

Looking for bilinear relations, one is naturally led to p-adic heights. Classically, these were defined as bilinear pairings on $J(\mathbf{Q})$ but since it is crucial that the non-abelian method of Chabauty–Kim factors through a non-abelian Selmer variety rather than the abelian variety J, we instead prefer to utilize a more general approach due to Nekovář [Nek93, §2]. Namely, he constructs a continuous bilinear pairing

$$h\colon\ \mathrm{H}^1_f(G_{\mathbf{Q}}, V) \times \mathrm{H}^1_f(G_{\mathbf{Q}}, V^*(1)) \longrightarrow \mathbf{Q}_p, \tag{34}$$

depending on some auxiliary choices, including the choice of a splitting of the Hodge filtration

$$s\colon V_{\mathrm{dR}}/\,\mathrm{Fil}^0\, V_{\mathrm{dR}} \longrightarrow V_{\mathrm{dR}}\,. \tag{35}$$

The global height h decomposes as a sum of local heights h_v, where v runs through the finite primes of $\mathbf{Q}$. Briefly, the idea is to lift a pair in $\mathrm{H}^1_f(G_{\mathbf{Q}}, V) \times \mathrm{H}^1_f(G_{\mathbf{Q}}, V^*(1))$ to a mixed extension of p-adic Galois representations with graded pieces $\mathbf{Q}_p, V$ and $\mathbf{Q}_p(1)$ and to define h_v on it. As explained in [BD18, Section 5], we can construct such a representation from a torsor $P \in \mathrm{H}^1_f(G_{\mathbf{Q}}, \mathrm{U})$, where U is attached to a Néron–Severi class as in §4.2, by twisting a certain quotient of the universal enveloping algebra of the $\mathbf{Q}_p$-unipotent étale fundamental group by P. There is an analogous local construction for $P \in \mathrm{H}^1_f(G_{\mathbf{Q}_v}, \mathrm{U})$.

We will assume throughout that $r = g$ and that the p-adic closure of $J(\mathbf{Q})$ has finite index in $J(\mathbf{Q}_p)$ [14]. Then there are isomorphisms

$$\mathrm{H}^1_f(G_{\mathbf{Q}}, V) \xrightarrow{\mathrm{res}_p} \mathrm{H}^1_f(G_{\mathbf{Q}_p}, V) \xrightarrow{\log} \mathrm{H}^0(X_{\mathbf{Q}_p}, \Omega^1)^\vee.$$

By Poincaré duality we obtain maps

$$\pi\colon \mathrm{H}^1_f(G_K, \mathrm{U}) \longrightarrow \mathrm{H}^0(X_{\mathbf{Q}_p}, \Omega^1)^\vee \otimes \mathrm{H}^0(X_{\mathbf{Q}_p}, \Omega^1)^\vee$$

for $K \in \{\mathbf{Q}, \mathbf{Q}_p\}$.

For ease of exposition, we shall assume for all $v \neq p$ that $h_v = 0$ for torsors coming from $X(\mathbf{Q}_v)$. The local height h_p will be discussed in more detail below. The main point is that it factors through $\mathrm{D_{cris}}$, so we obtain the following refinement of diagram (24):

$$\begin{array}{ccccc} X(\mathbf{Q}) & \longrightarrow & X(\mathbf{Q}_p) & \overset{\mathrm{per}_p}{\dashrightarrow} & \\ \downarrow \rho & & \downarrow \rho & & \\ \mathrm{H}^1_f(G_{\mathbf{Q}}, \mathrm{U}) & \xrightarrow{\mathrm{res}} & \mathrm{H}^1_f(G_{\mathbf{Q}_p}, \mathrm{U}) & \xrightarrow{\mathrm{D_{cris}}} & \mathrm{MF}^{\phi} \\ & \searrow^{\pi} & \downarrow \pi & & \downarrow h_p \\ & & \mathrm{H}^0(X_{\mathbf{Q}_p}, \Omega^1)^\vee \otimes \mathrm{H}^0(X_{\mathbf{Q}_p}, \Omega^1)^\vee & \xrightarrow[h]{} & \mathbf{Q}_p, \end{array} \tag{36}$$

where h_p is now defined on the image of $\mathrm{H}^1_f(G_{\mathbf{Q}_p}, \mathrm{U})$.

If (ψ_i) is a basis of the dual space of $\mathrm{H}^0(X_{\mathbf{Q}_p}, \Omega^1)^\vee \otimes \mathrm{H}^0(X_{\mathbf{Q}_p}, \Omega^1)^\vee$, then there are constants $\alpha_i \in \mathbf{Q}_p$ such that $h = \sum_i \alpha_i \psi_i$. We deduce that the locally analytic function

$$Q\colon X(\mathbf{Q}_p) \longrightarrow \mathbf{Q}_p; \qquad x \mapsto \sum_i \alpha_i \psi_i(\pi(\rho(x))) - h_p(\mathrm{per}_p(x)) \tag{37}$$

vanishes along $X(\mathbf{Q})$; furthermore, one can show that it has only finitely many zeroes (see [BD20]). We can use this function for the explicit computation of $X(\mathbf{Q})$ if we have algorithms to

(i) compute the α_i for a suitable explicitly computable basis ψ_i.

[14] If the latter condition fails, we may apply classical Chabauty as in §1.3.

(ii) expand the function $x \mapsto h_p(\mathrm{per}_p(x))$ into convergent power series on residue disks.

We can easily solve (i) given $x_1, \ldots, x_m \in X(\mathbf{Q})$ such that

$$\{\pi(\rho(x_i))\}_{i=1,\ldots,m} \quad \text{is a basis for} \quad \mathrm{H}^0(X_{\mathbf{Q}_p}, \Omega^1)^\vee \otimes \mathrm{H}^0(X_{\mathbf{Q}_p}, \Omega^1)^\vee;$$

in this case we only need to compute $h_p(\mathrm{per}_p(x_i))$ and $\pi(\rho(x_i))$. If we choose an $\mathrm{End}_0(J)$-equivariant splitting in (35), then the global height is also $\mathrm{End}_0(J)$-equivariant, thus reducing the number of points x_i required. Nevertheless, there need not exist enough points x_i, in which case we can solve (i) using generators of $J(\mathbf{Q}) \otimes \mathbf{Q}$ and a construction of p-adic heights on J due to Coleman and Gross [CG89].

Remark 5.1. It is possible to write down functions vanishing in $X(\mathbf{Q})$ with finitely many zeroes when $r < g + \mathrm{rk}_{\mathrm{NS}} - 1$ using p-adic heights [BD18, Proposition 5.9]. More generally, one can extend Nekovář's construction to construct such functions when $r < g^2$, conditional on the conjecture of Bloch–Kato, see [BD20, §4]. This has only been made explicit in the special case of the Kulesz–Matera–Schost family of bielliptic genus 2 curves, see the (unconditional) Theorem 1.2 of [BD20].

5.2. Local heights

In the remainder of this section we focus on (ii). We first discuss in more detail the local height h_p, following [Nek93, §4]. Let $P \in \mathrm{H}^1_f(G_{\mathbf{Q}_v}, \mathrm{U})$ and denote by M_P the mixed extension of $\pi(P)$ mentioned above. Then $h_p(M_P)$ is constructed using $\mathrm{D}_{\mathrm{cris}}(M_P)$, which is a mixed extension of filtered ϕ-modules with graded pieces $\mathbf{Q}_p$, $V_{\mathrm{dR}} := \mathrm{H}^1_{\mathrm{dR}}(X_{\mathbf{Q}_p})^\vee = \mathrm{D}_{\mathrm{cris}}(V)$ and $\mathbf{Q}_p(1)$.

For simplicity, we only describe h_p on the image of $X(\mathbf{Q}_p)$. The family $(\mathrm{D}_{\mathrm{cris}}(M_{\rho(x)}))_x$ interpolates in the following sense: There is a filtered connection $\mathcal{A}_Z = \mathcal{A}_Z(b)$ with Frobenius structure such that we have

$$\mathrm{D}_{\mathrm{cris}}(M_{\rho(x)}) \simeq x^* \mathcal{A}_Z \quad \text{for all } x \in X(\mathbf{Q}_p). \tag{38}$$

Suppose that we have isomorphisms

$$\begin{array}{rcccc} s^\phi(b,x) & : & \mathbf{Q}_p \oplus V_{\mathrm{dR}} \oplus \mathbf{Q}_p(1) & \xrightarrow{\sim} & x^* \mathcal{A}_Z \\ s^{\mathrm{Fil}}(b,x) & : & \mathbf{Q}_p \oplus V_{\mathrm{dR}} \oplus \mathbf{Q}_p(1) & \xrightarrow{\sim} & x^* \mathcal{A}_Z \end{array}$$

where s^ϕ is Frobenius-equivariant, and s^{Fil} respects the filtrations, and suppose we can write them as

$$s^\phi(b,x) = \begin{pmatrix} 1 & 0 & 0 \\ \boldsymbol{\alpha}_\phi(b,x) & 1 & 0 \\ \gamma_\phi(b,x) & \boldsymbol{\beta}_\phi^{\mathsf{T}}(b,x) & 1 \end{pmatrix} \quad s^{\mathrm{Fil}}(b,x) = \begin{pmatrix} 1 & 0 & 0 \\ 0 & 1 & 0 \\ \gamma_{\mathrm{Fil}}(b,x) & \boldsymbol{\beta}_{\mathrm{Fil}}^{\mathsf{T}}(b) & 1 \end{pmatrix}. \tag{39}$$

Note that we make a choice of basis differentials on the affine open Y (see §5.3) so that $s^\phi(b,x)$ and $s^{\mathrm{Fil}}(b,x)$ are of this form. The splitting s in (35) induces idempotents

$$\begin{array}{rcccc} s_1 & : & V_{\mathrm{dR}} & \longrightarrow & s(V_{\mathrm{dR}}/\mathrm{Fil}^0 V_{\mathrm{dR}}) \\ s_2 & : & V_{\mathrm{dR}} & \longrightarrow & \mathrm{Fil}^0 V_{\mathrm{dR}}. \end{array}$$

With respect to our choices, Nekovář's local height at p is

$$h_p(\mathrm{per}_p(x)) = \gamma_\phi(b,x) - \gamma_{\mathrm{Fil}}(b,x) - \boldsymbol{\beta}_\phi^\intercal(b,x)\cdot s_1(\boldsymbol{\alpha}_\phi)(b,x) - \boldsymbol{\beta}_{\mathrm{Fil}}^\intercal(b)\cdot s_2(\boldsymbol{\alpha}_\phi)(b,x). \tag{40}$$

So in order to solve (ii) we need to compute the entries of (39), which means computing the Hodge filtration and the Frobenius structure on $\mathcal{A}_Z$. For (i), we also need to explicitly compute the composition $\pi\circ\rho$. With respect to the dual basis of our chosen basis differentials on Y, the map $\pi\circ\rho$ is given by

$$\begin{aligned} \pi\circ\rho : Y(\mathbf{Q}_p) &\to H^0(X_{\mathbf{Q}_p},\Omega^1)^\vee \otimes H^0(X_{\mathbf{Q}_p},\Omega^1)^\vee \\ x &\mapsto \boldsymbol{\alpha}_\phi(b,x)^\intercal \cdot \begin{pmatrix} I_g \\ \hline 0_g \end{pmatrix} \otimes (\boldsymbol{\beta}_\phi^\intercal(b,x) - \boldsymbol{\beta}_{\mathrm{Fil}}^\intercal(b))\cdot \begin{pmatrix} 0_g \\ \hline I_g \end{pmatrix}. \end{aligned} \tag{41}$$

Note in particular that the first factor is the Abel–Jacobi map $\mathrm{AJ}_b(x)$, sending x to the functional $\omega \mapsto \int_b^x \omega$.

5.3. Computing the Hodge filtration

We work in an affine open subset Y of X. Set $d = \#(X\setminus Y)(\overline{\mathbf{Q}})$ and choose differentials $\omega_0,\ldots,\omega_{2g+d-2} \in H^0(Y_{\overline{\mathbf{Q}}},\Omega^1)$ on Y such that the following conditions are satisfied:

1. The differentials $\omega_0,\ldots\omega_{2g-1}$ are of the second kind (residue zero) on X and form a symplectic basis of $H^1_{\mathrm{dR}}(X_{\mathbf{Q}})$ with respect to the cup product pairing. We let $\boldsymbol{\omega}$ denote the column vector $(\omega_0,\ldots\omega_{2g-1})^\intercal$.
2. The differentials $\omega_{2g},\ldots,\omega_{2g+d-2}$ are of the third kind (all poles have order one) on X.

Universal properties give that the rank $2g+2$ vector bundle $\mathcal{A}_Z$ has a connection, a Hodge filtration, and a Frobenius structure, as discussed in [BDM+19, §4,5]. Here, we give algorithms that describe these objects.

Recall that we have a non-trivial class Z in $\mathrm{NS}(J)$ mapping to 0 in $\mathrm{NS}(X)$. This is equivalent to the choice of an endomorphism of $\mathrm{H}^1_{\mathrm{dR}}(X)$ satisfying several conditions (see [BDM+19, §4.4]), and we describe a method to compute this in the case of modular curves in Section 6. We denote the matrix of the correspondence Z on $\mathrm{H}^1_{\mathrm{dR}}(X/\mathbf{Q})$ also by Z, where we act on column vectors.

Choose a trivialization

$$s_0 : \mathcal{O}_Y \otimes (\mathbf{Q}_p \oplus V_{\mathrm{dR}} \oplus \mathbf{Q}_p(1)) \to \mathcal{A}_Z|_Y$$

such that, with respect to this trivialization, the connection ∇ on $\mathcal{A}_X$ is given by

$$\nabla = d + \Lambda,$$

where

$$\Lambda = -\begin{pmatrix} 0 & 0 & 0 \\ \boldsymbol{\omega} & 0 & 0 \\ \eta & \boldsymbol{\omega}^\intercal Z & 0 \end{pmatrix},$$

where η is a differential of the third kind on X that is uniquely determined by the following two properties:

1. It is in the space spanned by $\omega_{2g}, \ldots, \omega_{2g+d-2}$, and
2. The connection ∇ extends to a holomorphic connection on all of X.

The Hodge filtration on $\mathcal{A}_Z$ is determined completely from the Hodge filtration on its graded pieces, via universal properties. Here is an algorithm to compute the Hodge filtration:

Algorithm 5.2 (Computing the Hodge filtration on $\mathcal{A}_Z$).

1. Let $L/\mathbf{Q}$ denote a finite extension over which all the points of $X \setminus Y$ are defined. Compute local coordinates at each $x \in (X\backslash Y)(L)$.
2. For each $x \in (X\backslash Y)(L)$, compute power series for $\boldsymbol{\omega}_x$, the expansion of the vector of differentials $\boldsymbol{\omega}$ at x to large enough precision, which means at least mod $t_x^{d_x}$, where d_x is the order of the largest pole occurring.
3. Compute the vector $\boldsymbol{\Omega}_x$, defined by
$$d\boldsymbol{\Omega}_x = -\boldsymbol{\omega}_x.$$
4. Compute η as the unique linear combination of $\omega_{2g}, \ldots, \omega_{2g+d-2}$ such that
$$d\boldsymbol{\Omega}_x^{\intercal} Z \boldsymbol{\Omega}_x - \eta$$
has residue zero at all $x \in (X\backslash Y)(L)$. To do this, carry out the following:
   (a) Using local coordinates at each $x \in (X\backslash Y)(L)$, rewrite $\omega_{2g}, \ldots, \omega_{2g+d-2}$.
   (b) Solve for η by comparing residues.
5. Solve the system of equations for g_x in $L(\!(t_x)\!)/L[\![t_x]\!]$ such that
$$dg_x = \boldsymbol{\Omega}_x^{\intercal} Z d\boldsymbol{\Omega}_x - \eta.$$
6. Compute the vector of constants $\mathbf{b}_{\mathrm{Fil}}^{\intercal} = (b_g, \ldots, b_{2g-1}) \in \mathbf{Q}^g$ and the function γ_{Fil} characterized by $\gamma_{\mathrm{Fil}}(b) = 0$ and
$$g_x + \gamma_{\mathrm{Fil}} - \mathbf{b}_{\mathrm{Fil}}^{\intercal} N^{\intercal} \boldsymbol{\Omega}_x - \boldsymbol{\Omega}_x^{\intercal} Z N N^{\intercal} \boldsymbol{\Omega}_x \in L[\![t_x]\!] \tag{42}$$
where N is the $2g \times g$ matrix which has the zero matrix of dimension g and the identity matrix of dimension g as blocks. Set $\boldsymbol{\beta}_{\mathrm{Fil}} = \boldsymbol{\beta}_{\mathrm{Fil}}(b) = (0, \ldots, 0, b_g, \ldots, b_{2g-1})^{\intercal}$.

Remark 5.3. We note that [BD20, Lemma 6.5] simplifies some of the calculations in the case of a hyperelliptic curve X: in this case, we have that $\eta = 0$ and $\boldsymbol{\beta}_{\mathrm{Fil}} = (0, \ldots, 0)^{\intercal}$.

5.4. Computing the Frobenius structure

The Frobenius structure on $\mathcal{A}_Z$ can be determined explicitly in terms of double Coleman integrals, as discussed in [BDM+19, §5]. Here is an algorithm to compute it:

Algorithm 5.4 (Computing the Frobenius structure on $\mathcal{A}_Z$).

1. Use Tuitman's algorithm [Tui16, Tui17] to compute the matrix of Frobenius F and a vector $\boldsymbol{f}$ of overconvergent functions such that
$$\phi^* \boldsymbol{\omega} = d\boldsymbol{f} + F\boldsymbol{\omega},$$
where ϕ is a certain lift of Frobenius.

2. Let b_0, x_0 be Teichmüller representatives of b, x respectively. Compute the matrix

$$\mathrm{A} = \mathrm{I}(x, x_0)^{+} \cdot \mathrm{I}(b_0, b)^{-},$$

where we define for any pair $x_1, x_2 \in X(\mathbf{Q}_p)$ the parallel transport matrices

$$\mathrm{I}^{\pm}(x_1, x_2) = \begin{pmatrix} 1 & 0 & 0 \\ \int_{x_1}^{x_2} \boldsymbol{\omega} & 1 & 0 \\ \int_{x_1}^{x_2} \eta + \int_{x_1}^{x_2} \boldsymbol{\omega}^{\intercal} Z \boldsymbol{\omega} & \pm \int_{x_1}^{x_2} \boldsymbol{\omega}^{\intercal} Z & 1 \end{pmatrix},$$

where η is as computed in Algorithm 5.2 (see also Remark 5.3).
3. Explicitly solve the system

$$\begin{cases} d\mathbf{g}^{\intercal} = d\mathbf{f}^{\intercal} Z F, \\ dh = \boldsymbol{\omega}^{\intercal} F^{\intercal} Z \mathbf{f} + d\mathbf{f}^{\intercal} Z \mathbf{f} - \mathbf{g}^{\intercal} \boldsymbol{\omega} + \phi^{*}\eta - p\eta, \\ h(b_0) = 0. \end{cases}$$

Then compute the matrix

$$\mathrm{M}(b_0, x_0) = \begin{pmatrix} 1 & 0 & 0 \\ (I - F)^{-1}\mathbf{f} & 1 & 0 \\ \frac{1}{1-p}\left(\mathbf{g}^{\intercal}(I - F)^{-1}\mathbf{f} + h\right) & \mathbf{g}^{\intercal}(F - p)^{-1} & 1 \end{pmatrix}(x_0).$$

4. Finally, compute the matrix

$$s_0^{-1}(b, x) \circ s^{\phi}(b, x) = \mathrm{A} \cdot \mathrm{M}(b_0, x_0) = \begin{pmatrix} 1 & 0 & 0 \\ \boldsymbol{\alpha}_{\phi}(b, x) & 1 & 0 \\ \gamma_{\phi}(b, x) & \boldsymbol{\beta}_{\phi}^{\intercal}(b, x) & 1 \end{pmatrix}.$$

Remark 5.5. If X is a hyperelliptic curve, say the smooth projective model of the affine curve $Y\colon y^2 = f(x)$, where f is monic and has no repeated roots, then we can use Kedlaya's algorithm [Ked01] or Harrison's generalization [Har12] in Step (1) above. In fact, Tuitman's approach generalizes the approach of Kedlaya and Harrison. Note that the `SageMath` implementation of Kedlaya's algorithm takes the convention that Frobenius acts on columns, while the `Magma` implementation of Tuitman's algorithm as used here takes the convention that Frobenius acts on rows and thus differs by a transpose.

Remark 5.6. Computing the action of Frobenius in Step (1) gives us a way to compute Coleman integrals: in particular, if $b_0 = \phi(b_0)$ and $x_0 = \phi(x_0)$ are Teichmüller points, we compute the Coleman integral as

$$\int_{b_0}^{x_0} \boldsymbol{\omega} = (1 - F)^{-1}\left(\boldsymbol{f}(x_0) - \boldsymbol{f}(b_0)\right).$$

6. Examples

We illustrate the practicality of the method of Chabauty–Kim discussed in Section 5 by applying it to three new examples of curves whose rational points were previously unknown. They are all curves of the form

$$X_0(N)^+ := X_0(N)/w_N \tag{43}$$

where N is prime and w_N is the Atkin–Lehner involution, and therefore they have a unique rational cusp. The non-cuspidal rational points of $X_0(N)^+$ classify unordered pairs of elliptic curves that are related by an N-isogeny. The (non-) existence of non-CM points is of great interest, because Elkies [Elk04] shows that every non-CM $\mathbf{Q}$-curve is isogenous to one parametrised by a rational point on some $X_0(N)^+$.

We consider the cases $N = 67$, 73, and 103. For each value of N, the curve $X_0(N)^+$ is of genus 2 and its Jacobian has real multiplication. Thus, the rank of the Néron–Severi group is equal to 2, and the method outlined in Section 5 produces exactly one non-trivial locally analytic function on $X_0(N)^+(\mathbf{Q}_p)$ that vanishes on the set of rational points $X_0(N)^+(\mathbf{Q})$. Hence, unlike in the Chabauty–Coleman example at the end of Section 1, we need in addition the Mordell–Weil sieve (see §6.7) to extract the set of rational points from the larger quadratic Chabauty set.

We discuss the computation for $N = 67$ in some detail and briefly summarize the cases $N = 73$ and $N = 103$. These computations use the computer algebra system `Magma` [BCP97] and were started by Best, Bianchi, and Triantafillou at the workshop "Arithmetic Statistics and Diophantine Stability" at the Fondation des Treilles in July 2018. More details can be found at

`https://ngtriant.github.io/papers/BBBLMTV_Data.pdf`

6.1. An explicit model for $X_0(67)^+$

As is explained in [Mur92, Gal96], an affine model for the genus 2 curve $X_0(67)^+$ can be found explicitly as follows. Let f be the unique, up to conjugation, newform of level 67 and weight 2, which is furthermore invariant under the Atkin–Lehner involution w_{67}. The complex vector space spanned by f and its Galois conjugate f^c is isomorphic to the space of regular differentials on $X_0(67)^+$, and we may choose a basis g_1 and g_2 for this space such that $g_1 = q + \cdots$ and $g_2 = q^2 + \cdots$. Note that f and f^c can be computed up to arbitrary q-adic precision using `Magma` [BCP97]. Then $x = \frac{g_1}{g_2}$ and $y = \frac{q}{g_2}\frac{dx}{dq}$ are related by an equation of the form $y^2 = p(x)$, for some monic polynomial $p(x)$ of degree 6 whose coefficients can be determined from the q-expansions. Such an equation gives a model for $X_0(67)^+$. While g_2 is unique, g_1 is not. A certain choice of g_1 yields

$$Y\colon y^2 = x^6 + 2x^5 + x^4 - 2x^3 + 2x^2 - 4x + 1 \qquad [\,=: f_{67}(x)\,].$$

See [Mur92] for more details. For other examples of computations of models of higher genus modular curves, see [Gal96]. The projective closure X adds two points

FIGURE 1. The reduction of $\mathcal{X}_0(N)^+$ at N.

at infinity, ∞^+ and ∞^-, corresponding to $(1:1:0)$ and $(1:-1:0)$ respectively. By an explicit search, we quickly find several points in $X(\mathbf{Q})$. Indeed,

$$X(\mathbf{Q}) \supset \{\infty^+, \infty^-, (0, \pm 1), (-1, \pm 3), (1, \pm 1), (-2, \pm 7)\}. \tag{44}$$

Leprevost [Lep99] also found these points and conjectured that we have equality in (44).

Our goal is to use the machinery set up in Section 5, combined with the Mordell–Weil sieve, to show that $X(\mathbf{Q})$ consists precisely of these 10 points.

Using the explicit model Y, several arithmetic properties of $X_0(67)^+$ can be deduced. For instance, `Magma`'s implementation of 2-descent shows that the rank[15] of $J_0(67)^+(\mathbf{Q})$ is exactly 2. Alternatively, one can avoid the use of a model and draw the same conclusion from the Gross–Zagier–Kolyvagin–Logachev theorem [GZ86, KL89], by computing that (provably [Ste00, Chapter 3]) $L(f,1) = 0$ and (numerically [Cre97, Dok04]) $L'(f,1) \neq 0$.

6.2. The reduction of $X_0(N)^+$

Recall that the method outlined in Section 5 uses some global and local p-adic heights in the sense of Nekovář. Although these depend on some auxiliary choices that we have not made yet at this stage, we have already remarked that we can always ignore all the local heights at primes $v \neq p$ of potential good reduction. More generally, by work of Betts–Dogra [BD19, Proposition 1.2.1], the map $X(\mathbf{Q}_v) \to \mathbf{Q}_p$ induced by the local height at $v \neq p$ takes at most as many values as the number of irreducible components of a regular semi-stable model at v. Note that $X_0(N)^+$ has good reduction at all primes away from N. Using an argument analogous to [BDM+19, Theorem 6.6], we can show that for all primes N there is a regular semi-stable model $\mathcal{X}_0(N)^+$ of $X_0(N)^+$ whose special fibre is isomorphic to a projective line intersecting itself g times, where g is the genus of $X_0(N)^+$ (see Figure 1). The proof which is contained in [BDM+19], constructs such a model starting from the semi-stable Deligne–Rapoport model of $X_0(N)$, and shows that the quotient by the Atkin–Lehner involution remains semi-stable using a lemma of Raynaud. The self-intersections correspond to conjugate pairs of supersingular j-invariants in $\mathbf{F}_{N^2} \setminus \mathbf{F}_N$ (see [DR73, V, §1] and [Ogg75, §3]). In particular, the special fibre of $\mathcal{X}_0(N)^+$ consists of only one component, so the work of Betts–Dogra implies that there are no non-trivial contributions at $v \neq p$.

[15]Since the newforms of weight 2, level 67, which are invariant under w_{67} form a single Galois orbit, the Mordell–Weil rank over $\mathbf{Q}$ is necessarily a multiple of the genus, i.e., of 2.

6.3. Preliminary choices

A prime p and a base point b. Since by §6.2 the curve $X_0(N)^+$ has good reduction at all primes away from N, we could let our fixed p be any prime different from N; we pick $p = 11$. This choice may seem slightly peculiar to the reader familiar with the classical Chabauty–Coleman method, where it is often advantageous to choose the smallest possible prime of good reduction. The prime 11 has two main advantages for our purposes. First, the polynomial f_{67} has no linear factors over $\mathbf{Q}_{11}$. As a result, the lift of Frobenius that we use in §6.5 extends to all of $X(\mathbf{Q}_{11})$. While it is possible to deal with disks containing a point with y-coordinate equal to zero by working with a different lift of Frobenius or by using the trick discussed in [BDM+19, §5.5], our choice of p makes both the exposition and the computation significantly shorter. The second advantage of the prime 11 is somewhat post-hoc, coming from the final Mordell–Weil sieve step. It turns out that $J_0(N)^+(\mathbf{F}_q)$ has order divisible by 11^2 for several small primes q (including $q = 31$ and $q = 137$), which makes the Mordell–Weil sieve particularly efficient for proving that points of $X(\mathbf{Q}_{11})$ are *not* in $X(\mathbf{Q})$.

We choose $b = (1, 1)$ for the base point. Note that b lies in both our affine patch and in the affine patch at infinity. One advantage of b over other possible base points is that b will be a Teichmüller point for a convenient lift of Frobenius.

A basis for the de Rham cohomology of $X_0(67)^+$. It is well known that $\mathrm{H}^0(Y_{\overline{\mathbf{Q}}}, \Omega^1)$ has basis given by (the classes of) the differentials

$$\left\{ \frac{dx}{y}, \frac{x\,dx}{y}, \frac{x^2\,dx}{y}, \frac{x^3\,dx}{y}, \frac{x^4\,dx}{y} \right\}. \tag{45}$$

Moreover, inside of $\mathrm{H}^0(Y_{\overline{\mathbf{Q}}}, \Omega^1)$, we can identify $\mathrm{H}^1_{\mathrm{dR}}(X)$ with those differentials which have residue 0 at both points of $X \smallsetminus Y = \{\infty^+, \infty^-\}$. By working with the expansion of each differential in (45) in terms of the uniformizer $t_{\infty^\pm} = x^{-1}$ at $\infty^\pm$, we construct a new basis $\omega_0, \dots, \omega_4$ satisfying properties (1) and (2) of §5.3. In particular, we may take

$$\omega_0 = -\frac{dx}{y}, \quad \omega_1 = (-1 - x)\frac{dx}{y}, \quad \omega_2 = (-2 + x - x^3 - x^4)\frac{dx}{y},$$
$$\omega_3 = \frac{1}{2}\left(1 - x^2 - x^3\right)\frac{dx}{y}, \quad \omega_4 = (-x - x^2)\frac{dx}{y}.$$

From now on, $\boldsymbol{\omega}$ will denote the column vector $(\omega_0, \dots, \omega_3)^\mathsf{T}$.

6.3.1. A Néron–Severi class. The choice of a Néron–Severi class Z as in Section 4 is equivalent to the choice of an endomorphism of $\mathrm{H}^1_{\mathrm{dR}}(X)$ satisfying a list of conditions (see [BDM+19, §4.4]). Let ℓ be a prime of good reduction for X. In order to compute the action of the Hecke operator $T_\ell \in \mathrm{End}(\mathrm{H}^1_{\mathrm{dR}}(X/\mathbf{Q}_\ell))$ on the whole of $\mathrm{H}^1_{\mathrm{dR}}(X/\mathbf{Q}_\ell)$, rather than just on $\mathrm{Fil}^0\,\mathrm{H}^1_{\mathrm{dR}}(X/\mathbf{Q}_\ell)$, we use the Eichler–Shimura formula

$$T_\ell = \mathrm{Fr}_\ell^\mathsf{T} + \ell \cdot (\mathrm{Fr}_\ell^\mathsf{T})^{-1}.$$

The matrix of Frobenius Fr_ℓ with respect to the basis $\boldsymbol{\omega}$ may be computed using Tuitman's algorithm (we briefly postpone a discussion of this to Step (1) of §6.5, since this matrix for $\ell = p$, as well as one additional output of Tuitman's algorithm, are both needed at that step), and we identify the operator T_ℓ with its matrix representation with respect to $\boldsymbol{\omega}$. Note that the Eichler–Shimura formula holds for $X_0(67)$ and thus for $X_0(67)^+$, since the Atkin–Lehner involution commutes with T_ℓ at all $\ell \neq 67$.

To obtain from T_ℓ an endomorphism corresponding to a class $Z \in \mathrm{NS}(J)$ which maps to zero in $\mathrm{NS}(X)$, we first consider $\mathrm{Tr}(T_\ell) \cdot I_4 - 4T_\ell$, which has trace zero, and then multiply on the right by the inverse of the cup product matrix on $\boldsymbol{\omega}$. For example, choosing $\ell = 11$, we obtain the non-trivial endomorphism with matrix representation

$$Z = \begin{pmatrix} 0 & -8 & 12 & 8 \\ 8 & 0 & -8 & -12 \\ -12 & 8 & 0 & 0 \\ -8 & 12 & 0 & 0 \end{pmatrix}.$$

Since the Néron–Severi group has rank 2, choosing a different Hecke operator would only change the matrix Z by multiplication by a constant.

Remark 6.1. Using Tuitman's algorithm, we can compute the entries of T_ℓ (and therefore of Z) only up to some ℓ-adic precision. In our case, this is sufficient to carry out the steps of the quadratic Chabauty computation, since we have chosen $\ell = p$. It should, however, be possible to prove that Z is given exactly by the above matrix, and we may assume that this is the case, as doing so does not affect the computation in any crucial way.

6.4. Hodge Filtration on $\mathcal{A}_Z$

We now compute the Hodge filtration of the vector bundle $\mathcal{A}_Z$ attached to our choice of Néron–Severi class Z and base point b. Since the curve X is hyperelliptic, by Remark 5.3 we only need to compute γ_{Fil}, and we can do so using a simplified version of Algorithm 5.2. In particular, for each point at infinity $\infty^\pm$, we compute $\boldsymbol{\Omega}_{\infty^\pm}$ and $g_{\infty^\pm}$ by formal integration of Laurent series in the uniformizer $t_{\infty^\pm}$. Following the steps, we then find that γ_{Fil} has a pole of exact order 1 at $\infty^\pm$ with residue -8. Since γ_{Fil} must vanish at b, we conclude that

$$\gamma_{\mathrm{Fil}} = -8x + 8.$$

6.5. Frobenius structure on $\mathcal{A}_Z$

We compute the Frobenius structure on $\mathcal{A}_Z$ using Algorithm 5.4.

Step (1): We first fix a lift of Frobenius ϕ. We take $\phi(x) = x^p$, and extend to $\mathbf{Q}_{11}[x]$ by linearity. Since f_{67} has no zeros over $\mathbf{F}_{11}$, we extend this lift to a strict open neighborhood of the tube $]Y_{\mathbf{F}_p}[$, which consists of all points reducing to $Y_{\mathbf{F}_p}$, by expanding

$$\phi(y) = \sqrt{\phi(f_{67}(x))} = y^p \cdot \left(1 + \frac{\phi(f_{67}(x)) - f_{67}(x)^p}{y^{2p}}\right)^{1/2}.$$

as an *overconvergent* Laurent series in $\mathbf{Q}_p[\![x, y, y^{-1}]\!]$. This lift naturally extends to one-forms.

Next, we compute p-adic approximations of F and $\boldsymbol{f}$ using Tuitman's algorithm [Tui16, Tui17], a generalization of Kedlaya's algorithm which incorporates Lauder's fibration method [Lau06]. Roughly speaking, we first compute $\phi^*\omega_i$. Then, we reduce pole orders by iteratively subtracting differentials of overconvergent functions (constructed by solving linear systems) until $\phi^*\omega_i$ has been reduced to a cohomologous linear combination of basis differentials $\sum_j F_{ji}\omega_j$. The sum f_i of the functions from each step satisfies

$$\phi^*\omega_i = \sum_j F_{ji}\omega_j + df_i \,.$$

Note that in our working example, this F is the matrix Fr_ℓ that was computed in §6.3.1 since we chose $\ell = p = 11$ there as well.

Step (2): Since $b = (1, 1)$ is a Teichmüller point for ϕ, $I(b_0, b)^- = I(b, b)^-$ is an identity matrix. To compute the $I(x, x_0)^+$ on each residue disk, we expand the ω_i in terms of a uniformizer near each Teichmüller point x_0 and integrate formally. To compute $\int_{x_0}^{x} \boldsymbol{\omega}^\intercal Z\boldsymbol{\omega}$, we expand, formally integrate, multiply terms, and formally integrate again, as in steps (3) and (5) of Algorithm 5.2.

Step (3): The matrices Z and F are constants, so $\boldsymbol{g}^\intercal = \boldsymbol{f}^\intercal ZF$. We approximate h by iteratively "reducing" a p-adic approximation $(dh)_\sim$ to $dh = \boldsymbol{\omega}^\intercal F^\intercal Z\boldsymbol{f} + d\boldsymbol{f}Z\boldsymbol{f} - \boldsymbol{g}^\intercal\boldsymbol{\omega} + \phi^*\eta - p\eta$ as in Tuitman's algorithm until we find $a_j \in \mathbf{Q}_{11}$ and an overconvergent function $h_\sim(x)$ satisfying

$$(dh)_\sim = \sum_j a_j\omega_j + d(h_\sim)\,.$$

Then $h_\sim(x) - h_\sim(b)$ approximates $h(x)$. The remainder of Steps (3) and (4) of Algorithm 5.2 are straightforward. The terms $\boldsymbol{\alpha}_\phi(b, x), \boldsymbol{\beta}_\phi(b, x), \gamma_\phi(b, x)$ cannot be expressed compactly, so we omit them here.

6.6. The local p-adic height and a finite set of p-adic points containing $X(\mathbf{Q})$

We have now assembled all ingredients to compute the quadratic Chabauty function from (37), whose finite set of zeroes contains $X(\mathbf{Q})$. To find the constants a_i in (37), we use the discussion at the end of §5.1.

Set $K := \mathbf{Q}(\sqrt{5}) = \mathrm{End}_0(J_0(67)^+)$ and $K_p = K \otimes_{\mathbf{Q}} \mathbf{Q}_p$. If we pick a K-equivariant splitting s of the Hodge filtration in formula (40), then the global height h factors through the tensor product $H^0(X_{\mathbf{Q}_p}, \Omega^1)^\vee \otimes_{K_p} H^0(X_{\mathbf{Q}_p}, \Omega^1)^\vee$. We now choose auxiliary points $x_1 = (-2, 7), x_2 = (-1, 3) \in X(\mathbf{Q})$. Since $\mathrm{AJ}_b(x_1) = [\omega \mapsto \int_b^{x_1} \omega]$ is nonzero, $\mathrm{AJ}_b(x_1)$ is a K_p-basis for $H^0(X_{\mathbf{Q}_p}, \Omega^1)^\vee$. Using (41), we compute $(\pi \circ \rho)(x_i)$ in this basis.

We compute $h(\pi(\rho(x_i))) = h_p(\mathrm{per}_p(x_i))$ using (40), the results of §6.4, §6.5 and the splitting s associated to the K-equivariant basis $(\omega_0, \omega_1, \omega_2, \omega_3 - \omega_1)$. Writing ψ_1 for the projection onto the "rational part" and ψ_2 for the projection

Disks	x-coordinates of candidate points
$](0,\pm 1)[$	0 $0+7\cdot 11+0\cdot 11^2+3\cdot 11^3+3\cdot 11^4+\cdots$
$](1,\pm 1)[$	1 $1+6\cdot 11+6\cdot 11^2+8\cdot 11^3+7\cdot 11^4+\cdots$
$](6,\pm 5)[$	$6+5\cdot 11+8\cdot 11^2+2\cdot 11^3+4\cdot 11^4+\cdots$ $6+7\cdot 11+0\cdot 11^2+5\cdot 11^3+1\cdot 11^4+\cdots$
$](-2,\pm 7)[$	-2 $9+10\cdot 11+1\cdot 11^2+8\cdot 11^3+0\cdot 11^4+\cdots$
$](-1,\pm 3)[$	-1 $10+3\cdot 11+9\cdot 11^2+10\cdot 11^3+1\cdot 11^4+\cdots$
$]\infty^{\pm}[$	∞ $2\cdot 11^{-1}+4+10\cdot 11+9\cdot 11^2+8\cdot 11^3+7\cdot 11^4+\cdots$

TABLE 1. A set of 24 points of $X_0(67)^+(\mathbf{Q}_{11})$ containing $X_0(67)^+(\mathbf{Q})$.

onto the "$\sqrt{5}$ part," we find that the function sending $x \in X(\mathbf{Q}_p)$ to

$$\begin{aligned} Q(x) := h_p(\mathrm{per}_p(x)) &- (5\cdot 11+2\cdot 11^2+5\cdot 11^3+0\cdot 11^4+\cdots)\cdot \psi_1(\pi(\rho(x))) \\ &+ (4\cdot 11+0\cdot 11^2+4\cdot 11^3+0\cdot 11^4+\cdots)\cdot \psi_2(\pi(\rho(x))) \end{aligned} \tag{46}$$

vanishes for all $x \in X(\mathbf{Q})$.

We expand Q as a power series on each residue disk, find the roots, and repeat the computation on an affine patch containing the points at infinity to find a finite subset of $X(\mathbf{Q}_{11})$ which contains $X(\mathbf{Q})$. Using a Newton polygon argument, we find that every root of Q is simple. In addition to the 10 known rational points, we find 14 additional 11-adic zeros of Q (listed in Table 1). To show that these points are not rational, we turn to the Mordell–Weil sieve, described in the following subsection.

6.7. The Mordell–Weil sieve

We assume we are given a smooth projective curve $X/\mathbf{Q}$, p a prime of good reduction, a set $X_{\mathrm{known}} \subseteq X(\mathbf{Q})$ and a set $X_{\mathrm{extra}} \subseteq X(\mathbf{Q}_p)$ known to some finite p-adic precision, distinct from any of the X_{known} to that precision. The goal of the Mordell–Weil sieve, which we describe in this section, is to describe extra conditions that the points of $X(\mathbf{Q})$ satisfy that the points in X_{extra} do not. See also [Sik15, BS10, BBM17].

We will show that any rational point must be sufficiently close p-adically to an element of X_{known}. To do this, we prove that for each $x \in X(\mathbf{Q})$, there is some

$y \in X_{\mathrm{known}}$ such that $[x-y] \in J(\mathbf{Q})$ is p-adically close to the identity in $J(\mathbf{Q})$. We can get a handle on being p-adically close to $0 \in J(\mathbf{Q})$ using the p-adic filtration of $J(\mathbf{Q}_p)$ by

$$J_i = \left\{ x \in J(\mathbf{Q}_p) \colon x \equiv 0 \pmod{p^i} \right\}.$$

The important property of this filtration that we will make use of is that

$$J_0/J_1 \simeq J(\mathbf{F}_p),\ J_i/J_{i+1} \simeq \mathbf{F}_p^{\dim J},$$

so that p-adically close rational points must have difference in the Jacobian divisible by a large power of p. Then for any $D \in J(\mathbf{Q})$ we have $\# J(\mathbf{F}_p) \cdot p^i \cdot D \in J_{i+1}$.

The Mordell–Weil sieve locates small cosets within $J(\mathbf{Q})$ (that is, cosets of large index), that contain the image of $X(\mathbf{Q})$ under the Abel-Jacobi map $i_b \colon X \to J$ sending x to $[x-b]$. The sieve plays off local information at a finite set of primes v against the global Mordell–Weil group structure to find restrictions on $i_b(X(\mathbf{Q}))$. First we fix a prime v of good reduction and consider the following commutative diagram:

$$\begin{array}{ccc} X(\mathbf{Q}) & \xrightarrow{\ i_b\ } & J(\mathbf{Q}) \\ {\scriptstyle \mathrm{red}_{X,v}}\downarrow & & \downarrow{\scriptstyle \mathrm{red}_{J,v}} \\ X(\mathbf{F}_v) & \xrightarrow[\ i_{b,v}\ ]{} & J(\mathbf{F}_v) \end{array} \tag{47}$$

The commutativity of the diagram implies that the image of $X(\mathbf{Q})$ along $\mathrm{red}_{J,v} \circ i_b$ is contained in the image of $i_{b,v}$. The advantage of this observation is that the bottom row of the diagram deals with finite objects and information about these may be computed effectively. In particular, we can find $\operatorname{im} i_{b,v}$ given equations for X. In our setting of a hyperelliptic curve, algorithms for this go back to [Can87], and in general one can make use of work of Khuri-Makdisi [KM07]. Pulling the computed image $\operatorname{im} i_{b,v}$ back to $J(\mathbf{Q})$ gives a union of cosets for the kernel of $\mathrm{red}_{J,v}$ that contains the image of $X(\mathbf{Q})$. We will want to pick v so that the kernel of this map provides non-trivial information about cosets of the target subgroup, which means that the index of the kernel is divisible by p. The Mordell–Weil sieve diagram can be amended by using several primes v of good reduction or working with residue classes of $J(\mathbf{Q})$; it is also possible to make use of primes of bad reduction and to go deeper into the filtration $(J_i)_i$.

For simplicity, we suppose that $r = g$ and we fix a basis $D_1, \ldots, D_g$ of $J(\mathbf{Q})/J(\mathbf{Q})_{\mathrm{torsion}}$. If $x \in X(\mathbf{Q}_p)$ were to be rational, and we expressed

$$i_b(x) = \sum_{j=1}^{g} m_j D_j,\ m_j \in \mathbf{Z},$$

then we would have, via the linearity of the Coleman integral of regular 1-forms on the Jacobian,

$$\int_b^x \omega_i = \sum_{j=1}^{g} m_j \int_0^{D_j} \omega_i, \quad \text{for each } i \in \{1, \ldots, g\} \tag{48}$$

where we identify ω_i with the holomorphic differential it induces on J via ι_b. This can be used to determine the m_j for given $x \in X(\mathbf{Q}_p)$ modulo p^n for any n. We are done if we can show for every $x \in X_{\text{extra}}$ that the resulting coset of $J(\mathbf{Q})/p^n J(\mathbf{Q})$ does not meet the pullback of $i_{b,v}$ under $\text{red}_{J,v}$ for some v.

6.7.1. $X_0(67)^+$. We now give some details of this computation for $X_0(67)^+$, using the model

$$y^2 = x^6 + 2x^5 + x^4 - 2x^3 + 2x^2 - 4x + 1\,;$$

we have for X_{known} the 10 points found in (44). The quadratic Chabauty computation described above also results in a set X_{extra} of 11-adic points of cardinality 14, known to finite precision, whose elements are roots of the function Q in (46), but which do not appear to be rational. See Table 1 for their x-coordinates.

As above, we take $b = (1,1)$. With this choice, $D_1 = i_b(\infty^-)$ and $D_2 = i_b(\infty^+)$ are generators for $J(\mathbf{Q})$. For $x \in X_{\text{known}}$ we can find exact coefficients for $i_b(x)$ in terms of this basis. In particular, $i_b(X_{\text{known}})$ is given by pairs

$$(m_1, m_2) \in \big\{(1,0), (0,1), (-6,4), (7,-3), (3,-1), \\ (-2,2), (1,1), (0,0), (8,-5), (-7,6)\big\}.$$

Since we are working with $p = 11$, we look for primes v such that $\text{ord}_{11}(\#J(\mathbf{F}_v))$ is large.

We find that

$$J(\mathbf{F}_{31}) \simeq (\mathbf{Z}/(3 \cdot 11))^2 \text{ and } J(\mathbf{F}_{137}) \simeq \mathbf{Z}/3 \oplus \mathbf{Z}/(3 \cdot 11^2 \cdot 19)$$

and the image of $J(\mathbf{Q})/11^2 J(\mathbf{Q})$ inside these groups surjects onto the 11-parts. We pull back the images of $i_{b,31}$ and $i_{b,137}$ to cosets for $J(\mathbf{Q})/11^2 J(\mathbf{Q})$. Using (48) we compute $i_b(x)$ modulo 11^2 for all $x \in X_{\text{extra}}$, assuming x is rational, and we find that this does not meet our cosets for 31 or 137.

6.7.2. Further examples. In the case of $N = 73$ we run computations analogous to the ones described above, using the prime $p = 37$ for the quadratic Chabauty procedure, and applying the Mordell–Weil sieve with the prime 9511 to rule out the extra 37-adic points. Likewise for $N = 103$ we can perform quadratic Chabauty at $p = 3$. This gives 6 "extra" 3-adic points, which can be shown to be non-rational by applying the Mordell–Weil sieve using the prime 397.

6.7.3. Conclusion. In summary, we have shown:

Theorem 6.2. *The number of rational points on the Atkin–Lehner quotient modular curves $X_0(N)^+$ for $N \in \{67, 73, 103\}$ are as follows:*

$$\#X_0(67)^+(\mathbf{Q}) = 10\,, \quad \#X_0(73)^+(\mathbf{Q}) = 10\,, \quad \#X_0(103)^+(\mathbf{Q}) = 8\,.$$

Recall that an exceptional point on a modular curve is a rational point that is neither a cusp nor a CM point. According to [Gal96], Theorem 6.2 shows that $X_0(67)^+(\mathbf{Q})$ contains no exceptional points and that $X_0(73)^+(\mathbf{Q})$ and $X_0(103)^+(\mathbf{Q})$ contain precisely one exceptional point each, up to the hyperelliptic involution.

Furthermore, we may conclude that the table in [Box21, §4.6] contains all quadratic points on $X_0(67)$ and the table in [Box21, §4.7] contains all quadratic points on $X_0(73)$, complementing [Box21, Theorem 1.1].

Finally, our theorem implies that the list of j-invariants of $\mathbf{Q}$-curves attached to non-cuspidal rational points on $X_0(N)^+$ given in [BGX20, §4.1] is complete for $N \in \{67, 73, 103\}$.

Acknowledgements

We would like to thank Barry Mazur, whose kind enthusiasm and encouragement were a great motivation for the writing of this paper. We thank Levent Alpoge, Netan Dogra, Kiran Kedlaya, Minhyong Kim, Bjorn Poonen, Michael Stoll, Akshay Venkatesh, and John Voight for many useful conversations and suggestions relating to the material in this article, and Bas Edixhoven and Guido Lido for sharing an early version of their preprint [EL19] with us and for providing several useful suggestions and corrections on an earlier version of this article. Most of the computations described in Section 6 were done at the Workshop "Arithmetic Statistics and Diophantine Stability" at the Fondation des Treilles in July 2018. We are grateful to Barry Mazur and Karl Rubin for organizing this event. We thank the referees for several useful remarks.

JB was supported in part by NSF grant DMS-1702196, the Clare Boothe Luce Professorship (Henry Luce Foundation), and Simons Foundation grant #550023. AB was supported by Simons Foundation grant #550023. FB was supported in part by EPSRC and by Balliol College through a Balliol Dervorguilla scholarship and by an NWO Vidi grant. BL was supported by an NSF postdoctoral fellowship. SM was supported in part by DFG grant MU 4110/1-1 and an NWO Vidi grant. NT was supported by an NSF Graduate Research Fellowship under grant #1122374, by Simons Foundation grant #550033, and by the Research and Training Group in Algebra, Algebraic Geometry, and Number Theory at the University of Georgia (NSF grant #1344994) during various stages of this project. JV was supported by Francis Brown and ERC-COG 724638 'GALOP', the Carolyn and Franco Gianturco Fellowship at Linacre College (Oxford), and NSF Grant No. DMS-1638352, during various stages of this project.

References

[Bal15] J. Balakrishnan. Coleman integration for even-degree models of hyperelliptic curves. *LMS J. Comput. Math.*, 18(1):258–265, 2015.

[BBBM19] J.S. Balakrishnan, A. Besser, F. Bianchi, and J.S. Müller. Explicit quadratic Chabauty over number fields. *ArXiv preprint*, arXiv:1910.04653, 2019.

[BBK10] J. Balakrishnan, R. Bradshaw, and K. Kedlaya. Explicit Coleman integration for hyperelliptic curves. In *Algorithmic number theory (ANTS-IX)*, volume 6197 of *Lecture Notes in Comput. Sci.*, pages 16–31. Springer, Berlin, 2010.

[BBM17] Jennifer S. Balakrishnan, Amnon Besser, and J. Steffen Müller. Computing integral points on hyperelliptic curves using quadratic Chabauty. *Math. Comp.*, 86(305):1403–1434, 2017.

[BCP97] W. Bosma, J. Cannon, and C. Playoust. The Magma algebra system I: The user language. *J. Symb. Comp*, 24(3-4):235–265, 1997.

[BD18] J. Balakrishnan and N. Dogra. Quadratic Chabauty and rational points I: p-adic heights. *Duke Math. J.*, 167(11):1981–2038, 2018.

[BD19] A. Betts and N. Dogra. Ramification of étale path torsors and harmonic analysis on graphs. *ArXiv preprint*, arXiv:1909.05734, 2019.

[BD20] J. Balakrishnan and N. Dogra. Quadratic Chabauty and rational points II: Generalised height functions on Selmer varieties. *Int. Math. Res. Not. IMRN*, (rnz362), 2020.

[BDCKW18] Jennifer S. Balakrishnan, Ishai Dan-Cohen, Minhyong Kim, and Stefan Wewers. A non-abelian conjecture of Tate-Shafarevich type for hyperbolic curves. *Math. Ann.*, 372(1-2):369–428, 2018.

[BDM$^+$19] J. Balakrishnan, N. Dogra, S. Müller, J. Tuitman, and J. Vonk. Explicit Chabauty–Kim for the split Cartan modular curve of level 13. *Annals of Math.*, 189(3), 2019.

[BE20] D. Bertrand and B. Edixhoven. Pink's conjecture on unlikely intersections and families of semi-abelian varieties. *J. Éc. polytech. Math.*, 7:711–742, 2020.

[Bet17] A. Betts. The motivic anabelian geometry of local heights on abelian varieties. *Arxiv preprint*, arXiv:1706.04850, 2017.

[BGX20] Francesc Bars, Josep Gonzlez, and Xavier Xarles. Hyperelliptic parametrizations of $\mathbb{Q}$-curves. *The Ramanujan Journal*, 2020.

[Box21] J. Box. Quadratic points on modular curves with infinite Mordell-Weil group. *Math. Comp.*, 90(327):321–343, 2021.

[BPSS] A. Booker, D. Platt, J. Sijsling, and A. Sutherland. Genus 3 hyperelliptic curves. `http://math.mit.edu/~drew/lmfdb_genus3_hyperelliptic.txt`.

[BS10] Nils Bruin and Michael Stoll. The Mordell-Weil sieve: proving non-existence of rational points on curves. *LMS J. Comput. Math.*, 13:272–306, 2010.

[BT19] Benjamin Bakker and Jacob Tsimerman. The Ax-Schanuel conjecture for variations of Hodge structures. *Invent. Math.*, 217(1):77–94, 2019.

[BT20] J. Balakrishnan and J. Tuitman. Explicit Coleman integration for curves. *Math. Comp.*, 89(326):2965–2984, 2020.

[Can87] David G. Cantor. Computing in the Jacobian of a hyperelliptic curve. *Math. Comp.*, 48(177):95–101, 1987.

[CG89] Robert F. Coleman and Benedict H. Gross. p-adic heights on curves. In *Algebraic number theory*, volume 17 of *Adv. Stud. Pure Math.*, pages 73–81. Academic Press, Boston, MA, 1989.

[Cha41] C. Chabauty. Sur les points rationels des courbes algébriques de genre supérieur à l'unité. *C.R. Acad. Sci.*, 212:882–884, 1941.

[CK10] J. Coates and M. Kim. Selmer varieties for curves with CM Jacobians. *Kyoto J. Math.*, 50(4):827–852, 2010.

[CMSV19] Edgar Costa, Nicolas Mascot, Jeroen Sijsling, and John Voight. Rigorous computation of the endomorphism ring of a Jacobian. *Math. Comp.*, 88(317):1303–1339, 2019.

[Coh00] H. Cohen. *Advanced Topics in Computational Number Theory*, volume 193 of *Graduate Texts in Mathematics*. Springer, 2000.

[Col85] R. Coleman. Torsion points on curves and p-adic abelian integrals. *Annals of Math.*, 121:111–168, 1985.

[Cre97] J. Cremona. *Algorithms for modular elliptic curves.* Cambridge University Press, Cambridge, second edition, 1997.

[CS86] G. Cornell and J. Silverman. *Arithmetic Geometry.* Springer-Verlag, New York, 1986.

[DC20] Ishai Dan-Cohen. Mixed Tate motives and the unit equation II. *Algebra Number Theory*, 14(5):1175–1237, 2020.

[DCW15] Ishai Dan-Cohen and Stefan Wewers. Explicit Chabauty-Kim theory for the thrice punctured line in depth 2. *Proc. Lond. Math. Soc. (3)*, 110(1):133–171, 2015.

[DCW16] Ishai Dan-Cohen and Stefan Wewers. Mixed Tate motives and the unit equation. *Int. Math. Res. Not. IMRN*, (17):5291–5354, 2016.

[Del85] Pierre Deligne. Représentations ℓ-adiques. In *Astérisque 124*. 1985.

[Del89] P. Deligne. Le groupe fondamental de la droite projective moins trois points. In *Galois groups over* $\mathbf{Q}$, volume 16 of *Math. Inst. Res. Inst. Publ.*, pages 79–297. Springer-Verlag, 1989.

[Dog19] N. Dogra. Unlikely intersections and the Chabauty–Kim method over number fields. *ArXiv preprint*, arXiv:1903.05032v2, 2019.

[Dok04] Tim Dokchitser. Computing special values of motivic L-functions. *Experiment. Math.*, 13(2):137–149, 2004.

[DR73] P. Deligne and M. Rapoport. Les schémas de modules de courbes elliptiques. In W. Kuyk, editor, *Modular forms in one variable II*, volume 349 of *LNM*, pages 143–316. Springer-Verlag, 1973.

[EH17] J. Ellenberg and D. Hast. Rational points on solvable curves over $\mathbb{Q}$ via non-abelian Chabauty. *ArXiv preprint*, arXiv:1706.00525, 2017.

[EL19] B. Edixhoven and G. Lido. Geometric quadratic Chabauty. *ArXiv Preprint*, arXiv:1910.10752, 2019.

[Elk04] Noam D. Elkies. On elliptic K-curves. In *Modular curves and abelian varieties*, volume 224 of *Progr. Math.*, pages 81–91. Birkhäuser, Basel, 2004.

[Fal83] G. Faltings. Endlichkeitssätze für abelsche Varietäten über Zahlkörpern. *Invent. Math.*, 73(3):349–366, 1983.

[FL82] Jean-Marc Fontaine and Guy Laffaille. Construction de représentations p-adiques. *Annales scientifiques de l'É.N.S.*, 14(4):547–608, 1982.

[FM12] Benson Farb and Dan Margalit. *A Primer on Mapping Class Groups.* Princeton University Press, 2012.

[Gal96] S. D. Galbraith. Equations for modular curves. *Oxford DPhil thesis*, 1996.

[GLLM15] Fritz Grunewald, Michael Larsen, Alexander Lubotzky, and Justin Malestein. Arithmetic quotients of the mapping class group. *Geometric and Functional Analysis*, 25:1493–1542, 2015.

[Gro97] Alexander Grothendieck. Brief an G. Faltings. In *Geometric Galois actions, 1*, volume 242 of *London Math. Soc. Lecture Note Ser.*, pages 49–58. Cambridge University Press, 1997.

[GZ86] B. Gross and D. Zagier. Heegner points and derivatives of L-series. *Invent. Math.*, 84(2):225–320, 1986.

[Had11] M. Hadian. Motivic fundamental groups and integral points. *Duke Math. J.*, 160:503–565, 2011.

[Har12] M. C. Harrison. An extension of Kedlaya's algorithm for hyperelliptic curves. *J. Symb. Comp.*, 47(1):89 – 101, 2012.

[Ked01] Kiran S. Kedlaya. Counting points on hyperelliptic curves using Monsky-Washnitzer cohomology. *J. Ramanujan Math. Soc.*, 16(4):323–338, 2001.

[Ked07] K. Kedlaya. p-Adic cohomology: from theory to practice. *Arizona Winter School Notes*, 2007.

[Ked09] Kiran S. Kedlaya. p-adic cohomology. In *Algebraic geometry—Seattle 2005. Part 2*, volume 80 of *Proc. Sympos. Pure Math.*, pages 667–684. Amer. Math. Soc., Providence, RI, 2009.

[Kim05] M. Kim. The motivic fundamental group of $\mathbf{P}^1\backslash\{0,1,\infty\}$ and the theorem of Siegel. *Invent. Math.*, 161:629–656, 2005.

[Kim09] M. Kim. The unipotent Albanese map and Selmer varieties for curves. *Publ. RIMS*, 45:89–133, 2009.

[Kim10] M. Kim. Massey products for elliptic curves of rank 1. *J. Amer. Math. Soc.*, 23(3):725–747, 2010.

[KL89] V. A. Kolyvagin and D. Yu. Logachëv. Finiteness of the Shafarevich-Tate group and the group of rational points for some modular abelian varieties. *Algebra i Analiz*, 1(5):171–196, 1989.

[KM04] Kamal Khuri-Makdisi. Linear algebra algorithms for divisors on an algebraic curve. *Math. Comp.*, 73:333–357, 2004.

[KM07] Kamal Khuri-Makdisi. Asymptotically fast group operations on Jacobians of general curves. *Math. Comp.*, 76(260):2213–2239, 2007.

[Lau04] Alan G. B. Lauder. Deformation theory and the computation of zeta functions. *Proc. London Math. Soc. (3)*, 88(3):565–602, 2004.

[Lau06] Alan G. B. Lauder. A recursive method for computing zeta functions of varieties. *LMS J. Comput. Math.*, 9:222–269, 2006.

[Lep99] Franck Leprévost. The modular points of a genus 2 quotient of $X_0(67)$. In *Applications of curves over finite fields (Seattle, WA, 1997)*, volume 245 of *Contemp. Math.*, pages 181–187. Amer. Math. Soc., Providence, RI, 1999.

[Loo97] Eduard Looijenga. Prym representations of mapping class groups. *Geom. Dedicata*, 64(1):69–83, 1997.

[LV20] B. Lawrence and A. Venkatesh. Diophantine problems and p-adic period mappings. *Invent. Math.*, 221(3):893–999, 2020.

[Mas20] Nicolas Mascot. Hensel-lifting torsion points on Jacobians and Galois representations. *Math. Comp.*, 89(323):1417–1455, 2020.

[Mor22] L. J. Mordell. On the rational solutions of the indeterminate equations of the third and fourth degrees. *Proc. Cambridge Phil. Soc.*, 21:179–192, 1922.

[MP12] W. McCallum and B. Poonen. The method of Chabauty and Coleman. In *Explicit methods in number theory*, volume 36 of *Panor. Synthèses*, pages 99–117. Soc. Math. France, Paris, 2012.

[Mum83] David Mumford. *Tata lectures on theta*. Birkhäuser, 1983.

[Mur92] Naoki Murabayashi. On normal forms of modular curves of genus 2. *Osaka J. Math.*, 29(2):405–418, 1992.

[Nek93] J. Nekovar. On p-adic height pairings. In *Séminaire de Théorie des Nombres, Paris 1990-1991*, pages 127–202. Birkhäuser, 1993.

[Ogg75] A. P. Ogg. Automorphismes de courbes modulaires. In *Séminaire Delange-PisotPoitou (16e année: 1974/75), Théorie des nombres, Fasc. 1, Exp. No. 7*, page 8. 1975.

[Sik13] Samir Siksek. Explicit Chabauty over number fields. *Algebra Number Theory*, 7(4):765–793, 2013.

[Sik15] Samir Siksek. Chabauty and the Mordell-Weil sieve. In *Advances on superelliptic curves and their applications*, volume 41 of *NATO Sci. Peace Secur. Ser. D Inf. Commun. Secur.*, pages 194–224. IOS, Amsterdam, 2015.

[Sik17] Samir Siksek. Quadratic Chabauty for modular curves. *Arxiv preprint*, 2017.

[ST20] Nick Salter and Bena Tshishiku. Arithmeticity of the monodromy of some Kodaira fibrations. *Compos. Math.*, 156(1):114–157, 2020.

[Ste00] W. Stein. *Explicit approaches to modular abelian varieties*. ProQuest LLC, Ann Arbor, MI, 2000. Thesis (Ph.D.)–University of California, Berkeley.

[SV14] J. Sijsling and J. Voight. On computing Belyĭ maps. *Publications mathématiques de Besançon*, (1):73–131, 2014.

[Tui16] Jan Tuitman. Counting points on curves using a map to $\mathbf{P}^1$. *Math. Comp.*, 85(298):961–981, 2016.

[Tui17] J. Tuitman. Counting points on curves using a map to $\mathbf{P}^1$, II. *Finite Fields Appl.*, 45:301–322, 2017.

J.S. Balakrishnan and A.J. Best
Department of Mathematics & Statistics
Boston University
111 Cummington Mall
Boston, MA 02215, USA
e-mail: jbala@bu.edu
alex.j.best@gmail.com

F. Bianchi and J.S. Müller
Bernoulli Institute
University of Groningen
Nijenborgh 9
9747 AG Groningen, The Netherlands
e-mail: francesca.bianchi@rug.nl
steffen.muller@rug.nl

B. Lawrence
Department of Mathematics
University of Chicago
5734 S University Ave
Chicago, IL 60637, USA
e-mail: brianrl@math.uchicago.edu

N. Triantafillou
Department of Mathematics
University of Georgia
Athens, GA 30602, USA
e-mail: nicholas.triantafillou@uga.edu

J. Vonk
Institute for Advanced Study
1 Einstein Drive
Princeton, NJ 08540, USA
e-mail: vonk@ias.edu

Progress in Mathematics, Vol. 338, 75–89

The Syntomic Regulator for K_2 of Curves with Arbitrary Reduction

Amnon Besser

Abstract. We give a formula for the syntomic regulator on K_2 of a proper curve X over a p-adic field K. This generalizes the results of [Bes00c] where the curve was assumed to have good reduction. The formula is essentially the same with Coleman integration replaced by by Vologodsky integration.

Mathematics Subject Classification (2010). Primary: 19F27, 11S80. Secondary: 11G20, 11G20.

Keywords. p-adic cohomology, syntomic cohomology, regulators, algebraic K-theory, p-adic integration.

1. Introduction

In this paper we compute the syntomic regulator, more precisely the one defined by Nekovář and Nizioł [NN16], on the algebraic K-theory group K_2 of a smooth complete curve X defined over a field K which is a finite extension of $\mathbb{Q}_p$. In [Bes00c] we considered the case that X had good reduction over K and proved a formula for the cup product of the regulator with a form of the second type ω involving Coleman integration. The goal of this paper is to show that the same formula holds without the good reduction assumption, at least if we impose some restrictions on ω, when Coleman integration is replaced by Vologodsky integration [Vol03].

Let k be the residue field of K with cardinality $q = p^f$. We fix once and for all a uniformizer π of K, giving rise to a fixed choice of a branch of the p-adic logarithm $\log = \log_\pi$ with the property $\log(\pi) = 0$. To explain the formula we obtain we first recall that Nekovář and Nizioł define, for a smooth variety X over K, syntomic cohomology groups $H^i_{\mathrm{syn}}(X,j)$, in such a way that there is a spectral sequence

$$E_2^{p,q} = H^p_{\mathrm{st}}(G, H^q_{\text{ét}}(X \otimes \overline{K}, \mathbb{Q}_p(j))) \Rightarrow H^{p+q}_{\mathrm{syn}}(X,j)\ . \tag{1.1}$$

Here, $H^q_{\text{ét}}(X \otimes \overline{K}, \mathbb{Q}_p(j))$ is regarded as a representation of $G = \mathrm{Gal}(\overline{K}/K)$ and, for a de Rham representation V of G, $H_{\mathrm{st}}(V)$ is semi-stable cohomology which may be interpreted as higher exts in the category of potentially semi-stable representation and may be computed in terms of the complex $C_{\mathrm{st}}^{\bullet}(V)$ of [Nek93, 1.19]. Nekovář and Nizioł further construct regulator maps

$$\mathrm{reg} : K_i(X) \to H_{\mathrm{syn}}^{2j-i}(X, j) \;. \tag{1.2}$$

Assume now that X is a complete curve. Using the complex $C_{\mathrm{st}}^{\bullet}$ one can show (see Lemma 2.1) that $H_{\mathrm{st}}^0(G, H_{\text{ét}}^2(X \otimes \overline{K}, \mathbb{Q}_p(2))) = 0$ and $H_{\mathrm{st}}^1(G, H_{\text{ét}}^1(X \otimes \overline{K}, \mathbb{Q}_p(2)))$ surjects on $H_{\mathrm{dR}}^1(X/K)$ and one therefore gets a map

$$\alpha : H_{\mathrm{syn}}^2(X, 2) \to H_{\mathrm{dR}}^1(X/K) \;. \tag{1.3}$$

Let ω be a form of the second kind on X giving rise to a cohomology class $[\omega] \in H_{\mathrm{dR}}^1(X/K)$. We then get a map

$$\mathrm{reg}_{\omega} : K_2(X) \xrightarrow{\mathrm{reg}} H_{\mathrm{syn}}^2(X, 2) \xrightarrow{\alpha} H_{\mathrm{dR}}^1(X/K) \xrightarrow{\cup[\omega]} K \;. \tag{1.4}$$

It is this map that we would like to compute explicitly.

The sought after explicit description is given in terms of Vologodsky integration and the triple index. The latter, developed in Sections 7 and 8 of [BdJ12] and recalled here in Section 2, associates to formal integrals F, G, H of meromorphic differentials at a point x, together with some auxiliary choices of integrals, a number

$$\langle F, G; H \rangle_x \in K \;.$$

When F, G, H are Vologodsky integrals of meromorphic differentials on X their restriction to the formal neighborhood of any point in X is of the above form and one may compute the global triple index

$$\langle F, G; H \rangle_{\mathrm{gl}} = \sum_{x \in X} \langle F, G; H \rangle_x \;.$$

When the auxiliary choices are themselves restrictions of Vologodsky integrals, this expression depends only on F, G, H and not on the particular auxiliary choices.

Recall that K_2 of the function field $K(X)$ is generated by symbols $\{f, g\}$, where f and g are rational functions on X, subject to some relations. Suppose that f and g are such functions. Then $\log(f)$ and $\log(g)$ ($\log = \log_{\pi}$, the branch fixed before) are the Vologodsky integrals of the meromorphic forms $\mathrm{dlog}(f)$ and $\mathrm{dlog}(g)$ respectively. Picking a Vologodsky integral F_{ω} for ω we define

$$\rho_{\omega}(f, g) := \langle \log(f), F_{\omega}; \log(g) \rangle_{\mathrm{gl}} \;, \tag{1.5}$$

which turns out to be independent of the choice of F_{ω}. We conjecture the following:

Conjecture 1. *The map ρ_{ω} extends to a well-defined map $K_2(K(X)) \to K$ and the composed map*

$$K_2(X) \to K_2(K(X)) \xrightarrow{\rho_{\omega}} K$$

equals reg_{ω}.

Our main result verifies this conjecture under the following additional assumption on the form ω, which we hope to remove in future work.

Assumption 1.1. The cohomology class $[\omega]$, based changed to a semi-stable model, is in the kernel of the monodromy operator N.

Theorem 1.2. *Conjecture* 1 *is true under the additional assumption* 1.1.

While not phrased in this way in [Bes00c] this result was proved there in the good reduction case (see [Lan11, Proposition 1.13] as well as (3.7)). Just as in [Bes00c], it implies the following.

Corollary 1.3. *If ω is holomorphic, then the main Theorem holds with $\rho_\omega(f,g)$ replaced with $\int_{(f)} \log(g)\omega$*

The author would like to thank the Georgia Institute of Technology, where most of the work described here was done. He would also like to thank the referee for many useful comments that improved dramatically both the readability and the accuracy of the text. The author is currently supported by grant number 912/18 from the Israel Science Foundation.

2. Syntomic cohomology of curves with semi-stable reduction

In this section we first review the construction of the map α from (1.3) and reduce the proof of the main theorem to the case of a curve with semi-stable reduction. Then we study the syntomic cohomology of such curves and reduce further to a simplified formula, (2.16), applying to individual symbols.

We begin by recalling the complex $C^\bullet_{\mathrm{st}}(V)$ of [Nek93, 1.19] computing $H^\bullet_{\mathrm{st}}(G,V)$ for a de Rham representation V of $G = \mathrm{Gal}(\overline{K}/K)$. It is given by

$$\mathrm{D}_{\mathrm{st}}(V) \xrightarrow{(\varphi-1,N,-i)} \mathrm{D}_{\mathrm{st}}(V) \oplus \mathrm{D}_{\mathrm{st}}(V) \oplus \mathrm{DR}(V)/F^0 \xrightarrow{N+1-p\varphi+0} \mathrm{D}_{\mathrm{st}}(V)\,. \quad (2.1)$$

Here, D_{st} and DR are the functors defined by Fontaine: $\mathrm{D}_{\mathrm{st}}(V)$ is a K_0-vector space, where K_0 is the maximal unramified extension of $\mathbb{Q}_p$ inside K, equipped with a linear nilpotent operator N (called monodromy) and a semi-linear (with respect to the unique lift of Frobenius on K_0 operator φ (Frobenius), satisfying the relation

$$N\varphi = p\varphi N\,.$$

Let X be a smooth proper curve over K.

Lemma 2.1. *We have*

1. $H^0_{\mathrm{st}}(H^2_{\text{ét}}(X\otimes\overline{K},\mathbb{Q}_p(2))) = 0$.
2. *A surjection* $H^1_{\mathrm{st}}(H^1_{\text{ét}}(X\otimes\overline{K},\mathbb{Q}_p(2))) \to H^1_{\mathrm{dR}}(X/K)$.

Before proving the lemma we recall several facts about the cohomology of curve with semi-stable reduction X. Our reference is [dS06], relying in turn on [LS95], and we refer to the above for the precise definition of semi-stable reduction in our context. Let

$$D = \mathrm{D}_{\mathrm{st}}(H^1_{\text{ét}}(X\otimes\overline{K},\mathbb{Q}_p))\,. \quad (2.2)$$

By the semi-stable conjecture of Fontaine, proved by Tsuji [Tsu99] this is isomorphic to Hyodo–Kato cohomology of the special fiber (see below). Consequently, by [dS06, (0.5)], there is a weight decomposition $D = D^0 + D^1 + D^2$ where the linear Frobenius $\phi = \varphi^f$ operates on D^i with eigenvalues which are Weil numbers of weight i. The operator N vanishes on D^0 and D^1 and maps D^2 isomorphically onto D^0.

Proof of Lemma 2.1. Assume first that X has semi-stable reduction. We have $H^2_{\text{ét}}(X \otimes \overline{K}, \mathbb{Q}_p(2)) \cong \mathbb{Q}_p(1)$ and φ acts on $\mathrm{D}_{\text{st}}(\mathbb{Q}_p(1))$ in a non-trivial manner proving the first statement. For the second statement consider the commutative diagram

$$\begin{array}{ccccc} C'_{\text{st}} : D & \xrightarrow{(p^{-2}\varphi-1,N)} & D \oplus D & \xrightarrow{N+1-p^{-1}\varphi} & D \\ \downarrow & & \downarrow & & \downarrow \\ C''_{\text{st}} : D & \xrightarrow{(q^{-2}\phi-1,N)} & D \oplus D & \xrightarrow{N+1-q^{-1}\phi} & D \end{array}$$

In this diagram the first line is the complex obtained from (2.1) by removing the de Rham component (and twisting the operators appropriately) and the second is its analogue with the semi-linear φ replaced by the linear one. By weight considerations the operator $q^{-2}\phi - 1$ in invertible on D. This easily implies that the two left most vertical maps in the diagram are isomorphisms and one easily obtains an isomorphism on H^0 and an injection on H^1 of the rows. As for the cohomology of C''_{st}, clearly $H^0(C''_{\text{st}}) = 0$. Let us compute $H^1(C''_{\text{st}})$. If $Nx = (1 - q^{-1}\phi y)$, these two equal terms lie in D^0 and by doing an eigenspace decomposition to y we see that $y \in (1 - q^{-1})^{-1}Nx + D^2$. On the other hand, if $(x, y) = (q^{-2}\phi - 1, N)z$ then $z = (q^{-2}\phi - 1)^{-1}x$ and $y = Nz \in D^0$ so $y = (1 - q^{-1})^{-1}Nx$. Thus, the map $(x, y) \mapsto y - (1 - q^{-1})^{-1}Nx$ gives an isomorphism $H^1(C''_{\text{st}}) \cong D^2$. It is now easy to get the isomorphism $H^1_{\text{st}}(H^1_{\text{ét}}(X \otimes \overline{K}, \mathbb{Q}_p(2))) \cong H^1_{\text{dR}}(X/K) \oplus H^1(C'_{\text{st}})$, with $H^1(C'_{\text{st}}) \subset D^2$. In general, we reduce to the above situation after a finite base extension. □

Remark 2.2. It would be interesting to refine the results here by computing the component of the regulator landing in $H^1(C'_{\text{st}})$. We have nothing to say about this problem at the moment.

As in the introduction we get the map α from (1.3) and the regulator map from (1.4).

Lemma 2.3. *Conjecture* 1 *will be proved if it is true for X which has semi-stable reduction.*

Proof. By making a finite base change the curve has semi-stable reduction. Both sides of the formula behave well with respect to this base change. □

Assume from now onward that X has semi-stable reduction, thus, recalling the setup of [Bes17, Section 4], X is the generic fiber of a proper $\mathcal{O}_K$ scheme $\mathcal{X}$

with semi-stable reduction

$$T = \cup_i T_i \ . \tag{2.3}$$

In particular, locally near an intersection point $T_i \cap T_j$ there are coordinates x, y satisfying

$$xy = \pi \ , \ T_i = (x) \ , \ T_j = (y) \tag{2.4}$$

(here, (f) denotes the divisor of the rational function f). For simplicity we will assume that components T_i and T_j intersect at most one point. We can easily get to this by blowing up and the main theorem will apply without this assumption.

Let $\Gamma(X)$ be the dual graph of T with vertices V and edges E (this is of course an abuse of notation as it really depends on the particular model). The vertices correspond to the components T_v while the edges are ordered pairs of intersecting components (T_v, T_w) oriented from v to w, so that an edge e has tail $e^+ = v$ and head $e^- = w$. For such an edge we denote by $-e$ the same edge with reverse orientation.

The reduction map $X \to T$ allows us to split X into rigid analytic domains $U_v = \mathrm{red}^{-1} T_v$ which are wide open spaces in the sense of Coleman. These then intersect along annuli corresponding bijectively to the unoriented edges of $\Gamma(X)$. Indeed, in terms of the coordinates x, y appearing in (2.4) the annulus corresponding to the edge (T_i, T_j) gets mapped via x (or y) to the rigid analytic space $A(|\pi|, 1)$ with

$$A(r, s) := \{z \in \overline{K} \ , \ r < |z| < s\} \ . \tag{2.5}$$

An orientation of an annulus fixes a sign for the residue along this annulus and we match oriented edges with oriented annuli as in [Bes17, Definition 4.6]. We use the same notation for the edge and for the associated oriented annulus.

For the vector space D from (2.2) with its Frobenius and monodromy we have, by Tsuji's Theorem alluded to before,

$$D \cong H^1_{\mathrm{HK}}(\mathcal{X}_k) \ , \ \text{with } \iota : H^1_{\mathrm{HK}}(\mathcal{X}_k) \otimes_{K_0} K \xrightarrow{\sim} H^1_{\mathrm{dR}}(X/K) \ , \tag{2.6}$$

the Hyodo–Kato cohomology and Hyodo–Kato isomorphism [dS06, (0.3)], depending on the choice of the uniformizer π.

We now assume that

$$\mathcal{Y} \subset \mathcal{X} \text{ is a relative normal crossings divisor with generic fiber } Y. \tag{2.7}$$

By [NN16] the syntomic cohomology $H^*_{\mathrm{syn}}(X - Y, j)$ is equal to the cohomology $H^*_{l-\mathrm{syn}}((\mathcal{X} - \mathcal{Y}, \mathcal{X}), j)$, which is the cohomology of the total complex associated with the following diagram of complexes.

$$\begin{array}{ccc} R\Gamma_{\mathrm{HK}}(\mathcal{X} - \mathcal{Y}, \mathcal{X}) & \xrightarrow{(1-\varphi/p^j, \iota)} & R\Gamma_{\mathrm{HK}}(\mathcal{X} - \mathcal{Y}, \mathcal{X}) \oplus R\Gamma_{\mathrm{dR}}(X - Y/K)/F^j \\ \downarrow N & & \downarrow N+0 \\ R\Gamma_{\mathrm{HK}}(\mathcal{X} - \mathcal{Y}, \mathcal{X}) & \xrightarrow{1-\varphi/p^{j-1}} & R\Gamma_{\mathrm{HK}}(\mathcal{X} - \mathcal{Y}, \mathcal{X}) \end{array} \tag{2.8}$$

with $R\Gamma_{\mathrm{HK}}(\mathcal{X} - \mathcal{Y}, \mathcal{X})$ a functorial complex computing the Hyodo–Kato cohomology and ι lifting the Hyodo–Kato isomorphism (more precisely its restriction to the underlying K_0-vector space). We now consider a simplified cohomology theory

$$\begin{aligned}&H^*_{m-\mathrm{syn}}((\mathcal{X}-\mathcal{Y},\mathcal{X}),j)\\&\quad = H^*(MF(F^jR\Gamma_{\mathrm{dR}}(X-Y/K)\xrightarrow{1-\phi/q^j} R\Gamma_{\mathrm{HK}}(\mathcal{X}-\mathcal{Y},\mathcal{X})\otimes_{K_0}K))\,.\end{aligned}$$

This is the analogue of the modified syntomic cohomology introduced in the good reduction case in [Bes00b, Definition 8.4], but it is not as useful because the Frobenius cannot be computed by an explicit map on affine dagger spaces even in affine situations, which is why we will need further modifications later on.

Proposition 2.4. *There is a map, natural in pairs* $(\mathcal{X},\mathcal{Y})$ *and compatible with products*

$$H^*_{l-\mathrm{syn}}(*,*)\to H^*_{m-\mathrm{syn}}(*,*). \tag{2.9}$$

When $\mathcal{X}$ *is of relative dimension* 1 *and* $\mathcal{Y}$ *is empty this gives a factoring*

$$H^2_{l-\mathrm{syn}}(\mathcal{X},2)\to H^2_{m-\mathrm{syn}}(\mathcal{X},2)\to H^1_{\mathrm{dR}}(X/K)$$

of the map α *from* (1.3).

Proof. To obtain the required map we first project on the first line of diagram (2.8), giving a map

$$\begin{aligned}&H^*_{l-\mathrm{syn}}(*,*)\to H^*(MF(R\Gamma_{\mathrm{HK}}(\mathcal{X}-\mathcal{Y},\mathcal{X})\\&\qquad\qquad\xrightarrow{(1-\varphi/p^j,\iota)} R\Gamma_{\mathrm{HK}}(\mathcal{X}-\mathcal{Y},\mathcal{X})\oplus R\Gamma_{\mathrm{dR}}(X/K)/F^j)).\end{aligned}$$

The linear Frobenius ϕ is a power of the semi-linear Frobenius φ and one has a map from the mapping fiber of $1-\varphi/p^j$ to that of $1-\phi/q^j$ as in [Bes00b, (8.1)], giving a map to

$$H^*(MF(R\Gamma_{\mathrm{HK}}(\mathcal{X}-\mathcal{Y},\mathcal{X})\xrightarrow{(1-\phi/q^j,\iota)} R\Gamma_{\mathrm{HK}}(\mathcal{X}-\mathcal{Y},\mathcal{X})\oplus R\Gamma_{\mathrm{dR}}(X/K)/F^j)).$$

The linear Frobenius extends by linearity to K-vector spaces. Extending scalars from K_0 to K gives a map to

$$\begin{aligned}&H^*(MF(R\Gamma_{\mathrm{HK}}(\mathcal{X}-\mathcal{Y},\mathcal{X})\otimes_{K_0}K\\&\qquad\xrightarrow{(1-\phi/q^j,\iota)} R\Gamma_{\mathrm{HK}}(\mathcal{X}-\mathcal{Y},\mathcal{X})\otimes_{K_0}K\oplus R\Gamma_{\mathrm{dR}}(X/K)/F^j)).\end{aligned}$$

After tensoring with K the map ι is a quasi-isomorphism, and, inverting it in a functorial way, we find that this last cohomology is isomorphic to $H^*_{m-\mathrm{syn}}((\mathcal{X}-\mathcal{Y},\mathcal{X}),j)$ via the map induced from the following diagram.

$$\begin{array}{ccc}F^jR\Gamma_{\mathrm{dR}}(X-Y/K) & \xrightarrow{\quad 1-\phi/q^j\quad} & R\Gamma_{\mathrm{HK}}(\mathcal{X}-\mathcal{Y},\mathcal{X})\otimes_{K_0}K))\\ \downarrow & & \downarrow{\scriptstyle(\mathrm{Id},0)}\\ R\Gamma_{\mathrm{HK}}(\mathcal{X}-\mathcal{Y},\mathcal{X})\otimes_{K_0}K & \xrightarrow{(1-\phi/q^j,\iota)} & R\Gamma_{\mathrm{HK}}(\mathcal{X}-\mathcal{Y},\mathcal{X})\otimes_{K_0}K\oplus R\Gamma_{\mathrm{dR}}(X/K)/F^j.\end{array}$$

Functoriality is clear and compatibility with products is standard (see [BLZ16] for some details on products). Finally, the statement about curves can easily be recovered by tracing the behavior of the de Rham component on both cohomologies. □

We will denote by $\mathrm{reg}^{\mathrm{m}}$ the composition of the regulator map with the map (2.9). We will also denote by $\mathrm{reg}^{\mathrm{m}}$ the composition

$$\mathrm{reg}^{\mathrm{m}} : K_2(X) \to H^2_{\mathrm{syn}}(X,2) = H^2_{l-\mathrm{syn}}(\mathcal{X},2) \to H^2_{m-\mathrm{syn}}(\mathcal{X},2) \to H^1_{\mathrm{dR}}(X/K)$$

which, by the proposition above, equals the map $\alpha \circ \mathrm{reg}$. We are reduced to computing $\mathrm{reg}^{\mathrm{m}}$, or, composing with cup product with $[\omega]$, the map $\mathrm{reg}^{\mathrm{m}}_{\omega} = (\cup[\omega]) \circ \mathrm{reg}^{\mathrm{m}}$.

Suppose now that f and g are rational functions on X and suppose further that the divisors of f, g are supported on Y satisfying (2.7). As f and g are invertible on $X - Y$ they define classes in $K_1(X-Y)$ having regulators in $H^1_{\mathrm{syn}}(X - Y, 1)$. These have images, $\mathrm{reg}^{\mathrm{m}}(f)$ and $\mathrm{reg}^{\mathrm{m}}(g)$ respectively, in $H^1_{m-\mathrm{syn}}((\mathcal{X} - \mathcal{Y}, \mathcal{X}), 1)$. The cup product $\mathrm{reg}^{\mathrm{m}}(f) \cup \mathrm{reg}^{\mathrm{m}}(g) \in H^2_{m-\mathrm{syn}}((\mathcal{X}-\mathcal{Y},\mathcal{X}),2)$ is the image of $\mathrm{reg}(\{f,g\})$ under (2.9). By our assumptions we have a map $H^2_{m-\mathrm{syn}}((\mathcal{X} - \mathcal{Y}, \mathcal{X}), 2) \to H^1_{\mathrm{dR}}(X-Y)$ and we abuse the notation to get an element

$$\mathrm{reg}^{\mathrm{m}}(f) \cup \mathrm{reg}^{\mathrm{m}}(g) \in H^1_{\mathrm{dR}}(X-Y) \ . \tag{2.10}$$

At this point we recall the theory of the double index, as developed in the Coleman setup in [Bes00c, Section 4] and in the Vologodsky setup in [Bes05, Section 3], as well as the triple index developed in [BdJ12, Sections 7 and 8]. The double index associates an element $\langle F,G\rangle_x$, or $\langle F,G\rangle_e$, of K to a pair (F,G) of integrals, either of meromorphic differentials at a point x or of rigid analytic differentials on an (oriented) annulus e. Note that in both cases F and G may involve a constant multiple of $\log(z)$, where z is a uniformizer either at the point x or on the annulus e. The double index $\langle F,G\rangle$ is uniquely characterized by being bilinear in F and G and by equaling $\operatorname{Res} F dG$ whenever this makes sense. The residue are either at a point or on the oriented annulus (the notion of orientation was recalled after (2.5)). In both scenarios, the uniqueness implies that because locally for a morphism α (of either germs of a point on a curve or annuli) of degree r,

$$\operatorname{Res} \alpha^*\omega = r \operatorname{Res} \omega \ , \tag{2.11}$$

for a meromorphic (resp. rigid analytic) form ω, we have

$$\langle \alpha^* F, \alpha^* G\rangle = r\langle F,G\rangle. \tag{2.12}$$

When F and G are Vologodsky integrals of meromorphic differentials on a curve X they are of the type described above at every point and their global double index is defined as the sum of the local indices at all points. It depends only on dF and dG and, by [Bes05, Section 3] defines a pairing

$$\langle\ ,\ \rangle_{\mathrm{gl}} : H^1_{\mathrm{dR}}(X-Y/K) \times H^1_{\mathrm{dR}}(X/K) \to K \ ,$$

which is compatible with the cup product via the restriction map $H^1_{\mathrm{dR}}(X/K) \to H^1_{\mathrm{dR}}(X-Y/K)$.

The triple index is defined for 3 functions as above F, G, H, together with "auxiliary" data consisting of integrals $\int RdS$ for any $R \neq S$ among F,G,H, satisfying the relation

$$\int RdS + \int SdR = RS \ . \tag{2.13}$$

To ease notation we usually only list the primary data, so the triple index is denoted by $\langle F, G; H\rangle$. If we want to be more precise we can write (this notation does not appear in [BdJ12]) $\langle F, G; H\rangle_{U,W}$, with $dU = HdF$ and $dW = HdG$ and the understanding that the integrals $\int FdH$ and $\int GdH$ are determined by (2.13) while the remaining integrals do not affect the index by [BdJ12, Lemma 7.4]. As proved in [BdJ12, Proposition 7.3], the triple index is uniquely defined by the following properties:

1. It is trilinear.
2. It is symmetric in the first two variables.
3. It satisfies the triple identity
$$\langle F, G; H\rangle + \langle F, H; G\rangle + \langle G, H; F\rangle = 0\,.$$
4. It reduces to the double index when this makes sense as follows:
$$\langle F, G; H\rangle = \left\langle F, \int GdH \right\rangle \tag{2.14}$$
when $\operatorname{Res} dG = 0$.

The statement of these properties require some care because we neglected to auxiliary data. However, in each of these properties there is a clear choice of these auxiliary data and it is with respect to these that the properties hold. For example, symmetry means more precisely that
$$\langle F, G; H\rangle_{U,W} = \langle G, F; H\rangle_{W,U}\,.$$
It is important to note that the properties impose the behavior of the index with respect to changing the auxiliary data [BdJ12, Lemma 7.4]
$$\begin{aligned} \langle F, G; H\rangle_{U+C,W} &= \langle F, G; H\rangle_{U,W} + C \operatorname{Res} dG\,, \\ \langle F, G; H\rangle_{U,W+C} &= \langle F, G; H\rangle_{U,W} + C \operatorname{Res} dF\,. \end{aligned} \tag{2.15}$$
From this property it follows that the global triple index, defined as for the global double index by summing over all points or all annuli, depends only on F, G and H and not on the auxiliary data, provided these are taken to be Vologodsky integrals (or Coleman integrals on overconvergent domains). This completes the description of the triple index appearing in (1.5).

We now set
$$\rho'_\omega(f, g) = \langle \mathrm{reg}^{\mathrm{m}}(f) \cup \mathrm{reg}^{\mathrm{m}}(g), \omega\rangle_{\mathrm{gl}}$$
and we have

Proposition 2.5. *Conjecture* 1 *follows if we prove the formula*
$$\rho'_\omega(f, g) = \rho_\omega(f, g). \tag{2.16}$$

Proof. Suppose $\gamma \in K_2(X)$ restricts to $\sum\{f_i, g_i\} \in K_2(K(X))$. After change of basis, blowup and enlarging $\mathcal{Y}$ we may assume that the assumptions we made are satisfied on all f_i and g_i. The restriction of γ to $X - Y$ equals $\sum f_i \cup g_i$, where f_i

and g_i are considered elements of $K_1(X-Y)$. By the compatibility of the regulator with products and its functoriality we have

$$\begin{aligned}\operatorname{reg}(\gamma) \cup \omega &= \langle \operatorname{reg}(\gamma)|_{X-Y}, \omega\rangle_{\mathrm{gl}} = \left\langle \sum \operatorname{reg}^{\mathrm{m}}(f_i) \cup \operatorname{reg}^{\mathrm{m}}(g_i), \omega \right\rangle_{\mathrm{gl}} \\ &= \sum \rho'_\omega(f_i, g_i) = \sum \rho_\omega(f_i, g_i)\,,\end{aligned}$$

proving the theorem. □

The remainder of this work will be devoted to proving (2.16)

3. Using smooth part syntomic cohomology

The cohomology $H^*_{m-\mathrm{syn}}(*, *)$ already looks somewhat simpler than general syntomic cohomology. Unfortunately, it is still not easily amenable to explicit computations. The reason is that its definition involves the Hyodo–Kato isomorphism in some form, and this, in all known construction, does not seem to have an easy expression liftable to the derived level. To overcome this we use a trick, which is not quite satisfying as will become apparent below, but can still be made to do the required job. It consists of restricting to the smooth part of the special fiber

Definition 3.1. We define for any surjective $\mathcal{O}_K$ scheme $\mathcal{X}$ with smooth generic fiber the "smooth part" syntomic-cohomology groups $\hat{H}^i(\mathcal{X}, j)$ to be the cohomologies of

$$MF(F^j R\Gamma_{\mathrm{dR}}(X/K) \xrightarrow{1-\phi/q^j} R\Gamma_{\mathrm{rig}}(\mathcal{X}_s^{\mathrm{sm}}))\,,$$

where $\mathcal{X}_s^{\mathrm{sm}}$ is the smooth part of the special fiber of $\mathcal{X}$ and $R\Gamma_{\mathrm{rig}}$ is the associated rigid complex computing rigid cohomology [Bes00b]. The map in the mapping fiber is the composition

$$F^j R\Gamma_{\mathrm{dR}}(X/K) \to R\Gamma_{\mathrm{dR}}(X/K) \xrightarrow{sp} R\Gamma_{\mathrm{rig}}(\mathcal{X}_s^{\mathrm{sm}}) \xrightarrow{1-\phi/q^j} R\Gamma_{\mathrm{rig}}(\mathcal{X}_s^{\mathrm{sm}})\,,$$

where the specialization map sp is computed by taking a limit over compactifications of $\mathcal{X}$ of restriction maps, noting that X^{an} contains is a strict neighborhood of the tube of $\mathcal{X}_s^{\mathrm{sm}}$ in such a compactification (a similar argument to the one in [Bes00b, Section 5]).

Proposition 3.2. *For pairs $(\mathcal{X}, \mathcal{Y})$ as in (2.7) there is a functorial map*

$$H^*_{m-\mathrm{syn}}((\mathcal{X}-\mathcal{Y}, \mathcal{X}), *) \to \hat{H}^*(\mathcal{X}-\mathcal{Y}, *)\,, \tag{3.1}$$

compatible with products.

Proof. The map is induced by the diagram

$$\begin{array}{ccc} F^j R\Gamma_{\mathrm{dR}}(X-Y/K) & \xrightarrow{1-\phi/q^j} & R\Gamma_{\mathrm{HK}}(\mathcal{X}-\mathcal{Y}, \mathcal{X}) \otimes_{K_0} K)) \\ \downarrow & & \downarrow \\ F^j R\Gamma_{\mathrm{dR}}(X-Y/K) & \xrightarrow{1-\phi/q^j} & R\Gamma_{\mathrm{rig}}((\mathcal{X}_s - \mathcal{Y}_s)^{\mathrm{sm}}) \end{array}$$

where the vertical map on the right is the restriction on the part where the log structure is trivial. □

We denote the composition of the map (3.1) with the regulator map $\mathrm{reg}^{\mathrm{m}}$ by $\hat{\mathrm{reg}}$.

The advantage of the smooth part cohomology is that it is quite explicit. In fact, we have

Proposition 3.3. *Under assumptions* (2.7) *we have*

$$\hat{H}^i(\mathcal{X}-\mathcal{Y},j) = H^i\left(MF(F^jR\Gamma_{\mathrm{dR}}(X-Y/K) \xrightarrow{1-\phi/q^j} \bigoplus_v \Omega^\bullet((U_v-Y)^\dagger))\right). \tag{3.2}$$

Here, $(U_v-Y)^\dagger$ *is the dagger space corresponding to* U_v-Y*, which can be thought of as removing from* U_v *"discs of infinitesimally smaller than* 1 *radius" inside the residue discs of the points of* Y*. In this mapping fiber,* ϕ *is given explicitly on* $\Omega^\bullet((U_v-Y)^\dagger)$ *via a lift of Frobenius* ϕ_v*.*

Proof. It is enough to note that the rigid complexes of the components $T_v - \cup_{w\neq v}T_w - \mathcal{Y}_s$ of the smooth locus $(\mathcal{X}_s-\mathcal{Y}_s)^{\mathrm{sm}}$ are given by $\Omega^\bullet((U_v-Y)^\dagger)$ and that Frobenius on these complexes is given by a lift of Frobenius, as proved in [Bes00b, Section 10]. □

Unfortunately, this cohomology theory has two serious disadvantages which need to be overcome in order to be able to use it for computations. The first is that the map (3.1) is not an isomorphism and in fact is not even an injection. To see this, we need to recall a few facts that will also be needed in the sequel. Via the Hyodo–Kato isomorphism (2.6) the weight decomposition on D gives a corresponding weight decomposition on $H^1_{\mathrm{dR}}(X/K)$ [dS06, (0.6)] and consequently on

$$H^1_{\mathrm{dR}}(X-Y/K) = H^0 \oplus H^1 \oplus H^2$$

whose dependence on π we suppress as it remains fixed. We have an exact sequence

$$0 \to H^1(\Gamma(X),K) \to H^1_{\mathrm{dR}}(X-Y/K) \to \bigoplus_{v\in V} H^1_{\mathrm{dR}}(U_v-Y) \tag{3.3}$$

(see [Bes17, (4.5)]), which, via the comparison between de Rham and overconvergent cohomologies [BC94], in light of our assumption (2.7), becomes isomorphic to the sequence

$$0 \to H^1(\Gamma(X),K) \to H^1_{\mathrm{dR}}((X-Y)^\dagger/K) \to \bigoplus_{v\in V} H^1_{\mathrm{dR}}((U_v-Y)^\dagger).$$

The $H^1(\Gamma(X), K)$ is graph cohomology, defined by the complex

$$\begin{aligned} &C^0(\Gamma, A) \xrightarrow{d} C^1(\Gamma, A) \\ &C^0(\Gamma, A) = \{f : V(\Gamma) \to K\}, \\ &C^1(\Gamma, A) = \{f : E(\Gamma) \to K,\ f(-e) = -f(e)\} \\ &\quad df(e) = f(e^+) - f(e^-). \end{aligned}$$

Lemma 3.4. *There is a short exact sequence*

$$0 \to H^1(\Gamma(X), K) \to H^2_{m-\mathrm{syn}}((\mathcal{X} - \mathcal{Y}, \mathcal{X}), 2) \to \hat{H}^2(\mathcal{X} - \mathcal{Y}, 2).$$

Proof. This follows from the definitions of the relevant cohomologies together with the fact that $F^2 H^2_{\mathrm{dR}}(X - Y/K) = 0$ and from the exact sequence (3.3). □

Via the sequence (3.3) graph cohomology is identified with H^0. The induced operator N factors via

$$\mathrm{Im}\left(H^1_{\mathrm{dR}}(X - Y/K) \to \bigoplus_{v \in V} H^1_{\mathrm{dR}}(U_v - Y)\right) \to H^1(\Gamma(X), K)$$

as the map

$$([\omega_v])_v \mapsto (e = (v, w) \mapsto \mathrm{Res}_e(\omega)).$$

The Frobenius equivariant projection $\chi : H^1_{\mathrm{dR}}(X - Y/K) \to H^1(\Gamma(X), K)$ can in fact be lifted to level of cocycles by sending a form ω to the cocycle

$$e = (v, w) \mapsto F_v - F_w$$

where the Vologodsky integral F of ω restricts to the function F_v on U_v. By the main theorem of [BZ17], the cocycle $\chi(\omega)$ is the unique harmonic representative of the corresponding cohomology class. By Definition 4.11 and Corollary 5.3 of [Bes17] we have for any two forms on $X - Y$,

$$\langle \eta, \omega \rangle_{\mathrm{gl}} = \sum_v \langle \eta, \omega \rangle_{(U_v - Y)^\dagger} + \chi(\eta) \cdot N(\omega) - \chi(\omega) \cdot N(\eta) \tag{3.4}$$

where the dot refers to the "scalar product" of cocycles (sum of the product of their values on all edges). Note that in [Bes17] the global index was over $U_v - Y$ but it is always possible to replace indices on an annulus by the sum of the indices on the enclosed points. From this formula (and also by weight considerations) it is easily seen that the annihilator of H^0 (which we recall is the image of graph cohomology) under the cup product is $\mathrm{Ker}\, N$. Recall that the regulator $\hat{\mathrm{reg}}$ was defined just after (3.1) to be the composition of the regulator with the map (3.1) to $\hat{H}^*(\mathcal{X} - \mathcal{Y}, *)$. By Lemma 3.4 the kernel of the map (3.1) is precisely H^0. Thus, the global pairing of $\mathrm{reg}^{\mathrm{m}}(f) \cup \mathrm{reg}^{\mathrm{m}}(g)$ with ω can only possibly be computed from $\hat{\mathrm{reg}}(f) \cup \hat{\mathrm{reg}}(g)$ if $N\omega = 0$. We will prove in this section the following.

Proposition 3.5. *Suppose* $N\omega = 0$. *Then formula* (2.16) *holds.*

The second disadvantage of $\hat{H}^i(\mathcal{X}, j)$ is that elements of $F^j H^i_{\mathrm{dR}}(X/K)$ have "too many lifts" to it. Consider $\hat{H}^1(\mathcal{X}-\mathcal{Y}, 1)$, given explicitly, as follows from (3.2)

$$\begin{aligned}\hat{H}^1(\mathcal{X}-\mathcal{Y},1) = \{(\eta,(h_v))\ ,\ \eta \in H^0(X,\Omega^1_{X/K}\log(Y))\ ,\\ h_v \in \mathcal{O}((U_v - Y)^\dagger)\ ,\ dh_v = (1-\phi_v^*/q)\eta\}\,.\end{aligned} \tag{3.5}$$

Thus, the preimage of a fixed η is an affine space of dimension $|V|$ due to all the possible constants of integration in the choices of the h_v. On the other hand, the preimage of η in $H^1_{m-\mathrm{syn}}((\mathcal{X}-\mathcal{Y},\mathcal{X}),1)$ is affine of dimension 1. It is therefore natural to try to identify the image of a lift of η to $H^1_{m-\mathrm{syn}}((\mathcal{X}-\mathcal{Y},\mathcal{X}),1)$ in $\hat{H}^1(\mathcal{X}-\mathcal{Y},1)$. Here we need the easier, but slightly more precise problem of identifying the image of $\mathrm{reg}^{\mathrm{m}}(f)$ in $\hat{H}^1(\mathcal{X}-\mathcal{Y},1)$. The answer is not surprising.

Proposition 3.6. *For each $v \in V$ The function $h_v = (1-\phi_v^*/q)\log(f)$ belongs to $\mathcal{O}(U_v)$ and satisfies $dh_v = (1-\phi_v^*/q)\,\mathrm{dlog}(f)$. The element* $\hat{\mathrm{reg}}(f) \in \hat{H}^1(\mathcal{X}-\mathcal{Y},1)$ *is given, in terms of the representation* (3.5), *by* $(\mathrm{dlog}(f),(h_v)_v)$.

Proof. The first claim is well known and the second is obvious, leaving the verification that the suggested element is indeed the required image of the regulator, which it is up to an additive constant for each v. For each v we find, after possible base change, a K-rational point in U_v, giving a section $x : \mathcal{O}_K \to \mathcal{X}-\mathcal{Y}$. The diagram

$$\begin{array}{ccccccc} K_1(X-Y) & \longrightarrow & H^1_{m-\mathrm{syn}}((\mathcal{X}-\mathcal{Y},\mathcal{X}),1) & \longrightarrow & \hat{H}^1(\mathcal{X}-\mathcal{Y},1) & & \\ \downarrow{\scriptstyle x^*} & & \downarrow{\scriptstyle x^*} & & \downarrow{\scriptstyle x^*} & & \\ K_1(K) & \longrightarrow & H^1_{m-\mathrm{syn}}(\mathcal{O}_K,1) & \xrightarrow{\sim} & \hat{H}^1(\mathcal{O}_K,1) & \xrightarrow{\sim} & K \end{array}$$

commutes. The composed map on the bottom line is known to be the chosen branch of the logarithm and thus the required element pulls back via x to $\log(f(x))$. That this implies the result can be seen as in the description of pullbacks in finite polynomial cohomology in the proof of Theorem 1.1 in Section 9 of [Bes00a]. □

At this point we turn to a description of the right-hand side of (2.16) in terms of local indices.

Proposition 3.7. *For any ω of the second kind on X, with $N\omega = 0$, any Vologodsky integral F_ω of ω and any $f, g \in K(X)^\times$ we have*

$$\begin{aligned}\langle \log(f), F_\omega; \log(g)\rangle_{\mathrm{gl}} = \sum_v \langle \log(f), F_\omega; \log(g)\rangle_{(U_v - Y)^\dagger}\\ - \sum_e \chi(\omega)(e)\cdot \langle \log(f), \log(g)\rangle_e\,.\end{aligned}$$

Proof. This follows the same lines as the proof of Theorem 4.10 in [Bes17]. If we add to the left-hand side the sum over all $v \in V$ of the sum over all ends e of

$\langle \log(f), F_\omega; \log(g)\rangle_e$ we get

$$\sum_v \langle \log(f), F_\omega; \log(g)\rangle_{(U_v - Y)^\dagger}\,.$$

In what we added, every annulus appears twice – the annulus connecting U_v to U_w appears in the sum for v and the sum for w, with reverse orientations. Thus we obtain the sum over edges e of differences

$$\langle \log(f), (F_\omega)_v; \log(g)\rangle_e - \langle \log(f), (F_\omega)_w; \log(g)\rangle_e\,.$$

This notation, however, hides a potentially serious problem, namely: as discussed in Section 2 the triple indices depend also on auxiliary choices of integrals $\int \log(g)\omega$ and $\int \log(g)\,\mathrm{dlog}(f)$. These are made to be Vologodsky integrals. Writing these auxiliary choices into the notation we will get the difference

$$\begin{aligned}
&\langle \log(f), (F_\omega)_v; \log(g)\rangle_{(\int \log(g)\,\mathrm{dlog}(f))_v, (\int \log(g)\omega)_v, e}\\
&- \langle \log(f), (F_\omega)_w; \log(g)\rangle_{(\int \log(g)\,\mathrm{dlog}(f))_w, (\int \log(g)\omega)_w, e}\\
= \;&\langle \log(f), (F_\omega)_v; \log(g)\rangle_{(\int \log(g)\,\mathrm{dlog}(f))_v, (\int \log(g)\omega)_v, e}\\
&- \langle \log(f), (F_\omega)_v; \log(g)\rangle_{(\int \log(g)\,\mathrm{dlog}(f))_w, (\int \log(g)\omega)_w, e}\\
&+ \langle \log(f), (F_\omega)_v; \log(g)\rangle_{(\int \log(g)\,\mathrm{dlog}(f))_w, (\int \log(g)\omega)_w, e}\\
&- \langle \log(f), (F_\omega)_w; \log(g)\rangle_{(\int \log(g)\,\mathrm{dlog}(f))_w, (\int \log(g)\omega)_w, e}\,.
\end{aligned}$$

On the right-hand side of the equation, the first two lines differ only by the auxiliary data. By (2.15) this difference equals $c(e)\,\mathrm{Res}_e\,\mathrm{dlog}(f) + d(e)\,\mathrm{Res}_e\,\omega$, where $c(e)$ and $d(e)$ are the differences between the Vologodsky integrals $\int \log(g)\omega$ and $\int \log(g)\,\mathrm{dlog}(f)$ on v and w respectively. Similar arguments to those of [BZ17] show that both c and d are harmonic cocycles, and as $N\omega$ and $N\,\mathrm{dlog}\,f$ are exact the sum over all e of the relevant contributions vanish. By trilinearity of the triple index we now have that the difference of the last two lines equals

$$\langle \log(f), \chi(\omega)(e); \log(g)\rangle_e = \chi(\omega)(e)\cdot \langle \log(f), \log(g)\rangle_e\,,$$

and summing over all e completes the proof. □

Proof of Proposition 3.5. Suppose $\mathrm{reg}^{\mathrm{m}}(f) \cup \mathrm{reg}^{\mathrm{m}}(g) = [\eta] \in H^1_{\mathrm{dR}}(X - Y/K)$. By (3.4) and the assumption $N\omega = 0$ we have

$$\rho'_\omega(f,g) = \langle \eta, \omega\rangle_{\mathrm{gl}} = \sum_v \langle \eta, \omega\rangle_{(U_v - Y)^\dagger} - \chi(\omega)\cdot N(\eta)\,. \tag{3.6}$$

Let $\eta_v = \eta|_{U_v}$. Then $[\eta_v]$ is the v-component of $\widehat{\mathrm{reg}}(f) \cup \widehat{\mathrm{reg}}(g)$ and it is computed using Proposition 3.6 and the usual formulas for the cup product as in [Bes00c]. By Proposition 3.3 there we have

$$(1 - \phi_v/q^2)\eta_v = \eta_{0,v}(f,g) + d()$$

with [Bes00c, (3.2)]

$$\eta_{0,v}(f,g) = \frac{1}{q^2}\log f_0\,\mathrm{dlog}\,\phi_v^* g - \frac{1}{q}\log g_0\,\mathrm{dlog}\,f$$

where, by definition,

$$f_0 = \frac{f^q}{\phi_v^* f}.$$

We now have

$$\begin{aligned}
\langle \eta_v, \omega \rangle_{(U_v - Y)^\dagger} &= \sum_e \left\langle \log(f), \int F_\omega \,\mathrm{dlog}(g) \right\rangle_e \quad \text{by [Bes00c, Propositions 5.1, 5.3]} \\
&= \sum_e \langle \log(f), F_\omega; \log(g) \rangle_e \quad \text{by (2.14)} \\
&= \langle \log(f), F_\omega; \log(g) \rangle_{(U_v - Y)^\dagger} , \qquad (3.7)
\end{aligned}$$

where the sum over e is over the ends of $(U_v - Y)^\dagger$. We therefore deduce from (3.6)

$$\begin{aligned}
\rho'_\omega(f, g) &= \sum_v \langle \log(f), F_\omega; \log(g) \rangle_{(U_v - Y)^\dagger} - \chi(\omega) \cdot N(\eta) \\
&= \langle \log(f), F_\omega; \log(g) \rangle_{\mathrm{gl}} + \sum_e \chi(\omega)(e) \cdot \langle \log(f), \log(g) \rangle_e - \chi(\omega) \cdot N(\eta) ,
\end{aligned}$$

where the last equality follows from Proposition 3.7. Now, Lemma 3.8 below shows that the last two terms on the right-hand side of the last equality cancel each other, giving the required result. □

Lemma 3.8. *For each edge e of v we have* $\mathrm{Res}_e\, \eta_v = \langle \log(f), \log(g) \rangle_e$.

Proof. This is proved in [BBdJ]. For completeness, we reproduce the argument here. We have

$$\mathrm{Res}_e\, \eta_{0,v}(f, g) = \mathrm{Res}_e (1 - \phi_v / q^2) \eta_v = \left(1 - \frac{1}{q}\right) \mathrm{Res}_e\, \eta_v$$

by (2.11), and, using (2.12),

$$\begin{aligned}
\mathrm{Res}_e\, \eta_{0,v}(f, g) &= \frac{1}{q^2} \mathrm{Res}_e \log f_0 \,\mathrm{dlog}\, \phi_v^* g - \frac{1}{q} \mathrm{Res}_e \log g_0 \,\mathrm{dlog}\, f \\
&= \frac{1}{q^2} \langle \log f_0, \log \phi_v^* g \rangle_e - \frac{1}{q} \langle \log g_0, \log f \rangle_e \\
&= \frac{1}{q^2} \langle q \log f - \phi_v^* \log f, \phi_v^* \log g \rangle_e - \frac{1}{q} \langle q \log g - \phi_v^* \log g, \log f \rangle_e \\
&= \frac{1}{q} \langle \log f, \phi_v^* \log g \rangle_e - \frac{1}{q^2} \langle \phi_v^* \log f, \phi_v^* \log g \rangle_e - \langle \log g, \log f \rangle_e \\
&\quad + \frac{1}{q} \langle \phi_v^* \log g, \log f \rangle_e \\
&= \langle \log f, \log g \rangle_e - \frac{1}{q^2} \langle \phi_v^* \log f, \phi_v^* \log g \rangle_e = (1 - \frac{1}{q}) \langle \log f, \log g \rangle_e .
\end{aligned}$$

□

Propositions 2.5 and 3.5 taken together complete the proof of Theorem 1.2.

References

[BBdJ] J. S. Balakrishnan, A. Besser, and R. de Jeu. Calculating syntomic regulators for K_2 of hyperelliptic curves. In preparation.

[BC94] F. Baldassarri and B. Chiarellotto. Algebraic versus rigid cohomology with logarithmic coefficients. In *Barsotti Symposium in Algebraic Geometry (Abano Terme,* 1991), volume 15 of *Perspect. Math.*, pages 11–50. Academic Press, San Diego, CA, 1994.

[BdJ12] A. Besser and R. de Jeu. The syntomic regulator for K_4 of curves. *Pacific Journal of Mathematics*, 260(2):305–380, 2012.

[Bes00a] A. Besser. A generalization of Coleman's p-adic integration theory. *Inv. Math.*, 142(2):397–434, 2000.

[Bes00b] A. Besser. Syntomic regulators and p-adic integration I: rigid syntomic regulators. *Israel Journal of Math.*, 120:291–334, 2000.

[Bes00c] A. Besser. Syntomic regulators and p-adic integration. II. K_2 of curves. In *Proceedings of the Conference on p-adic Aspects of the Theory of Automorphic Representations (Jerusalem,* 1998), volume 120, pages 335–359, 2000.

[Bes05] A. Besser. p-adic Arakelov theory. *J. Number Theory*, 111(2):318–371, 2005.

[Bes17] A. Besser. p-adic heights and Vologodsky integration. Preprint, 2017.

[BLZ16] A. Besser, D. Loeffler, and S. Zerbes. Finite polynomial cohomology for general varieties. *Annales mathématiques du Québec.*, 40(1):203–220, 2016.

[BZ17] A. Besser and S. Zerbes. Vologodsky integration on curves with semi-stable reduction. Preprint, 2017.

[dS06] E. de Shalit. Coleman integration versus Schneider integration on semistable curves. *Doc. Math.*, (Extra Vol.):325–334, 2006.

[Lan11] A. Langer. On the syntomic regulator for products of elliptic curves. *J. Lond. Math. Soc.* (2), 84(2):495–513, 2011.

[LS95] B. Le Stum. La structure de Hyodo–Kato pour les courbes. *Rend. Sem. Mat. Univ. Padova*, 94:279–301, 1995.

[Nek93] J. Nekovář. On p-adic height pairings. In *Séminaire de Théorie des Nombres, Paris,* 1990–91, pages 127–202. Birkhäuser Boston, Boston, MA, 1993.

[NN16] J. Nekovář and W. Nizioł. Syntomic cohomology and p-adic regulators for varieties over p-adic fields. *Algebra Number Theory*, 10(8):1695–1790, 2016. With appendices by Laurent Berger and Frédéric Déglise.

[Tsu99] T. Tsuji. p-adic étale cohomology and crystalline cohomology in the semi-stable reduction case. *Invent. Math.*, 137(2):233–411, 1999.

[Vol03] V. Vologodsky. Hodge structure on the fundamental group and its application to p-adic integration. *Mosc. Math. J.*, 3(1):205–247, 260, 2003.

Amnon Besser
Department of Mathematics
Ben-Gurion University of the Negev
P.O.B. 653
Be'er-Sheva 84105, Israel

Progress in Mathematics, Vol. 338, 91–120

Toric Regulators

Amnon Besser and Wayne Raskind

Mathematics Subject Classification (2010). Primary: 19F27. Secondary 14G20.

Keywords. Regulators, totally degenerate reduction, motivic cohomology, étale cohomology, syntomic cohomology.

Introduction

In the mid-19th century, Dirichlet (for quadratic fields) and then Dedekind defined a regulator map relating the units in the ring of integers of an algebraic number field of finite degree over $\mathbb{Q}$ with r_1 real embeddings and $2r_2$ complex embeddings to a lattice of codimension one in a Euclidean space of dimension $r_1 + r_2$. They then showed how a determinant formed from this map and other invariants of the field are related to values of zeta and L-functions, known as Dirichlet's class number formula ([LD68, Supplemente, V, §§183 and 184]). Since then, the term "regulator" has been applied to many such maps in number theory and algebraic geometry such as higher algebraic K-theory of number fields, Abel–Jacobi maps for algebraic cycles, and more generally, motivic cohomology. In most cases, the source of the regulator map is a group of interest that is deemed to be difficult to compute, and the target somewhat easier to compute. A very general form of this circle of ideas can be found in Beilinson's conjectures relating motivic cohomology of a smooth projective variety over a number field to real Deligne cohomology and values of L-functions [Bei85], and their refinement by Bloch–Kato [BK90]. There are p-adic analogues of these conjectures, where real Deligne cohomology is replaced by a suitable p-adic cohomology theory such as syntomic or log-syntomic cohomology, and a conjectural relationship with values of p-adic L-functions. In the special but important case of a variety with totally degenerate reduction over a p-adic field K (please see below for definitions), this paper seeks to tie many of the p-adic conjectures and some of the known results together under the guise of what we call *toric regulators*, which relate motivic cohomology with p-adic tori (quotient of a multiplicative torus by a finitely generated free abelian group). These tori may be compact or not.

In [RX07b, RX07a], the second named author and Xarles studied a class of varieties X over a p-adic field K with what they termed *totally degenerate reduction*. In [RX07b] they studied the étale cohomology of X with $\mathbb{Z}_l$-coefficients and showed that for *all* l these are, up to finite torsion and cotorsion, extensions of direct sums of Tate twists. In [RX07a] they used this result to define *p-adic intermediate Jacobians*, which are p-adically uniformized tori, together with Abel–Jacobi maps from the Chow group of algebraic cycles that are homologically equivalent to zero. The first example of these is the Tate elliptic curve E_q, which is given rigid analytically by $\mathbb{G}_m/q^{\mathbb{Z}}$ with q of absolute value less than 1 in K. In this case, the intermediate Jacobian is just $K^\times/q^{\mathbb{Z}}$, and the Abel–Jacobi map is essentially the identity. More generally, for a p-adically uniformized curve X, their work recovers the p-adic uniformization of the Jacobian in a purely algebro-geometric way. The Abel–Jacobi map defined in loc. cit. should agree with that provided by Manin–Drinfeld [MD73], although they do not prove that in their paper.

The work [RX07b, RX07a] raises the very natural question of defining toric regulators in higher motivic cohomology, and that is the main purpose of this paper. Assuming the totally degenerate reduction of X is "split" (see Definition 1.3) we will define *higher intermediate Jacobians* $H_{\mathcal{T}}^{k+1}(X,\mathbb{Z}(r))$, given by the quotient of an algebraic torus by periods, and construct, assuming a certain natural conjecture, regulator maps into them, *toric regulators*.

The toric regulator is a refinement of the regulator of Sreekantan [Sre10b], which is a map

$$r_{\mathcal{D}} : H_{\mathcal{M}}^{k+1}(X,\mathbb{Z}(r)) \to H_{\mathcal{D}}^{k+1}(X,\mathbb{Q}(r))$$

where the group on the right is the cohomology of a certain cone defined by Consani [Con98]. It is a finite-dimensional $\mathbb{Q}$-vector space. We construct a valuation map (see (2.3))

$$H_{\mathcal{T}}^{k+1}(X,\mathbb{Z}(r)) \to H_{\mathcal{D}}^{k+1}(X,\mathbb{Z}(r))$$

and conjecture that after tensoring with $\mathbb{Q}$, the Sreekantan regulator is the valuation of the toric regulator.

Another interesting feature of the theory of the toric regulator is the relation with the syntomic regulator. Following the case of CH^1 of curves, where the toric regulator is the Abel–Jacobi map and the syntomic regulator is its logarithm, one expects that "the syntomic regulator is the logarithm of the toric regulator." We formulate this assertion precisely and prove it. From the point of view of the syntomic theory, this adds the interesting assertion that the syntomic regulator can be "exponentiated." This sometimes allows us to guess formulas for the toric regulator, and we will describe one such guess, but without presenting the syntomic motivation, for brevity.

The Tate elliptic curve is in some sense the original toric intermediate Jacobian. The papers [RX07b, RX07a] may be viewed as a purely algebro-geometric way using p-adic Hodge theory to interpret and generalize Tate's analytic theory and its generalization to curves of higher genus (Mumford curves) and abelian varieties by Mumford [Mum72b, Mum72a]. Another example is provided by K_2

of a Mumford curve X. In this case, it turns out that a regulator into an algebraic torus has already been developed, and termed the *rigid analytic regulator* by Pál [Pál10b] (we prove this except at the prime p). We also compare the log of the rigid analytic regulator with the syntomic regulator, as computed in [Bes21]. For the product of two Mumford curves, we explain a conjectural formula for the toric regulator on K_1, whose motivation is syntomic.

A question left for future work is the relation between the toric regulator and L-functions. Because the syntomic regulator is the logarithm of the toric regulator and is related to special values of p-adic L-functions, we are looking for such special values that may be "exponentiated." There have been several instances of such a phenomenon, starting with the refined Birch and Swinnerton–Dyer conjecture of Mazur and Tate [MT87] and its descendants, especially in Darmon's work on Stark–Heegner points [BD94, BD96, BD98, Dar98, Dar01]. These conjectures concern rational points on elliptic curves in terms of the Tate parameterize at a prime of split multiplicative reduction, and so fit perfectly with the toric regulator in this case. These conjectures inspired in turn the refined p-adic Stark conjecture of Gross [Gro88]. There is one example in higher K-theory, which is due to Pál [Pál10a] and Kondo and Yasuda [KY12], providing an L-function and a regulator formula in the case of K_2 of the Drinfeld modular curve, in analogy with Beilinson's work in the classical case [Bei85] and [BD14, Bru10, Nik10] in the p-adic case on K_2 of a modular curve. Note that to relate this with the toric regulator, one has to either import the result to the number field case or extend the toric regulator to the function field case (the theory of [RX07b] at the prime p currently relies on p-adic Hodge theory).

Another possible source of examples is the Sreekantan regulator. This is expected to be connected with L-values [Sre10b], and the example of K_1 of the product of two Drinfeld modular curves has been worked out by Sreekantan [Sre10a]. Because of the relation between the toric and Sreekantan regulators mentioned above, the sought after L-functions should be such that their valuation is the corresponding L-function of Sreekantan.

This work began while the first author was on sabbatical at Arizona State University, continued while the second author visited the first at Oxford University, and then during two visits of the first author to Wayne State University. It was completed while the first author was on sabbatical at the Georgia Institute of Technology and then a member of IHÉS. We thank all of these institutions and the Raymond and Beverly Sackler Foundation, whose fellowship supported the stay at IHÉS. The first author is currently supported by a grant from the Israel Science Foundation number 912/18.

1. The toric regulator

Let K be a finite extension of $\mathbb{Q}_p$ with ring of integers R, uniformizer π and residue field F. Let $\bar{K}$ be an algebraic closure of K and let $G = \mathrm{Gal}(\bar{K}/K)$. Let $\bar{F}$ be an algebraic closure of F. Let X be a smooth, projective, geometrically connected variety over K which has totally degenerate reduction in the sense of [RX07b]. Let k and r be non-negative integers such that

$$k - 2r \leq -1 . \tag{1.1}$$

In this section we define the toric regulator

$$H_{\mathcal{M}}^{k+1}(X, \mathbb{Z}(r))_0 \xrightarrow{\operatorname{reg}_t} H_{\mathcal{T}}^{k+1}(X, \mathbb{Z}(r))$$

from the motivic cohomology of X to the toric higher intermediate Jacobian of X. The subindex 0 on motivic cohomology refers to homologically trivial classes. This is only relevant when $k+1 = 2r$, in which case $H_{\mathcal{M}}^{k+1}(X, \mathbb{Z}(r)) = CH^r(X)$ are Chow groups. In this particular case the toric regulator is constructed in detail in [RX07a]. We will therefore mostly concentrate on the case of strict inequality in (1.1).

Let l be a prime number. As a consequence of the construction of étale realization functors [Ivo07], one gets an étale regulator map (a rational coefficients version was known for a long time but we will need integral coefficients),

$$\operatorname{reg}_l : H_{\mathcal{M}}^{k+1}(X, \mathbb{Z}(r))_0 \to H^1(K, M_l(r)) \text{ with } M_l = H_{\text{ét}}^k(X \otimes_K \bar{K}, \mathbb{Z}_l) \tag{1.2}$$

(see for example [Rio06]). Note that we have the restriction (1.1) as otherwise the motivic cohomology groups vanish.

As X has totally degenerate reduction, we have the following result of Raskind and Xarles (see [RX07b, Cor. 1 and Theorem 3] as well as [RX07a, Theorem 3] summarizing both the $l \neq p$ and $l = p$ cases).

Theorem 1.1. *There exist finitely generated abelian groups T_j^i and for each l, a filtration $W_\bullet$ on the $\mathbb{Z}_l$-module M_l and isogenies*

$$\operatorname{Gr}_i^W M_l/\operatorname{tor} \to \begin{cases} T^{-i}_{\frac{k+i}{2}}/\operatorname{tor} \otimes \mathbb{Z}_l(\frac{-k-i}{2}) & k+i \text{ even} \\ 0 & \text{otherwise,} \end{cases}$$

which are isomorphisms for almost all l.

Let us recall the construction of the groups T_j^i [RX07b, Section 3]. By assumption, X is the generic fiber of a proper $\mathcal{O}_K$-scheme $\mathcal{X}$ with strictly semi-stable reduction and special fiber Y, which decomposes as a union $Y = \cup_{i=1}^n Y_i$. Set, for each subset of indices $I \subset \{1, \dots, n\}$,

$$Y_I = \bigcap\nolimits_{i \in I} Y_i .$$

Note that our assumptions imply that the Y_I are projective. Let $\bar{Y}_I := Y_I \otimes \bar{F}$ and let $\bar{Y}^{(m)}$ be the disjoint union of $\bar{Y}_I$ over all subset I of size m. We first define groups $C_j^{i,k} = CH^{i+j-k}(\bar{Y}^{(2k-i+1)})$ for each triple (i,j,k) such that $k \geq \max(0,i)$ and 0 otherwise. Then we set

$$C_j^i = \bigoplus\nolimits_k C_j^{i,k} .$$

To make a complex out of these groups, we define the following maps: for a subset of indices I of size $m+1$ and an integer $0 < r \leq m+1$, we define I_r to be the subset obtained from I by deleting the rth index. There is an obvious inclusion map $\rho_r : Y_I \to Y_{I_r}$ and we define maps,

$$\theta_{i,m} = \sum_{r=1}^{m+1} (-1)^{r-1} \rho_r^* : CH^i(\bar{Y}^{(m)}) \to CH^i(\bar{Y}^{(m+1)}) \,,$$

$$\delta_{i,m} = \sum_{r=1}^{m+1} (-1)^r \rho_{r*} : CH^i(\bar{Y}^{(m+1)}) \to CH^{i+1}(\bar{Y}^{(m)}) \,,$$

$$d' = \bigoplus_{k \geq \max(0,i)} \theta_{i+j-k,2k-i+1} \,,$$

$$d'' = \bigoplus_{k \geq \max(0,i)} \theta_{i+j-k,2k-i} \,,$$

and finally

$$d_j^i = d' + d'' : C_j^i \to C_j^{i+1} \,.$$

Then we define

$$T_j^i := H^i(C_j^\bullet) \,.$$

The monodromy map

$$N : T_j^i \to T_{j-1}^{i+2} \tag{1.3}$$

is induced by the map $C_j^i \to C_{j-1}^{i+2}$ which is the identity on the common factors. The composed map

$$N^i : T_{j+i}^{-i} \to T_j^i \tag{1.4}$$

is an isogeny for $i \geq 0$ [RX07b, Proposition 1], implying that N is injective for negative i and surjective for positive i after tensoring with $\mathbb{Q}$.

Remark 1.2. The numerical condition (1.1) on k and r imply that the map $T_r^{k-2r} \to T_{r-1}^{k+2-2r}$ is injective (after tensoring with $\mathbb{Q}$) while $T_r^{k+1-2r} \to T_{r-1}^{k+3-2r}$ is injective except when $k - 2r = -1$ in (1.1), i.e., the case of cycles.

There exists a pairing [RX07b, p. 274]

$$(\ ,\ ) : T_j^i \times T_{d-j}^{-i} \to \mathbb{Z} \tag{1.5}$$

coming from the intersection pairing and inducing a duality on the torsion free quotients [RX07b, Proposition 1]. We further have the relation

$$(Nx, y) = -(x, Ny) \,. \tag{1.6}$$

To proceed we make an additional restriction on the variety X.

Definition 1.3. The variety X is said to have *split* totally degenerate reduction if it has totally degenerate reduction and the absolute Galois group of the residue field F acts trivially on all the groups $C_j^{i,k}$.

A variety with a totally degenerate reduction gains split totally degenerate reduction after a finite unramified field extension. It is tedious but possible to keep track of the Galois action in the non-split case, and it would be important if we considered, e.g., towers of field extensions of K. We assume that X has split totally degenerate reduction from now onward.

Let us further use the following terminology.

Definition 1.4. We say that a collection of maps indexed by the rational primes l is an *almost injection* (respectively, *almost surjection*) if for all l the corresponding kernel (respectively, cokernel) is finite and it is 0 for almost all l. We say it is an almost isomorphism if it is both an almost injection and an almost surjection.

To proceed with the construction of the toric regulator, we will want, as in [RX07a], to isolate from $M_l(r)$ the subquotient which is an extension of $T_*^* \otimes \mathbb{Z}_l$ by $T_*^* \otimes \mathbb{Z}_l(1)$, which, with our indexing, is the the subquotient

$$W_{2r-k}M_l(r)/W_{2r-k-4}M_l(r).$$

In order to do this, we will use the groups H_g^1 of Bloch–Kato ([BK90], §3). Recall that when $l \neq p$ $H_g^1 = H^1$, while when $l = p$ we have for a $\mathbb{Q}_p$-representation V that $H_g^1(K,V) = \ker : H^1(K,V) \to H^1(K, V \otimes B_{\mathrm{dR}})$ and for a $\mathbb{Z}_p$-module g-cohomology classes are the ones that become g after tensoring with $\mathbb{Q}$. The regulator reg_l from (1.2) takes values in $H_g^1(K, M_l(r))$. This is tautological for $l \neq p$ and follows from the work of Nekovář and Nizioł [NN16] when $l = p$ (see Section 4 (4.1) and the following statement and (4.3)).

Proposition 1.5. *There exist an integer n_0 and well-defined maps*

$$\mathrm{reg}_l' : H_{\mathcal{M}}^{k+1}(X, \mathbb{Z}(r))_0 \to H_g^1(K, W_{2r-k}M_l(r))$$

such that for any prime $l \neq p$ and any $\alpha \in H_{\mathcal{M}}^{k+1}(X, \mathbb{Z}(r))$ we have

$$\mathrm{reg}_l'(\alpha) = n_0 x, \text{ with } x \in H_g^1(K, W_{2r-k}M_l(r)) \text{ , } \iota_{2r-k}(x) = n_0 \,\mathrm{reg}_l(\alpha) \text{ ,}$$

with

$$\iota_r : W_r M_l \to M_l$$

the obvious injection.

Proof. The quotient $M_l(r)/W_{2r-k}M_l(r)$ is, up to torsion, an iterated extension of copies of $\mathbb{Z}_l(j)$ for $j < 0$. The H^0 of these groups is trivial, and by ([BK90] Example 3.9), we have $H_g^1(\mathbb{Z}_l(j)) = H^0(\mathbb{Q}_l/\mathbb{Z}_l(j))$, which are finite and killed by a fixed integer that is independent of l. □

Projecting on the quotient by $W_{2r-k-4}M_l(r)$ we obtain a map

$$H_{\mathcal{M}}^{k+1}(X, \mathbb{Z}(r))_0 \to H_g^1(K, W_{2r-k}M_l(r)/W_{2r-k-4}M_l(r)) \text{ .} \tag{1.7}$$

Now we can proceed in a similar manner to the proof of Proposition 1.5. The Galois module $W_{2r-k}M_l(r)/W_{2r-k-4}M_l(r)$ gives us an extension class in

$$\mathrm{Ext}^1(W_{2r-k}M_l(r)/W_{2r-k-2}M_l(r), W_{2r-k-2}M_l(r)/W_{2r-k-4}M_l(r)) \text{ .}$$

This is almost isomorphic to $\mathrm{Ext}^1(T_r^{k-2r} \otimes \mathbb{Z}_l, T_{r-1}^{k+2-2r} \otimes \mathbb{Z}_l(1))$ and so after multiplying by an integer n_1 we get Galois modules M_l' with a short exact sequence

$$0 \to T_{r-1}^{k+2-2r} \otimes \mathbb{Z}_l(1) \to M_l' \to T_r^{k-2r} \otimes \mathbb{Z}_l \to 0 \,, \tag{1.8}$$

and after multiplying again by an integer n_2 we get a regulator map

$$\mathrm{reg}_l'' : H_{\mathcal{M}}^{k+1}(X, \mathbb{Z}(r))_0 \to H_g^1(K, M_l') \,. \tag{1.9}$$

We now consider boundary maps in the long cohomology sequence from (1.8). In degree 0 we use Kummer theory to get the map

$$\tilde{N}_l : T_r^{k-2r} \otimes \mathbb{Z}_l \to H_g^1(K, T_{r-1}^{k+2-2r} \otimes \mathbb{Z}_l(1)) \cong T_{r-1}^{k+2-2r} \otimes K^{\times(l)} \,, \tag{1.10}$$

where $K^{\times(l)}$ is the l-completion of $K^\times$. This is essentially the monodromy pairing considered by Raskind and Xarles [RX07a, p 6064] (although they only define it in some cases). Let us call this the *augmented monodromy* (at l).

Suppose now that $l \neq p$. The l-part of the tame inertia group is isomorphic to $\mathbb{Z}_l(1)$ as a Frobenius module and we identify the two for convenience. The following is well known.

Lemma 1.6. *The following diagram commutes*

$$\begin{array}{ccc} K^\times & \xrightarrow{\text{Kummer}} & H^1(K, \mathbb{Z}_l(1)) \\ \downarrow{\scriptstyle \text{val}} & & \downarrow \\ \mathbb{Z} & \longrightarrow & \mathbb{Z}_l \end{array}$$

where the vertical map on the right is obtained by restriction to $\mathbb{Z}_l(1)$.

From this lemma and the relation between the monodoromy on étale cohomology and on the T's the following is easy for $l \neq p$. For $l = p$ it will be proved in Proposition 4.5.

Corollary 1.7. *The map*

$$T_r^{k-2r} \otimes \mathbb{Z}_l \xrightarrow{\tilde{N}_l} T_{r-1}^{k+2-2r} \otimes K^{\times(l)} \xrightarrow{\text{val}} T_{r-1}^{k+2-2r} \otimes \mathbb{Z}_l$$

is just the monodromy map (1.3) *with the appropriate indexing tensored with* $\mathbb{Z}_l$.

By Remark 1.2 the composed map is almost injective.

Lemma 1.8. *The obvious map from* $K^\times$ *to the pushout of*

$$\begin{array}{ccc} \prod_l K^{\times(l)} & \xrightarrow{\text{val}} & \prod_l \mathbb{Z}_l \\ & & \uparrow \\ & & \mathbb{Z} \end{array}$$

is an isomorphism. For $l \neq p$ *we have the short exact sequence*

$$0 \to (F^\times)_{l\text{-torsion}} \to K^{\times(l)} \xrightarrow{\text{val}} \mathbb{Z}_l \to 0 \,.$$

Proof. This is well known and follows from the fact that the group of units in the ring of integers in K is compact and complete with respect to its subgroups of finite index. This is not the case for a "larger" nonarchimedean-valued field such as $\mathbb{C}_p$. Note that as $l \neq p$ we have $(F^\times)_{l\text{-torsion}} = (K^\times)_{l\text{-torsion}}$. □

Corollary 1.9. *There is an* augmented monodromy *map*

$$T_r^{k-2r} \xrightarrow{\tilde{N}} T_{r-1}^{k+2-2r} \otimes K^\times \tag{1.11}$$

which gives the augmented monodromy at l (1.10) after l-completion for each l.

We can finally define one of the main objects of this paper.

Definition 1.10. The higher toric intermediate Jacobian of X in degree $k+1$ and twist r is defined by

$$H_T^{k+1}(X, \mathbb{Z}(r)) := \operatorname{coker}\left(T_r^{k-2r} \xrightarrow{\tilde{N}} T_{r-1}^{k+2-2r} \otimes K^\times\right) .$$

Proposition 1.11. *The boundary map in the long exact cohomology sequence of* (1.8),

$$H_g^1(K, T_r^{k-2r} \otimes \mathbb{Z}_l) \to H^2(K, T_{r-1}^{k+2-2r} \otimes \mathbb{Z}_l(1)) ,$$

is almost injective.

Proof. Suppose first that $l \neq p$. By local Tate duality and by the duality induced by the pairing 1.5 and the relation with the monodromy operator given in (1.6), we see that this map is almost dual to the map

$$H^0(K, T_{d-r+1}^{2r-k-2} \otimes \mathbb{Z}_l) \to H^1(K, T_{d-r}^{2r-k} \otimes \mathbb{Z}_l(1))$$

obtained from the dual of (1.6), but this is again the augmented monodromy map, this time in the range where after applying the valuation it is almost surjective, hence is almost surjective by Lemma 1.8. For the case $l = p$ see the syntomic theory of Section 4, in particular, Theorem 4.6. □

Corollary 1.12. *The group $H_g^1(K, M_l')$ is almost isomorphic to*

$$\operatorname{coker}\left(T_r^{k-2r} \otimes \mathbb{Z}_l \xrightarrow{\tilde{N}_l} T_{r-1}^{k+2-2r} \otimes K^{\times(l)}\right) .$$

Definition 1.13. The toric regulator completed at l is the map reg_t^l defined as the composition

$$H_{\mathcal{M}}^{k+1}(X, \mathbb{Z}(r)) \xrightarrow{\mathrm{reg}_l''} H_g^1(K, M_l') \xrightarrow{n_3} \operatorname{coker}\left(T_r^{k-2r} \otimes \mathbb{Z}_l \xrightarrow{\tilde{N}_l} T_{r-1}^{k+2-2r} \otimes K^{\times(l)}\right)$$

of the regulator reg_l'' from (1.9) and multiplication by an integer n_3 which is done to eliminate the difference between the two groups in Corollary 1.12.

In the following section we will state a conjectural formula, Conjecture 1, for the valuation of the toric regulator completed at l. Assuming this conjecture we obtain the main object of this work as follows.

Theorem 1.14. *Let X be a variety with split totally degenerate reduction such that Parshin's conjecture holds for each Y_I (see Remark 2.4 for Parshin's conjecture, in particular for the fact that it is true in all known cases of totally degenerate reduction). Suppose conjecture 1 is true. Then there exists a unique map, called the* toric regulator,

$$H_{\mathcal{M}}^{k+1}(X,\mathbb{Z}(r)) \xrightarrow{\mathrm{reg}_t} H_{\mathcal{T}}^{k+1}(X,\mathbb{Z}(r)) ,$$

such that for each prime l the composed map

$$H_{\mathcal{M}}^{k+1}(X,\mathbb{Z}(r)) \xrightarrow{\mathrm{reg}_t} H_{\mathcal{T}}^{k+1}(X,\mathbb{Z}(r)) = \operatorname{coker}\left(T_r^{k-2r} \xrightarrow{\tilde{N}} T_{r-1}^{k+2-2r} \otimes K^\times\right)$$
$$\xrightarrow{\otimes \mathbb{Z}_l} \operatorname{coker}\left(T_r^{k-2r} \otimes \mathbb{Z}_l \xrightarrow{\tilde{N}_l} T_{r-1}^{k+2-2r} \otimes K^{\times(l)}\right)$$

is the toric regulator completed at l of Definition 1.13.

Proof. By Lemma 1.8 it suffices to show the existence of a map

$$r_{\mathcal{D}} : H_{\mathcal{M}}^{k+1}(X,\mathbb{Z}(r)) \to \operatorname{coker}\left(T_r^{k-2r} \xrightarrow{N} T_{r-1}^{k+2-2r}\right) \otimes \mathbb{Q}$$

such that for any prime l the diagram

$$\begin{array}{ccc} H_{\mathcal{M}}^{k+1}(X,\mathbb{Z}(r)) & \xrightarrow{\mathrm{reg}_t^l} & \operatorname{coker}\left(T_r^{k-2r} \otimes \mathbb{Z}_l \xrightarrow{\tilde{N}_l} T_{r-1}^{k+2-2r} \otimes K^{\times(l)}\right) \\ \downarrow{\scriptstyle r_{\mathcal{D}}} & & \downarrow{\scriptstyle \mathrm{val}} \\ \operatorname{coker}\left(T_r^{k-2r} \xrightarrow{N} T_{r-1}^{k+2-2r}\right) \otimes \mathbb{Q} & \longrightarrow & \operatorname{coker}\left(T_r^{k-2r} \otimes \mathbb{Q}_l \xrightarrow{N\otimes\mathbb{Q}_l} T_{r-1}^{k+2-2r} \otimes \mathbb{Q}_l\right) \end{array} \tag{1.12}$$

commutes. In Section 2 below we will show how the work of Sreekantan [Sre10b] gives precisely such a map and Conjecture 1 will say that the diagram (1.12) commutes. This proves the theorem. □

Remark 1.15. We hope to prove Conjecture 1 in future work. It is somewhat problematic that the toric regulator is only defined after multiplication by an integer, thereby erasing finer roots of unity information. Fortunately, in various situations no such multiplication is required. We could hope that in fact the resulting regulator has a canonical root, but we have no evidence to support this.

2. The relation with the regulator of Sreekantan

In [Sre10b] Sreekantan constructs a new type of regulator and conjectures relations with special values of L-functions in the function field case. We will conjecture that, in a very precise sense, the Sreekantan regulator is exactly the toric regulator followed by the valuation map. As explained in Section 1, this is an important step in actually showing the existence of the toric regulator.

Let us recall the setup for Sreekantan's work. Let X be smooth and proper over K with semi-stable reduction. In his setup the reduction is not assumed to be completely degenerate. Sreekantan starts with a variety over a global field but this is for the sake of getting results about L-functions and he is really only interested in the completion at a finite prime for computing the regulator.

Sreekantan defines what he calls *Deligne cohomology groups at finite primes.* For this he first recalls work of Consani, who defines a certain complex based on groups $K^{i,j,k}$ [Con98, (3.13)]. In fact, it is easy to check (see also Observation 1 on p. 273 of [RX07b], noting that Consani's convention is the same as that of [BGS97] and [GNA90]) that we have

$$K^{i,j,k} = C^{i,k}_{\frac{d+j-i}{2}} \otimes \mathbb{Q}\,.$$

There are differentials and a monodromy operator which are the same as those of [RX07b] and Consani defines (immediately following (3.1))

$$K^{i,j} = \bigoplus_k K^{i,j,k} = C^{i}_{\frac{d+j-i}{2}} \otimes \mathbb{Q}\,.$$

We make the following definition after Consani.

Definition 2.1 (Consani [Con98]). The Consani complex with twist r is the complex $\mathcal{C}(r) = \mathrm{Cone}(N : K^{*-2r,*-d} \to K^{*-2r+2,*-d}) = \mathrm{Cone}(N : C^{*-2r}_{r} \to C^{*-2r+2}_{r-1}) \otimes \mathbb{Q}\,.$

For the normalization of the cone here see [Con98, p. 331].

Definition 2.2. The Deligne cohomology group $H^{k+1}_{\mathcal{D}}(X, \mathbb{Q}(r))$ is the $k+1-2r$ cohomology of the Consani complex $\mathcal{C}(r)$.

Comparing with Sreekantan the reader will observe that we have removed the v-notation, which was to indicate working with the completion of the global X at the finite prime v.

Proposition 2.3. *Suppose that all Y_I satisfy Parshin's conjecture (i.e., have torsion higher K-theory). Then we have an isomorphism*

$$H^{k+1}_{\mathcal{D}}(X, \mathbb{Q}(r)) \cong CH^{r-1}(Y, 2r-k-2) \otimes \mathbb{Q} \tag{2.1}$$

with Bloch's higher Chow groups.

Proof. This is a combination of results in [Con98]: Lemma 3.1 (see also p. 341 where the lemma is used but without precise reference in our context) and equation (2.3), which relies in turn on Conjecture 2.1, which is just Parshin's conjecture. □

Remark 2.4. Parshin's conjecture, that proper varieties over finite fields have torsion higher K-theory, is sometimes stated in the literature only for projective varieties, but, as remarked previously, the Y_I are projective in the totally degenerate reduction case. Furthermore, the totally degenerate assumptions puts severe restrictions on the Y_I, which may allow proving Parshin's conjecture for them, thereby obtaining (2.1) unconditionally. In fact, Parshin's conjecture holds if the $\bar{Y}_I$ are toric or cellular varieties, which is the case for all known examples of varieties with totally degenerate reduction, althought it is not known to follow from this assumption. Note finally that Parshin's conjecture is known in dimensions ≤ 1 by work of Quillen [Qui72] and Harder [Har77, Corollary 3.2.3].

Proposition 2.5. *Suppose now that X has totally degenerate reduction. Then there is a long exact sequence*

$$\begin{aligned}\cdots \to T_r^{k-2r} \otimes \mathbb{Q} \quad &\to T_{r-1}^{k+2-2r} \otimes \mathbb{Q} \to H_{\mathcal{D}}^{k+1}(X, \mathbb{Q}(r)) \\ \to T_r^{k+1-2r} \otimes \mathbb{Q} &\to T_{r-1}^{k+3-2r} \otimes \mathbb{Q} \to \cdots .\end{aligned}$$

Proof. This follows immediately from the definition of the Consani complex as a cone. □

Definition 2.6. The Sreekantan regulator is a map

$$r_{\mathcal{D}} : H_{\mathcal{M}}^{k+1}(X, \mathbb{Z}(r)) \to H_{\mathcal{D}}^{k+1}(X, \mathbb{Q}(r))$$

which is the composition of the boundary map (effectively in motivic homology)

$$H_{\mathcal{M}}^{k+1}(X, \mathbb{Z}(r)) \cong CH^r(X, 2r-k-1) \xrightarrow{\partial} CH^{r-1}(Y, 2r-k-2) \otimes \mathbb{Q}\,,$$

with the isomorphism (2.1).

By Remark 1.2, unless $k+1 = 2r$, the map $T_r^{k+1-2r} \to T_{r-1}^{k+3-2r}$ is injective after tensoring with $\mathbb{Q}$, giving an isomorphism

$$H_{\mathcal{D}}^{k+1}(X, \mathbb{Q}(r)) \cong \operatorname{coker}\left(T_r^{k-2r} \to T_{r-1}^{k+2-2r}\right) \otimes \mathbb{Q}\,. \tag{2.2}$$

Thus, the following conjecture is quite natural.

Conjecture 1. *Suppose that X has split totally degenerate reduction and that Parshin's conjecture holds for all Y_I. Assume $k+1 < 2r$. Then, with $r_{\mathcal{D}}$ as in Definition* 2.6*, diagram* (1.12) *commutes.*

As noted in Section 1, the existence of the toric regulator follows from this conjecture. In addition, we get the following obvious relation between the toric regulator and the Sreekantan regulator: Define the valuation map

$$\operatorname{val} : H_{\mathcal{T}}^{k+1}(X, \mathbb{Z}(r)) \to H_{\mathcal{D}}^{k+1}(X, \mathbb{Q}(r)) \tag{2.3}$$

as the composition of the valuation map and the isomorphism (2.2).

Corollary 2.7. *Assuming Conjecture* 1 *the following diagram commutes.*

$$\begin{array}{ccc} H_{\mathcal{M}}^{k+1}(X, \mathbb{Z}(r)) & \xrightarrow{\operatorname{reg}_t} & H_{\mathcal{T}}^{k+1}(X, \mathbb{Z}(r)) \\ & \searrow^{r_{\mathcal{D}}} & \downarrow \operatorname{val} \\ & & H_{\mathcal{D}}^{k+1}(X, \mathbb{Q}(r))\,. \end{array}$$

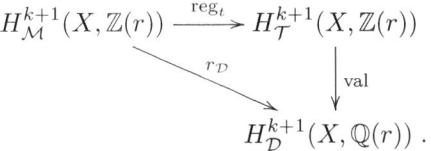

Remark 2.8. Note that Sreekantan does not deal with the case of cycles at all. If we try to argue by analogy, we should expect that in the case $k+1=2r$, the composed map $CH^r(X) \to H^{2r}_{\mathcal{D}}(X,\mathbb{Q}(r)) \to T^0_r$ factors via the cycle class map. Therefore, the analogous Sreekantan regulator maps the homologically trivial cycles $CH^r(X)_0$ again to $\operatorname{coker}\left(T_r^{-1} \to T_{r-1}^1\right)$, but as this last map is (almost) bijective, the corresponding regulator is trivial (unlike the toric regulator).

3. K_2 of curves and the rigid analytic regulator of Pál

In this section we assume that X is of dimension 1. We recall from [Bes17, Section 4] the setup that will be used here and also in some parts of Section 4. By assumption X is the generic fiber of a proper $\mathcal{O}_K$ scheme $\mathcal{X}$ with (Zariski) semi-stable reduction

$$Y = \bigcup_i Y_i \, . \tag{3.1}$$

In particular, locally near an intersection point $Y_i \cap Y_j$ there are coordinates x, y satisfying

$$xy = \pi \; , \; Y_i = (x) \; , \; Y_j = (y) \tag{3.2}$$

(here, (f) denotes the divisor of the rational function f).

For simplicity we will assume that components Y_i and Y_j intersect at at most one point.

Let $\Gamma(X)$ be the dual graph of Y with vertices V and edges E (this is of course an abuse of notation as it really depends on the particular model). The vertices correspond to the components Y_v while the edges are ordered pairs of intersecting components (Y_v, Y_w) oriented from v to w, so that an edge e has tail $e^+ = v$ and head $e^- = w$. For such an edge we denote by $-e$ the same edge with reverse orientation.

The reduction map $X \to Y$ allows us to split X into rigid analytic domains $U_v = \mathrm{red}^{-1} Y_v$ which are wide open spaces in the sense of Coleman. These then intersect along annuli corresponding bijectively to the unoriented edges of $\Gamma(X)$. Indeed, in terms of the coordinates x, y appearing in (3.2) the annulus corresponding to the edge (Y_i, Y_j) gets mapped via x (or y) to the rigid analytic space $A(|\pi|, 1)$ with

$$A(r,s) := \{z \in \bar{K} \; , \; r < |z| < s\} \, . \tag{3.3}$$

An orientation of an annulus fixes a sign for the residue along this annulus and we match oriented edges with oriented annuli as in [Bes17, Definition 4.6]. We use the same notation for the edge and for the associated oriented annulus.

At this point we recall some facts about graph cohomology and harmonic cochains on graphs. For a slightly expanded version the reader may consult [Bes17, Section 4]. For a graph $\Gamma = (V, E)$ and an abelian group A we define 0 and 1 cochains on Γ with values in A by

$$C^0(\Gamma, A) = \{f : V \to A\} \, , \; C^1(\Gamma, A) = \{f : E \to A \, , \; f(-e) = -f(e)\} \, .$$

We have a differential

$$d : C^0(\Gamma, A) \to C^1(\Gamma, A),\ df(e) = f(e^+) - f(e^-)$$

and the graph cohomology $H^1(\Gamma, A)$ is the cokernel of d. We have a pointwise product of $c, d \in C^1(\Gamma, A)$, assuming that A is a ring, defined by

$$c \cdot d = \sum_{e \in E(G)/\pm} c(e) \cdot d(e) ,$$

where the sum is over unoriented edges. The kernel of the dual differential

$$d^* : C^1(G, A) \to C^0(G, A),\ d^* f(v) = \sum_{e^+ = v} f(e) ,$$

is the space $\mathcal{H}(\Gamma, A)$ of *harmonic cochains* with values in A. The obvious injection $\mathcal{H}(\Gamma, A) \hookrightarrow C^1(\Gamma, A)$ induces a map

$$\mathcal{H}(\Gamma, A) \to H^1(\Gamma, A) , \tag{3.4}$$

which is an isomorphism if A is a $\mathbb{Q}$-vector space. With A a ring again, the space $\mathcal{H}(\Gamma, A)$ is orthogonal to $\operatorname{Im} d$, hence induces a pairing

$$\mathcal{H}(\Gamma, A) \times H^1(\Gamma, A) \to A . \tag{3.5}$$

Remark 3.1. There is an unoriented version of the above, which will be needed for what follows. This requires fixing an orientation of each edge. Then we can redefine 1-cochains with values in A as functions from unoriented edges to A. The differential and the dual differential are defined as above but using the fixed orientation on each edge and the pointwise product is unchanged. It is trivial that this construction produces isomorphic graph cohomologies.

Proposition 3.2. *Let X be as above, with dual graph Γ. We have isomorphisms*

$$T_1^{-1} \cong \mathcal{H}(\Gamma, \mathbb{Z}) ,$$
$$T_0^1 \cong H^1(\Gamma, \mathbb{Z}) .$$

With respect to these isomorphisms the monodromy map $N : T_1^{-1} \to T_0^1$ corresponds to the map (3.4) and the product $T_1^{-1} \times T_0^1 \to \mathbb{Z}$ to the pairing (3.5) induced by the pointwise product.

Proof. We have

$$C_1^{-1} = \bigoplus_{k \geq 0} C_1^{-1,k} = \bigoplus_{k \geq 0} CH^{-k}(\bar{Y}^{(2k+2)}) = CH^0(\bar{Y}^{(2)}) ,$$
$$C_1^{0} = \bigoplus_{k \geq 0} C_1^{0,k} = \bigoplus_{k \geq 0} CH^{1-k}(\bar{Y}^{(2k+1)}) = CH^1(\bar{Y}^{(1)}) ,$$
$$C_1^{-2} = \bigoplus_{k \geq 0} C_1^{-2,k} = \bigoplus_{k \geq 0} CH^{-1-k}(\bar{Y}^{(2k+3)}) = 0 ,$$
$$C_0^{1} = \bigoplus_{k \geq 1} C_0^{1,k} = \bigoplus_{k \geq 1} CH^{1-k}(\bar{Y}^{(2k)}) = CH^0(\bar{Y}^{(2)}) ,$$

$$C_0^2 = \bigoplus_{k\geq 2} C_0^{2,k} = \bigoplus_{k\geq 2} CH^{2-k}(\bar{Y}^{(2k-1)}) = 0\,,$$
$$C_0^0 = \bigoplus_{k\geq 0} C_0^{0,k} = \bigoplus_{k\geq 0} CH^{-k}(\bar{Y}^{(2k+1)}) = CH^0(\bar{Y}^{(1)})\,. \tag{3.6}$$

As all the components of $\bar{Y}^{(1)}$ are projective lines, their CH^1's, as well as the CH^0 of the components of $\bar{Y}^{(2)}$ are isomorphic to $\mathbb{Z}$ via the degree map, and the differential $C_1^{-1} \to C_1^0$ is the alternating sum of pushforward maps, which are clearly just the identity map on the appropriate $\mathbb{Z}$ summands. The chosen numbering of the components gives an orientation on all the edges. The unoriented edge $\{i,j\}$ gets the orientation (i,j) with $i<j$. With this the isomorphisms (3.6) are clear in the unoriented version of Remark 3.1. Similarly, The CH^0 of each component of $\bar{Y}^{(1)}$ is isomorphic to $\mathbb{Z}$ and the map $C_0^0 \to C_0^1$ is an alternating sum of pullbacks, which are again the identities on the corresponding $\mathbb{Z}$ components, giving the identification of the monodromy operator, again in the unoriented version. The identification of the pairing is clear. □

Corollary 3.3. *We have $H^2_{\mathcal{T}}(X,\mathbb{Z}(2)) \cong \mathcal{H}(\Gamma, K^\times)$ and the toric regulator in this case is therefore a map*

$$\mathrm{reg}_t : H^2_{\mathcal{M}}(X,\mathbb{Z}(2)) \to \mathcal{H}(\Gamma, K^\times)\,. \tag{3.7}$$

We now recall the (somewhat reformulated) definition of the Pál rigid analytic regulator [Pál10b]. We start by working over $\mathbb{C}_p$ and with closed annuli $A[r,s]$ instead of open ones (3.3).

Definition 3.4. View the annulus $e = A[r,s]$ as embedded in $\mathbb{P}^1$ in the obvious way, and let $D = D(r)$ be the disc $\{|z| < r\}$ inside $\mathbb{P}^1$. For rational functions f,g on $\mathbb{P}^1$ which have no poles or zeros on e, set

$$t_e(f,g) = \prod_{x\in D} t_x(f,g) \in \mathbb{C}_p^\times\,,$$

where t_x is the tame symbol at the point x. Let f,g be invertible rigid analytic functions on $A[r,s]$. Let f_n and g_n be sequences of rational functions on $\mathbb{P}^1$ that converge to f and g respectively on $A[r,s]$. Then set

$$t_e(f,g) = \lim_{n\to\infty} t_e(f_n,g_n) \in \mathbb{C}_p\,.$$

For an open annulus $e = A(r,s)$ define $t_e(f,g)$ for $f,g \in \mathcal{O}(e)^\times$ to be $t_{e'}(f,g)$ for any smaller closed annulus. Finally, for an oriented (open) annulus e and invertible rigid analytic functions f,g on e, define $t_e(f,g)$ by choosing an orientation preserving identification of e with $A(r,s)$.

Theorem 3.5. *The quantity $t_e(f,g)$ is well defined and in $\mathbb{C}_p^\times$. Furthermore we have the following:*

1. $t_{-e}(f,g) = t_e(f,g)^{-1}$.
2. $t_e(f,1-f) = 1$.

3. *Let U be the complement in $\mathbb{P}^1(\mathbb{C}_p)$ of the union of a finite number of disjoint closed balls $D[r_i]$ and let e_i be annuli $A(r_i, r_i+\varepsilon)$ where ε is chosen sufficiently small so that the e_i's are disjoint. Let f, g be invertible meromorphic functions on U which are invertible on the e_i. Then the following residue theorem holds:*

$$\prod_{x\in U} t_x(f,g) \cdot \prod_i t_{e_i}(f,g) = 1 \,.$$

Proof. For the closed disc $A[r,s]$ what we defined here is what Pál defines, in the course of proving [Pál10b, Theorem 2.2], as t_D for the boundary component $D(r)$ of the connected rational subdomain $A[r,s]$. In particular, it is well defined. Lemma 3.4 in [Pál10b] shows that the choice of a closed annulus inside an open annulus does not matter and Pál also shows [Pál10b, Theorem 3.11] that the construction commutes with morphisms of domains, which shows that it is independent of the choice of parameterization. Thus, the index is well defined. Switching the orientation corresponds to using $D = \{s < |z|\}$ and [Pál10b, Theorem 3.2 (iii)] immediately gives (1) (Pál works here already with elements of K_2 but the result certainly specializes to what we have here). The formula in Theorem 2.2 (iv) there immediately implies (2) and the residue theorem is easily deduced from (iii) of Theorem 3.2. □

Lemma 3.6. *Let* $\deg_e : \mathcal{O}(e)^\times \to \mathbb{Z}$ *be defined by* $\deg_e f = \operatorname{res}_e d\log f$. *Then, for a constant* $c \in \mathbb{C}_p^\times$, $t_e(c,f) = c^{\deg_e(f)}$.

Proof. This holds for rational functions f with a degree function equaling the number of zeros in the disc $D(r)$ minus the number of poles, which is clearly the same as our degree, by [Pál10b]. It then follows in general by continuity. □

To define the Pál regulator we first consider functions $f, g \in \mathbb{C}_p(X)^\times$ having no poles or zeros on any $e \in E$. We can then define $\operatorname{reg}_P(\{f,g\}) \in C^1(\Gamma, \mathbb{C}_p^\times)$ by

$$\operatorname{reg}_P(\{f,g\})(e) = t_e(f,g) \,.$$

Next, if $\alpha = \sum \{f_i, g_i\}$ is a formal combination of symbols and all the functions f_i, g_i are invertible on all $e \in E$, and all its tame symbols are 1, then

$$\operatorname{reg}_p(\alpha) = \prod \operatorname{reg}_P(\{f_i, g_i\})$$

is defined and the residue theorem implies that it is in $\mathcal{H}(\Gamma, \mathbb{C}_p)$. Suppose now that all the f_i, g_i are in $K(X)^\times$ and all tame symbols are 1, but without assuming they are invertible on the e's. By making a finite field extension we can ensure that all points where any of these functions have values in $\{0, 1, \infty\}$ are defined over K and, maintaining a semi-stable model by possibly blowing up, we get for the new graph Γ' that none of these points is in any of $e \in E'$ and we obtain $\operatorname{reg}_p(\alpha) \in \mathcal{H}(\Gamma', \mathbb{C}_p^\times)$. Now, (2) of Theorem 3.5 shows that reg_P factors via $K_2(K(X))$. Note that Γ' is obtained from Γ by subdividing edges and it is easily seen that $\mathcal{H}(\Gamma', \mathbb{C}_p^\times) \cong \mathcal{H}(\Gamma, \mathbb{C}_p^\times)$. Finally, [Pál10b, Proposition 4.2] shows that, for functions in $K(X)$,

reg_P in fact takes values in some finite extension and compatibility with Galois actions shows that in fact takes values in $K^\times$. To summarize, we get a map

$$\mathrm{reg}_P : K_2(X) \to \mathcal{H}(\Gamma, K^\times) .$$

Theorem 3.7. *The map* reg_P *equals the map* reg_t *from* (3.7) *except possibly at the prime* p*. In other words,* reg_P *gives* reg_t^l *upon completion at* l *for all primes* $l \neq p$ *(conjecturally also for* $l = p$*).*

Proof. The proof is inspired by the work of Asakura [Asa06, page 279]. The toric regulator completed at a prime l is the map

$$\begin{aligned} H^2_{\mathcal{M}}(X, \mathbb{Z}(2)) &\to H^1(K, H^1_{\text{ét}}(X \otimes_K \bar{K}, \mathbb{Z}_l(2))) \\ &\to H^1(K, T_1^{-1} \otimes \mathbb{Z}_l(1)) \to T_1^{-1} \otimes K^{\times(l)} \end{aligned}$$

and so we first need to understand the map $H^1_{\text{ét}}(X \otimes_K \bar{K}, \mathbb{Z}_l(1)) \to T_1^{-1} \otimes \mathbb{Z}_l$ from which the second left most map above is derived by twisting once and taking Galois cohomology.

Proposition 3.8. *Suppose* $l \neq p$*. Let* J *be the Jacobian of* X*. The map*

$$H^1_{\text{ét}}(X \otimes_K \bar{K}, \mathbb{Z}_l(1)) \cong T_l(J) \to T_1^{-1} \otimes \mathbb{Z}_l$$

is the limit of the maps

$$J[l^n] \to C^1(\Gamma, \mathbb{Z}/l^n)$$

defined as follows: let $[D] \in J[l^n]$ *by the class of a divisor* D*. Then* $l^n D$ *is the divisor of a function* f *and the map sends* $[D]$ *to* $(e \mapsto \mathrm{res}_e(d\log(f)) \pmod{l^n})$*.*

Proof. We begin by observing that in the map described here, we could just as well replace $[D]$ by a torsion class in $\mathrm{Pic}(X/\bar{K})$, represented by some Čech cocycle, with respect to a covering $\{V_k\}$, which is then the boundary of $V_k \mapsto g_k$ and map to the residue of $d\log g_k$ on an annulus, as this is independent of k modulo l^n.

The map we need is

$$H^1_{\text{ét}}(X \otimes_K \bar{K}, \mathbb{Z}/l^n(1)) = H^1_{\text{ét}}(Y, \mathbb{R}\Psi\mathbb{Z}/l^n(1)) \to H^1_{\text{ét}}(Y, \mathbb{R}^1\Psi\mathbb{Z}/l^n(1)[1]) ,$$

with $\mathbb{R}^1\Psi$ the functor of "nearby cycles", followed by the identification

$$\mathbb{R}^1\Psi\mathbb{Z}/l^n(1) \cong a_{2*}\mathbb{Z}/l^n ,$$

where, for any k, $a_k : \bar{Y}^{(k)} \hookrightarrow \bar{Y}$ is the obvious injection. To understand this identification we follow [Sai03, Proposition 1.2] (alternatives are [Jan88, Proposition 3.20] or the orginal [RZ82]). We have morphisms $X \xrightarrow{j} \mathcal{X} \xleftarrow{i} Y$. Then, according to Saito, we have isomorphisms,

$$\theta' : a_{k*}\mathbb{Z}/l^n \xrightarrow{\sim} i^*\mathbb{R}^k j_*\mathbb{Z}/l^n(k) \tag{3.8}$$

(the reader would notice a shift in notation due to our definition of $\bar{Y}^{(k)}$), and a short exact sequence

$$0 \to \mathbb{R}^k\Psi\mathbb{Z}/l^n \xrightarrow{\bar{\theta}} i^*\mathbb{R}^{k+1}j_*\mathbb{Z}/l^n(1) \to \mathbb{R}^{k+1}\Psi\mathbb{Z}/l^n(1) \to 0 , \tag{3.9}$$

whose definition we will recall in a moment. To make sense out of these formulas, note that we identify sheaves on Y with sheaves on $\bar{Y}$ with an action of the Galois group. Since there is no $\bar{Y}^{(3)}$, the isomorphism above gives $i^*\mathbb{R}^3 j_*\mathbb{Z}/l^n = 0$ and the short exact sequence for $k = 2$ gives $\mathbb{R}^2\Psi\mathbb{Z}/l^n = 0$. Then for $k = 1$ the short exact sequence yields an isomorphism

$$\mathbb{R}^1\Psi\mathbb{Z}/l^n(1) \xrightarrow{\bar{\theta}} i^*\mathbb{R}^2 j_*\mathbb{Z}/l^n(2)$$

from which we get the required identification by composing with the inverse of θ'.

Next we recall how the thetas are defined, focusing on a neighborhood of a point in $Y^{(2)}$ locally given by an equation $xy = \pi$. The special fiber Y is defined there by π and the two components of Y passing through the point are given by the additional equations $x = 0$ and $y = 0$. Let i_i be the embedding of Y_i into X, $j_i : X - Y_i \to X$ the obvious open immersion, and $a^i : Y_i \to Y$. Saito defines a map

$$\theta_i : a^i_*\mathbb{Z}/l^n \to i_i^*\mathbb{R}^1 j_{i*}\mathbb{Z}/l^n(1)$$

to be the map sending 1 to the Kummer image of the local generator of $\mathcal{O}(-Y_i)$. On our local neighborhood, these are clearly the Kummer images of x and y respectively, which we will denote by (x) and (y). Then, following Saito, the map

$$\theta' = \sum \theta_i : a_{1*}\mathbb{Z}/l^n \to i^*\mathbb{R}^1 j_*\mathbb{Z}/l^n(1)$$

is an isomorphism and the isomorphisms (3.8) are deduced from it by taking cup products. Finally, the map

$$i^*\mathbb{R}^k j_*\mathbb{Z}/l^n \to \mathbb{R}^k\Psi\mathbb{Z}/l^n$$

is surjective and the map $\bar{\theta}$ is induced by the map $i^*\mathbb{R}^k j_*\mathbb{Z}/l^n \to i^*\mathbb{R}^{k+1} j_*\mathbb{Z}/l^n(1)$ obtained by cup product with the Kummer class of π, which, on our neighborhood, is $(x)+(y)$. We finally obtain the map we want, locally on our chosen neighborhood as the composition

$$\mathbb{R}^1\Psi\mathbb{Z}/l^n(1) \xrightarrow{\text{lift}} i^*\mathbb{R}^1 j_*\mathbb{Z}/l^n(1) \xrightarrow{((x)+(y))\cup} i^*\mathbb{R}^2 j_*\mathbb{Z}/l^n(2) \xleftarrow{(x)\cup(y)\cup} a_{2*}\mathbb{Z}/l^n \tag{3.10}$$

where the last map is an isomorphism that needs to be inverted. From this it is clear that the Kummer images of x and y go to ± 1 respectively. Rigidifying and taking the residue of the corresponding $d\log$'s on the resulting annulus would obviously give the same. So, finally, our required map would map a torsion class in the Picard group to the cocycle (g_k) and then apply (3.10) to the Kummer image of g_k. This is now clearly the same as the map claimed in the proposition. □

Switching to K_2 for convenience, we can now concentrate on a single annulus e and compute the map

$$K_2(X)/l^n \to H^1(K, \operatorname{Pic}(X)[l^n] \otimes \mu_{l^n}) \xrightarrow{\operatorname{res}_e} H^1(K, \mu_{l^n}) \to K^\times/(K^\times)^{l^n} \tag{3.11}$$

where μ denotes roots of unity and res_e is the "e component" of the map in Proposition 3.8. Let $A = \mathcal{O}(e)$. Furthermore, identify e with some $A(r, s) \subset \mathbb{P}^1$. Let B be the ring of rational functions on $\mathbb{P}^1$ regular on e and let $D = D(r)$. We have a map $\mathrm{Spec}(A) \to X$ and using the pullback via this map we can write the map (3.11) as the composition of $K_2(X)/l^n \to K_2(A)/l^n$ with

$$K_2(A)/l^n \to H^1(K, \mathrm{Pic}(A)[l^n] \otimes \mu_{l^n}) \xrightarrow{\mathrm{res}_e} H^1(K, \mu_{l^n}) \to K^\times/(K^\times)^{l^n} . \quad (3.12)$$

For $f \in B$ we have

$$\mathrm{res}_e \, d\log(f) = \sum_{x \in D} \mathrm{res}_x \, d\log(f)$$

and therefore the restriction of (3.12) to $K_2(B)/l^n$ is the sum over $x \in D$ of the maps

$$K_2(B)/l^n \to H^1(K, \mathrm{Pic}(B)[l^n] \otimes \mu_{l^n}) \xrightarrow{\mathrm{res}_x} H^1(K, \mu_{l^n}) \to K^\times/(K^\times)^{l^n} .$$

According to [Asa06, Lemma 4.3 (2)] for each $x \in D$ this last map is just the tame symbol at x modulo $(K^\times)^{l^n}$. It is easy to see that if $f, g \in A$ are sufficiently well approximated by $f_i, g_i \in B$, then the image of $\{f_i, g_i\}$ in $K_2(A)/l^n$ is the same as that of $\{f, g\}$. The theorem follows easily. □

4. Relation with the syntomic regulator

In this section we will analyze the construction of the toric regulator at the prime $l = p$, after tensoring with $\mathbb{Q}$. We will see that the logarithm of the toric regulator may be computed using the syntomic regulator of Nekovář and Nizioł. We will then check our description of the toric regulator for K_2 of curves via the Pál regulator using the computation of the syntomic regulator in this case by the first named author [Bes21].

Let X be a smooth variety over a p-adic field K. Then, Nekovář and Nizioł [NN16, Theorem 5.9] show that the regulator map,

$$\mathrm{reg}_p : H^{k+1}_{\mathcal{M}}(X, \mathbb{Z}(r))_0 \to H^1(K, H^k_{\text{ét}}(X \otimes_K \bar{K}, \mathbb{Q}_p(r))) , \quad (4.1)$$

factors via the subgroup $H^1_{\mathrm{st}}(K, H^k_{\text{ét}}(X \otimes_K \bar{K}, \mathbb{Q}_p(r)))$, where, for a de Rham representation V of G, $H^1_{\mathrm{st}}(V)$ is semi-stable cohomology, which may be interpreted as the group of Yoneda extensions in the category of potentially semi-stable representation and may be computed in terms of the complex $C^\bullet_{\mathrm{st}}(V)$ of [Nek93, 1.19],

$$C^\bullet_{\mathrm{st}}(V) : \mathrm{D}_{\mathrm{st}}(V) \xrightarrow{(\varphi-1, N, -i)} \mathrm{D}_{\mathrm{st}}(V) \oplus \mathrm{D}_{\mathrm{st}}(V) \oplus \mathrm{DR}(V)/F^0 \xrightarrow{N+1-p\varphi+0} \mathrm{D}_{\mathrm{st}}(V) . \quad (4.2)$$

Here, D_{st} and DR are the functors defined by Fontaine: $\mathrm{D}_{\mathrm{st}}(V)$ is a K_0-vector space, where K_0 is the maximal unramified extension of $\mathbb{Q}_p$ inside K, equipped with a linear nilpotent operator N (called monodromy) and a semi-linear (with respect to the unique lift of Frobenius on K_0) operator φ (Frobenius), satisfying the relation

$$N\varphi = p\varphi N ,$$

and $\mathrm{DR}(V)$ is a filtered K-vector space, and there is an isomorphism

$$\mathrm{D}_{\mathrm{st}}(V) \otimes K \to \mathrm{DR}(V) .$$

We remark that the map between H_{st}, computed in terms of the complex $C^\bullet_{\mathrm{st}}$, and Galois cohomology, is a manifestation of the Bloch–Kato exponential map [BK90, Definition 3.10].

Assume now that X has totally degenerate reduction. Let V be the representation $V = H^k_{\text{ét}}(X \otimes_K \bar{K}, \mathbb{Q}_p(r))$. We first note that as the reduction of X is semi-stable, V is semi-stable as well. It follows [Nek93, 1.24 (3)] that

$$H^1_{\mathrm{st}}(K,V) = H^1_g(V)\,. \tag{4.3}$$

Let us compute $H^1_{\mathrm{st}}(K,V)$. Consider $D = \mathrm{D}_{\mathrm{st}}(V)$, which is a K_0-vector space with a Frobenius and a monodromy operator. By the semi-stable conjecture of Fontaine, proved by Tsuji [Tsu99], it is equal to the Hyodo–Kato cohomology $H^k(Y^\times/W^\times)$. It follows from [RX07b] that we have a slope decomposition

$$D = \bigoplus_{i+j=k} T_j^{i-j} \otimes K_0(r-j)$$

and the monodromy operator is compatible with the one defined on the T's. For simplicity let us renumber this as follows: We have a vector space decomposition

$$D = \bigoplus_i D^i\ ,\ D^i \cong T^i \otimes K_0\ ,\ T^i = T_{i+r}^{k-2r-2i}\ ,$$

where, with respect to the rational structure provided by T^i, the Frobenius φ acts by $p^i\sigma$, so that $(D^i)^{\varphi=p^i} = T^i \otimes \mathbb{Q}_p$. The monodromy maps are induced by the ones defined before $N : T^i \to T^{i-1}$. Note that by remark 1.2 the map $N : T^0 \to T^{-1}$ is injective after tensoring with $\mathbb{Q}$. The filtration is compatible with the filtration on the individual terms, where $\mathrm{DR}(\mathbb{Q}_p(i)) = F^{-i} \supset F^{-i-1} = 0$. We can now compute the semi-stable cohomology of V.

Proposition 4.1. *We have a short exact sequence*

$$0 \to \mathrm{DR}(V)/F^0 \to H^1_{\mathrm{st}}(K,V) \to (T^{-1}/NT^0) \otimes \mathbb{Q}_p \to 0 \tag{4.4}$$

and a map $H^1_{\mathrm{st}}(V) \to \mathrm{DR}(V)/(F^0 + T^0 \otimes \mathbb{Q}_p)$ *such that the composition*

$$\mathrm{DR}(V)/F^0 \to H^1_{\mathrm{st}}(K,V) \to \mathrm{DR}(V)/(F^0 + T^0 \otimes \mathbb{Q}_p) \tag{4.5}$$

is the projection. Both of these maps are functorial.

Proof. Let us begin by computing the cohomology of the complex $C^{\bullet\prime}_{\mathrm{st}}(V)$, which is obtained from $C^\bullet_{\mathrm{st}}(V)$ by dropping the de Rham component,

$$C^{\bullet\prime}_{\mathrm{st}}(V) : \mathrm{D}_{\mathrm{st}}(V) \xrightarrow{(\varphi-1,N)} \mathrm{D}_{\mathrm{st}}(V) \oplus \mathrm{D}_{\mathrm{st}}(V) \xrightarrow{N+1-p\varphi} \mathrm{D}_{\mathrm{st}}(V)\,.$$

We can decompose this last complex according to slopes

$$C^{\bullet\prime}_{\mathrm{st}}(V) = \bigoplus_i C^{\bullet i}_{\mathrm{st}}(V)$$

with

$$C^{\bullet i}_{\mathrm{st}}(V) : D^i \xrightarrow{(p^i\sigma-1,N)} D^i \oplus D^{i-1} \xrightarrow{N+1-p^i\sigma} D^{i-1}\,.$$

Since $p^i\sigma - 1$ is bijective unless $i = 0$, we immediately see that $H^0(C_{\mathrm{st}}^{\bullet i}) = 0$ unless $i = 0$, and also for $i = 0$ since $N : T^0 \to T^{-1}$ is injective after tensoring with $\mathbb{Q}$. Consider next H^1. Suppose $(x, y) \in D^i \oplus D^{i-1}$ represents an element in $H^1(C_{\mathrm{st}}^{\bullet i})$. If $i \neq 0$ we may use the bijectivity of $p^i\sigma - 1$ to assume $x = 0$ and the equation on y becomes $(p^i\sigma - 1)y = 0$ so $y = 0$ as well. Thus, $H^1(C_{\mathrm{st}}^{\bullet i}) = 0$ unless $i = 0$. In this last case we can write explicitly

$$H^1(C_{\mathrm{st}}^{\bullet 0}) = \frac{\{(x, y),\ x \in D^0,\ y \in D^{-1},\ Nx = (\sigma - 1)y\}}{\{((\sigma - 1)z, Nz),\ z \in D^0\}}$$

and we have the following.

Lemma 4.2. *We have an isomorphism* $(T^{-1}/NT^0) \otimes \mathbb{Q}_p \xrightarrow{\sim} H^1(C_{\mathrm{st}}^{\bullet 0})$ *given by* $u \mapsto (0, u)$.

Proof. The map is clearly well defined and injective. Surjectivity amounts to the statement that any element in $H^1(C_{\mathrm{st}}^{\bullet 0})$ has a representative $(0, y)$, i.e., that any representative (x, y) has $x \in \operatorname{Im} \sigma - 1$. This is true because N is defined over $\mathbb{Q}$ and is injective. □

Remark 4.3. The above lemma may be interpreted for an extension

$$0 \to T^{-1} \otimes \mathbb{Q}_p(1) \to V \to T^0 \otimes \mathbb{Q}_p \to 0 \tag{4.6}$$

as saying that the map $H^1_{\mathrm{st}}(K, T^{-1} \otimes \mathbb{Q}_p(1)) \to H^1_{\mathrm{st}}(K, V)$ is surjective.

We have an obvious short exact sequence of complexes

$$0 \to \mathrm{DR}(V)/F^0[1] \to C_{\mathrm{st}}^{\bullet}(V) \to C_{\mathrm{st}}^{\bullet\prime}(V) \to 0$$

and the associated long exact sequence together with the computation of the cohomology of $C_{\mathrm{st}}^{\bullet\prime}(V)$ immediately gives the short exact sequence (4.4). To define the map $H^1_{\mathrm{st}}(V) \to \mathrm{DR}(V)/(F^0 + T^0 \otimes \mathbb{Q}_p)$ start with a representative $(x, y, d) \in \mathrm{D}_{\mathrm{st}}(V) \oplus \mathrm{D}_{\mathrm{st}}(V) \oplus \mathrm{DR}(V)/F^0$ and use the computation of the cohomology of $C_{\mathrm{st}}^{\bullet\prime}(V)$ to see that it is equivalent to a representative with $x = 0$. The d component of this representative is now unique up to an element of $T^0 \otimes \mathbb{Q}_p$ and this gives the map. The composed map (4.5) is clearly the projection. □

Consider now the case $V = \mathbb{Q}_p(1)$, so that $T^{-1} = \mathbb{Z}$ while $T^0 = 0$. Clearly, in this case the map $H^1_{\mathrm{st}}(K, \mathbb{Q}_p(1)) \to \mathrm{DR}(V)/F^0 = K$ splits the short exact sequence (4.4) and we have an isomorphism $H_{\mathrm{st}}(K, \mathbb{Q}_p(1)) \cong K \oplus \mathbb{Q}_p$. Nekovář proves the following result.

Proposition 4.4 ([Nek93, 1.35]). *The composed map*

$$K^\times \xrightarrow{\text{Kummer}} H_{\mathrm{st}}(K, \mathbb{Q}_p(1)) \cong K \oplus \mathbb{Q}_p$$

is given by $x \to (\log(x), v(x))$.

Proposition 4.5. *For the short exact sequence* (4.4) *the composed map*

$$T^0 \otimes \mathbb{Q}_p \cong H^0_{\mathrm{st}}(K, T^0 \otimes \mathbb{Q}_p) \to H^1_{\mathrm{st}}(K, T^{-1} \otimes \mathbb{Q}_p(1)) \to T^{-1} \otimes K^{\times(l)} \xrightarrow{v} T^{-1} \otimes \mathbb{Q}_p$$

is just the monodromy map.

Proof. By Proposition 4.4 we can compute the map by projecting on the cohomology of the complexes $C_{\mathrm{st}}^{\bullet\prime}$, where the result is easy. □

Theorem 4.6. *The toric regulator at p exists. Furthermore we have the following commutative diagram, where $V = H^k_{\acute{e}t}(X \otimes_K \bar{K}, \mathbb{Q}_p(r))$,*

$$\begin{array}{ccccc} H^{k+1}_{\mathcal{M}}(X,\mathbb{Z}(r))_0 & \longrightarrow & H^1_{\mathrm{st}}(K,V) & \longrightarrow & \mathrm{DR}(V)/(F^0 + T^0 \otimes \mathbb{Q}_p) \\ \downarrow & & & & \downarrow \\ H^{k+1}_{\mathcal{T}}(X,\mathbb{Z}(r)) & & \xrightarrow{\quad\log\quad} & & T^{-1} \otimes K/T^0 \otimes \mathbb{Q}_p. \end{array}$$

Here, the vertical map on the right side is a projection relative to the subspace $\oplus_{i\le -2} T^i \otimes K$.

Proof. Consider the quotient $V' = V/W_{2r-k}V$. As this does not have non-positive D^i's, and as $\mathrm{DR}(V') = F^0$ (since this is true for all the Tate subquotients), we easily see that $H^1_{\mathrm{st}}(K,V') = 0$. Thus, we may again do the factoring, as in Section 1, of the regulator into $H^1_{\mathrm{st}}(K, W_{2r-k}V)$, and then project to $H^1_{\mathrm{st}}(K,V'')$ with $V'' = W_{2r-k}V/W_{2r-k-4}V$. Furthermore, the map $H^1_{\mathrm{st}}(K,V'') \to H^1_{\mathrm{st}}(K, T^0 \otimes \mathbb{Q}_p)$ is 0 because $\mathrm{DR}(\mathbb{Q}_p) = F^0$ and by Remark 4.3. Thus, the toric regulator exists at p as in Section 1. The commutativity of the diagram in the theorem is now straightforward from the following commutative diagram

$$\begin{array}{ccccc} H^1_{\mathrm{st}}(K, T^{-1} \otimes \mathbb{Q}_p(1)) & \longrightarrow & H^1_{\mathrm{st}}(K,V'') & \longrightarrow & 0 \\ \downarrow & & \downarrow & & \\ T^{-1} \otimes K & \longrightarrow & \mathrm{DR}(V'')/(F^0 + T^0 \otimes \mathbb{Q}_p) & & \end{array}$$

and the fact that the composition of the Kummer map with the vertical map on the left is just the log map by Nekovář's result 4.4. □

Nekovář and Nizioł define the syntomic regulator,

$$\mathrm{reg} : H^{k+1}_{\mathcal{M}}(X,\mathbb{Z}(r)) \to H^{k+1}_{\mathrm{syn}}(X,r)\ , \tag{4.7}$$

into syntomic cohomology groups. These groups are constructed in such a way that there is a spectral sequence

$$E_2^{p,q} = H^p_{\mathrm{st}}(K, H^q_{\acute{e}t}(X \otimes \bar{K}, \mathbb{Q}_p(r))) \Rightarrow H^{p+q}_{\mathrm{syn}}(X,r)\ . \tag{4.8}$$

One easily deduces from this spectral sequence and the syntomic regulator a map

$$\mathrm{reg} : H^{k+1}_{\mathcal{M}}(X,\mathbb{Z}(r))_0 \to H^1_{\mathrm{st}}(K, H^k_{\acute{e}t}(X \otimes_K \bar{K}, \mathbb{Q}_p(r))\ ,$$

which is the same as the map (4.1). The syntomic regulator is computed without étale cohomology, using a mixture of de Rham and (log) crystalline cohomology constructions, and so is more computable, at least in principle, using a kind of "p-adic differential geometry" approach. Indeed, in the good reduction case the syntomic regulator has been defined for a long time and has been computed in several cases, primarily by the first named author [Bes00, BdJ03, BdJ12, Bes12].

Recently, some of these results have been extended to the semi-stable reduction case [Bes21].

We can summarize the results and comments of this section to this point by the motto "The log of the toric regulator is computed from the syntomic regulator". To end this section we illustrate this by revisiting the case of K_2 of a totally degenerate curve and showing how the syntomic regulator is indeed the logarithm of the Pál rigid analytic regulator.

Let X be as in Section 3 and consider $k = 1$, $r = 2$ again. We have

$$V = H^1_{\text{ét}}(X \otimes_K \bar{K}, \mathbb{Q}_p(2)) .$$

Recall that in this case $T^0 = 0$. We have $\mathrm{DR}(V) \cong H^1_{\mathrm{dR}}(X/K)$ and $F^0 \,\mathrm{DR}(V) = F^2 H^1_{\mathrm{dR}}(X/K) = 0$. According to Theorem 4.6 the logarithm of the toric regulator map agrees with the composed map

$$\mathrm{reg}_p : H^2_{\mathcal{M}}(X, \mathbb{Z}(2)) \xrightarrow{\mathrm{reg}_{\mathrm{syn}}} H^1_{\mathrm{st}}(K, V) \to H^1_{\mathrm{dR}}(X/K) \to T^{-1} \otimes K . \tag{4.9}$$

We identify the vector space on the right-hand side with the space $\mathcal{H}(\Gamma, K)$ of K-valued harmonic cochains on the dual graph. As expected from our motto, we get the following result

Theorem 4.7. *Let X be as above. Then the diagram*

$$\begin{array}{ccc} H^2_{\mathcal{M}}(X, \mathbb{Z}(2)) & \xrightarrow{\mathrm{reg}_P} & T^{-1} \otimes K^{\times} \\ & \searrow^{\mathrm{reg}_p} & \downarrow \log \\ & & T^{-1} \otimes K \end{array}$$

commutes, with reg_P *the Pál regulator and* reg_p *the p-adic regulator from* (4.9).

Proof. Suppose that an element α of $K_2(X)$ restricts to an element $\sum\{f_i, g_i\}$ in K_2 of the function field of X. In [Bes21] the first named author proved a formula for the cup product with a cohomology class $[\omega]$ of the image of α under

$$H^2_{\mathcal{M}}(X, \mathbb{Z}(2)) \xrightarrow{\mathrm{reg}_{\mathrm{syn}}} H^1_{\mathrm{st}}(K, V) \to H^1_{\mathrm{dR}}(X/K) .$$

One checks easily that the projection $H^1_{\mathrm{st}}(K, V) \to H^1_{\mathrm{dR}}(X/K)$ defined there coincides with the one we have been using. This formula was valid when $[\omega]$ is in the kernel of the monodromy operator N (X can be any curve with semi-stable reduction). To explain the formula and to complete the proof we first analyze $H^1_{\mathrm{dR}}(X/K)$ in a bit more detail. We have the weight decomposition

$$H^1_{\mathrm{dR}}(X/K) = T^{-2} \otimes K \oplus T^{-1} \otimes K = H^1(\Gamma, K) \oplus \mathcal{H}(\Gamma, K) .$$

With respect to this decomposition the monodromy operator vanishes on $H^1(\Gamma, K)$ and maps $\mathcal{H}(\Gamma, K)$ on $H^1(\Gamma, K)$ via (3.4). The cup product makes both summands isotropic and gives the pointwise product (3.5) otherwise. Consequently, if χ and pr denote the projections on $H^1(\Gamma, K)$ and $\mathcal{H}(\Gamma, K)$ respectively, and we have $[\omega] \in \operatorname{Ker} N = H^1(\Gamma, K)$, $\beta \in H^1_{\mathrm{dR}}(X/K)$, then

$$[\omega] \cup \mathrm{pr}\, \beta = \chi([\omega]) \cdot \mathrm{pr}(\beta) .$$

We can finally introduce the formula of [Bes21], which expresses the cup product in term of expressions assigned to the individual symbols $\{f, g\}$. The expression that we need is the one in Proposition 3.7, which is the same as the expression for the regulator by the main theorem [Bes21, Theorem 1.2]. This expression is

$$\sum_v \langle \log(f), F_\omega; \log(g) \rangle_{(U_v - Z)^\dagger} - \sum_e \chi(\omega)(e) \cdot \langle \log(f), \log(g) \rangle_e \,.$$

Here, Z is a subset containing all the singularities of f and g, but in any case, when all the components of the reduction are projective lines, the first term vanishes because all the "triple indices" appearing in the sum vanish by [BdJ12, Proposition 8.4]. Thus, the projection of the regulator on $\mathcal{H}(\Gamma, K)$ is the map obtained by sending $\{f, g\}$ to

$$e \mapsto \langle \log(f), \log(g) \rangle_e \,.$$

Therefore, the following lemma completes the proof. □

Lemma 4.8. *For rigid analytic functions on the annulus e we have*

$$\langle \log(f), \log(g) \rangle_e = \log(t_e(f, g)).$$

Proof. Suppose, after identifying e with $A(r, s)$, that f and g are rational functions on $\mathbb{P}^1$. By [Bes00, Proposition 4.10] we have

$$\langle \log(f), \log(g) \rangle_e = \sum_{x \in D(r)} \langle \log(f), \log(g) \rangle_x \,.$$

At each point in $D = D(r)$ we have, essentially by definition,

$$\langle \log(f), \log(g) \rangle_x = \log t_x(f, g) \,.$$

Thus, the result is true for rational f and g and then is true in general by the definition of the Pál regulator and by continuity. □

We close this section with a conjecture, which is suggested by the relation between the toric regulator with both the syntomic and the Sreekantan regulator (but note that in this conjecture we do not need to make any assumptions about the reduction, other than being semi-stable).

Conjecture 2. *The composition*

$$H^{k+1}_{\mathcal{M}}(X, \mathbb{Z}(r))_0 \xrightarrow{reg} H^1_{\text{st}}(K, H^k_{\acute{e}t}(X \otimes_K \bar{K}, \mathbb{Q}_p(r)) \to H^1(C^{\bullet\prime}_{\text{st}}(H^k_{\acute{e}t}(X \otimes_K \bar{K}, \mathbb{Q}_p(r)))$$

factors via the Sreekantan regulator.

5. A conjectural formula for K_1 of surfaces

One nice feature of the relation with the syntomic regulator developed in the preceding section is that just as we are able to test formulas for the toric regulator by taking their logarithms and comparing with the syntomic regulator, we can look at formulas for the syntomic regulator and attempt to exponentiate them to get conjectural formulas for the toric regulator. In this section we present a conjecture for the toric regulator for K_1 of surfaces, which is suggested by the corresponding result of the first named author in the syntomic case [Bes12]. Unfortunately, the formula we need is the analogue of the one in [Bes12] for the semi-stable reduction case, and it is still conjectural. As it will further take some work to introduce the results we will not present the motivation here and describe the conjecture without it.

We begin by recalling, for easy reference, some facts about Mumford curves, which are well known to the experts. Let X be such a curve, given as $\mathcal{G}\backslash\mathbb{H}$ where $\mathbb{H}$ is the Drinfeld upper half-plane and $\mathcal{G}$ is some Schottky group. Let $\Gamma = (V, E)$ be the corresponding dual graph, which is the quotient $\mathcal{G}\backslash\mathbb{T}$, with $\mathbb{T}$ the tree of $\mathbb{H}$. Let l be a prime. The filtration on $M = H^1_{\text{ét}}(X \otimes \bar{K}, \mathbb{Z}_l(1))$ takes the form of a short exact sequence

$$0 \to H^1(\Gamma, \mathbb{Z}) \otimes \mathbb{Z}_l(1) \to M \to \mathcal{H}(\Gamma, \mathbb{Z}) \otimes \mathbb{Z}_l \to 0 . \tag{5.1}$$

We now describe the augmented monodromy map

$$\tilde{N} : \mathcal{H}(\Gamma, \mathbb{Z}) \to H^1(\Gamma, K^\times)$$

Proposition 5.1. *Let $\alpha \in \mathcal{H}(\Gamma, \mathbb{Z})$. Then there exists a unique $\omega_\alpha \in \Omega^1(X/K)$ such that for each $e \in E$ we have $\operatorname{res}_e(\omega_\alpha) = \alpha(e)$. Furthermore, on each U_v there exists $f_\alpha^v \in \mathcal{O}(U_v)^\times$ such that $d\log(f_\alpha^v) = \omega_\alpha|_{U_v}$.*

This appears in [Col00] without a clear indication of the proof (Coleman shows the existence of ω_α and mentions it is locally a $d\log$, presumably relying on the theory of theta functions as we will do). We show how it follows from [GvdP80] (or [MD73]). We start by defining theta functions.

Definition 5.2 ([GvdP80, II, (2.3) and (2.3.5)]). Let $\gamma \in \mathcal{G}$. The theta function u_γ is defined, up to a multiplicative constant, as follows: Pick any $x \in \mathbb{H}$ and let $w_{\gamma,x}$ be any function on $\mathbb{P}^1$ whose divisor is $\gamma(x) - x$. Then

$$u_\gamma := \prod_{\delta \in \mathcal{G}} \delta^* w_{\gamma,x} .$$

This function is independent of the choice of x as shown in [GvdP80, II, (2.3.4)].

One can normalize the function by normalizing w in an obvious way, but we will not do this. The function u_γ has no zeros or poles. It has a constant factor of automorphy, meaning that

$$\mu(\delta, \gamma) := \delta^* u_\gamma / u_\gamma \tag{5.2}$$

is constant. As a consequence the one form $\omega_\gamma := d\log(u_\gamma)$ is $\mathcal{G}$ invariant and holomorphic, descending to a holomorphic one form on X. It is furthermore easy to see that μ is bi-multiplicative and therefore induces a map

$$\mu : \mathcal{G}^{\text{ab}} \times \mathcal{G}^{\text{ab}} \to K^\times .$$

We compute the residues of ω_γ. On $\mathbb{H}$ it is clear that for an edge $\tilde{e}$ of $\mathbb{T}$ the residue of $d\log(w_{\gamma,x})$ on $\tilde{e}$ is non-zero if and only if $\tilde{e}$ sits on the path between the vertex v_x, corresponding to the domain where x resides, and $\gamma(v_x)$, and it is ± 1

depending on its orientation compared with that of the path. Averaging on $\mathcal{G}$, we immediately get that the residue of ω_γ on an edge $e \in E(\Gamma)$ is

$$\sum_{\delta \in \mathcal{G}} \mathrm{res}_{\delta\tilde{e}}\, d\log(w_{\gamma,x}) \ , \ \ \tilde{e} \text{ any lift of } e.$$

Now we recall that as the graph Γ is finite, the space $\mathcal{H}(\Gamma, \mathbb{Z})$ of harmonic forms on Γ can be identified with the first homology of Γ with coefficients in $\mathbb{Z}$, which is, by definition,

$$H_1(\Gamma, \mathbb{Z}) := \ker(\mathbb{Z}[E] \xrightarrow{d} \mathbb{Z}[V]) \ , \ d(e) = (e^+) - (e^-) \ .$$

Furthermore, the Hurewicz isomorphism

$$\mathrm{Hur} : \mathcal{G}^{\mathrm{ab}} = H_1(\mathcal{G}, \mathbb{Z}) \to H_1(\Gamma, \mathbb{Z}) \tag{5.3}$$

can be evaluated on $\gamma \in \mathcal{G}$ by choosing a vertex v and pushing down the path from v to $\gamma(v)$ from $\mathbb{T}$ to Γ. This immediately gives

Proposition 5.3. *The harmonic cocycle* $e \mapsto \mathrm{res}_e\, u_\gamma$ *is* $\mathrm{Hur}(\gamma)$.

Proof of Proposition 5.1. By [GvdP80, Chapter VI (4.2) Proposition] the functions ω_γ span $\Omega^1(X)$. The result follows easily. □

Let $\alpha \in \mathcal{H}(\Gamma, \mathbb{Z})$. For each $e \in E$ the functions $f_\alpha^{e^+}$ and $f_\alpha^{e^-}$ of Proposition 5.1 are both defined on e and have, by assumption, the same image under the $d\log$ map. Therefore, their quotient on e maps to 0 under $d\log$ and is therefore a constant $c_\alpha(e) \in K^\times$. The map $e \mapsto c_\alpha(e)$ is a cocycle on Γ with values in $K^\times$ and its image in $H^1(\Gamma, K^\times)$ is uniquely determined by α. By pairing with harmonic cocycles we get a bilinear form

$$\mu' : \mathcal{H}(\Gamma, \mathbb{Z}) \times \mathcal{H}(\Gamma, \mathbb{Z}) \to K^\times \ .$$

Proposition 5.4. *Identifying* $\mathcal{H}(\Gamma, \mathbb{Z})$ *with* $\mathcal{G}^{\mathrm{ab}}$ *via the Hurewicz isomorphism* (5.3) *we have the identification* $\mu' = \mu$ *with the bilinear form* μ *defined in* (5.2).

Proof. Using Proposition 5.3 what we need to prove is that for the form $\omega = \omega_\gamma = d\log(u_\lambda) = d\log(u)$, the homomorphism $\delta \mapsto \delta^* u/u$ in $H^1(\mathcal{G}, K^\times)$ represents the same cohomology class as the map $e \mapsto d\log^{-1}\omega|_{U_{e^+}} - d\log^{-1}\omega|_{U_{e^-}}$ in $H^1(\Gamma, K^\times)$. But both are clearly the image of ω under the boundary map in the short exact sequence

$$1 \to K^\times \to \mathcal{O}^\times \xrightarrow{d\log} \Omega \to 0$$

either as sheaves on X or, taking global sections on $\mathbb{H}$, as $\mathcal{G}$-modules. □

Because μ is symmetric by [MD73, Theorem 1] we get.

Corollary 5.5. *The form* μ' *is symmetric.*

Corollary 5.6. *We have* $\tilde{N}(\alpha) = c_\alpha$.

Proof. This is because μ gives the periods for the p-adic uniformization of the Jacobian of X by [MD73]. □

We will from now onward normalize the choice of f_α^v in such a way that

$$c_\alpha \in \mathcal{H}(\Gamma, K^\times)\,. \tag{5.4}$$

This may require multiplying by some integer.

Suppose now we have two Mumford curves X_i, $i = 1, 2$, with corresponding graphs Γ_i, and we consider the surface $X = X_1 \times X_2$. We want a formula for the toric regulator

$$\operatorname{reg}_t : H^3_{\mathcal{M}}(X, \mathbb{Z}(2)) \to H^3_{\mathcal{T}}(X, \mathbb{Z}(2))$$

given on elements of the form

$$\Theta = \sum (C_j, g_j)\,, \tag{5.5}$$

where $C_j \subset X$ are curves and g_j is a rational functions on C_j such that the divisors of the g_j cancel on X.

Let us first compute $H^3_{\mathcal{T}}(X, \mathbb{Z}(2))$. Putting aside the uninteresting terms corresponding to $H^0 \otimes H^2$, the main contribution to $H^2_{\text{ét}}(X, \mathbb{Z}_l(2))$ is

$$M = M_1 \otimes M_2\,,\ M_i = H^1_{\text{ét}}(X_i \otimes \bar{K}, \mathbb{Z}_l(1))\,.$$

Taking the tensor product of the short exact sequences of the form (5.1) corresponding to M_i we get on M a 3-step filtration and we may consider the interesting quotient M' having the following short exact sequence,

$$\begin{aligned} 0 &\to \big(\mathcal{H}(\Gamma_1, \mathbb{Z}) \otimes H^1(\Gamma_2, \mathbb{Z}) \oplus \mathcal{H}(\Gamma_2, \mathbb{Z}) \otimes H^1(\Gamma_1, \mathbb{Z})\big) \otimes \mathbb{Z}_l(1) \\ &\to M' \to \mathcal{H}(\Gamma_1, \mathbb{Z}) \otimes \mathcal{H}(\Gamma_2, \mathbb{Z}) \to 0\,, \end{aligned}$$

and we may apply a projection P on one of the summands on the left to get an extension

$$0 \to \big(\mathcal{H}(\Gamma_1, \mathbb{Z}) \otimes H^1(\Gamma_2, \mathbb{Z})\big) \otimes \mathbb{Z}_l(1) \to M'' \to \mathcal{H}(\Gamma_1, \mathbb{Z}) \otimes \mathcal{H}(\Gamma_2, \mathbb{Z}) \to 0.$$

The associated augmented monodromy is

$$\mathrm{id}_{\mathcal{H}(\Gamma_1,\mathbb{Z})} \otimes \tilde{N}_2 : \mathcal{H}(\Gamma_1, \mathbb{Z}) \otimes \mathcal{H}(\Gamma_2, \mathbb{Z}) \to \mathcal{H}(\Gamma_1, \mathbb{Z}) \otimes H^1(\Gamma_2, \mathbb{Z}) \otimes K^\times \tag{5.6}$$

and we get a regulator

$$P \operatorname{reg}_t : H^3_{\mathcal{M}}(X, \mathbb{Z}(2)) \to P H^3_{\mathcal{T}}(X, \mathbb{Z}(2))\,.$$

We use the duality between graph cohomology and harmonic cocycles to view the resulting intermediate Jacobian $PH^3_{\mathcal{T}}(X, \mathbb{Z}(2))$, which is the cokernel of (5.6), as bilinear forms $H^1(\Gamma_1, \mathbb{Z}) \times \mathcal{H}(\Gamma_2, \mathbb{Z}) \to K^\times$ modulo those forms which are obtained from bilinear forms $H^1(\Gamma_1, \mathbb{Z}) \times H^1(\Gamma_2, \mathbb{Z}) \to \mathbb{Z}$ by composing in the second coordinate with $\tilde{N}_2 : \mathcal{H}(\Gamma_2, \mathbb{Z}) \to H^1(\Gamma_2, \mathbb{Z}) \otimes K^\times$ (to be precise, we need to compose with the dual of $\tilde{N}_2$ but this is the same by Corollary 5.5).

To construct the required form out of the element (5.5), we are going to further assume that for each index j, the curve C_j has semi-stable reduction, the projections $\pi_i : C_j \to X_i$ are finite for $i = 1, 2$ and that they give maps of graphs between the corresponding dual graphs. Let $\alpha \in \mathcal{H}(\Gamma_2, \mathbb{Z})$, $\beta \in H^1(\Gamma_1, \mathbb{Z})$. Identify β with a harmonic representative (which may require again multiplying by a fixed integer), and let $(f_\alpha^v)_{v \in V_2}$ be the corresponding functions as in Proposition 5.1, normalized as in (5.4). We pick an orientation for the edges of Γ_2. Consider a curve C_j with a rational function g_j on it. The map induced by π_2 on graphs determines an orientation on the edges of Γ_{C_j}. Furthermore, we get pullbacked function $h^w = \pi_2^* f_\alpha^{\pi_2(w)}$ for each $w \in V_{C_j}$. Define

$$\mathrm{reg}_?(\Theta)(\beta,\alpha)_j = \sum_{e \in E_{C_j}} t_e(g_j, h^{e^+})^{\beta(\pi_1(e))} ,$$

$$\mathrm{reg}_?(\Theta)(\beta,\alpha) = \sum_j \mathrm{reg}_?(\Theta)(\beta,\alpha)_j .$$

Note that the only place where the orientation of the graph enters is when deciding on the function h^{e^+}. Let us check that this gives a well-defined element of $PH^3_{\mathcal{T}}(X, \mathbb{Z}(2))$, fixing a single $C = C_j$ and $g = g_j$. Because of the harmonicity condition there is no ambiguity in β. Since we also imposed harmonicity in the construction of f_α^v, these function, and consequently the functions h^w, are defined up to a single multiplicative factor c. Correspondingly, the terms $t_e(g, h^{e^+})$ will change by $t_e(g,c) = c^{-\deg_e g}$ by Lemma 3.6, with $e \mapsto \deg_e(g) = \mathrm{res}_e\, d\log g$. This last quantity belongs to $dC^0(\Gamma_C, \mathbb{Z})$ by [BZ17, Lemma 2.1] and thus the regulator does not change.

Now we check what happens if we change the orientation. Suppose we change the orientation for one $e \in E_2$. That changes the orientation for all edges in $e' \in \pi_2^{-1} e$ and for each of these $h^{e'+}$ is replaced by $h^{e'+}/\tilde{N}_2(\alpha)(e)$. Thus, the change is given by some quadratic form evaluated on β and $\tilde{N}_2(\alpha)$ as required.

Conjecture 3. *We have* $P\,\mathrm{reg}_t = \mathrm{reg}_?$.

Note that this conjecture only computes (a part of) the toric regulator for very special elements. It does cover, though, certain cases that have been studied in connection with special values of L-functions [BDR15a, BDR15b, Sre10a] and for this reason we find it interesting in spite of its limited scope.

References

[Asa06] M. Asakura. Surjectivity of p-adic regulators on K_2 of Tate curves. *Invent. Math.*, 165(2):267–324, 2006.

[BD94] M. Bertolini and H. Darmon. Derived heights and generalized Mazur–Tate regulators. *Duke Math. J.*, 76(1):75–111, 1994.

[BD96] M. Bertolini and H. Darmon. Heegner points on Mumford–Tate curves. *Invent. Math.*, 126(3):413–456, 1996.

[BD98] M. Bertolini and H. Darmon. Heegner points, p-adic L-functions, and the Cerednik–Drinfeld uniformization. *Invent. Math.*, 131(3):453–491, 1998.

[BD14] M. Bertolini and H. Darmon. Kato's Euler system and rational points on elliptic curves I: a p-adic Beilinson formula. *Israel J. Math.*, 199(1):163–188, 2014.

[BdJ03] A. Besser and R. de Jeu. The syntomic regulator for the K-theory of fields. *Ann. Sci. École Norm. Sup.* (4), 36(6):867–924, 2003.

[BdJ12] A. Besser and R. de Jeu. The syntomic regulator for K_4 of curves. *Pacific Journal of Mathematics*, 260(2):305–380, 2012.

[BDR15a] M. Bertolini, H. Darmon, and V. Rotger. Beilinson–Flach elements and Euler systems I: Syntomic regulators and p-adic Rankin L-series. *J. Algebraic Geom.*, 24(2):355–378, 2015.

[BDR15b] M. Bertolini, H. Darmon, and V. Rotger. Beilinson–Flach elements and Euler systems II: the Birch–Swinnerton–Dyer conjecture for Hasse–Weil–Artin L-series. *J. Algebraic Geom.*, 24(3):569–604, 2015.

[Bei85] A.A. Beilinson. Higher regulators and values of L-functions. *J. Sov. Math.*, 30:2036–2070, 1985.

[Bes00] A. Besser. Syntomic regulators and p-adic integration. II. K_2 of curves. In *Proceedings of the Conference on p-adic Aspects of the Theory of Automorphic Representations (Jerusalem,* 1998), volume 120, pages 335–359, 2000.

[Bes12] A. Besser. On the syntomic regulator for K_1 of a surface. *Israel Journal of Mathematics*, 190:29–66, 2012.

[Bes17] A. Besser. p-adic heights and Vologodsky integration. Preprint, 2017.

[Bes21] A. Besser. The syntomic regulator for K_2 of curves with arbitrary reduction. This volume, pages 75–89.

[BGS97] S. Bloch, H. Gillet, and C. Soulé. Algebraic cycles on degenerate fibers. In *Arithmetic geometry (Cortona,* 1994), Sympos. Math., XXXVII, pages 45–69. Cambridge Univ. Press, Cambridge, 1997.

[BK90] S. Bloch and K. Kato. L-functions and Tamagawa numbers of motives. In *The Grothendieck Festschrift I*, volume 86 of *Prog. in Math.*, pages 333–400, Boston, 1990. Birkhäuser.

[Bru10] F. Brunault. Régulateurs p-adiques explicites pour le K_2 des courbes elliptiques. In *Actes de la Conférence "Fonctions L et Arithmétique"*, Publ. Math. Besançon Algèbre Théorie Nr., pages 29–57. Lab. Math. Besançon, Besançon, 2010.

[BZ17] A. Besser and S. Zerbes. Vologodsky integration on curves with semi-stable reduction. Preprint, 2017.

[Col00] R.F. Coleman. The monodromy pairing. *Asian J. Math.*, 4(2):315–330, 2000.

[Con98] C. Consani. Double complexes and Euler L-factors. *Compositio Math.*, 111(3):323–358, 1998.

[Dar98] H. Darmon. Stark-Heegner points over real quadratic fields. In *Number theory (Tiruchirapalli,* 1996), volume 210 of *Contemp. Math.*, pages 41–69. Amer. Math. Soc., Providence, RI, 1998.

[Dar01] H. Darmon. Integration on $\mathcal{H}_p \times \mathcal{H}$ and arithmetic applications. *Ann. of Math.* (2), 154(3):589–639, 2001.

[GNA90] F. Guillén and V. Navarro Aznar. Sur le théorème local des cycles invariants. *Duke Math. J.*, 61(1):133–155, 1990.

[Gro88] B.H. Gross. On the values of abelian L-functions at $s = 0$. *J. Fac. Sci. Univ. Tokyo Sect. IA Math.*, 35(1):177–197, 1988.

[GvdP80] L. Gerritzen and M. van der Put. *Schottky groups and Mumford curves*, volume 817 of *Lect. Notes in Math.* Springer, Berlin Heidelberg New York, 1980.

[Har77] G. Harder. Die Kohomologie S-arithmetischer Gruppen über Funktionenkörpern. *Invent. Math.*, 42:135–175, 1977.

[Ivo07] F. Ivorra. Réalisation ℓ-adique des motifs mixtes. *Doc. Math.*, (12):607–671, 2007.

[Jan88] U. Jannsen. *Mixed motives and algebraic K-theory*, volume 1400 of *Lect. Notes in Math.* Springer, Berlin Heidelberg New York, 1988.

[KY12] S. Kondo and S. Yasuda. Zeta elements in the K-theory of Drinfeld modular varieties. *Math. Ann.*, 354(2):529–587, 2012.

[LD68] P.G. Lejeune Dirichlet. *Vorlesungen über Zahlentheorie / von P.G. Lejeune Dirichlet. Hrsg. und mit Zusätzen vers. von R. Dedekind; Abth.* 2. Chelsea, New York, 4., umgearbeitete und vermehrte Auflage edition, 1968. Corrected reprint of the edition published in Braunschweig by F. Vieweg und Sohn, in 1894.

[MD73] Yu. Manin and V. Drinfeld. Periods of p-adic Schottky groups. *J. Reine Angew. Math.*, 262/263:239–247, 1973. Collection of articles dedicated to Helmut Hasse on his seventy-fifth birthday.

[MT87] B. Mazur and J. Tate. Refined conjectures of the "Birch and Swinnerton–Dyer type". *Duke Math. J.*, 54(2):711–750, 1987.

[Mum72a] D. Mumford. An analytic construction of degenerating abelian varieties over complete rings. *Compositio Math.*, 24:239–272, 1972.

[Mum72b] D. Mumford. An analytic construction of degenerating curves over complete local rings. *Compositio Math.*, 24:129–174, 1972.

[Nek93] J. Nekovář. On p-adic height pairings. In *Séminaire de Théorie des Nombres, Paris,* 1990–91, pages 127–202. Birkhäuser Boston, Boston, MA, 1993.

[Nik10] M. Niklas. *Rigid Syntomic Regulators and the p-adic L-function of a modular form.* PhD thesis, University of Regensburg, 2010.

[NN16] J. Nekovář and W. Nizioł. Syntomic cohomology and p-adic regulators for varieties over p-adic fields. *Algebra Number Theory,* 10(8):1695–1790, 2016. With appendices by Laurent Berger and Frédéric Déglise.

[Pál10a] A. Pál. The rigid analytical regulator and K_2 of Drinfeld modular curves. *Publ. Res. Inst. Math. Sci.*, 46(2):289–334, 2010.

[Pál10b] A. Pál. A rigid analytical regulator for the K_2 of Mumford curves. *Publ. Res. Inst. Math. Sci.*, 46(2):219–253, 2010.

[Qui72] D. Quillen. On the cohomology and K-theory of the general linear groups over a finite field. *Ann. of Math.* (2), 96:552–586, 1972.

[Rio06] J. Riou. Realization functors. Preprint, 2006.

[RX07a] W. Raskind and X. Xarles. On p-adic intermediate Jacobians. *Trans. Amer. Math. Soc.*, 359(12):6057–6077 (electronic), 2007.

[RX07b] W. Raskind and X. Xarles. On the étale cohomology of algebraic varieties with totally degenerate reduction over p-adic fields. *J. Math. Sci. Univ. Tokyo,* 14(2):261–291, 2007.

[RZ82] M. Rapoport and Th. Zink. Über die lokale Zetafunktion von Shimuravarietäten. Monodromiefiltration und verschwindende Zyklen in ungleicher Charakteristik. *Invent. Math.*, 68(1):21–101, 1982.

[Sai03] T. Saito. Weight spectral sequences and independence of l. *J. Inst. Math. Jussieu*, 2(4):583–634, 2003.

[Sre10a] R. Sreekantan. K_1 of products of Drinfeld modular curves and special values of L-functions. *Compos. Math.*, 146(4):886–918, 2010.

[Sre10b] R. Sreekantan. Non-Archimedean regulator maps and special values of L-functions. In *Cycles, motives and Shimura varieties*, Tata Inst. Fund. Res. Stud. Math., pages 469–492. Tata Inst. Fund. Res., Mumbai, 2010.

[Tsu99] T. Tsuji. p-adic étale cohomology and crystalline cohomology in the semi-stable reduction case. *Invent. Math.*, 137(2):233–411, 1999.

Amnon Besser
Department of Mathematics
Ben-Gurion University of the Negev
P.O.B. 653
Be'er-Sheva 84105, Israel
e-mail: `bessera@math.bgu.ac.il`

Wayne Raskind
Department of Mathematics
Wayne State University
Detroit, MI 48202, USA
e-mail: `raskind@wayne.edu`

Current address:
Institute for Defense Analyses
Center for Communications Research
805 Bunn Drive
Princeton, NJ 08540, USA
e-mail: `wraskind@idaccr.org`

Progress in Mathematics, Vol. 338, 121–140

Higher Displays Arising from Filtered de Rham–Witt Complexes

Oli Gregory and Andreas Langer

Abstract. For a smooth projective scheme X over a ring R on which p is nilpotent that meets some general assumptions we prove that the crystalline cohomology is equipped with the structure of a higher display which is a relative version of Fontaine's strongly divisible lattices. Frobenius-divisibility is induced by the Nygaard filtration on the relative de Rham–Witt complex. For a nilpotent PD-thickening S/R we also consider the associated relative display and can describe it explicitly by a relative version of the Nygaard filtration on the de Rham–Witt complex associated to a lifting of X over S. We prove that there is a crystal of relative displays if moreover the mod p reduction of X has a smooth and versal deformation space.

Mathematics Subject Classification (2010). 14F30, 14F40.

Keywords. Relative de Rham–Witt complex, higher displays, crystal of relative displays.

1. Introduction

For a ring R in which p is nilpotent, we constructed in [LZ07] an exact tensor category of displays which contains the displays associated to p-divisible groups [Zin02] as a full subcategory. If $R = k$ is a perfect field, a display is a finitely generated free $W(k)$-module M endowed with an injective Frobenius-linear map $F : M \to M$. In general, displays can be regarded as a relative version of Fontaine's strongly divisible lattices [Fon83]. For a smooth projective scheme X over Spec R we developed a strategy in [LZ07] to equip the crystalline cohomology $H^n_{\text{cris}}(X/W(R))$ with the structure of a display. For a precise statement see below. The following assumptions were essential in the construction of such "geometric" displays:

There exists a compatible system of smooth liftings $X_n/W_n(R)$ for $n \in \mathbb{N}$ of X/R such that the following properties hold:

(A1) The cohomology groups $H^i(X_n, \Omega^j_{X_n/W_n(R)})$ are for each n, i and j locally free $W_n(R)$-modules of finite type.

The first named author is supported by the ERC Consolidator Grant 681838 "K3CRYSTAL".

(A2) For each n the de Rham spectral sequence degenerates at E_1

$$E_1^{i,j} = H^j(X_n, \Omega^i_{X_n/W_n(R)}) \Rightarrow \mathbb{H}^{i+j}(X_n, \Omega^\bullet_{X_n/W_n(R)})$$

(compare $(*)$ and $(**)$ in ([LZ07], p. 150) and assumptions 5.2, 5.3 in [LZ07]). For example, these assumptions are satisfied by K3 surfaces, abelian schemes and smooth relative complete intersections (see [LZ07], Introduction).

Let $I_R := VW(R)$ and $W\Omega^\bullet_{X/R}$ be the relative de Rham–Witt complex as constructed in [LZ04]. For $r \geq 0$ define the complex $\mathcal{N}^r W\Omega^\bullet_{X/R}$ as follows:

$$(W\Omega^0_{X/R})_{[F]} \xrightarrow{d} (W\Omega^1_{X/R})_{[F]} \xrightarrow{d} \cdots \xrightarrow{d} (W\Omega^{r-1}_{X/R})_{[F]} \xrightarrow{dV} W\Omega^r_{X/R} \xrightarrow{d} \cdots$$

This is a complex of $W(R)$-modules where $(W\Omega^i_{X/R})_{[F]}$, for $i < r$, is considered as a $W(R)$-module via restriction of scalars along $W(R) \xrightarrow{F} W(R)$. It was conjectured in ([LZ07], Conj. 5.8) that the predisplay structure (Appendix, Def. A.5) on $P_0 := H^n_{\mathrm{cris}}(X/W(R))$, defined by the data $P_r := \mathbb{H}^n(X, \mathcal{N}^r W\Omega^\bullet_{X/R})$ and maps $\hat{\alpha}_r : I_R \otimes P_r \to P_{r+1}$, $\hat{\iota}_r : P_{r+1} \to P_r$ and $\hat{F}_r : P_r \to P_0$ induced by the corresponding maps of complexes $\hat{\alpha}_r : I_R \otimes \mathcal{N}^r W\Omega^\bullet_{X/R} \to \mathcal{N}^{r+1} W\Omega^\bullet_{X/R}$, $\hat{\iota}_r : \mathcal{N}^{r+1} W\Omega^\bullet_{X/R} \to \mathcal{N}^r W\Omega^\bullet_{X/R}$ and $\hat{F}_r : \mathcal{N}^r W\Omega^\bullet_{X/R} \to W\Omega^\bullet_{X/R}$ given in ([LZ07], (5)) and the diagram below and in the same order between the vertically written complexes

$$\begin{array}{ccccccc}
I_S \otimes (W\Omega^0_{X/R})_{[F]} & \longrightarrow & (W\Omega^0_{X/R})_{[F]} & \xrightarrow{p} & (W\Omega^0_{X/R})_{[F]} & \xrightarrow{\mathrm{id}} & W\Omega^0_{X/R} \\
\downarrow{\scriptstyle \mathrm{id}\otimes d} & & \downarrow{\scriptstyle d} & & \downarrow{\scriptstyle d} & & \downarrow{\scriptstyle d} \\
\cdots & & \cdots & & \cdots & & \cdots \\
\downarrow{\scriptstyle \mathrm{id}\otimes d} & & \downarrow{\scriptstyle d} & & \downarrow{\scriptstyle d} & & \downarrow{\scriptstyle d} \\
I_S \otimes (W\Omega^{r-1}_{X/R})_{[F]} & \longrightarrow & (W\Omega^{r-1}_{X/R})_{[F]} & \xrightarrow{p} & (W\Omega^{r-1}_{X/R})_{[F]} & \xrightarrow{\mathrm{id}} & W\Omega^{r-1}_{X/R} \\
\downarrow{\scriptstyle \mathrm{id}\otimes dV} & & \downarrow{\scriptstyle d} & & \downarrow{\scriptstyle dV} & & \downarrow{\scriptstyle d} \\
I_S \otimes W\Omega^r_{X/R} & \xrightarrow{\tilde{F}} & (W\Omega^r_{X/R})_{[F]} & \xrightarrow{V} & W\Omega^r_{X/R} & \xrightarrow{F} & W\Omega^r_{X/R} \\
\downarrow{\scriptstyle \mathrm{id}\otimes d} & & \downarrow{\scriptstyle dV} & & \downarrow{\scriptstyle d} & & \downarrow{\scriptstyle d} \\
I_S \otimes W\Omega^{r+1}_{X/R} & \xrightarrow{mult} & W\Omega^{r+1}_{X/R} & \xrightarrow{\mathrm{id}} & W\Omega^{r+1}_{X/R} & \xrightarrow{pF} & W\Omega^{r+1}_{X/R} \\
\downarrow{\scriptstyle \mathrm{id}\otimes d} & & \downarrow{\scriptstyle d} & & \downarrow{\scriptstyle d} & & \downarrow{\scriptstyle d} \\
I_S \otimes W\Omega^{r+2}_{X/R} & \xrightarrow{mult} & W\Omega^{r+2}_{X/R} & \xrightarrow{\mathrm{id}} & W\Omega^{r+2}_{X/R} & \xrightarrow{p^2F} & W\Omega^{r+2}_{X/R} \\
\downarrow{\scriptstyle \mathrm{id}\otimes d} & & \downarrow{\scriptstyle d} & & \downarrow{\scriptstyle d} & & \downarrow{\scriptstyle d} \\
\cdots & & \cdots & & \cdots & & \cdots
\end{array}$$

define a display structure on P_0 (Appendix, Def. A.7). (Note that in the definition of $\hat{\alpha}_r$, $\tilde{F}(V\xi \otimes \omega) := \xi F\omega$.)

In this note we prove this conjecture under the assumption $r \leq n < p$, which is a standard hypothesis in integral p-adic Hodge theory, and prove the following:

Theorem 1.1.

(a) *Let R be a local ring in which p is nilpotent, and let X be a smooth projective scheme over Spec R. Assume that there exists a compatible system of liftings $X_n/\mathrm{Spec}\, W_n(R)$ satisfying* (A1) *and* (A2). *Then the data $(P_r, \hat{\iota}_r, \hat{\alpha}_r, \hat{F}_r)$ form a display structure $\mathcal{P}_R$ on $H^n_{cris}(X/W(R))$ for $r \leq n < p$.*

(b) *Assume that there exists in addition a frame $A \to R$ and a smooth projective pA-adic lifting $\mathcal{Y}/\mathrm{Spf}\, A$ of X such that the $\mathcal{Y}_n := \mathcal{Y} \times_{\mathrm{Spf}\, A} \mathrm{Spf}\, A/p^n$ satisfy the analogous assumptions* (A1) *and* (A2). *Then the display structure on $H^n_{cris}(X/W(R))$ obtained by base change from the window structure on $H^n_{cris}(X/A)$ (compare* [LZ07] *Theorem* 5.5 *and Corollary* 5.6) *is isomorphic to $\mathcal{P}_R$.*

Remark. Theorem 1.1(a) was shown for reduced rings R in ([LZ07], Theorem 5.7).

In our second main result we derive a relative version of Theorem 1.1 on relative displays using a modified version of the complexes $\mathcal{N}^r W\Omega^\bullet_{X/R}$. For this, let $S \to R$ be a homomorphism of rings in which p is nilpotent and such that the kernel $\mathfrak{a}$ is equipped with nilpotent divided powers. In [LZ19] and [Gre17] we considered the Witt frames $\mathcal{W}_S$, $\mathcal{W}_{S/R}$, $\mathcal{W}_R$ (Appendix, Def. A.3, A.4) and the canonical homomorphisms of frames $\mathcal{W}_S \xrightarrow{u} \mathcal{W}_{S/R}$ and $\mathcal{W}_{S/R} \to \mathcal{W}_R$. Let $X/\mathrm{Spec}\, R$ be as before and let X_S be a smooth projective lifting of X over Spec S, admitting liftings $(X_S)_n$ over Spec $W_n(S)$ that satisfy the assumptions (A1) and (A2). Let $\tilde{\mathfrak{a}}$ be the logarithmic Teichmüller ideal in $W(S)$, as defined in ([Zin02], 1.4 and Appendix, Def. A.4). Then $\mathcal{J} := \tilde{\mathfrak{a}} \oplus VW(S)$ is the kernel of the composite map $W(S) \to S \to R$ and is again equipped with a PD-structure (see [Zin02], 2.3). Define the complex $\mathcal{N}^r_{rel/R} W\Omega^\bullet_{X_S/S}$ as follows:

$$\begin{aligned}&(W\mathcal{O}_{X_S})_{[F]} \oplus \tilde{\mathfrak{a}}^r W\mathcal{O}_{X_S}\\ &\quad \xrightarrow{d\oplus d} (W\Omega^1_{X_S/S})_{[F]} \oplus \tilde{\mathfrak{a}}^{r-1} W\Omega^1_{X_S/S} \xrightarrow{d\oplus d} \cdots\\ &\qquad \cdots \xrightarrow{d\oplus d} (W\Omega^{r-1}_{X_S/S})_{[F]} \oplus \tilde{\mathfrak{a}} W\Omega^{r-1}_{X_S/S} \xrightarrow{dV+d} W\Omega^r_{X_S/S} \xrightarrow{d} \cdots\end{aligned}$$

The maps $\hat{\alpha}_r$, $\hat{\iota}_r$, $\hat{F}_r$ on $\mathcal{N}^r W\Omega^\bullet_{X_S/S}$ that define the predisplay structure on $H^n_{\mathrm{cris}}(X/W(S)) = H^n_{\mathrm{cris}}(X_S/W(S))$ can easily be extended to maps on the complexes $\mathcal{N}^r_{rel/R} W\Omega^\bullet_{X_S/S}$, where multiplication by p on $W\Omega^j_{X_S/S}$ is replaced by the map

$$\pi : W\Omega^j_{X_S/S} \oplus \tilde{\mathfrak{a}}^{r-j} W\Omega^j_{X_S/S} \to W\Omega^j_{X_S/S} \oplus \tilde{\mathfrak{a}}^{r-j-1} W\Omega^j_{X_S/S}$$

which is multiplication by p on $W\Omega^j_{X_S/S}$ and the inclusion on the other summand. The divided Frobenius $\hat{F}_r$ is defined on the subcomplex $\mathcal{N}^r W\Omega^\bullet_{X_S/S}$ as before and on $\tilde{\mathfrak{a}}^{r-j} W\Omega^j_{X_S/S}$ it is defined to be the zero map. In analogy to ([LZ07], (5)) we get induced maps of complexes

$$\hat{\alpha}_r : \mathcal{J} \otimes \mathcal{N}^r_{rel/R} W\Omega^\bullet_{X_S/S} \to \mathcal{N}^{r+1}_{rel/R} W\Omega^\bullet_{X_S/S}$$
$$\hat{\iota}_r : \mathcal{N}^{r+1}_{rel/R} W\Omega^\bullet_{X_S/S} \to \mathcal{N}^r_{rel/R} W\Omega^\bullet_{X_S/S}$$
$$\hat{F}_r : \mathcal{N}^r_{rel/R} W\Omega^\bullet_{X_S/S} \to W\Omega^\bullet_{X_S/S}.$$

It is then easy to see that the above data form a predisplay $\mathcal{P}_{S/R}$ on the hypercohomology of these complexes over the relative Witt frame $\mathcal{W}_{S/R}$ (see ([LZ19], Def. 2 and Appendix, Def. A.5). Then one has:

Theorem 1.2.

(a) *Let $\mathcal{P}_S$ be the display over S constructed in Theorem* 1.1(a). *Then the associated display $u_*\mathcal{P}_S$ under the homomorphism of frames $\mathcal{W}_S \xrightarrow{u} \mathcal{W}_{S/R}$ is isomorphic to $\mathcal{P}_{S/R}$. In particular, the predisplay $\mathcal{P}_{S/R}$ is a display over $\mathcal{W}_{S/R}$ in the sense of* ([LZ19], *Def.* 3 *and Appendix, Def. A.*7).

(b) *Assume that R is artinian local with perfect residue field k. Assume that $X_0 := X \times_R k$ has a smooth versal formal deformation space and that the assumptions* (32) *and* (33) *of* [LZ19] (*analogous to* (A1) *and* (A2)) *are satisfied. Then the display $\mathcal{P}_{S/R}$ only depends – up to isomorphism – on X, not on the lifting X_S. The collection $(\mathcal{P}_{S/R})_{S\to R}$ where $S \to R$ are PD-morphisms defines a crystal of relative displays.*

Remark. Relative displays were first considered by Zink [Zin02] and Lau [Lau14] in their classification of p-divisible groups. Zink associated to a formal p-divisible group $\mathcal{G}$ and a nilpotent PD-thickening $S \to R$ a relative display $\mathscr{D}_{S/R}(\mathcal{G})$ (which he calls a triple in [Zin02]) which is – up to isomorphism – the unique relative display lifting the display associated to $\mathcal{G}$ over R to S. Using Dieudonné displays, i.e., displays defined over the small Witt ring [Zin01], Zink extended the classification to all p-divisible groups over an artinian local ring with perfect residue field, see also [Lau14] and [Mes07]. Over slightly more general rings, called admissible rings, Lau [Lau14] constructed a unique functor from p-divisible groups to crystals of relative displays. The relative display is obtained by base change from a window structure associated to the universal p-divisible group over the deformation ring. The proof of Theorem 1.2(b) was inspired by this construction.

We assume that the reader is familiar with the basic definitions and properties of the relative de Rham–Witt complex [LZ04] including the comparison to crystalline cohomology and the explicit description of the de Rham–Witt complex of a polynomial algebra. In an appendix we recall the basic definitions of the theory of higher displays including frames, windows and relative displays.

2. Proof of the theorems

Theorem 1.1(a) is a consequence of the following result which was conjectured in ([LZ07], Conj. 4.1) and was recently proved by the second named author in ([Lan18], Thm. 0.2) under the assumption $r < p$:

Theorem 2.1. *Let R be a ring on which p is nilpotent and let $X/\operatorname{Spec} R$ be smooth projective, and $X_n/\operatorname{Spec} W_n(R)$ a compatible system of smooth liftings. Let $\mathcal{F}^r\Omega^\bullet_{X_n/W_n(R)}$ be the following complex:*

$$I_R \otimes \mathcal{O}_{X_n} \xrightarrow{pd} I_R \otimes \Omega^1_{X_n/W_n(R)} \xrightarrow{pd} \cdots \xrightarrow{pd} I_R \otimes \Omega^{r-1}_{X_n/W_n(R)} \xrightarrow{d} \Omega^r_{X_n/W_n(R)} \xrightarrow{d} \cdots$$

where $I_R := VW_{n-1}(R)$, and let $\mathcal{N}^rW_n\Omega^\bullet_{X/R}$ be the Nygaard complex, given as

$$(W_{n-1}\Omega^0_{X/R})_{[F]} \xrightarrow{d} (W_{n-1}\Omega^1_{X/R})_{[F]} \xrightarrow{d} \cdots \xrightarrow{d} (W_{n-1}\Omega^{r-1}_{X/R})_{[F]} \xrightarrow{dV} W_n\Omega^r_{X/R} \xrightarrow{d} \cdots$$

Then $\mathcal{F}^r\Omega^\bullet_{X_n/W_n(R)}$ and $\mathcal{N}^rW_n\Omega^\bullet_{X/R}$ are isomorphic in the derived category of $W_n(R)$-modules for $r < p$.

2.1. Proof of Theorem 1.1(a)

Under the assumptions (A1) and (A2) it follows from [LZ07] Propositions 3.2 and the projection formula Proposition 3.1 that one has a degenerating spectral sequence $E_1^{i,j} \Rightarrow \mathbb{H}^{i+j}(X_n, \mathcal{F}^r\Omega^\bullet_{X_n/W_n(R)})$ where

$$E_1^{i,j} = \begin{cases} I_R \otimes H^j(X_n, \Omega^i_{X_n/W_n(R)}) & \text{for } i < r \\ H^j(X_n, \Omega^i_{X_n/W_n(R)}) & \text{for } i \geq r \end{cases}$$

It follows from the proof of Theorem 2.1 ([Lan18], Theorem 0.2) that the isomorphisms $\mathcal{F}^r\Omega^\bullet_{X_n/W_n(R)} \cong \mathcal{N}^rW_n\Omega^\bullet_{X/R}$ are compatible for varying n and yield an isomorphism $\mathcal{F}^r\Omega^\bullet_{X_\bullet/W_\bullet(R)} \cong \mathcal{N}^rW_\bullet\Omega^\bullet_{X/R}$ of procomplexes in $D_{\text{pro,Zar}}(X)$. This induces an isomorphism resp. a decomposition

$$P_r = \mathbb{H}^n(X, \mathcal{F}^r\Omega^\bullet_{X_\bullet/W_\bullet(R)}) \cong I_RL_0 \oplus I_RL_1 \oplus \cdots \oplus I_RL_{r-1} \oplus L_r \oplus \cdots \oplus L_n$$

where $L_i := H^{n-i}(X, \Omega^i_{X_\bullet/W_\bullet(R)})$. Since the divided Frobenius $\hat{F}_r$ is defined on $\mathbb{H}^n(X, \mathcal{F}^r\Omega^\bullet_{X_\bullet/W_\bullet(R)})$ via Theorem 2.1, we can define $\Phi_r : L_r \to P_0$ by $\Phi_r := \hat{F}_r|_{L_r}$.

To show that $(P_r, \hat{F}_r, \hat{\iota}_r, \hat{\alpha}_r)$ defines a display on $H^n_{\text{cris}}(X/W(R))$ is equivalent to the condition that

$$\bigoplus_{i=0}^n \Phi_i : \bigoplus_{i=0}^n L_i \to \bigoplus_{i=0}^n L_i$$

is a σ-linear isomorphism, or equivalently that $\det(\oplus_{i=0}^n \Phi_i) \in W(R)^\times$. This is reduced by base change to the case that $R = k$ is a perfect field in the same way as in the proof of ([LZ07] Thm. 5.5), and then follows from ([Fon83], p. 91) and ([Kat87], Prop. 2.5).

2.2. Proof of Theorem 1.2(a)

We are going to explicitly construct displays over the relative Witt frames. For this, let $S \to R$ be a homomorphism of rings in which p is nilpotent and such that the kernel $\mathfrak{a}$ is equipped with divided powers. Then the kernel of $W(S) \to R$ is $\tilde{\mathfrak{a}} \oplus I_S$ where $I_S := VW(S)$ and $\tilde{\mathfrak{a}}$ is the logarithmic Teichmüller ideal. Then we consider the relative Witt frame $\mathcal{W}_{S/R}$ as defined in [LZ19].

We first prove a relative version of Theorem 2.1:

Theorem 2.2. *Let $X_S/\operatorname{Spec} S$ be a smooth projective lifting of $X/\operatorname{Spec} R$ and assume that X_S admits a compatible system of smooth liftings $(X_S)_n$ over $\operatorname{Spec} W_n(S)$. Set $I_S := VW_{n-1}(S)$. Let $\mathcal{F}il^r_{rel/R}\Omega^\bullet_{(X_S)_n/W_n(S)}$ be the following complex:*

$$I_S\mathcal{O}_{(X_S)_n} \oplus \tilde{\mathfrak{a}}^r\mathcal{O}_{(X_S)_n} \xrightarrow{pd\oplus d} I_S\Omega^1_{(X_S)_n/W_n(S)} \oplus \tilde{\mathfrak{a}}^{r-1}\Omega^1_{(X_S)_n/W_n(S)} \xrightarrow{pd\oplus d} \cdots$$
$$\cdots \xrightarrow{pd\oplus d} I_S\Omega^{r-1}_{(X_S)_n/W_n(S)} \oplus \tilde{\mathfrak{a}}\Omega^{r-1}_{(X_S)_n/W_n(S)} \xrightarrow{d+d} \Omega^r_{(X_S)_n/W_n(S)} \xrightarrow{d} \cdots$$

and let $\mathcal{N}^r_{rel/R}W_n\Omega^\bullet_{X_S/S}$ be the following complex:

$$(W_{n-1}\mathcal{O}_{X_S})_{[F]} \oplus \tilde{\mathfrak{a}}^r W_n\mathcal{O}_{X_S} \xrightarrow{d\oplus d} (W_{n-1}\Omega^1_{X_S/S})_{[F]} \oplus \tilde{\mathfrak{a}}^{r-1}W_n\Omega^1_{X_S/S} \xrightarrow{d\oplus d} \cdots$$
$$\cdots \xrightarrow{d\oplus d} (W_{n-1}\Omega^{r-1}_{X_S/S})_{[F]} \oplus \tilde{\mathfrak{a}}W_n\Omega^{r-1}_{X_S/S} \xrightarrow{dV+d} W_n\Omega^r_{X_S/S} \xrightarrow{d} \cdots$$

Then for $r < p$ the complexes $\mathcal{F}il^r_{rel/R}\Omega^\bullet_{(X_S)_n/W_n(S)}$ and $\mathcal{N}^r_{rel/R}W_n\Omega^\bullet_{X_S/S}$ are quasi-isomorphic.

Proof. First notice that we may write $\mathcal{F}il^r_{rel/R}\Omega^\bullet_{(X_S)_n/W_n(S)}$ and $\mathcal{N}^r_{rel/R}W_n\Omega^\bullet_{X_S/S}$ as the mapping cones of certain morphisms of complexes:

$$\mathcal{F}il^r_{rel/R}\Omega^\bullet_{(X_S)_n/W_n(S)} = \operatorname{Cone}(\tilde{\mathfrak{a}}^{(r)}\Omega^{<r}_{(X_S)_n/W_n(S)}[-1] \xrightarrow{f} \mathcal{F}^r\Omega^\bullet_{(X_S)_n/W_n(S)})$$

and

$$\mathcal{N}^r_{rel/R}W_n\Omega^\bullet_{X_S/S} = \operatorname{Cone}(\tilde{\mathfrak{a}}^{(r)}W_n\Omega^{<r}_{X_S/S}[-1] \xrightarrow{g} \mathcal{N}^r W_n\Omega^\bullet_{X_S/S})$$

where $\tilde{\mathfrak{a}}^{(r)}\Omega^{<r}_{(X_S)_n/W_n(S)}$ and $\tilde{\mathfrak{a}}^{(r)}W_n\Omega^{<r}_{X_S/S}$ are the truncated complexes of the complexes $\tilde{\mathfrak{a}}^{(r)}\Omega^\bullet_{(X_S)_n/W_n(S)}$

$$\tilde{\mathfrak{a}}^r\mathcal{O}_{(X_S)_n} \xrightarrow{-d} \tilde{\mathfrak{a}}^{r-1}\Omega^1_{(X_S)_n/W_n(S)} \xrightarrow{-d} \cdots \xrightarrow{-d} \tilde{\mathfrak{a}}\Omega^{r-1}_{(X_S)_n/W_n(S)} \xrightarrow{-d} \tilde{\mathfrak{a}}\Omega^r_{(X_S)_n/W_n(S)} \xrightarrow{-d} \cdots$$

and $\tilde{\mathfrak{a}}^{(r)}W_n\Omega^\bullet_{X_S/S}$

$$\tilde{\mathfrak{a}}^r W_n\mathcal{O}_{X_S} \xrightarrow{-d} \tilde{\mathfrak{a}}^{r-1}W_n\Omega^1_{X_S/S} \xrightarrow{-d} \cdots \xrightarrow{-d} \tilde{\mathfrak{a}}W_n\Omega^{r-1}_{X_S/S} \xrightarrow{-d} \tilde{\mathfrak{a}}W_n\Omega^r_{X_S/S} \xrightarrow{-d} \cdots$$

respectively. The morphisms f and g are respectively given by

$$\begin{array}{ccccccccccccc}
0 & \longrightarrow & \tilde{\mathfrak{a}}^r\mathcal{O} & \xrightarrow{-d} & \cdots & \xrightarrow{-d} & \tilde{\mathfrak{a}}^2\Omega^{r-2} & \xrightarrow{-d} & \tilde{\mathfrak{a}}\Omega^{r-1} & \longrightarrow & 0 & \longrightarrow & \cdots \\
\downarrow & & \downarrow{\scriptstyle 0} & & & & \downarrow{\scriptstyle 0} & & \downarrow{\scriptstyle d} & & \downarrow & & \\
I_S\mathcal{O} & \xrightarrow{pd} & I_S\Omega^1 & \xrightarrow{pd} & \cdots & \xrightarrow{pd} & I_S\Omega^{r-1} & \xrightarrow{d} & \Omega^r & \xrightarrow{d} & \Omega^{r+1} & \xrightarrow{d} & \cdots
\end{array} \tag{2.3}$$

and

$$\begin{array}{ccccccccccc} 0 & \longrightarrow & \tilde{\mathfrak{a}}^r W_n\mathcal{O} & \xrightarrow{-d} \cdots \xrightarrow{-d} & \tilde{\mathfrak{a}}^2 W_n\Omega^{r-2} & \xrightarrow{-d} & \tilde{\mathfrak{a}} W_n\Omega^{r-1} & \longrightarrow & 0 & \longrightarrow & \cdots \\ \downarrow & & \downarrow{\scriptstyle 0} & & \downarrow{\scriptstyle 0} & & \downarrow{\scriptstyle d} & & \downarrow & & \\ (W_{n-1}\mathcal{O})_{[F]} & \xrightarrow{d} (W_{n-1}\Omega^1)_{[F]} \xrightarrow{d} & \cdots & \xrightarrow{d} & (W_{n-1}\Omega^{r-1})_{[F]} & \xrightarrow{dV} & W_n\Omega^r & \xrightarrow{d} & W_n\Omega^{r+1} & \xrightarrow{d} & \cdots \end{array} \tag{2.4}$$

(we briefly omitted the subscripts for typographical reasons). We will construct a morphism of distinguished triangles in the derived category

$$\begin{array}{ccccccc} \tilde{\mathfrak{a}}^{(r)}\Omega^{<r}_{(X_S)_n/W_n(S)}[-1] & \xrightarrow{f} & \mathcal{F}^r\Omega^\bullet_{(X_S)_n/W_n(S)} & \longrightarrow & \mathcal{F}il^r_{rel/R}\Omega^\bullet_{(X_S)_n/W_n(S)} & \xrightarrow{+1} & \\ \downarrow & & \downarrow & & \downarrow & & \\ \tilde{\mathfrak{a}}^{(r)}W_n\Omega^{<r}_{X_S/S}[-1] & \xrightarrow{g} & \mathcal{N}^rW_n\Omega^\bullet_{X_S/S} & \longrightarrow & \mathcal{N}^r_{rel/R}W_n\Omega^\bullet_{X_S/S} & \xrightarrow{+1} & \end{array} \tag{2.5}$$

where the middle vertical arrow is the quasi-isomorphism of Theorem 2.1 and the left vertical map is defined as follows:

Assume first that there exists a closed embedding $(X_S)_n \xrightarrow{i} Z_n$ into a projective smooth $W_n(S)$-scheme which is a Witt lift of $Z_n \times_{\mathrm{Spec}\ W_n(S)} \mathrm{Spec\ S} = Z_S$ in the sense of [LZ04] Definition 3.3. Such a Witt lift always exists locally ([LZ04] Prop. 3.2 and remarks after Def. 3.3) and induces maps $\mathcal{O}_{Z_n} \to W_n(\mathcal{O}_{Z_S}) \to W_n(\mathcal{O}_{X_S})$. Let $\mathcal{O}_{D_n}$ be the PD-envelope of i and let $\mathcal{J}^{[r]}$ be the divided power ideal with $\mathcal{J} = \ker(\mathcal{O}_{Z_n} \to \mathcal{O}_{(X_S)_n})$. The comparison with crystalline cohomology yields a chain of quasi-isomorphisms [BO78] Theorem 7.1 and [LZ04] Theorem 3.5

$$\Omega^\bullet_{(X_S)_n/W_n(S)} \xleftarrow{\cong} \Omega^\bullet_{D_n/W_n(S)} \xrightarrow{\cong} W_n\Omega^\bullet_{X_S/S}\,. \tag{2.6}$$

We construct a complex $\tilde{\mathfrak{a}}^{(r)}\Omega^{<r}_{D_n/W_n(S)}$ together with a diagram of maps

$$\tilde{\mathfrak{a}}^{(r)}\Omega^{<r}_{(X_S)_n/W_n(S)} \leftarrow \tilde{\mathfrak{a}}^{(r)}\Omega^{<r}_{D_n/W_n(S)} \to \tilde{\mathfrak{a}}^{(r)}W_n\Omega^{<r}_{X_S/S}\,. \tag{2.7}$$

The argument is very similar to the proof of [Lan18] Theorem 0.2. Consider diagram 2.8 for $r < p$ (on top of page 128).

The complex $\mathcal{J}^{[s]} \to \mathcal{J}^{[s-1]}\Omega^1_{D_n} \to \cdots \to \Omega^s_{D_n} \to \Omega^{s+1}_{D_n}$ is exact in degrees $< s$ and quasi-isomorphic to $\Omega^{\geq s}_{(X_S)_n/W_n(S)}[-s]$ by [BO78] Theorem 7.2. Since all $\mathcal{J}^{[l]}\Omega^k_{D_n}$ are – locally – free $\mathcal{O}_{(X_S)_n}$-modules by [BO78] Prop. 3.32, the complexes $\mathcal{J}^{[s-\bullet]}\Omega^\bullet_{D_n}$ remain exact in degrees $< s$ after $\otimes_{W_n(S)}S$ and $\otimes_{W_n(S)}R$; they coincide then with the corresponding complexes for the embeddings $X_S \to Z_S$, resp $X \to Z_S \times_{\mathrm{Spec}\ S} \mathrm{Spec}\ R$, since $\tilde{\mathfrak{a}}$ is a direct summand of $\ker(W_n(S) \to R)$ the lower horizontal sequence in (2.8) is exact in degrees $< r$ and quasi-isomorphic to $\tilde{\mathfrak{a}}\Omega^{\geq r-1}_{(X_S)_n/W_n(S)}[-(r-1)]$. By an easy induction argument (replace R by $S/\mathfrak{a}^l$) one

$$
\begin{array}{ccccccccc}
\tilde{\mathfrak{a}}^r\mathcal{O}_{D_n} & & & & & & & & \\
& \overset{-d}{\searrow} & & & & & & & \\
\tilde{\mathfrak{a}}^{r-1}\mathfrak{J} & \xrightarrow{-d} & \tilde{\mathfrak{a}}^{r-1}\Omega^1_{D_n} & & & & & & \\
& & & \overset{-d}{\searrow} & & & & & \\
\vdots & & \vdots & & \ddots & & & & \\
& & & & & \overset{-d}{\searrow} & & & \\
\tilde{\mathfrak{a}}^2\mathfrak{J}^{[r-2]}\mathcal{O}_{D_n} & \xrightarrow{-d} & \tilde{\mathfrak{a}}^2\mathfrak{J}^{[r-3]}\Omega^1_{D_n} & \xrightarrow{-d} & \cdots \xrightarrow{-d} & \tilde{\mathfrak{a}}^2\Omega^{r-2}_{D_n} & & & \\
& & & & & & \overset{-d}{\searrow} & & \\
\tilde{\mathfrak{a}}\mathfrak{J}^{[r-1]}\mathcal{O}_{D_n} & \xrightarrow{-d} & \tilde{\mathfrak{a}}\mathfrak{J}^{[r-2]}\Omega^1_{D_n} & \xrightarrow{-d} & \cdots & \cdots \xrightarrow{-d} & \tilde{\mathfrak{a}}\Omega^{r-1}_{D_n} & \xrightarrow{-d} \tilde{\mathfrak{a}}\Omega^r_{D_n} & \xrightarrow{-d} \cdots
\end{array}
\tag{2.8}
$$

sees that the other horizontal sequences in (2.8) are – up to the diagonal term – exact as well. The degree wise sum of the two lower sequences is quasi-isomorphic to

$$\left(\tilde{\mathfrak{a}}^2\Omega^{r-2}_{(X_S)_n/W_n(S)} \xrightarrow{-d} \tilde{\mathfrak{a}}\Omega^{r-1}_{(X_S)_n/W_n(S)} \xrightarrow{-d} \tilde{\mathfrak{a}}\Omega^{r}_{(X_S)_n/W_n(S)} \xrightarrow{-d} \cdots\right)[-(r-2)] .$$

Finally, the degree wise sum of all horizontal sequences yields a complex denoted by $\tilde{\mathfrak{a}}^{(r)}\Omega^\bullet_{D_n/W_n(S)}$ which is quasi-isomorphic to $\tilde{\mathfrak{a}}^{(r)}\Omega^\bullet_{(X_S)_n/W_n(S)}$. From the above we conclude that the complex $\tilde{\mathfrak{a}}\mathcal{F}il^r\Omega^\bullet_{D_n}$

$$\tilde{\mathfrak{a}}\mathcal{J}^{[r]}\mathcal{O}_{D_n} \xrightarrow{-d} \tilde{\mathfrak{a}}\mathcal{J}^{[r-1]}\Omega^1_{D_n} \xrightarrow{-d} \cdots \xrightarrow{-d} \tilde{\mathfrak{a}}\Omega^r_{D_n} \xrightarrow{-d} \tilde{\mathfrak{a}}\Omega^{r+1}_{D_n} \xrightarrow{-d} \cdots$$

is quasi-isomorphic to

$$(\tilde{\mathfrak{a}}\Omega^r_{(X_S)_n/W_n(S)} \xrightarrow{-d} \tilde{\mathfrak{a}}\Omega^{r+1}_{(X_S)_n/W_n(S)} \xrightarrow{-d} \tilde{\mathfrak{a}}\Omega^{r+2}_{(X_S)_n/W_n(S)} \xrightarrow{-d} \cdots)[-r] .$$

The natural embedding of $\tilde{\mathfrak{a}}\mathcal{F}il^r\Omega^\bullet_{D_n}$ into the lower horizontal complex in diagram (2.8) defines an injective map

$$\tilde{\mathfrak{a}}\mathcal{F}il^r\Omega^\bullet_{D_n} \to \tilde{\mathfrak{a}}^{(r)}\Omega^\bullet_{D_n/W_n(S)} .$$

We denote the mapping cone of this map by $\tilde{\mathfrak{a}}^{(r)}\Omega^{<r}_{D_n/W_n(S)}$. The notation is justified because the complex vanishes in degrees $\geq r$. We see that $\tilde{\mathfrak{a}}^{(r)}\Omega^{<r}_{D_n/W_n(S)}$ is quasi-isomorphic to $\tilde{\mathfrak{a}}^{(r)}\Omega^{<r}_{(X_S)_n/W_n(S)}$.

Note that under the canonical map $\mathcal{O}_{D_n} \to W_n(\mathcal{O}_{X_S})$ the image of $\mathfrak{J}$ is contained in $VW_{n-1}(\mathcal{O}_{X_S})$ and hence the image of $\mathfrak{a}\cdot\mathfrak{J}$ is zero in $W_n(\mathcal{O}_{X_S})$. Hence the map $\mathcal{O}_{D_n} \to W_n(\mathcal{O}_{X_S})$, compatible with Frobenius, induces a well-defined map

$$\tilde{\mathfrak{a}}^{(r)}\Omega^\bullet_{D_n/W_n(S)} \to \tilde{\mathfrak{a}}^{(r)}W_n\Omega^\bullet_{X_S/S} .$$

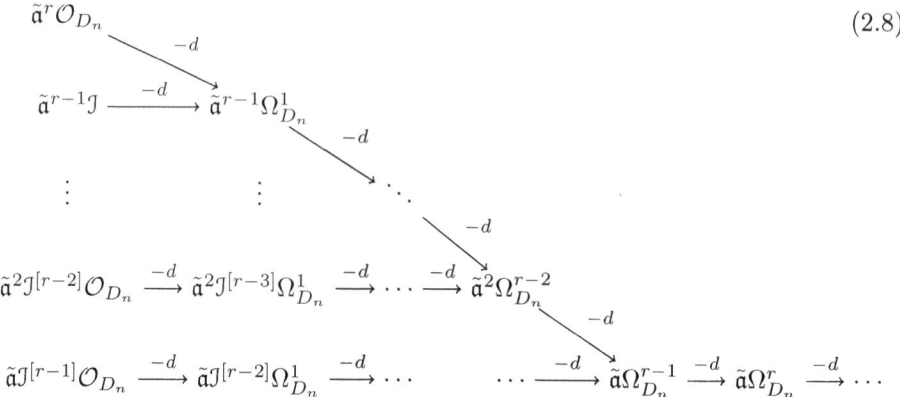

If one has two embeddings $(X_S)_n \xrightarrow{i} Z_n$, $(X_S)_n \xrightarrow{i'} Z'_n$ into Witt lifts then by considering the product embedding $(X_S)_n \xrightarrow{(i,i')} Z_n \times Z'_n$ we get a well-defined map in the derived category

$$\tilde{\mathfrak{a}}^{(r)}\Omega^{\bullet}_{(X_S)_n/W_n(S)} \to \tilde{\mathfrak{a}}^{(r)}W_n\Omega^{\bullet}_{X_S/S}$$

which induces a map

$$\tilde{\mathfrak{a}}^{(r)}\Omega^{<r}_{(X_S)_n/W_n(S)} \to \tilde{\mathfrak{a}}^{(r)}W_n\Omega^{<r}_{X_S/S}$$

on the truncated complexes. This defines (2.7).

To show that it is a quasi-isomorphism is a Zariski-local question on X_S, so it suffices to check the case where $X_S = \text{Spec B}$ is affine and B is étale over a polynomial algebra $A := S[T_1, \dots, T_d]$. Set $A_n := W_n(S)[T_1, \dots, T_d]$ and let $\phi_n : A_n \to A_{n-1}$ be the map extending $F : W_n(S) \to W_{n-1}(S)$ given by setting $\phi_n(T_i) = T_i^p$, and let $\delta_n : A_n \to W_n(A)$ be the unique $W_n(S)$-algebra homomorphism which sends each T_i to its Teichmüller representative. Then the data (A_n, ϕ_n, δ_n) is a Frobenius lift of A to $W(S)$ (see §3 of [LZ04]). Since $A \to B$ is étale, there exists a unique set of liftings B_n of B which are each étale over A_n, and homomorphisms $\psi_n : B_n \to B_{n-1}$, $\epsilon_n : B_n \to W_n(B)$ which are compatible with ϕ_n, δ_n. The morphism $\tilde{\mathfrak{a}}^{(r)}\Omega^{\bullet}_{B_n/W_n(S)} \to \tilde{\mathfrak{a}}^{(r)}W_n\Omega^{\bullet}_{B/S}$ is the one induced by the ϵ_n.

First let us treat the special case that $B = A = S[T_1, \dots, T_d]$; the quasi-isomorphism is easy to establish because in this case the de Rham–Witt complex has a rather explicit description (originally due to Illusie in the case of a perfect field, and by §2 of [LZ04] in our generality). Indeed, the de Rham–Witt complex decomposes into a direct sum of an integral part and an acyclic fractional part ([LZ04], (3.9)) and the fractional part is contained in the image of V resp. dV. We must check that the fractional part is still acyclic after multiplying by the logarithmic Teichmüller ideal $\tilde{\mathfrak{a}}$. But multiplying anything in the image of V by an element a means applying Frobenius to a, and Frobenius kills $\tilde{\mathfrak{a}}$ by [Zin02] Lemma 38, so we conclude that the fractional part of the de Rham–Witt complex is annihilated by $\tilde{\mathfrak{a}}$.

Now we return to the general case where $X_S = \text{Spec } B$ for B étale over $A = S[T_1, \dots, T_d]$. Choose an integer m such that $p^m W_n(S) = 0$ and set $\phi^m := \phi_{m+n} \circ \cdots \circ \phi_{n+1} : A_{m+n} \to A_n$. Then for each l we have an isomorphism (see the proof of [LZ04] Theorem 3.5)

$$\Omega^l_{B_n/W_n(S)} \cong B_n \otimes_{A_n} \Omega^l_{A_n/W_n(S)} \cong B_{m+n} \otimes_{A_{m+n},\phi^m} \Omega^l_{A_n/W_n(S)}$$

and, likewise, for each j, l an isomorphism

$$\tilde{\mathfrak{a}}^j\Omega^l_{B_n/W_n(S)} \cong B_n \otimes_{A_n} \tilde{\mathfrak{a}}^j\Omega^l_{A_n/W_n(S)} \cong B_{m+n} \otimes_{A_{m+n},\phi^m} \tilde{\mathfrak{a}}^j\Omega^l_{A_n/W_n(S)}$$

and this gives an isomorphism of complexes

$$\tilde{\mathfrak{a}}^{(r)}\Omega^{\bullet}_{B_n/W_n(S)} \cong B_{m+n} \otimes_{A_{m+n},\phi^m} \tilde{\mathfrak{a}}^{(r)}\Omega^{\bullet}_{A_n/W_n(S)}$$

if we give the right-hand side the differential $1 \otimes -d$.

Let

$$W_n\Omega^{\bullet}_{A/S} = W_n\Omega^{\text{int},\bullet}_{A/S} \oplus W_n\Omega^{\text{frac},\bullet}_{A/S}$$

be the decomposition into integral and fractional parts, as mentioned above. Note that this is a direct sum decomposition of complexes of A_{m+n}-modules via restriction of scalars $A_{m+n} \xrightarrow{\phi^m} A_n$. Then using the following facts proven in [LZ04] Theorem 3.5 and Proposition 1.7

1. base change of $W_n\Omega^\bullet$ for étale maps
2. $W_n\Omega^\bullet_{B/S} \cong B_{m+n} \otimes_{A_{m+n},\phi^m} W_n\Omega^\bullet_{A/S}$

we have

$$\begin{aligned} W_n\Omega^\bullet_{B/S} &= (B_{m+n} \otimes_{A_{m+n}} W_n\Omega^{\mathrm{int},\bullet}_{A/S}) \oplus (B_{m+n} \otimes_{A_{m+n}} W_n\Omega^{\mathrm{frac},\bullet}_{A/S}) \\ &=: W_n\Omega^{\mathrm{int},\bullet}_{B/S} \oplus W_n\Omega^{\mathrm{frac},\bullet}_{B/S}. \end{aligned}$$

Since $W_n\Omega^{\mathrm{frac},\bullet}_{A/S}$ is acyclic and B_{m+n} is a flat A_{m+n}-module, $W_n\Omega^{\mathrm{frac},\bullet}_{B/S}$ is acyclic too.

Then we define complexes $\tilde{\mathfrak{a}}^{(r)}W_n\Omega^\bullet_{B/S}$, $\tilde{\mathfrak{a}}^{(r)}W_n\Omega^{\mathrm{int},\bullet}_{B/S}$ and $\tilde{\mathfrak{a}}^{(r)}W_n\Omega^{\mathrm{frac},\bullet}_{B/S}$ in exactly the same manner as at the beginning of the proof. Evidently we get a direct sum decomposition

$$\tilde{\mathfrak{a}}^{(r)}W_n\Omega^\bullet_{B/S} = \tilde{\mathfrak{a}}^{(r)}W_n\Omega^{\mathrm{int},\bullet}_{B/S} \oplus \tilde{\mathfrak{a}}^{(r)}W_n\Omega^{\mathrm{frac},\bullet}_{B/S}.$$

Since $W_n\Omega^{\mathrm{int},\bullet}_{A/S} \cong \Omega^\bullet_{A_n/W_n(S)}$, we get an isomorphism

$$\begin{aligned} \tilde{\mathfrak{a}}^{(r)}W_n\Omega^{\mathrm{int},\bullet}_{B/S} &\cong B_{m+n} \otimes_{A_{m+n},\phi^m} \tilde{\mathfrak{a}}^{(r)}W_n\Omega^{\mathrm{int},\bullet}_{A/S} \\ &\cong B_{m+n} \otimes_{A_{m+n},\phi^m} \tilde{\mathfrak{a}}^{(r)}\Omega^\bullet_{A_n/W_n(S)} \cong \tilde{\mathfrak{a}}^{(r)}\Omega^\bullet_{B_n/W_n(S)}. \end{aligned}$$

Since $\tilde{\mathfrak{a}}$ annihilates the fractional part $W_n\Omega^\bullet_{A/S}$ as observed above, we get that $\tilde{\mathfrak{a}}^{(r)}W_n\Omega^{\mathrm{frac},\bullet}_{B/S}$ vanishes. Hence we obtain an isomorphism

$$\tilde{\mathfrak{a}}^{(r)}W_n\Omega^\bullet_{B/S} \cong \tilde{\mathfrak{a}}^{(r)}\Omega^\bullet_{B_n/W_n(S)}$$

and likewise for the truncated complexes

$$\tilde{\mathfrak{a}}^{(r)}W_n\Omega^{<r}_{B/S} \cong \tilde{\mathfrak{a}}^{(r)}\Omega^{<r}_{B_n/W_n(S)}$$

as desired. Since the construction of this isomorphism using PD-envelopes of embeddings into Witt lifts is compatible with the construction of the comparison map

$$\mathcal{F}^r\Omega^\bullet_{(X_S)_n/W_n(S)} \to \mathcal{N}^rW_n\Omega^\bullet_{X/S}$$

([Lan18], Theorem 0.2) the diagram (2.5) commutes on the left.

Let $\mathcal{F}il^r\Omega^\bullet_{D_n/W_n(S)}$ be the complex constructed in the proof of [Lan18, Theorem 0.2] (see page 1868). Analogously to the map

$$f : \tilde{\mathfrak{a}}^{(r)}\Omega^{<r}_{(X_S)_n/W_n(S)}[-1] \to \mathcal{F}^r\Omega^\bullet_{(X_S)_n/W_n(S)}$$

in (2.3), one can define a canonical map

$$\tilde{f} : \tilde{\mathfrak{a}}^{(r)}\Omega^{<r}_{D_n/W_n(S)}[-1] \to \mathcal{F}il^r\Omega^\bullet_{D_n/W_n(S)}$$

which is the zero map in degrees $\neq r$ and equal to d in degree r.

Let $\mathcal{F}il^r_{rel/R}\Omega^\bullet_{D_n/W_n(S)}$ be the mapping cone of $\tilde{f}$. One obtains a commutative diagram of complexes

$$\begin{array}{ccccccc}
\tilde{\mathfrak{a}}^{(r)}\Omega^{<r}_{(X_S)_n/W_n(S)}[-1] & \xrightarrow{f} & \mathcal{F}^r\Omega^\bullet_{(X_S)_n/W_n(S)} & \longrightarrow & \mathcal{F}il^r_{rel/R}\Omega^\bullet_{(X_S)_n/W_n(S)} & \xrightarrow{+1} & \cdots \\
\uparrow & & \uparrow & & \uparrow & & \\
\tilde{\mathfrak{a}}^{(r)}\Omega^{<r}_{D_n/W_n(S)}[-1] & \xrightarrow{\tilde{f}} & \mathcal{F}il^r\Omega^\bullet_{D_n/W_n(S)} & \longrightarrow & \mathcal{F}il^r_{rel/R}\Omega^\bullet_{D_n/W_n(S)} & \xrightarrow{+1} & \cdots \\
\downarrow & & \downarrow & & \downarrow & & \\
\tilde{\mathfrak{a}}^{(r)}W_n\Omega^{<r}_{X_S/S}[-1] & \xrightarrow{g} & \mathcal{N}^r W_n\Omega^\bullet_{X_S/S} & \longrightarrow & \mathcal{N}^r_{rel/R}W_n\Omega^\bullet_{X_S/S} & \xrightarrow{+1} & \cdots
\end{array} \tag{2.9}$$

where the right vertical arrows are canonical maps induced on the level of mapping cones by the commutative diagrams of complexes on the left. Since the vertical arrows on the left and in the middle of the diagram are quasi-isomorphisms by construction, the vertical arrows on the right-hand side are also quasi-isomorphisms. This proves that (2.5) is a morphism of distinguished triangles in the derived category where all vertical arrows are isomorphisms.

In the absence of a global embedding into a Witt lift one proceeds by simplicial methods as in the proof of [Lan18] Theorem 0.2, [LZ04] §3.2 and [Ill79] II.1. to obtain Theorem 2.2. For the convenience of the reader we recall the argument. Let $(X_S)_n(i)$, $i \in I$ be a covering of $(X_S)_n$ inducing a covering $X_S(i)$ of X_S and an embedding $(X_S)_n(i) \to (Z_S)_n(i)$ which is a Witt lift of $Z_S(i) := (Z_S)_n(i) \times_{W_n(S)} S$. We set

$$(X_S)_n(i_1, \ldots, i_r) = (X_S)_n(i_1) \cap \cdots \cap (X_S)_n(i_r)$$

(and likewise for X_S itself) and

$$(Z_S)_n(i_1, \ldots, i_r) = (Z_S)_n(i_1) \times_{W_n(S)} \cdots \times_{W_n(S)} (Z_S)_n(i_r)\,.$$

We denote by $D_n(i_1, \ldots, i_r)$ the PD-envelope of the canonical morphism

$$(X_S)_n(i_1, \ldots, i_r) \to (Z_S)_n(i_1, \ldots, i_r).$$

One gets simplicial schemes $X_S^\bullet \to (X_S)_n^\bullet \to D_n^\bullet \to (Z_S)_n^\bullet$ and an isomorphism in the derived category of simplicial complexes of sheaves on $X_S^\bullet$

$$\mathcal{F}il^r_{rel/R}\Omega^\bullet_{(X_S)_n^\bullet} \to \mathcal{N}^r_{rel/R}W_n\Omega^\bullet_{X_S^\bullet/S}\,.$$

Let $X_S^\bullet \xrightarrow{\theta} X_S$ be the natural augmentation. By applying $R\theta_*$ we get by cohomological descent in the Zariski topology the desired isomorphism in Theorem 2.2. □

We now prove Theorem 1.2(a). As in Theorem 2.1, the isomorphisms between the complexes in Theorem 2.2 are compatible for varying n. One first assumes the existence of a compatible system of embeddings into Witt lifts; in the general case one uses again simplicial methods as outlined in [Lan18] to obtain an isomorphism of procomplexes $\mathcal{F}il^r_{rel/R}\Omega^\bullet_{(X_S)_\bullet/W_\bullet(S)} \cong \mathcal{N}^r_{rel/R}W_\bullet\Omega^\bullet_{X_S/S}$. Let

$$(\mathcal{P}_{S/R})_r = \mathbb{H}^n(X_S, \mathcal{N}^r_{rel/R}W_\bullet\Omega^\bullet_{X_S/S}) = \mathbb{H}^n(X_S, \mathcal{F}il^r_{rel/R}\Omega^\bullet_{(X_S)_\bullet/W_\bullet(S)})\,.$$

Using the same argument as in the proof of Theorem 1.1(a), we see that the E_1-spectral sequence associated to the complex $\mathcal{F}il^r_{rel/R}\Omega^\bullet_{(X_S)_\bullet/W_\bullet(S)}$ degenerates. This implies a decomposition

$$(\mathcal{P}_{S/R})_r = \mathcal{J}_r L_0 \oplus \mathcal{J}_{r-1} L_1 \oplus \cdots \oplus \mathcal{J} L_{r-1} \oplus L_r \oplus \cdots \oplus L_n$$

where $L_i = H^{n-i}((X_S)_\bullet, \Omega^i_{(X_S)_\bullet/W_\bullet(S)})$ and $\mathcal{J}_i = \tilde{\mathfrak{a}}^i \oplus I_S$ (compare the construction of standard displays over the relative Witt frame $\mathcal{W}_{S/R}$ in [LZ19] and Appendix, Def. A.6).

The maps $\hat{F}_r : (\mathcal{P}_{S/R})_r \to (\mathcal{P}_{S/R})_0 = H^n_{\text{cris}}(X/W(S))$ induce maps $\Phi_r : L_r \to (\mathcal{P}_S)_0$ by $\Phi_r = \hat{F}_r | L_r$. To show that $((\mathcal{P}_{S/R})_r, \hat{F}_r, \hat{\iota}_r, \hat{\alpha}_r)$ defines a relative display on $H^n_{\text{cris}}(X/W(S))$ is equivalent to the condition that

$$\bigoplus_{i=0}^{n} \Phi_i : L_0 \oplus \cdots \oplus L_n \to (\mathcal{P}_S)_0$$

is a σ-linear isomorphism. The argument is the same as in the proof of Theorem 1.1(a). On the level of standard displays (i.e., displays given by standard data), the relative display associated to the display $\mathcal{P}_S$ is given by the inclusions (Appendix, Remark A.8)

$$I_S L_0 \oplus \cdots \oplus I_S L_{r-1} \oplus L_r \oplus \cdots \oplus L_n \hookrightarrow \mathcal{J}_r L_0 \oplus \cdots \oplus \mathcal{J} L_{r-1} \oplus L_r \oplus \cdots \oplus L_n \,.$$

Since the chain of quasi-isomorphisms between $\mathcal{F}^r\Omega^\bullet_{(X_S)_n/W_n(S)}$ and $\mathcal{N}^r W_n\Omega^\bullet_{X_S/S}$ and between $\mathcal{F}il^r_{rel/R}\Omega^\bullet_{(X_S)_n/W_n(S)}$ and $\mathcal{N}^r_{rel/R} W_n\Omega^\bullet_{X_S/S}$ are compatible under the inclusion maps (see the commutative diagram (2.5))

$$\mathcal{F}^r\Omega^\bullet_{(X_S)_n/W_n(S)} \hookrightarrow \mathcal{F}il^r_{rel/R}\Omega^\bullet_{(X_S)_n/W_n(S)}$$

and

$$\mathcal{N}^r W_n\Omega^\bullet_{X_S/S} \hookrightarrow \mathcal{N}^r_{rel/R} W_n\Omega^\bullet_{X_S/S}$$

we conclude that $u_*\mathcal{P}_S = \mathcal{P}_{S/R}$. This proves Theorem 1.2(a).

2.3. Proof of Theorem 1.1(b)

Let $A \to R$ be a frame for R such that the kernel $\mathfrak{a}$ is equipped with divided powers, by definition A is equipped with a lifting $\sigma : A \to A$ of the Frobenius $A/pA \to A/pA$. We consider the Cartier map $A \to W(A)$ into the Witt ring ([Ill79], 0.1.3.16). Then $A \to R$ factors through

$$A \to W(A) \to W(R) \to R \,.$$

The kernel $\mathcal{J}$ of $W(A) \to R$ is $\tilde{\mathfrak{a}} \oplus VW(A) = \tilde{\mathfrak{a}} \oplus I_A$, where $\tilde{\mathfrak{a}}$ is the logarithmic Teichmüller ideal, equipped again with divided powers. We then get a second frame $(W(A), \mathcal{J}, \sigma, \dot{\sigma})$ for R, where σ is the Frobenius on $W(A)$ and $\dot{\sigma} : \mathcal{J} \to W(A)$, $a + V\xi \mapsto \xi$. This is the definition of the relative Witt frame $\mathcal{W}_{A/R}$ (Appendix, Def. A.4).

Assuming the existence of liftings $\mathcal{Y}/\operatorname{Spf} A$ of X that satisfy (A1) and (A2), we get by base change liftings $\tilde{\mathcal{Y}}/\operatorname{Spf} W(A)$ that also satisfy (A1) and (A2). It is therefore enough to show Theorem 1.1(b) by working with the relative Witt frame $\mathcal{W}_{A/R}$ and the lifting $\tilde{\mathcal{Y}}$. Then a window over $W(A)$ ([LZ07], Def. 5.1 and Appendix, Def. A.2) is the same as a display $\mathcal{P}_{A/R}$ over the relative Witt frame $\mathcal{W}_{A/R}$. We denote now by $\mathcal{P}_{A/R}$ the display associated to the lifting $\tilde{\mathcal{Y}}$ that exists by ([LZ07], Thm. 5.5). Let $Y_{m,s} := \tilde{\mathcal{Y}} \times_{W(A)} W_s(A/p^m)$. Then

$$(\mathcal{P}_{A/R})_r = \varprojlim_{s,m} H^n_{\text{cris}}(X, \mathcal{J}^{[r]}_{X/W_s(A/p^m)})$$

is equipped with a divided Frobenius $F_r = \frac{F}{p^r}$ where F is the Frobenius on crystalline cohomology. Assume $A \to R$ factors through $A/p^m \to R$. By ([BO78], Thm. 7.2) the groups

$$H^n_{\text{cris}}(X, \mathcal{J}^{[r]}_{X/W(A/p^m)}) = \varprojlim_{s} H^n_{\text{cris}}(X, \mathcal{J}^{[r]}_{X/W_s(A/p^m)})$$

are the hypercohomology groups of the procomplexes $\mathcal{F}il^{[r]}\Omega^\bullet_{Y_{\bullet,m}/W_\bullet(A/p^m)}$ defined as follows:

$$(\tilde{\mathfrak{a}}^{[r]}_m \oplus p^{r-1}I_{A/p^m})\Omega^0_{Y_{\bullet,m}/W_\bullet(A/p^m)} \xrightarrow{d} (\tilde{\mathfrak{a}}^{[r-1]}_m \oplus p^{r-2}I_{A/p^m})\Omega^1_{Y_{\bullet,m}/W_\bullet(A/p^m)} \xrightarrow{d} \cdots$$
$$\cdots \xrightarrow{d} (\tilde{\mathfrak{a}}_m \oplus I_{A/p^m})\Omega^{r-1}_{Y_{\bullet,m}/W_\bullet(A/p^m)} \xrightarrow{d} \Omega^r_{Y_{\bullet,m}/W_\bullet(A/p^m)} \xrightarrow{d} \cdots$$

where $\tilde{\mathfrak{a}}_m$ is the logarithmic Teichmüller ideal associated to $\mathfrak{a}_m := \ker(A/p^m \to R)$ and $I_{A/p^m} := VW(A/p^m)$, and we have used that $\tilde{\mathfrak{a}}_m \cdot I_{A/p^m} = 0$ and for the ideal $\mathcal{J}_m = \ker(W(A/p^m) \to R)$ we have $\mathcal{J}^{[s]}_m = \tilde{\mathfrak{a}}^{[s]}_m \oplus p^{s-1}I_{A/p^m}$.

As A and $W(A)$ are p-torsion free, multiplication by p on $W(A)$ and the pro-group $W_\bullet(A/p^\bullet)$ is injective, hence the procomplexes $\mathcal{F}il^{[r]}\Omega^\bullet_{Y_{\bullet,\bullet}/W_\bullet(A/p^\bullet)}$ and $\mathcal{F}il^{[r]}_{rel/R}\Omega^\bullet_{Y_{\bullet,\bullet}/W_\bullet(A/p^\bullet)}$, defined as

$$(\tilde{\mathfrak{a}}^{[r]}_\bullet \oplus I_{A/p^\bullet})\Omega^0_{Y_{\bullet,\bullet}/W_\bullet(A/p^\bullet)} \xrightarrow{d\oplus pd} (\tilde{\mathfrak{a}}^{[r-1]}_\bullet \oplus I_{A/p^\bullet})\Omega^1_{Y_{\bullet,\bullet}/W_\bullet(A/p^\bullet)} \xrightarrow{d\oplus pd} \cdots$$
$$\cdots \xrightarrow{d\oplus pd} (\tilde{\mathfrak{a}}_\bullet \oplus I_{A/p^\bullet})\Omega^{r-1}_{Y_{\bullet,\bullet}/W_\bullet(A/p^\bullet)} \xrightarrow{d} \Omega^r_{Y_{\bullet,\bullet}/W_\bullet(A/p^\bullet)} \xrightarrow{d} \cdots$$

are isomorphic. By Theorem 2.2, the procomplexes $\mathcal{F}il^{[r]}_{rel/R}\Omega^\bullet_{Y_{\bullet,\bullet}/W_\bullet(A/p^\bullet)}$ and $\mathcal{N}^r_{rel/R}W_\bullet\Omega^\bullet_{Y_\bullet/(A/p^\bullet)}$, defined as

$$\tilde{\mathfrak{a}}^r_\bullet W_\bullet\mathcal{O}_{Y_\bullet} \oplus (W_\bullet\mathcal{O}_{Y_\bullet})_{[F]} \xrightarrow{d\oplus d} \tilde{\mathfrak{a}}^{r-1}_\bullet W_\bullet\Omega^1_{Y_\bullet/(A/p^\bullet)} \oplus (W_\bullet\Omega^1_{Y_\bullet/(A/p^\bullet)})_{[F]} \xrightarrow{d\oplus d} \cdots$$
$$\cdots \xrightarrow{d\oplus d} \tilde{\mathfrak{a}}_\bullet W_\bullet\Omega^{r-1}_{Y_\bullet/(A/p^\bullet)} \oplus (W_\bullet\Omega^{r-1}_{Y_\bullet/(A/p^\bullet)})_{[F]} \xrightarrow{d+dV} W_\bullet\Omega^r_{Y_\bullet/(A/p^\bullet)} \xrightarrow{d} \cdots$$

are quasi-isomorphic. This implies that

$$(\mathcal{P}_{A/R})_r = \mathbb{H}^n(Y_\bullet, \mathcal{N}^r_{rel/R}W_\bullet\Omega^\bullet_{Y_\bullet/(A/p^\bullet)})$$

and the divided Frobenius on $(\mathcal{P}_{A/R})_r$ is induced by the divided Frobenius on $\mathcal{N}^r_{rel/R}W_\bullet\Omega^\bullet_{Y_\bullet/(A/p^\bullet)}$. It is unique because A and $W(A)$ are p-torsion free.

The morphism of frames $\mathcal{W}_{A/R} \xrightarrow{\epsilon} \mathcal{W}_R$ induces a base change ϵ_* on displays. Let $\tilde{X} := \tilde{\mathcal{Y}} \times_{\text{Spf}\, W(A)} \text{Spec}\, W(R)$ be the induced lifting of X over $W(R)$. The standard display defined on $\tilde{L}_0 \oplus \cdots \oplus \tilde{L}_n$ with

$$\tilde{L}_i = H^{n-i}(\tilde{\mathcal{Y}}, \Omega^i_{\tilde{\mathcal{Y}}/\mathrm{Spf}\, W(A)})$$

is transformed into the standard display on $L_0 \oplus \cdots \oplus L_n$ with

$$L_i = \tilde{L}_i \otimes_{W(A)} W(R) = H^{n-i}(\tilde{X}, \Omega^i_{\tilde{X}/W(R)}) .$$

Note that under the composite map $\kappa : A \to W(A) \to W(R)$, $\kappa(a) \in I_R$ for $a \in \mathfrak{a}$, hence the image of $\tilde{\mathfrak{a}} \oplus I_A$ in $W(R)$ is I_R.

We have canonical reduction maps

$$\mathcal{F}il^{[r]}_{rel/R}\Omega^\bullet_{Y_{\bullet,\bullet}/W_\bullet(A/p^\bullet)} \to \mathcal{F}^r\Omega^\bullet_{\tilde{X}/W_\bullet(R)}$$

and

$$\mathcal{N}^r_{rel/R}W_\bullet\Omega^\bullet_{Y_\bullet/(A/p^\bullet)} \to \mathcal{N}^r W_\bullet\Omega^\bullet_{X/R} .$$

Under the base change of displays $\epsilon_*\mathcal{P}_{A/R}$ is a display over R with $(\epsilon_*\mathcal{P}_{A/R})_r$ given by the hypercohomology of $\mathcal{F}^r\Omega^\bullet_{\tilde{X}/W(R)}$. Since the quasi-isomorphisms between $\mathcal{F}il^{[r]}_{rel/R}\Omega^\bullet_{Y_{\bullet,\bullet}/W_\bullet(A/p^\bullet)}$ and $\mathcal{N}^r_{rel/R}W_\bullet\Omega^\bullet_{Y_\bullet/(A/p^\bullet)}$ and between $\mathcal{F}^r\Omega^\bullet_{\tilde{X}/W(R)}$ and $\mathcal{N}^r W_\bullet\Omega^\bullet_{X/R}$ are compatible under the canonical reduction maps, we see that the divided Frobenius F_r on $(\epsilon_*\mathcal{P}_{A/R})_r$ obtained by base change coincides with the divided Frobenius on the Nygaard complexes $\mathcal{N}^r W_\bullet\Omega^\bullet_{X/R}$. This finishes the proof of Theorem 1.1(b).

2.4. Proof of Theorem 1.2(b)

As in the theorem, we assume that R is an artinian local $W(k)$-algebra with residue field k, and that the special fibre X_0 is a smooth projective variety with smooth versal deformation space $\mathfrak{S}$. Write $\mathfrak{X}/\mathfrak{S}$ for the versal family. Then $\mathfrak{S} \cong \mathrm{Spf}\, A$, where $A = W(k)[\![t_1, \ldots, t_h]\!]$ is a formal power series algebra over $W(k)$.

Suppose now that X_S is a deformation of X/R over a PD-thickening $S \twoheadrightarrow R$ and let us write $\mathcal{P}_S(X_S)$ for the $\mathcal{W}_S$-display structure on $H^n_{\mathrm{cris}}(X/W(S))$. Write $u : \mathcal{W}_S \to \mathcal{W}_{S/R}$ for the frame homomorphism. Then we must prove that the relative display $\mathcal{P}_{S/R} = u_*\mathcal{P}_S$ does not depend on the lifting X_S. That is to say, given another deformation X'_S of X over S with associated display $\mathcal{P}_S(X'_S)$, the relative displays $\mathcal{P}_{S/R}(X_S) := u_*\mathcal{P}_S(X_S)$ and $\mathcal{P}_{S/R}(X'_S) := u_*\mathcal{P}_S(X'_S)$ coincide.

By the versality of $\mathfrak{S}$, the deformations X_S and X'_S are induced by two $W(k)$-algebra homomorphisms $A \underset{y}{\overset{x}{\rightrightarrows}} S$. Let $\mathcal{A}_{\mathrm{triv}} = (A, 0, A, \sigma, \sigma/p)$ be the trivial frame for A and write $\mathcal{P}^{\mathrm{triv}}_A$ for the $\mathcal{A}_{\mathrm{triv}}$-window structure on the versal family (given by [LZ07] Thm 5.5). Then $\mathcal{P}_{S/R}(X_S)$ and $\mathcal{P}_{S/R}(X'_S)$ are the base change of $\mathcal{P}^{\mathrm{triv}}_A$ along the two induced frame homomorphisms

$$\mathcal{A}_{\mathrm{triv}} \underset{y}{\overset{x}{\rightrightarrows}} \mathcal{W}_{S/R}$$

that arise from the commutative diagram

$$\begin{array}{ccc} A & \overset{x}{\underset{y}{\rightrightarrows}} & S \\ {\scriptstyle =}\downarrow & & \downarrow \\ A & \longrightarrow & R\,. \end{array}$$

That is $\mathcal{P}_{S/R}(X_S) = x_* \mathcal{P}_A^{\text{triv}}$ and $\mathcal{P}_{S/R}(X'_S) = y_* \mathcal{P}_A^{\text{triv}}$.

Now consider the following diagram

$$\begin{array}{ccccccccc} 0 & \longrightarrow & J & \longrightarrow & B := A\hat{\otimes}_{W(k)}A & \xrightarrow{\text{mult.}} & A & \longrightarrow & 0 \\ & & & & \downarrow & & \downarrow & & \\ & & & & S & \twoheadrightarrow & R\,. & & \end{array}$$

Write $D_B(J)$ for the PD-envelope of (B, J). Similarly, set $A_0 := W(k)[T_1, \dots, T_h]$, $B_0 := A_0 \otimes_{W(k)} A_0$, $J_0 := \ker(B_0 \xrightarrow{\text{mult.}} A_0)$ and write $D_{B_0}(J_0)$ for the PD-envelope of (B_0, J_0). Then $D_{B_0}(J_0)$ is the PD-polynomial algebra over B_0 in h variables. Since $A = W(k)[\![t_1, \dots, t_h]\!]$ is flat over $A_0 = W(k)[t_1, \dots, t_h]$, [BO78] Proposition 3.21 gives that $D_B(J)$ is the PD-polynomial algebra over B in h variables. In particular, $D_B(J)$ is a flat $D_{B_0}(J_0)$-module, so is certainly p-torsion free. We get a diagram

$$\begin{array}{ccc} D_B(J) & \longrightarrow & S \\ \downarrow & & \downarrow \\ A & \longrightarrow & R\,. \end{array}$$

Let $\widehat{D_B(J)} := \varprojlim_n D_B(J)/p^n$ denote the p-adic completion of $D_B(J)$. Then $\mathcal{A} = \left(\widehat{D_B(J)} \to A\right)$ is a frame for A. The sections $A \rightrightarrows A \otimes_{W(k)} A$ induce frame morphisms

$$\mathcal{A}_{\text{triv}} \rightrightarrows \mathcal{A} \to \mathcal{W}_{S/R}$$

given by the following diagram

$$\begin{array}{ccccc} A & \rightrightarrows & \widehat{D_B(J)} & \longrightarrow & S \\ \downarrow & & \downarrow & & \downarrow \\ A & \longrightarrow & A & \longrightarrow & R\,. \end{array}$$

Since the geometric construction of windows on $H^n_{\text{cris}}(\mathfrak{X}/A)$ ([LZ07], Thm. 5.5) is compatible with base change ([LZ07], Cor. 5.6), the base change of $\mathcal{P}_A^{\text{triv}}$ along both frame morphisms $\mathcal{A}_{\text{triv}} \rightrightarrows \mathcal{A}$ gives the same $\mathcal{A}$-window; it is the $\mathcal{A}$-window $\mathcal{P}_{\mathcal{A}}$ given by applying ([LZ07], Thm. 5.5) to the frame $\mathcal{A}$. We may now conclude the proof since the $\mathcal{W}_{S/R}$-displays $\mathcal{P}_{S/R}(X_S)$ and $\mathcal{P}_{S/R}(X'_S)$ are both given by the base change of $\mathcal{P}_{\mathcal{A}}$ along $\mathcal{A} \to \mathcal{W}_{S/R}$.

Appendix

In this appendix we recall the basic definitions of frames, windows and displays as given in [LZ07] and [LZ19].

Definition A.1. Let R be a ring such that p is topologically nilpotent in R. A frame (A, σ, α) for R consists of a torsion-free p-adic ring A with an endomorphism $\sigma : A \to A$ lifting the Frobenius on A/p and a surjective homomorphism $\alpha : A \to R$ such that the kernel $\mathfrak{a} = \ker \alpha$ has divided powers.

Definition A.2. Let $\mathcal{A} = (A, \sigma, \alpha)$ be a frame for R. An $\mathcal{A}$-window consists of

1. a finitely generated projective A-module P_0
2. a descending filtration of P_0 by A-submodules

$$P_{i+1} \subset P_i \subset \cdots \subset P_1 \subset P_0$$

3. σ-linear homomorphisms

$$F_i : P_i \to P_0$$

such that the following conditions are satisfied

1. $\mathfrak{a}P_i \subset P_{i+1}$; $P_{i+1}/\mathfrak{a}P_i$ is a finitely generated projective R-module E_{i+1} for $i \geq 0$. Let $E_0 = P_0/\mathfrak{a}P_0$.
2. The inclusions in 2) induce injective R-module homomorphisms

$$E_{i+1} \to E_i \to \cdots \to E_0$$

3. $\mathfrak{a}P_i = P_{i+1}$ for i large enough.
4. $F_i(x) = pF_{i+1}(x)$ for $x \in P_{i+1}$.
5. The union of the images $F_i(P_i)$ for $i \in \mathbb{Z}_{\geq 0}$ generate P_0 as an A-module.

It is then shown in [LZ07] page 181 that a window is isomorphic to a standard window, that is there are finitely generated projective A-modules $L_0, \ldots, L_d$ with $\bigoplus_{i=0}^{d} L_i = P_0$ and σ-linear homomorphisms $\Phi_i : L_i \to \bigoplus_{j=0}^{d} L_j$ such that the determinant of $\Phi_0 \oplus \cdots \oplus \Phi_d$ is a unit. Attached to this data we set for $i \geq 0$

$$P_i = \mathfrak{a}^i L_0 \oplus \mathfrak{a}^{i-1} L_1 \oplus \cdots \oplus \mathfrak{a} L_{i-1} \oplus L_i \oplus \cdots \oplus L_d$$

and define F_i on P_i, that is $F_i|_{\mathfrak{a}^{i-k} L_k}$ for $k < i$ resp. $F_i|_{L_k}$ for $k \geq i$, as follows:

$$F_i(ax) = \frac{\sigma(a)}{p^{i-k}} \Phi_k(x) \text{ for } 0 \leq k < i,\ x \in L_k,\ a \in \mathfrak{a}^{i-k}$$
$$F_i(x) = p^{k-i} \Phi_k(x) \text{ for } i \leq k,\ x \in L_k.$$

Then these data (P_i, F_i) and the obvious inclusions $P_{i+1} \to P_i$ form a window called a standard window.

To define higher displays we will use frames over the ring of Witt vectors for a given p-adic ring R.

Definition A.3. Let S be a p-adic ring and $W(S)$ its Witt vectors. The Witt frame $\mathcal{W}_S = (W(S), \mathcal{J} = I_S, S, \sigma, \dot{\sigma})$ consists of the data $I_S = VW(S)$, $W(S) \to S$ the augmentation map with kernel I_S, σ the Frobenius on $W(S)$ and $\dot{\sigma} : I_S \to W(S)$, $V\xi \mapsto \xi$.

Definition A.4. Let $S \to R$ be a surjective homomorphism of p-adic rings such that $\mathfrak{a}$ becomes nilpotent in S/pS. Then the relative Witt frame

$$\mathcal{W}_{S/R} = (W(S), \mathcal{J}, R, \sigma, \dot{\sigma})$$

consists of the ideal $\mathcal{J} = \ker(W(S) \to R)$ which is a direct sum $\mathcal{J} = \tilde{\mathfrak{a}} \oplus I_S$ where $I_S = VW(S)$ is the augmentation ideal in $W(S)$ and $\tilde{\mathfrak{a}} \subset W(S)$ is the ideal consisting of logarithmic Teichmüller representatives of elements of $\mathfrak{a}$ ([Zin02], 1.4). We recall the definition:

The divided powers on $\mathfrak{a}$ yield divided Witt polynomials

$$w'_n(\underline{a}) = \sum_{i=0}^{n} p^i a_i^{p^n - i} = \sum_{i=0}^{n} (p^{n-i} - 1)! \gamma_{p^{n-i}}(a_i)$$

for $\underline{a} = (a_0, a_1, \ldots) \in W(\mathfrak{a})$, and an isomorphism

$$\log : W(\mathfrak{a}) \xrightarrow{\simeq} \mathfrak{a}^{\mathbb{N}}$$
$$\underline{a} \mapsto (w'_0(\underline{a}), \ldots, w'_n(\underline{a}), \ldots).$$

Define $\tilde{\mathfrak{a}} = \log^{-1}(\mathfrak{a}, 0, 0, \ldots)$. This is an ideal in $W(S)$.

The map σ is the Frobenius on $W(S)$. We have $\sigma(\tilde{\mathfrak{a}}) = 0$, $I_S \cdot \tilde{\mathfrak{a}} = 0$ and define $\dot{\sigma} : \mathcal{J} \to W(S)$ by $\dot{\sigma}(a + V\xi) = \xi$ for $a \in \tilde{\mathfrak{a}}$, $\xi \in W(S)$.

Then both $\mathcal{W}_S$ and $\mathcal{W}_{S/R}$ are equipped with maps called "Verjüngung". In the case of $\mathcal{W}_S$ these are maps $\nu : I_S \otimes I_S \to I_S$, $V\xi_1 \otimes V\xi_2 \mapsto V(\xi_1\xi_2)$ with iterations $\nu^{(k)} : I_S^{\otimes k} \to I_S$, $V\xi_1 \otimes \cdots \otimes V\xi_k \mapsto V(\xi_1 \cdots \xi_k)$, and $\pi : I_S \to I_S$, $V\xi \mapsto pV\xi$. In the case of $\mathcal{W}_{S/R}$, the Verjüngung consists of the two maps $\nu : \mathcal{J} \otimes_{W(S)} \mathcal{J} \to \mathcal{J}$ and $\pi : \mathcal{J} \to \mathcal{J}$ with

$$\nu((a_1 + V\xi_1) \otimes (a_2 + V\xi_2)) = a_1 \cdot a_2 + V(\xi_1 \cdot \xi_2)$$

and

$$\pi(a + V\xi) = a + pV\xi$$

with obvious iterations $\nu^{(k)}$ that satisfy the properties (3) and (4) on page 460 of [LZ19].

In the following let $\mathcal{F}$ be one of the frames considered above, that is $\mathcal{F} = \mathcal{W}_S$ or $\mathcal{F} = \mathcal{W}_{S/R}$.

Definition A.5. An $\mathcal{F}$-predisplay $(P_i, \iota_i, \alpha_i, F_i)$ consists of the following data:

1. A sequence of $W(S)$-modules P_i for $i \geq 0$.
2. Two sets of $W(S)$-module homomorphisms

$$\iota_i : P_{i+1} \to P_i \ , \quad \alpha_i : \mathcal{J} \otimes_{W(S)} P_i \to P_{i+1}$$

   for $i \geq 0$.

3. A set of σ-linear maps for $i \geq 0$

$$F_i : P_i \to P_0$$

which satisfy the following properties:

1. Consider the following morphisms:

$$\begin{array}{ccc} \mathcal{J} \otimes P_i & \xrightarrow{\alpha_i} & P_{i+1} \\ {\scriptstyle \mathrm{id}_{\mathcal{J}} \otimes \iota_{i-1}} \downarrow & & \downarrow {\scriptstyle \iota_i} \\ \mathcal{J} \otimes P_{i-1} & \xrightarrow[\alpha_{i-1}]{} & P_i \end{array}$$

the compositions $\iota_i \circ \alpha_i$ and $\alpha_{i-1} \circ (\mathrm{id}_{\mathcal{J}} \otimes \iota_{i-1})$ are the multiplication maps $\mathcal{J} \otimes P_i \to P_i$ for all i.

2. $F_{i+1} \otimes \alpha_i = \tilde{F}_i$.

Here, for each σ-linear map $f : M \to N$ between $W(S)$-modules M, N, we define a new σ-linear map $\tilde{f} : \mathcal{J} \otimes M \to N$ by $\tilde{f}(\eta \otimes m) = \dot{\sigma}(\eta) f(m)$ for $\eta \in \mathcal{J}$.

In the following we define standard data of a display over a Witt frame. In the case of $\mathcal{W}_S$ the definition is given in [LZ07] Definition 2.5. We only recall the definition of standard data of a display over the relative Witt frame $\mathcal{W}_{S/R}$ given in [LZ19] pp. 460–461.

Definition A.6. A display given by standard data over the relative Witt frame $\mathcal{W}_{S/R}$ is given as follows:

It consists of finitely generated projective $W(S)$-modules $L_0, \ldots, L_d$ and σ-linear homomorphisms

$$\Phi_i : L_i \to L_0 \oplus \cdots \oplus L_d$$

such that $\Phi_0 \oplus \cdots \oplus \Phi_d : L_0 \oplus \cdots \oplus L_d \to L_0 \oplus \cdots \oplus L_d$ is a σ-linear isomorphism. Define $\mathcal{J}_i = \tilde{\mathfrak{a}}^i \oplus VW(S)$. Set

$$P_i = \mathcal{J}_i L_0 \oplus \mathcal{J}_{i-1} L_1 \oplus \cdots \oplus \mathcal{J} L_{i-1} \oplus L_i \oplus \cdots \oplus L_d\,.$$

The map ι_i is defined by the following diagram

$$\begin{array}{ccccccccccc} \mathcal{J}_{i+1} L_0 & \oplus & \mathcal{J}_i L_1 & \oplus \cdots \oplus & \mathcal{J} L_i & \oplus & L_{i+1} & \oplus \cdots \oplus & L_d \\ \downarrow {\scriptstyle \pi} & & \downarrow {\scriptstyle \pi} & & \downarrow {\scriptstyle \mathrm{id}} & & \downarrow {\scriptstyle \mathrm{id}} & & \downarrow {\scriptstyle \mathrm{id}} \\ \mathcal{J}_i L_0 & \oplus & \mathcal{J}_{i-1} L_1 & \oplus \cdots \oplus & L_i & \oplus & L_{i+1} & \oplus \cdots \oplus & L_d\,. \end{array}$$

The homomorphisms $\alpha_i : \mathcal{J} \otimes P_i \to P_{i+1}$ are defined as follows

$$\begin{array}{ccccccccccc} \mathcal{J} \otimes \mathcal{J}_i L_0 & \oplus & \mathcal{J} \otimes \mathcal{J}_{i-1} L_1 & \oplus \cdots \oplus & \mathcal{J} \otimes L_i & \oplus & \mathcal{J} \otimes L_{i+1} & \oplus \cdots \oplus & \mathcal{J} \otimes L_d \\ \downarrow {\scriptstyle \nu} & & \downarrow {\scriptstyle \nu} & & \downarrow {\scriptstyle \mathrm{mult}} & & \downarrow {\scriptstyle \mathrm{mult}} & & \downarrow {\scriptstyle \mathrm{mult}} \\ \mathcal{J}_{i+1} L_0 & \oplus & \mathcal{J}_i L_1 & \oplus \cdots \oplus & L_i & \oplus & L_{i+1} & \oplus \cdots \oplus & L_d\,. \end{array}$$

Finally we define σ-linear maps $F_i : P_i \to P_0$ as

$$\begin{array}{ccccccccccc} \mathcal{J}_i L_0 & \oplus \cdots \oplus & \mathcal{J} L_{i-1} & \oplus & L_i & \oplus & L_{i+1} & \oplus & L_{i+2} & \oplus \cdots \\ \downarrow{\scriptstyle \tilde{\Phi}_0} & & \downarrow{\scriptstyle \tilde{\Phi}_{i-1}} & & \downarrow{\scriptstyle \Phi_i} & & \downarrow{\scriptstyle p\Phi_{i+1}} & & \downarrow{\scriptstyle p^2\Phi_{i+2}} & \\ L_0 & \oplus \cdots \oplus & L_{i-1} & \oplus & L_i & \oplus & L_{i+1} & \oplus & L_{i+2} & \oplus \cdots \end{array}$$

where $\tilde{\Phi}_j$ is defined by $\tilde{\Phi}_j(\eta l_j) = \dot{\sigma}(\eta)\Phi_j(l_j)$ for $\eta \in \mathcal{J}_j$, $l_j \in L_j$, $j < i$.

These data meet the requirements of a predisplay.

Definition A.7. Let $\mathcal{F}$ be either of the Witt frames considered above. Then an $\mathcal{F}$-display is an $\mathcal{F}$-predisplay which is isomorphic to the display associated to standard data. The choice of such an isomorphism is called a normal decomposition.

For $S \to R$ a surjective PD-morphism one has the following morphisms of frames equipped with Verjüngung

$$\mathcal{W}_S \xrightarrow{\epsilon} \mathcal{W}_{S/R} \to \mathcal{W}_R \,.$$

Remark A.8. For a morphism of frames with Verjüngung $u : \mathcal{F} \to \mathcal{F}'$, we have a base change map of displays, that is the $\mathcal{F}'$-display $u_*\mathcal{P}$ obtained by base change of an $\mathcal{F}$-display $\mathcal{P}$ exists ([LZ19], Prop. 6).

For any $\mathcal{W}_S$-display $\mathcal{P}_S$ we can associate the base change $\mathcal{P}_{S/R} = \epsilon_*\mathcal{P}_S$, we also call the relative display for the morphism $S \to R$ associated to $\mathcal{P}_S$. It is clear from the definitions that if $\mathcal{P}_S$ has a normal decomposition with finitely generated projective $W(S)$-modules L_i, $i = 0, \ldots, d$, then using the same finitely generated projective modules L_i for the relative display $\mathcal{P}_{S/R}$, the inclusion map $I_S = VW(S) \to \tilde{\mathfrak{a}}^i + I_S = \mathcal{J}_i$ (for $\mathfrak{a} = \ker(S \to R)$) and the obvious extensions of ι_i, α_i, F_i to the P_i in Definition A.6 built from the L_i defines the standard data for $u_*\mathcal{P}_S =: \mathcal{P}_{S/R}$.

Remark A.9. If (P_i) is an $\mathcal{A}$-window for a frame $\mathcal{A} = (A \to R)$ as in definitions $A.1$–$A.2$, then there is an induced $\mathcal{W}_R$-display given by the composite ring homomorphism $\kappa : A \to W(A) \to W(R)$ that satisfies

$$\kappa(\sigma(a)) = F\kappa(a) \qquad ; a \in A$$
$$\kappa\left(\frac{\sigma(a)}{p}\right) = V^{-1}\kappa(a) \quad ; a \in \mathfrak{a}\,.$$

It is described explicitly for standard data associated to the window and also gives an invariant construction in [LZ07] page 182.

Remark A.10. For $S \to R$ as above and the induced morphism of frames $\mathcal{W}_S \to \mathcal{W}_R$, the base change from $\mathcal{W}_S$-displays to $\mathcal{W}_R$-displays coincides with the one given in [LZ07] Prop. 2.12.

References

[BO78] P. Berthelot, A. Ogus, *Notes on crystalline cohomology*, Princeton University Press (1978)

[Fon83] J.-M. Fontaine, *Cohomologie de de Rham, cohomologie cristalline et représentations p-adiques*, Lecture Notes in Math. **1016**, Algebraic Geometry (Tokyo/Kyoto, 1982), 86–108, Springer (1983)

[Gre17] O. Gregory, *Crystals of relative displays and Grothendieck–Messing deformation theory*, University of Exeter PhD thesis (2017)

[Ill79] L. Illusie, *Complexe de de Rham–Witt et cohomologie cristalline*, Ann. Sci. École Norm. Sup. (4) **12**, 501–661 (1979)

[Kat87] K. Kato, *On p-adic vanishing cycles (Applications of ideas of Fontaine–Messing)*, Advanced Studies in Pure Math. **10**, 207–251 (1987)

[Lan18] A. Langer, *p-adic Deformation of motivic Chow groups*, Doc. Math. **23**, 1863–1894 (2018)

[LZ04] A. Langer and T. Zink, *De Rham–Witt cohomology for a proper and smooth morphism*, J. Inst. Math. Jussieu 3, 231–314 (2004)

[LZ07] A. Langer and T. Zink, *De Rham–Witt cohomology and displays*, Documenta Mathematica **12**, 147–191 (2007)

[LZ19] A. Langer, T. Zink, *Grothendieck–Messing deformation theory for varieties of K3-type*, Tunisian Journal of Mathematics, Vol. 1, No. 4, 455–517 (2019)

[Lau14] E. Lau, *Relations between Dieudonné displays and crystalline Dieudonné theory*, Algebra & Number Theory **8**, 2201–2262 (2014)

[Mes07] W. Messing, *Travaux de Zink*, Séminaire Bourbaki 2005/2006, exp. 964, Astérisque 311, 341–364 (2007)

[Zin01] T. Zink, *A Dieudonné theory for p-divisible groups*, Class field theory – its centenary and prospect, 139–160, Adv. Stud. Pure Math. 30, Math. Soc. Japan (2001)

[Zin02] T. Zink, *The display of a formal p-divisible group*, Astérisque **278**, 127–248 (2002)

Oli Gregory and Andreas Langer
University of Exeter Mathematics
Exeter EX4 4QF
Devon, UK
e-mail: gregory@exeter.ac.uk
a.langer@exeter.ac.uk

Progress in Mathematics, Vol. 338, 141–177

Orbifold Submersion and Analytic Torsions

Xiaonan Ma

Abstract. In this paper, we establish the curvature theorem of determinant line bundles for an orbifold Kähler fibration as an extension of Bismut–Gillet–Soulé's curvature theorem. Then we introduce Bismut–Köhler analytic torsion form for an orbifold Kähler fibration. Finally we calculate the behaviour of the Quillen metric by orbifold submersions as an extension of Berthomieu–Bismut's result.

Mathematics Subject Classification (2010). 58J20, 32L10.

Keywords. Analytic torsion, Orbifold.

0. Introduction

Let ξ be a Hermitian vector bundle on a compact Hermitian complex manifold X. Let $\lambda(\xi)$ be the inverse of the determinant of the cohomology of ξ. Quillen defined first a metric on $\lambda(\xi)$ in the case that X is a Riemann surface. Quillen metric is the product of the L^2 metric on $\lambda(\xi)$ by the Ray–Singer analytic torsion of the Dolbeault complex. The logarithm of the Ray–Singer analytic torsion [39] is a linear combination of derivatives at zero of the zeta function of the Hodge Laplacians acting on smooth forms of various degrees. In [12], Bismut, Gillet, and Soulé have established a general theory on Quillen metric for any dimensional compact Kähler manifolds, in particular their anomaly formulas for Quillen metrics computes the variation of Quillen metric on the metrics on ξ and TX by using some Bott–Chern classes; for a holomorphic submersion, they proved their determinant line bundle from spectral theory has canonically a holomorphic structure, and is isomorphic canonically to the Knudsen–Mumford line bundle from sheaf theory, as holomorphic line bundles. They have shown that the Quillen metric is a smooth metric on the determinant line bundle $\lambda(\xi)$ of the cohomology groups of the fibers, even both L^2-metric and the analytic torsion could be discontinuous, their curvature formula calculates the curvature of $\lambda(\xi)$ with Quillen metric which refines the degree two part of the Riemann–Roch–Grothendieck theorem at the differential form level.

Later, Bismut and Köhler [13] (refer also [11], [22] in the special case) have extended the analytic torsion of Ray–Singer to the analytic torsion forms T for a holomorphic submersion. In particular, the equation on $\frac{\overline{\partial}\partial}{2i\pi}T$ gives a refinement of the Riemann–Roch–Grothendieck theorem at the level of differential forms. They have also established the corresponding anomaly formulas.

In [22], Gillet and Soulé had conjectured an arithmetic Riemann–Roch theorem in Arakelov geometry. The analytic torsion form is contained in their definition of direct image. In [23], they have established it for the first arithmetic Chern class and Bismut–Lebeau's embedding formula [15] for Quillen metric plays an important role in their proof. In [24], they have established the high degree version by using Bismut's embedding formula [6] for torsion forms. For the various equivariant extensions cf. [29], [5], [16], and the recent works [17, 18].

Note also that for a submersion $\pi : M \to B$ of compact Kähler manifolds and a holomorphic vector bundle ξ on M, by [28], there exists a canonical isomorphism σ from $\lambda_M(\xi)$, the determinant of the cohomology of ξ over M, to $\lambda(R^\bullet\pi_*\xi)$, here $R^\bullet\pi_*\xi$ is the direct image of ξ. In [2], Berthomieu and Bismut have obtained a formula for the Quillen norm of σ in terms of Bott–Chern classes on M and the analytic torsion forms of the fibration π. In our thesis [31, 32], we establish the family version of [2].

In [34], we define the analytic torsion for orbifolds and established the corresponding anomaly formula and embedding formula. This paper is a continuation of [34]. For an orbifold submersion, we will study the curvature formula for the Quillen metric and define the analytic torsion form, then extend Berthomieu–Bismut's result [2] for an orbifold submersion.

An complex orbifold can be always represented locally by $\mathbb{C}^n/G$ where the finite group G acts $\mathbb{C}$-linearly on $\mathbb{C}^n$. The simplest complex orbifold is a global orbifold M/G where G is a finite group acting holomorphically on a complex manifold M.

We will use the heat kernel method to solve our problem. Thanks to finite propagation speed of the solution of the hyperbolic equation [20], [35, Appendix D], we can use the local family index theory of Bismut [3]. Since, locally, we have to meet G-manifold, to generalize the results to the orbifold case, we must understand very well the situation of G-equivariant complex manifolds. After localized, we will apply the results of [5] and [33] to our situation.

Orbifold appears naturally in many important cases, for example: the symplectic reduction, the problem on moduli spaces. In [27], Kawasaki has extended the Riemann–Roch–Hirzebruch theorem to the orbifold case. Bismut and Labourie [14] also proved the Verlinde formula by using Kawasaki's theorem.

For applications of the analytic torsion in Arakelov geometry, cf. the book [42], in particular the recent works [17], [36], [37]. We also hope our results have corresponding versions in Arakelov geometry. For applications of analytic torsion on the moduli space of $K3$ surfaces, cf. Yoshikawa's works [43], [44], in particular, in [45], for general abelian Calabi–Yau orbifolds of dimension three, BCOV invariant was defined and the curvature theorem was proved for global orbifolds there.

Let us explain the contain of this paper in detail now. For a complex vector space F, we denote $\det F = \Lambda^{\max} F$ and denote by $(\det F)^{-1} := \det F^*$ its dual line.

Let ξ be a holomorphic orbifold vector bundle on an n-dimensional complex orbifold X. Let $H^{\bullet}(X,\xi)$ be the cohomology of sheaf of holomorphic sections of ξ over X.

The determinant of the cohomology of ξ over X is defined as

$$\lambda(\xi) := (\det H^{\bullet}(X,\xi))^{-1} = \otimes_{j=0}^{n}(\det H^{j}(X,\xi))^{(-1)^{j+1}}. \tag{0.1}$$

Let ΣX be the strata of X which has a natural orbifold structure. Let m_i be the multiplicity of the connected component X_i of $X \cup \Sigma X$ (cf. (1.2)). For α a differential form on $X \cup \Sigma X$, we denote simply

$$\int_{X\cup\Sigma X} \alpha = \sum_i \frac{1}{m_i}\int_{X_i} \alpha. \tag{0.2}$$

Let h^{TX}, h^{ξ} be Hermitian metrics on TX, ξ. Then as in the smooth case, in [34], we defined the analytic torsion and the Quillen metric on the complex line $\lambda(\xi)$ (cf. (3.4)) and established the anomaly formula in [34, Theorem 4.2], and the local term are certain integral of differential forms on $X \cup \Sigma X$, not on X. For example, $\mathrm{Td}^{\Sigma}(TX, h^{TX})$ is the Todd form on $X \cup \Sigma X$ associated with the holomorphic Hermitian connection on (TX, h^{TX}), which appears in Kawasaki's formulas [27]. Other Chern–Weil forms will be denoted in a similar way. In particular, the form $\mathrm{ch}^{\Sigma}(\xi, h^{\xi})$ (cf. (2.8)) on $X \cup \Sigma X$ is the Chern–Weil representative of the Chern character of $(\xi^{\mathrm{pr}}, h^{\xi})$, with ξ^{pr} the maximal proper orbifold subbundle of ξ.

As the space of $\mathscr{C}^{\infty}$ sections of an orbifold vector bundle is identified as the space of $\mathscr{C}^{\infty}$ sections of its maximal proper orbifold subbundle. In the whole paper, we can assume that ξ is a proper orbifold vector bundle.

Let $\pi : M \to B$ be a proper orbifold submersion of complex orbifolds. Then by Proposition 1.4, locally π is a quotient of a fibration with fiber of a compact orbifold X, by a finite group.

We assume that π is a Kähler fibration in the sense of Bismut–Gillet–Soulé, i.e., there is a smooth closed real $(1,1)$-form on M such that it induces a Kähler form along the fiber, cf. Definition 1.7. Let ξ be a holomorphic orbifold vector bundle on M. Let h^{ξ} be a Hermitian metric on ξ.

When the base B is a complex manifold, then the direct image $R^{\bullet}\pi_{*}\xi$ is well defined as an element in K-group of B. In this case, we establish in Theorem 2.3 the family local index theorem as an extension of Bismut's family local index theorem.

When B is a complex orbifold, as one of our main results, in Section 3.3, we define the determinant line bundle as a proper orbifold holomorphic line bundle on B by using the spectral analysis, also Knudsen–Mumford orbifold line bundle from sheaf theory, then Theorem 3.5 as an extension of [12, Theorem 3.14], shows the canonical isomorphism of these orbifold line bundles is holomorphic. In Theorem 3.6, we compute the curvature of the associated Chern connection as a consequence of the family local index theorem. Thus we extend Bismut–Gillet–Soulé's classical curvature theorem [12, Theorem 0.3] to the orbifold case.

We assume now that the direct image $R^k\pi_*\xi(0 \le k \le \dim X)$ are orbifold vector bundle on B. Then in Section 4, we introduce the analytic torsion form which is a differential form on $B \cup \Sigma B$, and we establish its anomaly formula.

Now we assume further that M, B are compact Kähler orbifolds. Let σ be the canonical section of $\lambda_M(\xi) \otimes \lambda^{-1}(R^\bullet\pi_*\xi)$.

Let h^{TM}, h^{TB} be Kähler metrics on TM and TB. Let h^{TX} be the metric on TX induced by h^{TM}. Let ω^M be the Kähler form of h^{TM}.

Let $H^\bullet(X, \xi|_X)$ be the cohomology of $\xi|_X$. Let $h^{H(X,\xi|_X)}$ be the L^2-metric on $H^\bullet(X, \xi|_X)$ constructed in Section 4 associated to h^{TX}, h^ξ. Let $T(\omega^M, h^\xi)$ be the analytic torsion forms on $B \cup \Sigma B$ constructed in Section 4, which extend the analytic torsion forms of Bismut–Köhler to the orbifold case. Let $\widetilde{\mathrm{Td}}^\Sigma(TM, TB, h^{TM}, h^{TB})$ be the Bott–Chern class on $M \cup \Sigma M$ constructed as in [10] such that

$$\frac{\overline{\partial}\partial}{2i\pi}\widetilde{\mathrm{Td}}^\Sigma(TM, TB, h^{TM}, h^{TB}) = \mathrm{Td}^\Sigma(TM, h^{TM}) - \pi^*(\mathrm{Td}^\Sigma(TB, h^{TB}))\,\mathrm{Td}^\Sigma(TX, h^{TX}). \quad (0.3)$$

Let $|| \quad ||_{\lambda_M(\xi)\otimes\lambda^{-1}(R^\bullet\pi_*\xi)}$ be the Quillen metric on the complex line $\lambda_M(\xi) \otimes \lambda^{-1}(R^\bullet\pi_*\xi)$ attached to the metrics $h^{TM}, h^\xi, h^{TB}, h^{H(X,\xi|_X)}$ on $TM, \xi, TB, R^\bullet\pi_*\xi$. The last purpose of this paper is to calculate the Quillen metric

$$||\sigma||_{\lambda_M(\xi)\otimes\lambda^{-1}(R^\bullet\pi_*\xi)}$$

as an extension of [2, Theorem 3.1]

Theorem 5.1. *The following identity holds,*

$$\log(||\sigma||^2_{\lambda_M(\xi)\otimes\lambda^{-1}(R^\bullet\pi_*\xi)}) = -\int_{B\cup\Sigma B} \mathrm{Td}^\Sigma(TB, h^{TB})T(\omega^M, h^\xi) + \int_{M\cup\Sigma M} \widetilde{\mathrm{Td}}^\Sigma(TM, TB, h^{TM}, h^{TB})\,\mathrm{ch}^\Sigma(\xi, h^\xi). \quad (0.4)$$

Let $m_{i,B}, m_{i,M}$ be the multiplicities of the connected components B_i, M_i of $B \cup \Sigma B, M \cup \Sigma M$. Then we can reformulate (0.4) as

$$\log(||\sigma||^2_{\lambda_M(\xi)\otimes\lambda^{-1}(R^\bullet\pi_*\xi)}) = -\sum_i \frac{1}{m_{i,B}}\int_{B_i} \mathrm{Td}^\Sigma(TB, h^{TB})T(\omega^M, h^\xi) + \sum_i \frac{1}{m_{i,M}}\int_{M_i} \widetilde{\mathrm{Td}}^\Sigma(TM, TB, h^{TM}, h^{TB})\,\mathrm{ch}^\Sigma(\xi, h^\xi). \quad (0.5)$$

This paper is organized as follows. The first four sections are concerned with some generalities of orbifolds and of analytic torsions. In Section 1, we recall the definition of orbifold, and construct the Bismut superconnection for a submersion of orbifolds. In Section 2, We extend Kawasaki's theorem to a relative situation. In Section 3, we construct the Quillen metrics for an orbifold, and prove their anomaly formulas. In Section 4, we construct the analytic torsion forms for a submersion of orbifolds. In Section 5, we extend the result of [2] to the orbifold case.

The first version of this paper was written in 1998 when I was visiting at ICTP. The first part was published in [34]. For the recent works on the analytic torsion for orbifold flat vector bundles, cf. recent works [21], [41].

In the whole paper, we use the superconnection formalism of Quillen [38]. If $E = E^+ \oplus E^-$ is a $\mathbb{Z}_2$-graded vector space, and $\tau = \pm 1$ defines the $\mathbb{Z}_2$-grading, for $A \in \mathrm{End}(E)$, we denote $\mathrm{Tr}_s[A]$ the supertrace of A, i.e.,

$$\mathrm{Tr}_s[A] = \mathrm{Tr}[\tau A]. \tag{0.6}$$

The reader is referred for more details to [4], [10], [2].

Acknowledgements. We are very much indebted to Professor Jean-Michel Bismut for very helpful discussions and suggestions. Thanks also to a referee for his useful comments.

1. Orbifolds and superconnections

In this section, we extend the Bismut superconnection of [3] to a Kähler fibration of orbifolds.

This section is organized as follows. In Section 1.1, we recall the definition of an orbifold following [34, §1.1]. In Section 1.2, we describe the Kähler fibration. In Section 1.3, we explain the construction of the Bismut superconnection $B_u (u > 0)$ [3] for a submersion of orbifolds.

1.1. Definition of an orbifold

We define at first a category $\mathcal{M}_s$ as follows: The objects of $\mathcal{M}_s$ are the class of pairs (G, M) where M is a connected smooth manifold and G is a finite group acting effectively on M. Let (G, M) and (G', M') be two objects, then a morphism $\Phi : (G, M) \to (G', M')$ is a family of open embedding $\varphi : M \to M'$ satisfying:

i) For each $\varphi \in \Phi$, there is an injective group homomorphism $\lambda_\varphi : G \to G'$ that makes φ be λ_φ-equivariant.

ii) For $g \in G', \varphi \in \Phi$, we define $g\varphi : M \to M'$ by $(g\varphi)(x) = g\varphi(x)$ for $x \in M$. If $(g\varphi)(M) \cap \varphi(M) \neq \phi$, then $g \in \lambda_\varphi(G)$.

iii) For $\varphi \in \Phi$, we have $\Phi = \{g\varphi : g \in G'\}$.

Definition 1.1. Let X be a paracompact Hausdorff space and let $\mathcal{U}$ be a cover of X consisting of connected open subsets. We assume $\mathcal{U}$ satisfies the condition:

$$\text{For any } x \in U \cap U',\ U,\ U' \in \mathcal{U}, \text{ there is } U'' \in \mathcal{U} \text{ such that } x \in U'' \subset U \cap U'. \tag{1.1}$$

Then an orbifold structure $\mathcal{V}$ on X is the following:

i) For $U \in \mathcal{U}$, $\mathcal{V}(U) = ((G_U, \widetilde{U}) \xrightarrow{\tau} U)$ is a ramified covering $\widetilde{U} \to U$ giving an identification $U \simeq \widetilde{U}/G_U$.

ii) For $U, V \in \mathcal{U}, U \subset V$, there is a morphism $\varphi_{VU} : (G_U, \widetilde{U}) \to (G_V, \widetilde{V})$ that covers the inclusion $U \subset V$.

iii) For $U, V, W \in \mathcal{U}, U \subset V \subset W$, we have $\varphi_{WU} = \varphi_{WV} \circ \varphi_{VU}$.

If $\mathcal{U}'$ is a refinement of $\mathcal{U}$ satisfying (1.1), then there is an orbifold structure $\mathcal{V}'$ such that $\mathcal{V} \cup \mathcal{V}'$ is an orbifold structure. We consider $\mathcal{V}$ and $\mathcal{V}'$ to be equivalent. Such an equivalence class is called an orbifold structure over X. So we may choose $\mathcal{U}$ arbitrarily fine.

In the above definition, we can replace $\mathcal{M}_s$ by a category of manifolds with an additional structure such as orientation, Riemannian metric or complex structure. We understand that the morphisms (and the groups) preserve the specified structure. So we can define oriented, Riemannian or complex orbifolds.

Let $(X, \mathcal{V})$ be an orbifold. For each $x \in X$, we can choose a small neighbourhood $(G_x, \widetilde{U}_x) \to U_x$ such that $\widetilde{x} \in \widetilde{U}_x$, the unique inverse image of x, is a fixed point of G_x. (Such G_x is unique up to isomorphisms for each $x \in X$, [40, p. 468].) Let $(1), (h_x^1), \dots, (h_x^{\rho_x})$ be the conjugacy classes in G_x. Let $Z_{G_x}(h_x^j)$ be the centralizer of h_x^j in G_x. One also notes $\widetilde{U}_x^{h_x^j}$ the fixed points of h_x^j over $\widetilde{U}_x$. Then we have a natural bijection

$$\Big\{(y, (h_y^j)) : y \in U_x, j = 1, \dots, \rho_y\Big\} \simeq \coprod_{j=1}^{\rho_x} \widetilde{U}_x^{h_x^j} / Z_{G_x}(h_x^j). \tag{1.2}$$

So we can define globally

$$\Sigma X = \{(x, (h_x^j)) : x \in X, G_x \neq \{1\}, j = 1, \dots, \rho_x\}. \tag{1.3}$$

Then ΣX has a natural orbifold structure defined by

$$\Big\{(Z_{G_x}(h_x^j)/K_x^j, \widetilde{U}_x^{h_x^j}) \to \widetilde{U}_x^{h_x^j}/Z_{G_x}(h_x^j)\Big\}_{(x, U_x, j)}. \tag{1.4}$$

Here K_x^j is the kernel of the representation $Z_{G_x}(h_x^j) \to \text{Diffeo}\,(\widetilde{U}_x^{h_x^j})$, the diffeomorphism group of $\widetilde{U}_x^{h_x^j}$. The number $m = |K_x^j|$ is called the multiplicity of ΣX in X at $(x, (h_x^j))$. Since the multiplicity is locally constant on ΣX, we may assign the multiplicity m_i to each connected component ΣX_i of ΣX.

Definition 1.2. An orbifold vector bundle ξ over an orbifold $(X, \mathcal{V})$ is defined as follows: ξ is an orbifold and for $U \in \mathcal{U}$, $(G_U^\xi, \widetilde{p}_U : \widetilde{\xi}_U \to \widetilde{U})$ is a G_U^ξ-equivariant vector bundle such that the morphism $\varphi_{\xi_U \xi_V}$ is a morphism of equivariant vector bundles, and $(G_U^\xi, \widetilde{\xi}_U)$ (resp. $(G_U^\xi/K_U, \widetilde{U}), K_U = \text{Ker}(G_U^\xi \to \text{Diffeo}(\widetilde{U}))$) (In general, G_U^ξ does not act effectively on $\widetilde{U}$, i.e., $K_U \neq \{1\}$) is the orbifold structure of ξ (resp. X). For $x \in X$, we denote the fiber of the vector bundle $\widetilde{\xi}_U$ at an inverse image of x in $\widetilde{U}$, as the vector space $\widetilde{\xi}_x$.

If G_U^ξ acts effectively on $\widetilde{U}$ for $U \in \mathcal{U}$, we call that ξ is a *proper* orbifold vector bundle.

For an orbifold vector bundle ξ, let $\widetilde{\xi_U^{\text{pr}}}$ be the maximal K_U-invariant subbundle of $\widetilde{\xi}_U \to \widetilde{U}$, then $(G_U, \widetilde{\xi_U^{\text{pr}}})$ defines a proper orbifold vector bundle ξ^{pr}.

A natural example is the (proper) orbifold tangent bundle TX which is defined by:

$$(G_U, T\widetilde{U} \to \widetilde{U}), \quad \text{for} \quad U \in \mathcal{U}$$

Let $\xi \to X$ be an orbifold vector bundle. A section $s : X \to \xi$ is called $\mathscr{C}^\infty$ (or $\mathscr{C}^k$) if for each $U \in \mathcal{U}$, $s_{|U}$ is covered by a G_U^ξ-invariant smooth (or $\mathscr{C}^k$) section $\widetilde{s}_U : \widetilde{U} \to \widetilde{\xi}_U$.

If X is oriented, we define the integral $\int_X \omega$ for a form over X (i.e., a section of $\Lambda(T^*X)$ over X): if $\mathrm{supp}(\omega) \subset U \in \mathcal{U}$, then

$$\int_X \omega = \frac{1}{|G_U|}\int_{\widetilde{U}} \widetilde{\omega}_U. \tag{1.5}$$

In the sequel, if G does not act effectively on the connected manifold M, we will identify the couple (G, M) as an element $(G/K, M)$ in $\mathcal{M}_s$, with $K = \mathrm{Ker}(G \to \mathrm{Diffeo}(M))$.

Definition 1.3. Let M, B be two orbifolds, a map $\pi : M \to B$ is said to define an orbifold submersion if there exist $\mathcal{U}, \mathcal{U}'$ open covers of M, B, such that $\pi(\mathcal{U}) \subset \mathcal{U}'$, and $(G_U, \widetilde{U})_{U\in\mathcal{U}}$, $(G_V, \widetilde{V})_{V\in\mathcal{V}}$ are the orbifold structures of M, B; for $U \in \mathcal{U}$, there is $\widetilde{\pi} : \widetilde{U} \to \widetilde{V}$ a G_U-equivariant submersion of $\widetilde{U}$ onto $\widetilde{V}$ that covers $\pi : U \to V = \pi(U)$, and $(G_U, \widetilde{V}) = (G_V, \widetilde{V})$ in $\mathcal{M}_s$; if $U_1 \subset U_2, U_1, U_2 \in \mathcal{U}$, then $\Phi_{\pi(U_2)\pi(U_1)}$ is induced by $\Phi_{U_2U_1}$.

Let $\pi : M \to B$ be an orbifold submersion of M onto B, then the related tangent bundle TM/B is defined by: over $\widetilde{U}$, $((G_U, T\widetilde{U}/\widetilde{V}) \to \widetilde{U})$.

Proposition 1.4. *If $\pi : M \to B$ is a proper orbifold submersion of M onto B, then for each $b \in B$, there exists a small neighborhood $(G_b, \widetilde{V}_b) \to V_b$, $\widetilde{M_b}$ an orbifold, such that π is induced by a G_b-equivariant orbifold submersion $\widetilde{\pi}_b : \widetilde{M_b} \to \widetilde{V}_b$ with compact fiber $\overline{X}$.*

Proof. Let $\mathcal{U}$ be a cover of M in Definition 1.3. For $U \in \mathcal{U}$, set

$$K_U = \mathrm{Ker}\{G_U \to \mathrm{Diffeo}(\widetilde{\pi}(\widetilde{U}))\}. \tag{1.6}$$

As π is proper, for $b \in B$, we can find $V \subset B$ open, $b \in V$, $(G_b, \widetilde{V}) \xrightarrow{\gamma} V$ be a ramified covering of V, and $\gamma^{-1}(b) = \{b_0\}$, such that there is $((G_{U_i}, \widetilde{U}_i) \to U_i)_{i\in I}$ ($I = \{1, \cdots, q\}$) induced by the orbifold structure of M, the map $\widetilde{\pi} : (G_{U_i}, \widetilde{U}_i) \to (G_{U_i}, \widetilde{V}) = (G_b, \widetilde{V})$ is a G_{U_i}-equivariant submersion of $\widetilde{U}_i$ onto $\widetilde{V}$, and $\pi^{-1}(V) = \cup_{i\in I} U_i$.

For $W_1 \subset W_2, W_1, W_2 \in \mathcal{U}$, by definition, there exist morphisms

$$\begin{aligned} \Phi_{W_2W_1} &: (G_{W_1}, \widetilde{W_1}) \to (G_{W_2}, \widetilde{W_2}),\\ \Phi_{\pi(W_2)\pi(W_1)} &: (G_{\pi(W_1)}, \widetilde{\pi}(\widetilde{W_1})) \to (G_{\pi(W_2)}, \widetilde{\pi}(\widetilde{W_2})) \text{ in } \mathcal{M}_s, \end{aligned} \tag{1.7}$$

such that $\Phi_{\pi(W_2)\pi(W_1)}$ is induced by $\Phi_{W_2W_1}$. We note that $\widetilde{\pi}(\widetilde{W_j})$ is a ramified covering of $\pi(W_j)$ for $j = 1, 2$.

Let $\widetilde{\mathcal{U}} = \{(\widetilde{W}, \varphi)$: there exist $i \in I$, such that $(G_W, \widetilde{W}) \to W \subset U_i$, $W \in \mathcal{U}$, and $\varphi \in \Phi_{V\pi(W)}\}$. Let $a_1 = (\widetilde{W}_1, \varphi_1)$, $a_2 = (\widetilde{W}_2, \varphi_2) \in \widetilde{\mathcal{U}}$, $W_1 \subset W_2$, for each

$\psi \in \Phi_{W_2 W_1}$, we also denote $\psi \in \Phi_{\pi(W_2)\pi(W_1)}$ the associated open embedding. Thus for $\psi \in \Phi_{W_2 W_1}$, we have the commutative diagram

$$\begin{array}{ccccccccc} \widetilde{W}_1 & \xrightarrow{g} & \widetilde{W}_1 & \xrightarrow{\widetilde{\pi}} & \widetilde{\pi}(\widetilde{W}_1) & \xrightarrow{g} & \widetilde{\pi}(\widetilde{W}_1) & \xrightarrow{\varphi_1} & \widetilde{V} \\ \downarrow \psi & & \downarrow \psi & & \downarrow \psi & & \downarrow \psi & & \\ \widetilde{W}_2 & \xrightarrow{\lambda_\psi(g)} & \widetilde{W}_2 & \xrightarrow{\widetilde{\pi}} & \widetilde{\pi}(\widetilde{W}_2) & \xrightarrow{\lambda_\psi(g)} & \widetilde{\pi}(\widetilde{W}_2) & \xrightarrow{\varphi_2} & \widetilde{V} \end{array} \tag{1.8}$$

Put

$$\Phi_{a_2 a_1} = \{\psi \in \Phi_{W_2 W_1} : \varphi_2 \psi = \varphi_1 \text{ as a map from } \widetilde{\pi}(\widetilde{W}_1) \text{ to } \widetilde{V}\}. \tag{1.9}$$

I.e., for $\psi \in \Phi_{a_2 a_1}$, the commutative diagram (1.8) is completed by the identity map $\mathrm{Id} : \widetilde{V} \to \widetilde{V}$.

Claim: *$\Phi_{a_2 a_1} : (K_{W_1}, (\widetilde{W}_1, \varphi_1)) \to (K_{W_2}, (\widetilde{W}_2, \varphi_2))$ is a morphism in $\mathcal{M}_s$.*

Proof of the claim. The K_{W_1}-action on $(\widetilde{W}_1, \varphi_1)$ is defines by its action on $\widetilde{W}_1$. i) For $\psi \in \Phi_{a_2 a_1} \subset \Phi_{W_2 W_1}$, the injective group homomorphism $\lambda_\psi : G_{W_1} \to G_{W_2}$ makes that ψ is λ_ψ-equivariant. Note that for $g \in G_{W_1}$, $\widetilde{x} \in \widetilde{W}_1$, by (1.8), we have

$$\psi(g\widetilde{x}) = \lambda_\psi(g)\psi(\widetilde{x}), \quad \psi\widetilde{\pi}(g\widetilde{x}) = \lambda_\psi(g)\widetilde{\pi}\psi(\widetilde{x}) = \lambda_\psi(g)\psi\widetilde{\pi}(\widetilde{x}). \tag{1.10}$$

Thus if $g \in K_{W_1}$, $\lambda_\psi(g) \in G_{W_2}$ fixes $\widetilde{\pi}(\psi(\widetilde{W}_1)) = \psi(\widetilde{\pi}(\widetilde{W}_1))$, an open set of $\widetilde{\pi}(\widetilde{W}_2)$. But G_{W_2} is compact and acts on $\widetilde{W}_2$ which is connected, thus we conclude that $\lambda_\psi(g)$ acts as identity on $\widetilde{\pi}(\widetilde{W}_2)$, i.e., $\lambda_\psi(g) \in K_{W_2}$. Thus λ_ψ induces an injective group homomorphism $\lambda_\psi : K_{W_1} \to K_{W_2}$.

ii) Assume now $(h\psi)(\widetilde{W}_1) \cap \psi(\widetilde{W}_1) \neq \phi$, and $h \in K_{W_2}$. The first condition implies $h \in \lambda_\psi(G_{W_1})$, i.e., there exists $g \in G_{W_1}$ such that $h = \lambda_\psi(g)$. But $h \in K_{W_2}$ means that $\lambda_\psi(g)$ acts as identity on $\widetilde{\pi}(\widetilde{W}_2)$, this implies that g acts as identity $\widetilde{\pi}(\widetilde{W}_1)$ by (1.8), i.e., $g \in K_{W_1}$. We conclude that $h \in \lambda_\psi(K_{W_1})$.

iii) For any $\psi', \psi \in \Phi_{a_2 a_1}$, there exists $g \in G_{W_2}$ such that $g\psi = \psi'$. By (1.9),

$$\lambda_{\varphi_2}(g)\varphi_2\psi = \varphi_2 g\psi = \varphi_2\psi' = \varphi_1 = \varphi_2\psi. \tag{1.11}$$

Thus $\lambda_{\varphi_2}(g)$ acts as identity on an open set $\varphi_2\psi(\widetilde{\pi}(\widetilde{W}_1))$ of $\widetilde{V}$, thus as identity on $\widetilde{V}$, this implies that g acts as identity on $\widetilde{\pi}(\widetilde{W}_2)$, i.e., $g \in K_{W_2}$, thus $\Phi_{a_2 a_1} = \{g\psi : g \in K_{W_2}\}$.

The proof of the claim is completed. □

For $i \in I$, we denote $\widetilde{U}_i = (\widetilde{U}_i, 1) \in \widetilde{\mathcal{U}}$. We define an equivalence relation $\sim$ on $\overline{M} = \cup_{i \in I} \widetilde{U}_i / K_{U_i}$: For $\widetilde{x} \in \widetilde{U}_i, \widetilde{y} \in \widetilde{U}_j$, $\widetilde{x} \sim \widetilde{y}$ if and only if there exist $(G_W, \widetilde{W}) \to W \subset U_i \cap U_j$, $\varphi_1, \varphi_2 \in \Phi_{V\pi(W)}$, $\widetilde{z} \in \widetilde{W}$ such that

$$\widetilde{x} \in \Phi_{\widetilde{U}_i a_1}(\{\widetilde{z}\}), \widetilde{y} \in \Phi_{\widetilde{U}_j a_2}(\{\widetilde{z}\}), \text{ for } a_1 = (\widetilde{W}, \varphi_1), a_2 = (\widetilde{W}, \varphi_2) \in \widetilde{\mathcal{U}}. \tag{1.12}$$

We can interpret (1.12) for $\widetilde{x}, \widetilde{z}$ by the following commutative diagram:

$$\begin{array}{cccccc} \widetilde{z}\in & \widetilde{W} & \xrightarrow{\widetilde{\pi}} & \widetilde{\pi}(\widetilde{W}) & & \\ & \Big\downarrow \Phi_{\widetilde{U}_i a_1} & & \Big\downarrow & \searrow^{\varphi_1} & \\ \widetilde{x}\in & \widetilde{U}_i & \xrightarrow{\widetilde{\pi}} & \widetilde{\pi}(\widetilde{U}_i)=\widetilde{V} & \xrightarrow{\mathrm{Id}} & \widetilde{V}\end{array} \tag{1.13}$$

Let $\widetilde{M_b} = \overline{M}/\sim$. Let $\mathcal{U}' = \{(\widetilde{W}, \varphi)/K_W : (\widetilde{W}, \varphi) \in \widetilde{\mathcal{U}}\}$, then $\mathcal{U}'$ is a covering of $\widetilde{M_b}$ which satisfies the condition (1.1). We get the conditions i), ii), iii) of Definition 1.1 from the claim. So $\mathcal{U}'$ defines an orbifold structure on $\widetilde{M_b}$.

Note that K_{U_i} is a normal subgroup of G_{U_i} and $G_b = G_{U_i}/K_{U_i}$, thus G_b acts naturally on $\widetilde{U}_i/K_{U_i}$, so G_b acts on $\widetilde{M_b}$. Let $\widetilde{\pi} : \widetilde{M_b} \to \widetilde{V}$ be induced by $\widetilde{\pi} : \widetilde{U}_i \to \widetilde{V}$, then $\widetilde{\pi}$ is an orbifold submersion, and $\widetilde{\pi}$ is G_b -equivariant.

Now the procedure is standard. Note that the kernel of $d\widetilde{\pi} : T\widetilde{M_b} \to T\widetilde{V}$ is an orbifold vector bundle. By choosing a horizontal subbundle $T^H\widetilde{M_b}$ of $T\widetilde{M_b}$ (for example, by taking the orthogonal complement of $\mathrm{Ker}(d\widetilde{\pi})$ with respect to a metric on $T\widetilde{M_b}$), such that

$$T\widetilde{M_b} = \mathrm{Ker}(d\widetilde{\pi}) \oplus T^H\widetilde{M_b}. \tag{1.14}$$

As $\widetilde{V}$ is a manifold, we know that $T^H\widetilde{M_b}$ is a usual vector bundle. Now the horizontal lift of any ball $B(p,r)$, with the center p and radius r, in $\widetilde{V}$ along the radius direction gives a trivialization

$$\widetilde{\pi}^{-1}(B(p,r)) = B(p,r) \times \overline{X}_p. \tag{1.15}$$

Note that for any point in V such that $G_p = \{1\}$, $\overline{X}_p = \pi^{-1}(\{p\})$, thus as a real orbifold, the fiber $\overline{X}$ has a canonical model.

The proof of Proposition 1.4 is completed. □

Let $(X, \mathcal{V})$ be a compact connected Riemannian orbifold. For $x, y \in X$, put

$$\begin{aligned} d(x,y) = \mathrm{Inf}\Big\{ \textstyle\sum_i \int_{t_{i-1}}^{t_i} |\frac{\partial}{\partial t}\widetilde{\gamma}_i(t)|dt \Big| & \gamma : [0,1] \to X, \gamma(0) = x, \gamma(1) = y, \text{such that} \\ & \text{there exist } t_0 = 0 < t_1 < \cdots < t_k = 1, U_i \in \mathcal{U}, \gamma([t_{i-1}, t_i]) \subset U_i, \\ & \widetilde{\gamma}_i : [t_{i-1}, t_i] \to \widetilde{U}_i \quad \mathscr{C}^\infty, \text{ that covers } \gamma_{|[t_{i-1},t_i]}. \Big\} \end{aligned}$$

Then (X, d) is a metric space.

1.2. Kähler fibrations

In the rest of this paper, we always work on complex orbifolds, especially, all morphisms considered in Section 1.1 are holomorphic. For an orbifold complex vector bundle, we denote the underlying real orbifold vector bundle by adding a subscript $\mathbb{R}$.

Definition 1.5. A Kähler form on a complex orbifold X is a real closed $(1,1)$-form ω on X such that ω induces a (orbifold) metric on TX.

Let $\pi : M \to B$ be a proper holomorphic orbifold submersion of M onto B. Let TM, TB be the holomorphic tangent bundles to M, B. From Proposition 1.4, the holomorphic relative tangent bundle TX of the fibration π is well defined as an orbifold vector bundle over M. Let J^{TX} be the complex structure on the real relative tangent bundle $T_{\mathbb{R}}X$.

Lemma 1.6. *For $\pi : M \to B$ a proper holomorphic orbifold submersion, for any $b \in B$, we can choose $\widetilde{M}_b$ in Proposition 1.4 such that π is induced by a G_b-equivariant holomorphic orbifold submersion $\widetilde{\pi}_b : \widetilde{M}_b \to \widetilde{V}_b$.*

Proof. As all morphisms in the proof of Proposition 1.4 are holomorphic, we get Lemma 1.6 from the proof of Proposition 1.4. □

Let h^{TX} be a Hermitian metric on TX. Let T^HM be an orbifold vector subbundle of TM, such that

$$TM = T^HM \oplus TX. \tag{1.16}$$

We now define the Kähler fibration as in [11, Definition 1.4].

Definition 1.7. The triple (π, h^{TX}, T^HM) is said to define a Kähler fibration if there exists a smooth real 2-form ω of complex type (1,1), which has the following properties:

a) ω is closed.
b) $T^H_{\mathbb{R}}M$ and $T_{\mathbb{R}}X$ are orthogonal with respect to ω,
c) If $X, Y \in T_{\mathbb{R}}X$, then $\omega(X,Y) = \langle X, J^{TX}Y\rangle_{g^{T_{\mathbb{R}}X}}$ with $g^{T_{\mathbb{R}}X}$ the metric on $T_{\mathbb{R}}X$ induced by h^{TX}.

Now we have an analogue of [11, Theorems 1.5 and 1.7].

Theorem 1.8. *Let ω be a real smooth 2-form on M of complex type $(1,1)$, which has the following two properties:*

a) ω is closed.
b) The bilinear map $X, Y \in T_{\mathbb{R}}X \to \omega(J^{TX}X, Y)$ defines a Hermitian product h^{TX} on TX.

For $x \in M$, set

$$T^H_xM = \{Y \in T_xM : \text{ for any } X \in T_xX, \omega(X, \overline{Y}) = 0\}.$$

Then T^HM is an orbifold subbundle of TM such that $TM = T^HM \oplus TX$. Also (π, h^{TX}, T^HM) is a Kähler fibration, and ω is an associated $(1,1)$-form.

A smooth real $(1,1)$-form ω' on M is associated with the Kähler fibration (π, h^{TX}, T^HM) if and only if there is a real smooth closed $(1,1)$-form η on B such that

$$\omega' - \omega = \pi^*\eta.$$

Proof. The proof is as same as in [11, Theorems 1.5 and 1.7]. □

1.3. The Bismut superconnection of a Kähler fibration

In this part, we will define the Bismut superconnection by proceeding as in [13, §1], [2, §2].

Let $\pi : M \to B$ be a proper holomorphic orbifold submersion of M onto B with fibre X. Let ω^M be a real closed (1,1) form on M taken as in Theorem 1.8. Let ξ be a complex orbifold vector bundle on M. Let h^ξ be a Hermitian metric on ξ. Let ∇^{TX}, ∇^ξ be the holomorphic Hermitian connections on $(TX, h^{TX}), (\xi, h^\xi)$.

We will temporarily assume that B is a complex manifold. Then π is a fibration of M on B which is modelled on orbifold X: There is an open covering $\mathcal{U}$ of B such that if $U \in \mathcal{U}$, $\pi^{-1}(U)$ is diffeomorphic to $U \times X$.

Definition 1.9. For $0 \leq k \leq \dim X$, $b \in B$, let E_b^k be the vector space of $\mathscr{C}^\infty$ sections of $(\Lambda^k(T^{*(0,1)}X) \otimes \xi)|_{X_b}$ over X_b. Set

$$E_b = \oplus_{k=0}^{\dim X} E_b^k, \quad E_b^+ = \oplus_{k \,\mathrm{even}} E_b^k, \quad E_b^- = \oplus_{k \,\mathrm{odd}} E_b^k. \tag{1.17}$$

As in [3, §1f)], [11, §1d)], we can regard the E_b's as the fibers of a smooth $\mathbb{Z}$-graded infinite-dimensional vector bundle over the base B. Smooth sections of E over B will be identified with smooth sections of $\Lambda(T^{*(0,1)}X) \otimes \xi$ over M.

Let dv_X be the Riemannian volume form on X associated with h^{TX}. Let $\langle\ \rangle_{\Lambda(T^{*(0,1)}X)\otimes\xi}$ be the Hermitian product induced by h^{TX}, h^ξ on $\Lambda(T^{*(0,1)}X)\otimes\xi$. The Hermitian product $\langle\ \rangle$ on E is defined by: If $s, s' \in E$, set

$$\langle s, s' \rangle = \left(\frac{1}{2\pi}\right)^{\dim X} \int_X \langle s, s' \rangle_{\Lambda(T^{*(0,1)}X)\otimes\xi}\, dv_X. \tag{1.18}$$

For $b \in B$, let $\overline{\partial}^{X_b}$ be the Dolbeault operator acting on E_b, and let $\overline{\partial}^{X_b *}$ be its formal adjoint with respect to the Hermitian product (1.18). Set

$$D^X = \overline{\partial}^{X_b} + \overline{\partial}^{X_b *}. \tag{1.19}$$

If $U \in T_\mathbb{R} B$, let U^H be the lift of U in $T_\mathbb{R}^H M$, so that $\pi_* U^H = U$.

Definition 1.10. If $U \in T_\mathbb{R} B$, if s is a smooth section of E over B, set

$$\nabla_U^E s = \nabla_{U^H}^{\Lambda(T^{*(0,1)}X)\otimes\xi} s. \tag{1.20}$$

Let $c(T_\mathbb{R} X)$ be the Clifford algebra of $(T_\mathbb{R} X, h^{TX})$. The bundle $\Lambda(T^{*(0,1)}X)\otimes\xi$ is a $c(T_\mathbb{R} X)$-Clifford module. In fact, if $U \in TX$, let $U' \in T^{*(0,1)}X$ correspond to U by the metric h^{TX}. If $U, V \in TX$, set

$$c(U) = \sqrt{2} U' \wedge, \quad c(\overline{V}) = -\sqrt{2} i_{\overline{V}}. \tag{1.21}$$

Let P^{TX} be the projection $TM \simeq T^H M \oplus TX \to TX$.

If U, V are smooth vector fields on B, set

$$T(U^H, V^H) = -P^{TX}[U^H, V^H]. \tag{1.22}$$

Then T is a tensor. By [11, Theorem 1.7], we know that as a 2-form, T is of complex type (1,1).

Let $f_1, \dots, f_{2m}$ be a base of $T_{\mathbb{R}}B$, and let $f^1, \dots, f^{2m}$ be the dual base of $T^*_{\mathbb{R}}B$.

Definition 1.11. Set

$$c(T) = \frac{1}{2} \sum_{1 \le \alpha, \beta \le 2m} f^\alpha f^\beta c\Big(T(f^H_\alpha, f^H_\beta)\Big). \tag{1.23}$$

Then $c(T)$ is a section of $(\Lambda^2(T^*_{\mathbb{R}}B) \widehat{\otimes} \operatorname{End}(\Lambda(T^{*(0,1)}X) \otimes \xi))^{\mathrm{odd}}$.

Definition 1.12. For $u > 0$, let B_u be the Bismut superconnection constructed in [3, §3], [11, §2a)],

$$B_u = \sqrt{u} D^X + \nabla^E - \frac{c(T)}{2\sqrt{2u}}. \tag{1.24}$$

Let N_V be the number operator defining the $\mathbb{Z}$-grading on $\Lambda(T^{*(0,1)}X) \otimes \xi$ and on E. N_V acts by multiplication by k on $\Lambda^k(T^{*(0,1)}X) \otimes \xi$. If $U, V \in T_{\mathbb{R}}B$, set

$$\omega^{H\overline{H}}(U, V) = \omega^M(U^H, V^H). \tag{1.25}$$

Definition 1.13. For $u > 0$, set

$$N_u = N_V + \frac{i\omega^{H\overline{H}}}{u}. \tag{1.26}$$

In general, B is not a complex manifold. By Proposition 1.4, we verify easily that the above objects go down to B (Ex, E is an orbifold bundle over B), so we can define the Bismut superconnection B_u $(u > 0)$ over B as locally over $\widetilde{V}_b$.

2. Family index theorem

In this section, we describe basic properties of the operator $\overline{\partial}^X$ on a complex orbifold, and we extend Kawasaki's theorem to a relative situation.

This section is organized as follows. In Section 2.1, we give the Hodge decomposition for $\overline{\partial}^X$ operator over a complex orbifold. In Section 2.2, we state the family version of Kawasaki's theorem.

We use the notation of Section 1.

2.1. $\overline{\partial}$-operator on a complex orbifold

Let X be a compact complex orbifold of complex dimension l. Let ξ be a holomorphic orbifold vector bundle on X.

Let $\mathcal{O}_X$ be the sheaf over X of local G_U-invariant holomorphic functions over $\widetilde{U}$, for $U \in \mathcal{U}$. Then by [19], $(X, \mathcal{O}_X)$ is an analytic space. The local G^ξ_U -invariant holomorphic sections of $\widetilde{\xi} \to \widetilde{U}$ define also a coherent analytic sheaf $\mathcal{O}_X(\xi)$ over X.

Let $\mathcal{D}^k(\xi)$ be the sheaf of $\mathscr{C}^\infty$ sections of $\Lambda^k(T^{*(0,1)}X) \otimes \xi$ over X. Then we have the operator $\overline{\partial}^X : \mathcal{D}^k(\xi) \to \mathcal{D}^{k+1}(\xi)$ and an exact sequence of $\mathcal{O}_X$-sheaves

$$0 \to \mathcal{O}_X(\xi) \to \mathcal{D}^1(\xi) \overset{\overline{\partial}^X}{\to} \cdots \overset{\overline{\partial}^X}{\to} \mathcal{D}^l(\xi) \to 0. \tag{2.1}$$

Put $\Omega^k(X,\xi) = \Gamma(X, \mathcal{D}^k(\xi))$, $\Omega^\bullet(X,\xi) = \oplus_k \Omega^k(X,\xi)$, then we have $(\Omega^\bullet(X,\xi), \overline{\partial}^X)$ the Dolbeault complex of $\mathscr{C}^\infty$ sections of $\Lambda(T^{*(0,1)}X) \otimes \xi$ over X:

$$0 \to \Omega^0(X,\xi) \overset{\overline{\partial}^X}{\to} \cdots \overset{\overline{\partial}^X}{\to} \Omega^l(X,\xi) \to 0. \tag{2.2}$$

The sheaves $\mathcal{D}^k(\xi)$ are fine [27], so their higher cohomology groups vanish. So

$$H^\bullet(\Omega^\bullet(X,\xi), \overline{\partial}^X) \simeq H^\bullet(X, \mathcal{O}_X(\xi)). \tag{2.3}$$

In the sequel, we also note $H^\bullet(X, \mathcal{O}_X(\xi))$ by $H^\bullet(X,\xi)$.

Let h^{TX}, h^ξ be Hermitian metrics on TX, ξ. Then D^X in (1.19) induced by h^{TX}, h^ξ is an elliptic operator and

$$D^{X,2} = \overline{\partial}^X \overline{\partial}^{X*} + \overline{\partial}^{X*} \overline{\partial}^X \tag{2.4}$$

preserves the $\mathbb{Z}$-grading on $\Omega^\bullet(X,\xi)$.

The following proposition is [34, Proposition 2.2].

Proposition 2.1 (The Hodge Decomposition Theorem). *There is a L^2-orthogonal direct sum decomposition of the ξ-value $(0,k)$-forms*

$$\Omega^k(X,\xi) = \operatorname{Ker}(D^X) \oplus \operatorname{Im}(\overline{\partial}^X) \oplus \operatorname{Im}(\overline{\partial}^{X*}). \tag{2.5}$$

From (2.3), (2.5), there is a canonical identification

$$\operatorname{Ker}(D^X) \simeq H^\bullet(X,\xi). \tag{2.6}$$

Definition 2.2. Let P^X be the vector space of smooth forms on X, which are sums of forms of type (k,k). Let $P^{X,0}$ be the vector space of the forms $\alpha \in P^X$ such that there exist smooth forms β, γ on X for which $\alpha = \partial\beta + \overline{\partial}\gamma$.

We define $P^{X \cup \Sigma X}$, $P^{X \cup \Sigma X, 0}$ in the same way.

2.2. Family index theorem

We use the notation of Section 1.3.

Let M be a complex orbifold. Let ΣM be the strata of M defined by (1.3). Let B be a complex manifold. Let $\pi : M \to B$ be a proper orbifold holomorphic submersion of M onto B with compact fibre X. Then $\pi' : M \cup \Sigma M \to B$ is also an orbifold submersion with compact fibre $X \cup \Sigma X$. Let m_i be the multiplicity of each connected component M_i ($m_i = 1$, if $M_i = M$) of $M \cup \Sigma M$. Let ξ be an orbifold vector bundle on M. Let ξ^{pr} be the maximal proper orbifold subbundle of ξ.

We assume that π is a Kähler fibration with respect to a real closed $(1,1)$-form ω^M on M. Let D_+^X, D_-^X be the restrictions of D^X to E^+, E^-.

Let B_u ($u > 0$) be the Bismut superconnection on E constructed in Section 1.3 which is attached to the (1,1) form ω^M on M and to the metric h^ξ on ξ.

If A is a (q,q) matrix, set

$$\begin{aligned}\mathrm{Td}(A) &= \det\Big(\frac{A}{1-e^{-A}}\Big), \quad \mathrm{Td}'(A) = \frac{\partial}{\partial u}\mathrm{Td}(A+u)|_{u=0},\\ \mathrm{ch}(A) &= \mathrm{Tr}[\exp(A)].\end{aligned} \tag{2.7}$$

The genera associated with Td and ch are called the Todd genus and the Chern character.

Let $\mathcal{U}$ be a cover of $(M,\mathcal{V})$ which defines the submersion π as in Definition 1.3. Recall that for $U\in\mathcal{U}$, we denote $\mathcal{V}(U)=((G_U,\widetilde{U})\to U)$. By [5, (2.20)], [33, (1.15)], [34, (1.6), (1.7)], the forms $\mathrm{Td}^{\Sigma}(TX,h^{TX})$, $\mathrm{ch}^{\Sigma}(\xi,h^{\xi})$ over $M\cup\Sigma M$ are defined by: on $\widetilde{U}^g/Z_{G_U}(g)$ $(g\in G_U)$, as $\mathrm{Td}_g(\widetilde{TX},h^{TX})$ and

$$\mathrm{ch}_g(\widetilde{\xi^{\mathrm{pr}}},h^{\xi}) = \mathrm{Tr}\Big[g\exp(\frac{i}{2\pi}R^{\widetilde{\xi^{\mathrm{pr}}}})\Big], \tag{2.8}$$

where $R^{\widetilde{\xi^{\mathrm{pr}}}}$ is the curvature of the holomorphic Hermitian connection on $(\widetilde{\xi^{\mathrm{pr}}},h^{\xi})$. Then $\mathrm{Td}^{\Sigma}(TX,h^{TX}),\mathrm{ch}^{\Sigma}(\xi,h^{\xi})$ are closed on $M\cup\Sigma M$, and their cohomology classes don't depend on the metrics h^{TX},h^{ξ}.

Let Φ be the homomorphism of $\Lambda^{\mathrm{even}}(T^*_{\mathbb{R}}B)$ into itself: $\alpha\to(2i\pi)^{-\deg\alpha/2}\alpha$.

The following result extends [11, Theorem 2.2].

Theorem 2.3. *For any $u>0$, the differential forms on B, $\mathrm{Tr}_s[\exp(-B_u^2)]$ are elements of P^B. They are closed and they are in the same cohomology class, which does not depend on $u>0$. Also uniformly on compact sets in B,*

$$\lim_{u\to 0}\Phi\,\mathrm{Tr}_s[\exp(-B_u^2)] = \sum_i \frac{1}{m_i}\int_{M_i/B}\mathrm{Td}^{\Sigma}(TX,h^{TX})\,\mathrm{ch}^{\Sigma}(\xi,h^{\xi}), \tag{2.9}$$

and the differential form in the right-hand side of (2.9) is also in the same cohomology class as $\Phi\,\mathrm{Tr}_s[\exp(-B_u^2)]$.

If B is compact, then the index bundle as an element in the K-group $K(B)$ is well defined:

$$\mathrm{Ind}(D_+^X) = \mathrm{Ker}(D_+^X) - \mathrm{Ker}(D_-^X)\in K(B). \tag{2.10}$$

The differential forms considered above represent in cohomology $\mathrm{ch}(\mathrm{Ker}(D_+^X) - \mathrm{Ker}(D_-^X))$.

Proof. Let $P_u(x,y,b)$ $(x,y\in\pi^{-1}(b), b\in B)$ be the kernel of the heat operator $\exp(-B_u^2)$ with respect to the Riemannian volume form $dv_X(y)$ on (TX,h^{TX}). By the method of [1, Theorem 9.50], we know $P_u(x,y,b)$ defines a smooth family of smoothing operators along the fibers X.

Proceeding as in [11, Theorem 2.2], $\mathrm{Tr}_s[\exp(-B_u^2)]\in P^B$. They are closed and they are in the same cohomology class.

In [34, §6.6], we observe that the finite propagation speed for hyperbolic equations [20, §7.8], [35, Appendix D.2] holds for orbifolds. By (1.24), and using finite propagation speed as in [6, §11b)], [7], one shows that the problem of calculating the limit of $\mathrm{Tr}_s[\exp(-B_u^2)]$ as $u\to 0$ is local on $X_b (b\in B)$.

By Definition 1.3 and the discussion between (1.2)–(1.4), for each $x \in M$, we can choose a chart $\tau : (G_x, \widetilde{U}_x) \to U_x$, such that $\tau^{-1}(x)$ is a point $\widetilde{x}$ and $\widetilde{\pi} : \widetilde{U}_x \to \pi(U_x)$ is a G_x-equivariant submersion. For $\epsilon > 0$, let $B(\widetilde{x}, \epsilon) \subset \widetilde{U}_x$ be the ball with the center $\widetilde{x}$ and radius ϵ. If ϵ is small enough, there exist $x_i \in \pi^{-1}(b)(i \in I = \{1, \dots, k\})$, such that $\{(G_{x_i}, B(\widetilde{x}_i, \frac{\epsilon}{2})) \to B(\widetilde{x}_i, \frac{\epsilon}{2})/G_{x_i}\}_{i \in I}$ is a cover of $\pi^{-1}(b)$. Let $\{\rho_{x_i}\}$ be a partition of unity subordinate to this cover. Then we can replace X by $(\widetilde{TX})_{x_i}/G_{x_i} = \mathbb{C}^l/G_{x_i}$ $(l = \dim X)$, with $0 \in (\widetilde{TX})_{x_i}$ representing x_i.

Note that if Q_U has a $\mathscr{C}^k$-kernel $\widetilde{Q}_U(\widetilde{y}_1, \widetilde{y}_2)$ over $\widetilde{U} \times \widetilde{U}$, then for $y_1, y_2 \in U$,

$$Q_U(y_1, y_2) = \frac{1}{|K_U^\xi|} \sum_{g \in G_U^\xi} (g, 1) \widetilde{Q}_U(g^{-1}\widetilde{y}_1, \widetilde{y}_2), \tag{2.11}$$

is the kernel of the operator

$$Q_U : \mathscr{C}^\infty(U, \xi|_U) \to \mathscr{C}^\infty(U, \xi|_U),$$

with $\tau(\widetilde{y}_i) = y_i (i = 1, 2)$.

Let ${}'\nabla^{\Lambda(T_{\mathbb{R}}^* B) \otimes \Lambda(T^{*(0,1)}X)}$ be the connection on $\Lambda(T_{\mathbb{R}}^* B) \otimes \Lambda(T^{*(0,1)}X)$ along the fibre X given as in [6, Definition 11.7].

For $u > 0$, let $\psi_u : \Lambda(T_{\mathbb{R}}^* B) \to \Lambda(T_{\mathbb{R}}^* B)$ be the map

$$\alpha \in \Lambda(T_{\mathbb{R}}^* B) \to u^{-\frac{\deg \alpha}{2}} \alpha \in \Lambda(T_{\mathbb{R}}^* B).$$

Taken $y \in \mathbb{C}^l$, set $Y = y + \overline{y}$. We identify

$$(\Lambda(T_{\mathbb{R}}^* B) \otimes \Lambda(T^{*(0,1)}X))_Y, \ \xi_Y \quad \text{with} \quad (\Lambda(T_{\mathbb{R}}^* B) \otimes \Lambda(T^{*(0,1)}X))_0, \ \xi_0$$

by parallel transport along the curve $t \in [0, 1] \to tY$ with respect to the connection

$$\psi_u {}'\nabla^{\Lambda(T_{\mathbb{R}}^* B) \otimes \Lambda(T^{*(0,1)}X)} \psi_u^{-1}, \ \nabla^\xi.$$

Let $dv_{T_{x_i}X}(y)$ be the Riemannian volume form on $((\widetilde{TX})_{x_i}, h_{x_i}^{TX}) \simeq \mathbb{R}^{2l}$. For $y \in \mathbb{C}^l$, $|y| < \epsilon/2$, set

$$dv_X(y) = k(y) dv_{T_{x_i}X}(y). \tag{2.12}$$

Let $\widetilde{P}_u(x, y, b)(x, y \in (\widetilde{TX})_{x_i})$ be the kernel of $\exp(-B_u^2)$ associated to $dv_{T_{x_i}X}(y)$. Then by (2.11), and using finite propagation speed as in [6, §11b)], we get

$$\begin{aligned} &\lim_{u \to 0} \int_{M/B} \rho_{x_i} \Phi \operatorname{Tr}_s[P_u(y, y, b)] dv_X(y) \\ &= \lim_{u \to 0} \int_{\widetilde{U}_{x_i}/V_{x_i}} \rho_{x_i} \frac{1}{|G_{x_i}|} \sum_{g \in G_{x_i}} \Phi \operatorname{Tr}_s[g \widetilde{P}_u(g^{-1}y, y, b)] k(y) dv_{T_{x_i}X}(y). \end{aligned} \tag{2.13}$$

By [3, Theorems 4.11–4.15] and [33, Proof of Theorem 2.12], (1.4), one finds that

$$\begin{aligned}
&\lim_{u\to 0}\int_{\widetilde{U}_{x_i}/V_{x_i}} \frac{1}{|G_{x_i}|}\sum_{g\in G_{x_i}} \rho_{x_i}\Phi\,\mathrm{Tr}_s\Big[g\exp(-B_u^2)(g^{-1}y,y,b)\Big]k(y)dv_{T_{x_i}X}(y)\\
&=\frac{1}{|G_{x_i}|}\sum_{g\in G_{x_i}}\int_{\widetilde{U}^g_{x_i}/V_{x_i}} \rho_{x_i}\,\mathrm{Td}_g(TX,g^{TX})\,\mathrm{ch}_g(\xi,h^\xi) \qquad (2.14)\\
&=\sum_j \frac{1}{m_j}\int_{X_j}\rho_{x_i}\,\mathrm{Td}^\Sigma(TX,g^{TX})\,\mathrm{ch}^\Sigma(\xi,h^\xi).
\end{aligned}$$

By (2.13), (2.14), we get (2.9).

Using the same argument of [3, Theorem 3.4] (also. [1, Chap. 9]), we get the last part of Theorem 2.3. □

3. Quillen metrics and curvature theorem

In this section, we construct the Quillen metrics on the inverse of the determinant of the cohomology of a holomorphic orbifold vector bundle, and establish the curvature formula. We extend the results of [12] to complex orbifolds.

This section is organized as follows. In Section 3.1, by [12], we construct the Quillen metrics. In Section 3.2, we recall our anomaly formulas. In Section 3.3, we establish the curvature formula.

In this section, we use the notation of Section 1.1. We remark that all the morphisms considered in Section 1.1 are holomorphic in the rest of the paper.

3.1. Quillen metrics

Let X be a compact complex orbifold of complex dimension l. Let ξ be a holomorphic orbifold vector bundle on X. Let h^{TX}, h^ξ be smooth Hermitian metrics on TX, ξ. Let $h^{H(X,\xi)}$ be the corresponding metric on $H^\bullet(X,\xi)$ induced by the restriction of the L_2-metric (1.18) to $\mathrm{Ker}(D^X)$ via the canonical isomorphism (2.6).

Let $\lambda(\xi)$ be the inverse of the determinant of the cohomology of ξ on X.

$$\det H^\bullet(X,\xi)=\otimes_{i=0}^{\dim X}(\det H^i(X,\xi))^{(-1)^i},\quad \lambda(\xi)=(\det H^\bullet(X,\xi))^{-1}. \qquad (3.1)$$

Let $|\ \ |_{\lambda(\xi)}$ be the metric on $\lambda(\xi)$ induced by $h^{H(X,\xi)}$. The metric $|\ \ |_{\lambda(\xi)}$ will be called the L_2-metric on $\lambda(\xi)$.

Let P be the orthogonal projection operator from $\Omega^\bullet(X,\xi)$ on $\mathrm{Ker}(D^X)$ with respect to the Hermitian product (1.18). Set $P^\perp=1-P$. Let N be the number operator defining the $\mathbb{Z}$-grading of $\Omega^\bullet(X,\xi)$, i.e., N acts by multiplication by k on $\Omega^k(X,\xi)$. For $s\in\mathbb{C}, \mathrm{Re}(s)>\dim X$, set

$$\theta^\xi(s)=-\,\mathrm{Tr}_s[N(D^{X,2})^{-s}P^\perp]. \qquad (3.2)$$

Then

$$\theta^\xi(s)=\frac{-1}{\Gamma(s)}\int_0^{+\infty}t^{s-1}\,\mathrm{Tr}_s\Big[N\exp(-tD^{X,2})P^\perp\Big]dt. \qquad (3.3)$$

From the small time asymptotic expansion of the heat kernel (cf. [34, Proposition 2.1]), (3.3), $\theta^{\xi}(s)$ extends to a meromorphic function of $s \in \mathbb{C}$ which is holomorphic at $s = 0$.

Following [38], [12], now we define the Quillen metric on the line $\lambda(\xi)$.

Definition 3.1. Let $\| \quad \|_{\lambda(\xi)}$ be the Quillen metric on the line $\lambda(\xi)$,

$$\| \quad \|_{\lambda(\xi)} = | \quad |_{\lambda(\xi)} \exp\Big(-\frac{1}{2}\frac{\partial \theta^{\xi}}{\partial s}(0)\Big). \tag{3.4}$$

3.2. Anomaly formulas for Quillen metrics

Let $h^{'TX}, h'^{\xi}$ be another couple of metrics on TX, ξ. We denote with a $'$ the objects attached to $h^{'TX}, h'^{\xi}$.

As in [10, §1f)], in [34, (1.8)], we constructed classes $\widetilde{\mathrm{Td}}^{\Sigma}(TX, h^{TX}, h'^{TX})$ and $\widetilde{\mathrm{ch}}^{\Sigma}(\xi, h^{\xi}, h'^{\xi})$ in $P^{X\cup\Sigma X}/P^{X\cup\Sigma X,0}$ such that

$$\begin{aligned} \frac{\overline{\partial}\partial}{2i\pi}\widetilde{\mathrm{Td}}^{\Sigma}(TX, h^{TX}, h'^{TX}) &= \mathrm{Td}^{\Sigma}(TX, h'^{TX}) - \mathrm{Td}^{\Sigma}(TX, h^{TX}),\\ \frac{\overline{\partial}\partial}{2i\pi}\widetilde{\mathrm{ch}}^{\Sigma}(\xi, h^{\xi}, h'^{\xi}) &= \mathrm{ch}^{\Sigma}(\xi, h'^{\xi}) - \mathrm{ch}^{\Sigma}(\xi, h^{\xi}). \end{aligned} \tag{3.5}$$

Let m_i be the multiplicity of each connected component X_i of $X \cup \Sigma X$.

The following result is [34, Theorem 0.1] which extends the anomaly formulas of [12, Theorem 1.23], to orbifolds.

Theorem 3.2. *Assume that the metrics h^{TX} and h'^{TX} are Kähler. Then*

$$\begin{aligned} \log\left(\frac{\| \quad \|'^2_{\lambda(\xi)}}{\| \quad \|^2_{\lambda(\xi)}}\right) = \sum_i \Big(\frac{1}{m_i}\int_{X_i} \widetilde{\mathrm{Td}}^{\Sigma}(TX, h^{TX}, h'^{TX})\,\mathrm{ch}^{\Sigma}(\xi, h^{\xi})\\ + \frac{1}{m_i}\int_{X_i} \mathrm{Td}^{\Sigma}(TX, h'^{TX})\widetilde{\mathrm{ch}}^{\Sigma}(\xi, h^{\xi}, h'^{\xi})\Big). \end{aligned} \tag{3.6}$$

3.3. The curvature of the determinant line bundle for a Kähler fibration

We now do the same assumption as in Section 1.3 and we use the same notations.

Let $\pi : M \to B$ be a proper holomorphic orbifold submersion of M onto B with compact fibre X. Let ξ be a holomorphic orbifold vector bundle on M. Let ω^M be a real, closed (1,1) form on M taken as in Theorem 1.8. Let h^{TX} be the metric on TX induced by ω^M. Let h^{ξ} be a Hermitian metric on ξ.

We will temporarily assume that B is a complex manifold. Let λ be the $\mathscr{C}^{\infty}$ determinant line bundle on B constructed as in [12, §1b)]. By proceeding as in [12, §1c)], we can define a holomorphic structure on the line bundle λ.

We explain the construction in detail here. Let $\nabla^{E''}$ be the anti-holomorphic part of the connection ∇^E in (1.20) on the infinite-dimensional vector bundle E on B. For $a > 0$, set

$$U^a = \{y \in B : a \notin \mathrm{Spec}(D_y^2)\}, \tag{3.7}$$

where $\mathrm{Spec}(D_y^2)$ is the spectrum of the operator D_y^2. Then on U^a, the sum of the eigenspaces of the operator D_y^2 acting on E_y^j of eigenvalues $< a$, $K_y^{a,j}$ forms a smooth finite-dimensional vector bundle. On U^a, λ coincides with the line bundle λ^a

$$\lambda^a = \otimes_{j=0}^{\dim X} (\det K^{b,j})^{(-1)^{j+1}}, \tag{3.8}$$

and for $0 < a < c$, over $U^a \cap U^c$, we identify λ^a and λ^c by

$$s \in \lambda^a \to s \otimes T(\overline{\partial}^{(a,c)}) \in \lambda^c, \tag{3.9}$$

with $\overline{\partial}^{(a,c)}$ the restriction of $\overline{\partial}^X$ to $K^{c,j}/K^{a,j}$ and the torsion $T(\overline{\partial}^{(a,c)})$ for the complex $(K^{c,j}/K^{a,j}, \overline{\partial}^{(a,c)})$ is defined in [10, Definition 1.1]. Let P^a be the orthogonal projection operator from E onto K^a. By [12, Theorem 1.3], the holomorphic structure on λ^a is defined by

$$\overline{\partial}^{\lambda^a} = \mathrm{Tr}_s[P^a \nabla^{E''} P^a] \tag{3.10}$$

and the identification λ^a and λ^c in (3.9) is holomorphic.

The sheaf $\mathscr{O}_M$ is coherent as explained in Section 2.1. By [35, Theorem 5.4.16], $(M, \mathscr{O}_M)$ is a normal complex space and $\mathscr{O}_M(\xi)$ is a $\mathscr{O}_M$-coherent analytic sheaf, thus by a theorem of Grauert [25], for all $i \geq 0$, the $\mathscr{O}_B$-module $R^i\pi_*\xi$ is coherent. If $i > \dim M$, then $R^i\pi_*\xi = 0$. The functor $R^\bullet\pi_*$ maps the derived category of $\mathscr{O}_M$-module to the derived category of $\mathscr{O}_B$-modules and sends coherent sheaves to complexes with coherent cohomology. As B is a complex manifold, for any $y \in B$, the local ring $\mathscr{O}_{B,y}$ is regular, hence all coherent analytic sheaves on B is perfect and more generally any complex with bounded coherent cohomology is perfect. Thus as in [12, Theorem 3.4], we can associate a (graded) invertible holomorphic sheaf $\det(R^\bullet\pi_*\xi)$ on B, and the associated Knudsen–Mumford determinant line bundle is

$$\lambda^{KM} = (\det(R^\bullet\pi_*\xi))^{-1}. \tag{3.11}$$

In particular, if $R^i\pi_*\xi$ is locally free for all i, we get

$$\lambda^{KM}(\xi) = \otimes_{i\geq 0} (\det(R^i\pi_*\xi))^{(-1)^{i+1}}. \tag{3.12}$$

Let $\mathscr{O}_B^\infty$ be the sheaf of $\mathscr{C}^\infty$ functions on B. Let $\mathscr{H}_{\overline{\partial}}^j(\xi)$ be the cohomology sheaves of the relative Dolbeault complex $(\mathcal{D}_X^\bullet(\xi), \overline{\partial}^X)$ in (2.1) as $\mathscr{O}_B^\infty$-modules. Let $\mathcal{D}_M^\bullet$ be the sheaf of Dolbeault complexes on M, then we can use the partition of unity argument for $\mathcal{D}_M^\bullet$, thus $\mathcal{D}_M^\bullet$ is fine, from the argument of [12, p. 342],

$$R^j\pi_*\xi = \mathscr{H}^j(\pi_*(\mathcal{D}_M^\bullet(\xi))). \tag{3.13}$$

The natural map $T^*M \to T^*X$ induces a map of complexes $\mathcal{D}_M^\bullet(\xi) \to \mathcal{D}_X^\bullet(\xi)$, thus a canonical map on cohomology sheaves

$$\varrho_j : (R^j\pi_*\xi) \otimes_{\mathscr{O}_B} \mathscr{O}_B^\infty \to \mathscr{H}_{\overline{\partial}}^j(\xi). \tag{3.14}$$

Again as B is a manifold, the algebraic argument in the proof of [12, Theorem 3.5] holds, thus we get the analogue of [12, Theorem 3.5]:

Theorem 3.3. *For all $j \geq 0$, the map ϱ_j is an isomorphism.*

Under the assumption of the Kähler fibration, as in [11, Theorem 2.8], we have

$$\overline{\partial}^M = \nabla^{E''} + \overline{\partial}^X. \tag{3.15}$$

From the arguments of the proof of [12, Corollary 3.9, Theorem 3.14], by Theorem 3.3 and (3.15), we get the analogue of [12, Theorem 3.14]:

Theorem 3.4. *The smooth isomorphism λ^{KM} and λ via* (3.14) *is an isomorphism of holomorphic line bundles.*

If B is not a complex manifold, then for each $b \in B$, we consider over $\widetilde{V}_b$ as in Lemma 1.6. By proceeding as in the proof of Proposition 1.4, we construct $\widetilde{\xi}$ a holomorphic orbifold vector bundle on $\widetilde{M}_b$ induced by ξ. Then the above construction gives a G_b-equivariant holomorphic line bundle $\widetilde{\lambda}$ on $\widetilde{V}_b$ and natural compatibilities for different local charts $(G_b, \widetilde{V}_b)$ in Lemma 1.6. Thus we get the determinant line bundle λ as a holomorphic orbifold line bundle on B.

From the algebraic side, the Knudsen–Mumford line bundle $\widetilde{\lambda}^{KM}$ on $\widetilde{V}_b$ is also well defined and G_b-action lifts naturally on it. Thus we get the Knudsen–Mumford line bundle λ^{KM} as a holomorphic orbifold line bundle over B. Moreover, the isomorphism ϱ_j in (3.14) over $\widetilde{V}_b$ is G_b-equivariant via the argument from [12, §3]. Thus we get

Theorem 3.5. *The smooth isomorphism λ^{KM} and λ via* (3.14) *is an isomorphism of holomorphic orbifold line bundles.*

For $\alpha \in \Lambda(T^*_{\mathbb{R}}B)$, $\alpha^{(j)}$ denotes the component of α in $\Lambda^j(T^*_{\mathbb{R}}B)$.

Let m_j be the multiplicity of the component X_j of $X \cup \Sigma X$ in Proposition 1.4. The following result extends the curvature theorem [10, Theorem 0.3], [12, Theorem 1.27] to orbifolds.

Theorem 3.6. *The Quillen metric $\| \ \|_\lambda$ on λ is a smooth metric on B. Let ∇^λ be the holomorphic Hermitian connection on the Hermitian orbifold line bundle $(\lambda, \| \ \|_\lambda)$, then*

$$(\nabla^\lambda)^2 = 2i\pi \left[\sum_j \frac{1}{m_j} \int_{X_j} \mathrm{Td}^\Sigma(TX, g^{TX}) \,\mathrm{ch}^\Sigma(\xi, h^\xi) \right]^{(2)}. \tag{3.16}$$

Proof. Note that for $b \in B$, the Quillen metric $\| \ \|_{\widetilde{\lambda}}$ on the G_b-equivariant holomorphic line bundle $\widetilde{\lambda}(\xi)$ over $\widetilde{V}_b$, is smooth and G_b-invariant. Thus $\| \ \|_\lambda$ on the orbifold line bundle $\lambda(\xi)$ is smooth over B.

We still need to compute the curvature of $(\widetilde{\lambda}, \| \ \|_{\widetilde{\lambda}})$ on $\widetilde{V}_b$, for $b \in B$.

As the argument of [12, Theorem 1.8] is purely functional analysis, by [8, Theorem 1.18] and [9, Theorem 1.19], we know $(\nabla^\lambda)^2$ over $\widetilde{V}_b$ is the constant term in the asymptotic of

$$\mathrm{Tr}_s[\exp(-B_u^2)]^{(2)} \text{ as } u \to 0. \tag{3.17}$$

Now by combining with Theorem 2.3 for the fibration $\widetilde{\pi}_b : \widetilde{M}_b \to \widetilde{V}_b$, we get (3.16). □

4. Analytic torsion forms and anomaly formulas

In this section, we construct the analytic torsion forms associated with an orbifold submersion, and we explain the anomaly formulas. This extends the results of [11], [13] to the orbifold case.

This section is organized as follows. In Section 4.1, we describe the transgression formulas of the superconnection forms, which depend on $u \in]0, +\infty[$. In Section 4.2, proceeding as in [3], [11], [1], we obtain the results on the asymptotics of these forms as $u \to 0$ and $u \to +\infty$. In Section 4.3, we construct the analytic torsion forms, which extend [13]. In Section 4.4, we give the anomaly formulas of the analytic torsion forms, which extend [13] to the orbifold case.

We use here the same notation as in Sections 1, 2.1.

4.1. Superconnection forms and double transgression formulas

Let $\pi : M \to B$ be a proper holomorphic orbifold submersion of M onto B with compact fibre X. Let $n = \dim M$. Let ξ be a holomorphic orbifold vector bundle on M.

By Lemma 1.6, for each $b \in B$, there exists a neighbourhood $(G_b, \widetilde{V}_b) \to V_b$, an orbifold $\widetilde{M}_b$, such that π is induced by a G_b-equivariant orbifold submersion $\widetilde{\pi}_b : \widetilde{M}_b \to \widetilde{V}_b$ with compact fibre X. By proceeding as in the proof of Proposition 1.4, we construct $\widetilde{\xi}$ a holomorphic orbifold vector bundle on $\widetilde{M}_b$ induced by ξ.

The direct image $R^\bullet\pi_*\xi$ is well defined as a $\mathcal{O}_B$-sheaf. Let $\mathcal{D}^j_M(\xi)$ be the sheaf of $\mathscr{C}^\infty$ sections of $\Lambda^j(T^{*(0,1)}M) \otimes \xi$ over M. We have an exact sequence of $\mathcal{O}_M$-sheaves:

$$0 \to \mathcal{O}_M(\xi) \to \mathcal{D}^1_M(\xi) \overset{\overline{\partial}^M}{\to} \cdots \overset{\overline{\partial}^M}{\to} \mathcal{D}^n_M(\xi) \to 0. \tag{4.1}$$

The sheaves $\mathcal{D}^j_M(\xi)$ are fine, as we can apply the partition of unity argument for $\mathcal{D}^\bullet_M(\xi)$, so $(\mathcal{D}^\bullet_M(\xi), \overline{\partial}^M)$ is a π_*-acyclic resolution of $\mathcal{O}_M(\xi)$. So the direct image $R^\bullet\pi_*\xi$ is defined by the presheaf, cf. (3.13):

$$V \to H^\bullet(\Gamma(\pi^{-1}(V), \mathcal{D}^\bullet_M(\xi)), \overline{\partial}^M).$$

But for $b \in B$, on V_b, the presheaf $V \to H^\bullet(\Gamma(\pi^{-1}(V), \mathcal{D}^\bullet_M(\xi)), \overline{\partial}^M)$ is exactly the G_b-invariant sections of $R^\bullet\widetilde{\pi}_{b*}\widetilde{\xi}$ over $\widetilde{V}_b$.

If on each $\widetilde{V}_b$, we define a G_b-equivariant coherent sheaf $R^\bullet\widetilde{\pi}_{b*}\widetilde{\xi}$, then by construction, we verify that this defines a proper coherent sheaf on B.

By the above discussion, the direct image $R^\bullet\pi_*\xi$ is an orbifold $\mathcal{O}_B$-coherent sheaf: over $\widetilde{V}_b$, it is defined by $R^\bullet\widetilde{\pi}_{b*}\widetilde{\xi}$.

We make the *basic assumption* that for $0 \le k \le \dim X$, $b \in B$, the sheaves $R^k\widetilde{\pi}_{b*}\widetilde{\xi}$ is locally free. Then $R^\bullet\pi_*\xi$ is a proper orbifold vector bundle over B. For $p \in \widetilde{V}_b$, let $H^\bullet(X_p, \widetilde{\xi}|_{X_p}) = \oplus_{k=0}^{\dim X} H^k(X_p, \widetilde{\xi}|_{X_p})$ be the cohomology of the sheaf of holomorphic sections of $\widetilde{\xi}$ restricted to X_p. Then the $H^\bullet(X_p, \widetilde{\xi}|_{X_p})$ are the fibres of a G_b-equivariant holomorphic $\mathbb{Z}$-graded vector bundle $H^\bullet(X_p, \widetilde{\xi}|_{X_p})$ on $\widetilde{V}_b$, and $H^\bullet(X_p, \widetilde{\xi}|_{X_p}) = R^\bullet\widetilde{\pi}_{b*}\widetilde{\xi}$. So the $H^\bullet(X_p, \widetilde{\xi}|_{X_p})$ defined an orbifold vector bundle $H^\bullet(X, \xi|_X)$.

Let ω^M be a real closed (1,1) form on M such that ω^M induces a Kähler metric on TX (cf. Theorem 1.8). Let h^ξ be a Hermitian metric on ξ.

We verify easily that the objects on M (for example: ω^M, ξ, h^ξ) lift on $\widetilde{M}_b$. We denote with a $\widetilde{\ }$ the objects we considered in Section 1.1 which are attached to $\widetilde{\pi}_b : \widetilde{M}_b \to \widetilde{V}_b$.

For $p \in \widetilde{V}_b$, set

$$K_p = \{f \in \widetilde{E}_p : \overline{\partial}^{X_p} f = 0, \overline{\partial}^{X_p *} f = 0\}. \tag{4.2}$$

By the Hodge theory (2.6),

$$K_p \simeq H^\bullet(X_p, \widetilde{\xi}|_{X_p}). \tag{4.3}$$

The identification (4.3) induces an identification of the corresponding smooth vector bundles on $\widetilde{V}_b$. Also K inherits a G_b-invariant Hermitian product from the L_2-Hermitian product on $\widetilde{E}$. Let $h^{H(X,\xi|_X)}$ be the corresponding smooth metric on $H^\bullet(X, \xi|_X)$.

Recall that $\widetilde{E}$ is a G_b-equivariant bundle over $\widetilde{V}_b$ and the contribution of ξ is only from its maximal proper orbifold subbundle ξ^{pr} of ξ.

Let B_u be the Bismut superconnection on E constructed in Section 1.3.

Let $\widetilde{P}_u(x, y, p)$ $(x, y \in \widetilde{\pi}^{-1}(p), p \in \widetilde{V}_b)$ be the kernel associated to the operator $\exp(-B_u^2)$ with respect to $dv_X(y)/(2\pi)^{\dim X}$, then we know $\widetilde{P}_u(x, y, p)$ defines a smooth family of smoothing operators.

We define $\mathrm{Tr}_s^\Sigma[\exp(-B_u^2)]$, $\mathrm{Tr}_s^\Sigma[N_u \exp(-B_u^2)]$ as forms over $B \cup \Sigma B$ by: If a connected component B_i of $B \cup \Sigma B$, is locally defined by $\Big((Z_{G_b}(h), \widetilde{V}_b^h) \to \widetilde{V}_b^h/Z_{G_b}(h)\Big)$ $(h \in G_b$, $\widetilde{V}_b^h$ is the fixed point of h over $\widetilde{V}_b)$, then over $\widetilde{V}_b^h/Z_{G_b}(h)$,

$$\begin{aligned} \mathrm{Tr}_s^\Sigma[\exp(-B_u^2)] &= \mathrm{Tr}_s[h \exp(-B_u^2)], \\ \mathrm{Tr}_s^\Sigma[N_u \exp(-B_u^2)] &= \mathrm{Tr}_s[h N_u \exp(-B_u^2)]. \end{aligned} \tag{4.4}$$

As in [11, Theorems 2.2 and 2.9], the forms

$$\Phi\, \mathrm{Tr}_s^\Sigma[\exp(-B_u^2)] \quad \text{and} \quad \Phi\, \mathrm{Tr}_s^\Sigma[N_u \exp(-B_u^2)]$$

lie in $P^{B\cup\Sigma B}$. By Theorem 2.3, we know that the forms $\Phi\,\mathrm{Tr}_s^\Sigma[\exp(-B_u^2)]$ are closed and that their cohomology class is constant and equal to $\mathrm{ch}^\Sigma(H^\bullet(X,\xi|_X))$.

Theorem 4.1. *For $u>0$, the following identity holds*

$$\frac{\partial}{\partial u}\Phi\,\mathrm{Tr}_s^\Sigma[\exp(-B_u^2)] = -\frac{1}{u}\frac{\overline{\partial}\partial}{2i\pi}\Phi\,\mathrm{Tr}_s^\Sigma[N_u\exp(-B_u^2)]. \tag{4.5}$$

Proof. In (4.4), the action h commutes with B_u, N_u. Now, by proceeding as in [11, Theorem 2.9], we get (4.5). □

If $(\alpha_u)_{u>0}$ is a family of smooth forms over $B\cup\Sigma B$, we write that as $u\to 0$ (resp. $u\to+\infty$), $\alpha_u = O(u^k)$, if for any compact subset $K\subset B\cup\Sigma B$, and any $j\in\mathbb{N}$, there is $C>0$ such that the sup of α_u and its derivative of order $\le j$ on K are dominated by Cu^k.

4.2. The asymptotics of the superconnection forms

Clearly, for $b\in B$, in Proposition 1.4, we can choose the ramified covering $(G_{U_i},\widetilde{U}_i)$ of $\pi^{-1}(V_b)$ as the type $(G_x,\widetilde{U}_x)$ such that $\widetilde{U}_x$ is a neighbourhood of $0\in\mathbb{C}^n (n=\dim M)$ and such that G_x acts linearly on $\mathbb{C}^n$. Now, we fixe a choice of

$$(G_{U_i},\widetilde{U}_i) = (G_{x_i},\widetilde{U}_{x_i})_{i\in I}(I=\{1,\dots,k\}),(G_b,\widetilde{V}_b)\to V_b. \tag{4.6}$$

Let $\widetilde{\pi}:(G_{U_i},\widetilde{U}_i)\to(G_b,\widetilde{V}_b)$ be the G_{U_i}-equivariant holomorphic submersion of $\widetilde{U}_i$ onto $\widetilde{V}_b$, and $\pi^{-1}(V_b)=\cup_{i\in I}U_i$. The map $\widetilde{\pi}$ induces naturally a morphism $\pi_i: G_{U_i}\to G_b$. Let $K_{x_i}=K_{U_i}=\mathrm{Ker}\{\pi_i: G_{U_i}\to G_b\}$. Then for $h\in G_b, g\in\pi_i^{-1}(h)$, $\widetilde{\pi}:\widetilde{U}_i^g\to\widetilde{V}_b^h$ is also a submersion. Let ρ_i be a partition of unity of $\pi^{-1}(V_1)$ subordinate to $\{U_i\}_{i\in I}$, for $b\in V_1\subset V_b$ compact.

Let $\beta=\inf_{i\in I}\{\text{injectivity radius of } x_i \text{ on } \widetilde{U}_{x_i}\}$. Take $\alpha\in]0,\beta/4]$.

Let f be a smooth even function defined on $\mathbb{R}$ with values in $[0,1]$, such that

$$f(t)=\begin{cases}1 & \text{for} \quad |t|\le\alpha/2\\ 0 & \text{for} \quad |t|\ge\alpha.\end{cases} \tag{4.7}$$

Set

$$g(t)=1-f(t). \tag{4.8}$$

Definition 4.2. For $u\in]0,1],a\in\mathbb{C}$, set

$$F_u(a)=\int_{-\infty}^{+\infty}\exp(ita\sqrt{2})\exp\left(\frac{-t^2}{2}\right)f(ut)\frac{dt}{\sqrt{2\pi}}, \tag{4.9}$$
$$G_u(a)=\int_{-\infty}^{+\infty}\exp(ita\sqrt{2})\exp\left(\frac{-t^2}{2}\right)g(ut)\frac{dt}{\sqrt{2\pi}}.$$

Clearly

$$F_u(a)+G_u(a)=\exp(-a^2). \tag{4.10}$$

The functions $F_u(a), G_u(a)$ are even holomorphic functions. So there exist holomorphic functions $\widetilde{F}_u(a)$, $\widetilde{G}_u(a)$ such that

$$F_u(a) = \widetilde{F}_u(a^2), \quad G_u(a) = \widetilde{G}_u(a^2). \tag{4.11}$$

Let μ be a form on $M \cup \Sigma M$, we define $\int_{X\cup\Sigma X} \mu$ as a form over $B \cup \Sigma B$: locally over $\widetilde{V}_b^h / Z_{G_b}(h) \subset B \cup \Sigma B$, we denote

$$\int_{X\cup\Sigma X} \mu = \sum_{i\in I} \frac{1}{|K_{U_i}|} \sum_{g\in\tau_{U_i}^{-1}(h)} \int_{\widetilde{U}_i^g/\widetilde{V}^h} \rho_i \mu. \tag{4.12}$$

Put

$$\begin{aligned} C_{-1} &= \int_{X\cup\Sigma X} \frac{\omega^M}{2\pi} \operatorname{Td}^\Sigma(TX, h^{TX}) \operatorname{ch}^\Sigma(\xi, h^\xi), \\ C_0 &= \int_{X\cup\Sigma X} \Big(-(\operatorname{Td}')^\Sigma(TX, h^{TX}) + \dim X \operatorname{Td}^\Sigma(TX, h^{TX})\Big) \operatorname{ch}^\Sigma(\xi, h^\xi). \end{aligned} \tag{4.13}$$

Set

$$\begin{aligned} \operatorname{ch}^\Sigma(H^\bullet(X, \xi|_X), h^{H(X,\xi|_X)}) &= \sum_{k=0}^{\dim X} (-1)^k \operatorname{ch}^\Sigma(H^k(X, \xi|_X), h^{H(X,\xi|_X)}), \\ \operatorname{ch}'^\Sigma(H^\bullet(X, \xi|_X), h^{H(X,\xi|_X)}) &= \sum_{k=0}^{\dim X} (-1)^k k \operatorname{ch}^\Sigma(H^k(X, \xi|_X), h^{H(X,\xi|_X)}). \end{aligned} \tag{4.14}$$

Theorem 4.3. *As $u \to 0$*

$$\Phi \operatorname{Tr}_s^\Sigma[\exp(-B_u^2)] = \int_{X\cup\Sigma X} \operatorname{Td}^\Sigma(TX, h^{TX}) \operatorname{ch}^\Sigma(\xi, h^\xi) + O(u). \tag{4.15}$$

There are forms $C'_j \in P^{B\cup\Sigma B} (j \geq -1)$ such that for $k \in \mathbb{N}$, as $u \to 0$

$$\Phi \operatorname{Tr}_s^\Sigma[N_u \exp(-B_u^2)] = \sum_{j=-1}^{k} C'_j u^j + O(u^{k+1}). \tag{4.16}$$

Also

$$\begin{aligned} C'_{-1} &= C_{-1}, \\ C'_0 &= C_0 \quad \textit{in} \quad P^{B\cup\Sigma B}/P^{B\cup\Sigma B,0}. \end{aligned} \tag{4.17}$$

Proof. Recall that in the construction of the orbifold $\widetilde{M}_b$, we use the local coordinate system $(K_{U_i}, \widetilde{U}_i) \to \widetilde{U}_i/K_{U_i}$.

By (2.8) and the definition of smooth sections for an orbifold vector bundle, only the maximal proper orbifold subbundle ξ^{pr} of ξ makes contributions in various steps, thus we will assume simply that ξ is a proper orbifold vector bundle on M.

Following (4.4), we will calculate the following limit as $u \to 0$,

$$I_i(h, u) = \int_X \rho_i(p, x) \Phi \operatorname{Tr}_s[h\widetilde{P}_u(x, x, p)] dv_X. \tag{4.18}$$

Lemma 4.4. *There exist $c > 0, C > 0$ such that for $u \in]0,1]$*

$$\left| \operatorname{Tr}_s[\rho_i h \widetilde{G}_u(B_u^2)] \right| \leq c \exp\left(\frac{-C}{u^2}\right). \tag{4.19}$$

Proof. By proceeding as in the proof of [2, Proposition 8.3], we have (4.19). □

Let $\widetilde{F}_u(B_u^2)(x_1, x_2)((x_1, x_2) \in X_p \times X_p)$ be the smooth kernel associated with $\widetilde{F}_u(B_u^2)$ with respect to $dv_X(x_2)/(2\pi)^{\dim X}$. Using (4.5), (4.9), and finite propagation speed [20, §7.8], [35, Appendix D. 2], it is clear that $\widetilde{F}_u(B_u^2)(x, x') = 0$ if $d(x, x') > \alpha$, and $\widetilde{F}_u(B_u^2)(x, x')$ depends only on the restriction of B_u^2 to $B^X(x, \alpha)$.

We replace X by $(\widetilde{TX})_{x_i}/K_{x_i} = \mathbb{C}^l/K_{x_i}$ $(l = \dim X)$, with $0 \in (\widetilde{TX})_{x_i}$ representing x_i, and that the extended fibration over $\mathbb{C}^l$ coincides with the given fibration over $B(0, 2\alpha) \subset \mathbb{C}^l$.

Let Δ^{TX} be the standard Laplacian on $((\widetilde{TX})_{x_i}, h_{x_i}^{TX})$. Let $\rho(Y)$ be a $\mathscr{C}^\infty$ function over $\mathbb{C}^l$ which is equal 1 if $|Y| \leq \alpha$, equal 0 if $|Y| \geq 2\alpha$. Let

$$L_u^1 = (1 - \rho^2(Y)) \left(-\frac{1}{2} u \Delta^{TX} \right) + \rho^2(Y) B_u^2. \tag{4.20}$$

Let $\widetilde{F}_u(L_u^1)(x, x')(x, x' \in \mathbb{C}^l)$ be the smooth kernel of $\widetilde{F}_u(L_u^1)$ with respect to $dv_{T_{x_i}X}(x')/(2\pi)^{\dim X}$. For $y \in \mathbb{C}^l$, $|y| < 2\alpha$, as in (2.12), set

$$dv_X(y) = k(y) dv_{T_{x_i}X}(y). \tag{4.21}$$

Then, for $|y| < 2\alpha$, $y \in (\widetilde{TX})_{x_i}^g$, we get

$$dv_{X^g}(y) = k(y) dv_{T_{x_i}X^g}(y). \tag{4.22}$$

By (2.11) and the above discussion, if α is enough small, for

$$(x, x') \in \operatorname{supp}(\rho_i) \times \operatorname{supp}(\rho_i),$$

we get

$$\widetilde{F}_u(B_u^2)(x, x') = k(x') \sum_{g \in K_{x_i}} (g, 1) \widetilde{F}_u(L_u^1)(g^{-1}\widetilde{x}, \widetilde{x}'). \tag{4.23}$$

Note that K_{x_i} acts on $\widetilde{\xi}$ as we explained above (4.18) that ξ is proper.

By (4.18), (4.19), (4.23), we get

$$\begin{aligned} \lim_{u \to 0} I_i(h, u) &= \lim_{u \to 0} \int_X \rho_i(p, x) \Phi \operatorname{Tr}_s[h \widetilde{F}_u(B_u^2)(x, x)] dv_X/(2\pi)^{\dim X} \\ &= \lim_{u \to 0} \int_{\mathbb{C}^l} \frac{1}{|K_{x_i}|} \sum_{g \in K_{x_i}} \rho_i(p, x) \Phi \operatorname{Tr}_s[h(g \widetilde{F}_u(L_u^1))(\widetilde{x}, \widetilde{x})] k(\widetilde{x}) dv_{T_{x_i}X}/(2\pi)^{\dim X} \\ &= \lim_{u \to 0} \int_{\mathbb{C}^l} \frac{1}{|K_{x_i}|} \sum_{g \in \tau_{U_i}^{-1}(h)} \rho_i(p, x) \Phi \operatorname{Tr}_s[g \widetilde{F}_u(L_u^1)(g^{-1}\widetilde{x}, \widetilde{x})] k(\widetilde{x}) dv_{T_{x_i}X}/(2\pi)^{\dim X}. \end{aligned} \tag{4.24}$$

We observe that for any $k \in \mathbb{N}$, $c > 0$, there is $C > 0, C' > 0$ such that for $u \subset]0,1]$,

$$\sup_{|\mathrm{Im}(a)| \le c} |a|^k \left| \widetilde{F}_u(a^2) - \exp(-a^2) \right| \le C' \exp(\frac{-C}{u^2}). \tag{4.25}$$

For each $g \in \tau_{U_i}^{-1}(h)$, by using (4.22), (4.25), and by proceeding as in [33, (2.42)–(2.51)], we get

$$\begin{aligned} &\lim_{u\to 0} \int_{\mathbb{C}^l} \rho_i(p,\widetilde{x}) \Phi \operatorname{Tr}_s[g\widetilde{F}_u(L_u^1)(g^{-1}\widetilde{x},\widetilde{x})] k(\widetilde{x}) dv_{T_{x_i}X}/(2\pi)^l \\ &= \lim_{u\to 0} \int_{(\widetilde{TX})^g_{x_i}} \int_{z \in \widetilde{N}_{X^g/X}} \rho_i(p,(\widetilde{x},\widetilde{z})) \Phi \operatorname{Tr}_s[g\widetilde{F}_u(L_u^1)(g^{-1}(\widetilde{x},\widetilde{z}),(\widetilde{x},\widetilde{z}))] \\ &\qquad k(\widetilde{x},\widetilde{z}) dv_{T_{x_i}X^g}(\widetilde{x}) dv_{N_{X^g/X},x_i}(\widetilde{z})/(2\pi)^l \\ &\qquad = \int_{(\widetilde{TX})^g_{x_i}} \rho_i(p,\widetilde{x}) \operatorname{Td}_g(\widetilde{TX}, h^{TX}) \operatorname{ch}_g(\widetilde{\xi}, h^{\xi}). \end{aligned} \tag{4.26}$$

By (4.12), (4.18), (4.19), (4.24), (4.26), we get (4.15).

By combining the techniques of proof of [11, Theorems 2.2, 2.3, 2.9 and 2.16] and the proof of (4.15), we get (4.16) and (4.17). □

Theorem 4.5. *As $u \to +\infty$*

$$\begin{aligned} \Phi \operatorname{Tr}_s^{\Sigma}[\exp(-B_u^2)] &= \operatorname{ch}^{\Sigma}(H^\bullet(X,\xi|_X), h^{H(X,\xi|_X)}) + O\left(\frac{1}{\sqrt{u}}\right), \\ \Phi \operatorname{Tr}_s^{\Sigma}[N_u \exp(-B_u^2)] &= \operatorname{ch}'^{\Sigma}(H^\bullet(X,\xi|_X), h^{H(X,\xi|_X)}) + O\left(\frac{1}{\sqrt{u}}\right). \end{aligned} \tag{4.27}$$

Proof. Equation (4.27) was stated in [13, Theorem 3.4], if M, B are complex manifolds. By proceeding as in [1, Theorem 9.23], we get also (4.27) in our situation. □

4.3. Analytic torsion forms

For $s \in \mathbb{C}, \mathrm{Re}(s) > 1$, set

$$\zeta_1(s) = -\frac{1}{\Gamma(s)} \int_0^1 u^{s-1} \Big(\Phi \operatorname{Tr}_s^{\Sigma}[N_u \exp(-B_u^2)] - \operatorname{ch}'^{\Sigma}(H^\bullet(X,\xi|_X), h^{H(X,\xi|_X)}) \Big) du.$$

Using (4.16), we see that $\zeta_1(s)$ extends to a holomorphic function of $s \in \mathbb{C}$ near $s = 0$.

For $s \in \mathbb{C}, \mathrm{Re}(s) < \frac{1}{2}$, set

$$\zeta_2(s) = -\frac{1}{\Gamma(s)} \int_1^{+\infty} u^{s-1} \Big(\Phi \operatorname{Tr}_s^{\Sigma}[N_u \exp(-B_u^2)] - \operatorname{ch}'^{\Sigma}(H^\bullet(X,\xi|_X), h^{H(X,\xi|_X)}) \Big) du.$$

Then $\zeta_2(s)$ is a holomorphic function of s.

Definition 4.6. Set

$$T(\omega^M, h^{\xi}) = \frac{\partial}{\partial s}(\zeta_1 + \zeta_2)(0). \tag{4.28}$$

Then $T(\omega^M, h^\xi)$ is a smooth form on $B \cup \Sigma B$. Using (4.16), (4.27), we get

$$\begin{aligned} T(\omega^M, h^\xi) = & -\int_0^1 \Big(\Phi \operatorname{Tr}_s^\Sigma [N_u \exp(-B_u^2)] - \frac{C'_{-1}}{u} - C'_0 \Big) \frac{du}{u} \\ & - \int_1^{+\infty} \Big(\Phi \operatorname{Tr}_s^\Sigma [N_u \exp(-B_u^2)] - \operatorname{ch}'^\Sigma (H^\bullet(X, \xi|_X), h^{H(X,\xi|_X)}) \Big) \frac{du}{u} \\ & + C'_{-1} + \Gamma'(1) \Big(C'_0 - \operatorname{ch}'^\Sigma (H^\bullet(X, \xi|_X), h^{H(X,\xi|_X)}) \Big). \end{aligned} \tag{4.29}$$

Theorem 4.7. *The form $T(\omega^M, h^\xi)$ lies in $P^{B\cup\Sigma B}$. Moreover*

$$\begin{aligned} \frac{\overline{\partial}\partial}{2i\pi} T(\omega^M, h^\xi) = & \operatorname{ch}^\Sigma \Big(H^\bullet(X, \xi|_X), h^{H(X,\xi|_X)} \Big) \\ & - \int_{X\cup\Sigma X} \operatorname{Td}^\Sigma(TX, h^{TX}) \operatorname{ch}^\Sigma(\xi, h^\xi). \end{aligned} \tag{4.30}$$

Proof. By Theorems 4.1, 4.3, 4.5, we get (4.30). □

4.4. Anomaly formulas for the analytic torsion forms

Let now (ω', h'^ξ) be another couple of objects similar to (ω, h^ξ). We denote with a $'$ the objects associated to (ω', h'^ξ).

Theorem 4.8. *The following identity holds in $P^{B\cup\Sigma B}/P^{B\cup\Sigma B,0}$,*

$$T(\omega', h'^\xi) - T(\omega, h^\xi) = \widetilde{\operatorname{ch}}^\Sigma \Big(H^\bullet(X, \xi|_X), h^{H(X,\xi|_X)}, h'^{H(X,\xi|_X)} \Big) \tag{4.31}$$
$$- \int_{X\cup\Sigma X} \Big[\widetilde{\operatorname{Td}}^\Sigma (TX, h^{TX}, h'^{TX}) \operatorname{ch}^\Sigma(\xi, h^\xi) + \operatorname{Td}^\Sigma(TX, h'^{TX}) \widetilde{\operatorname{ch}}^\Sigma(\xi, h^\xi, h'^\xi) \Big].$$

In particular, the class of $T(\omega, h^\xi) \in P^{B\cup\Sigma B}/P^{B\cup\Sigma B,0}$ depends only on (h^{TX}, h^ξ).

Proof. By (4.4), and by combining the proof of [33, Theorem 2.13], and Theorem 4.3, we have (4.31). □

5. The Quillen norm in the submersion case

Let $\pi : M \to B$ be a holomorphic orbifold submersion of M onto B with compact fibre X. Let ξ be a holomorphic orbifold vector bundle on M. In this section, we will calculate the Quillen norm of the canonical section of $\lambda_M(\xi) \otimes \lambda^{-1}(R^\bullet \pi_* \xi)$. This extends the result of [2, Theorem 3.1] to the orbifold case.

This section is organized as follows. In Section 5.1, we state a formula for the Quillen norm of the canonical section σ. In Section 5.2, we introduce a 1-form on $\mathbb{R}_+^* \times \mathbb{R}_+^*$ as in [2, §3a)]. In Section 5.3, we state eight intermediate results which we need for the proof of Theorem 5.1, whose proofs are delayed to Sections 5.5–5.8. In Section 5.4, we prove Theorem 5.1. In Section 5.5, we prove Theorems 5.7–5.11. In Section 5.6, we prove Theorem 5.12. In Section 5.7, we prove Theorem 5.13. In Section 5.8, we prove Theorem 5.14.

We use the notation of Sections 1, 4.

5.1. A formula for the Quillen norm of the canonical section σ

Let M, B be compact complex orbifolds. Let $\pi : M \to B$ be a holomorphic orbifold submersion of M onto B with compact fiber X. Let ξ be a holomorphic orbifold vector bundle on M.

We assume that the sheaves $R^k\pi_*\xi (0 \leq k \leq \dim X)$ are orbifold vector bundles on B. Set

$$\begin{aligned}\lambda_M(\xi) &= \otimes_j (\det H^j(M,\xi))^{(-1)^{j+1}},\\ \lambda(R^\bullet\pi_*\xi) &= \otimes_{j,k}(\det H^j(B, R^k\pi_*\xi))^{(-1)^{j+k+1}}.\end{aligned} \tag{5.1}$$

By [28], the line $\lambda_M(\xi)\otimes\lambda^{-1}(R^\bullet\pi_*\xi)$ has a canonical nonzero section σ.

Let h^{TM}, h^{TB} be Kähler metrics on TM and TB. Let h^{TX} be the metric induced by h^{TM} on TX. Let h^ξ be a Hermitian metric on ξ.

On M, we have the exact sequence of holomorphic Hermitian proper orbifold vector bundles (cf. Definition 1.2)

$$0 \to TX \to TM \to \pi^*TB \to 0. \tag{5.2}$$

By a construction of [10, §1f)], there is a uniquely defined class of forms

$$\widetilde{\mathrm{Td}}^\Sigma(TM, TB, h^{TM}, h^{TB}) \in P^{M\cup\Sigma M}/P^{M\cup\Sigma M,0},$$

such that

$$\begin{aligned}&\frac{\overline{\partial}\partial}{2i\pi}\widetilde{\mathrm{Td}}^\Sigma(TM, TB, h^{TM}, h^{TB})\\ &\quad= \mathrm{Td}^\Sigma(TM, h^{TM}) - \pi^*(\mathrm{Td}^\Sigma(TB, h^{TB}))\,\mathrm{Td}^\Sigma(TX, h^{TX}).\end{aligned} \tag{5.3}$$

Let ω^M be the Kähler form of h^{TM}. Let $\|\ \ \|_{\lambda_M(\xi)\otimes\lambda^{-1}(R^\bullet\pi_*\xi)}$ be the Quillen metric on the line $\lambda_M(\xi)\otimes\lambda^{-1}(R^\bullet\pi_*\xi)$ attached to the metrics h^{TM}, h^ξ, h^{TB}, $h^{H(X,\xi|_X)}$ on TM, ξ, TB, $R^\bullet\pi_*\xi$.

Recall that the integral $\int_{B\cup\Sigma B}$ is defined in (4.12).

Now we state the main result of this section, which extends [2, Theorem 3.1].

Theorem 5.1. *The following identity holds,*

$$\begin{aligned}\log\left(\|\sigma\|^2_{\lambda_M(\xi)\otimes\lambda^{-1}(R^\bullet\pi_*\xi)}\right) &= -\int_{B\cup\Sigma B}\mathrm{Td}^\Sigma(TB, h^{TB})T(\omega^M, h^\xi)\\ &\quad+\int_{M\cup\Sigma M}\widetilde{\mathrm{Td}}^\Sigma(TM, TB, h^{TM}, h^{TB})\,\mathrm{ch}^\Sigma(\xi, h^\xi).\end{aligned} \tag{5.4}$$

Proof. The remainder of this section is devoted to the proof of Theorem 5.1. □

Remark 5.2. By Theorem 4.8, to prove Theorem 5.1 for any Kähler metrics h^{TM}, h^{TB}, we only need to establish (5.4) for one given metrics h^{TM}, h^{TB}. So by replacing h^{TM} by $h^{TM} + \pi^*h^{TB}$, we may and we will assume that $\widetilde{h}^{TM}$ is a Kähler metric on TM and

$$h^{TM} = \widetilde{h}^{TM} + \pi^*h^{TB}. \tag{5.5}$$

5.2. A fundamental closed 1-form

Let N_V, N_H be the number operators of $\Lambda(T^{*(0,1)}X)$, $\Lambda(T^{*(0,1)}B)$. As in [2, §4], the operators N_V and N_H act naturally on $\Lambda(T^{*(0,1)}M)$. Of course, $N = N_V + N_H$ defines the total grading of $\Lambda(T^{*(0,1)}M) \otimes \xi$ and $\Omega^\bullet(M, \xi)$.

Definition 5.3. For $T > 0$, let h_T^{TM} be the Kähler metric on TM

$$h_T^{TM} = \frac{1}{T^2}\widetilde{h}^{TM} + \pi^* h^{TB}. \tag{5.6}$$

Let $\langle\ \ \rangle_T$ be the Hermitian product (1.18) on $\Omega^\bullet(M, \xi)$ attached to the metrics h_T^{TM}, h^ξ. Let D_T^M be the corresponding operator constructed in (1.19) acting on $\Omega^\bullet(M, \xi)$. Let $*_T$ be the Hodge operator with respect to the metric h_T^{TM}. Then $*_T$ acts on $\Lambda(T_{\mathbb{R}}^* M) \otimes \xi$.

Theorem 5.4. *Let $\alpha_{u,T}$ be the 1-form on $\mathbb{R}_+^* \times \mathbb{R}_+^*$*

$$\alpha_{u,T} = \frac{2du}{u}\,\mathrm{Tr}_s\left[N \exp(-u^2 D_T^{M,2})\right] + dT\,\mathrm{Tr}_s\left[*_T^{-1}\frac{\partial *_T}{\partial T}\exp(-u^2 D_T^{M,2})\right]. \tag{5.7}$$

Then $\alpha_{u,T}$ is closed.

Proof. The proof of Theorem 5.4 is identical to the proof of [2, Theorem 4.3 and (4.30)]. □

Take $\epsilon, A, T, 0 < \epsilon \leq 1 \leq A < +\infty$, $1 \leq T_0 < +\infty$. Let $\Gamma = \Gamma_{\epsilon,A,T_0}$ be the oriented contour in $\mathbb{R}_+^* \times \mathbb{R}_+^*$

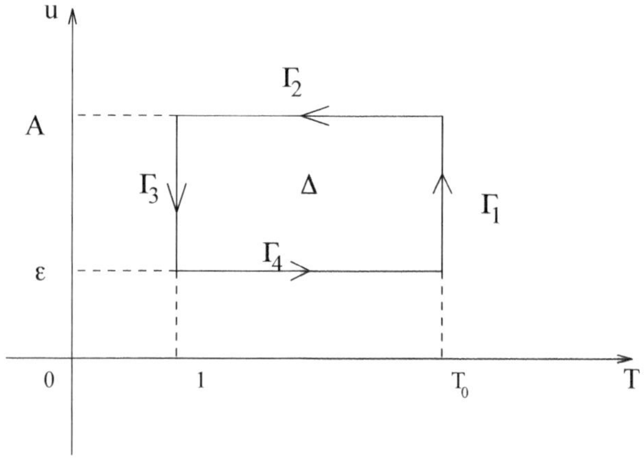

The contour Γ is made of the four oriented pieces $\Gamma_1, \ldots, \Gamma_4$ indicated above. For $1 \leq k \leq 4$, set

$$I_k^0 = \int_{\Gamma_k} \alpha. \tag{5.8}$$

Theorem 5.5. *The following identity holds,*

$$\sum_{k=1}^{4} I_k^0 = 0. \tag{5.9}$$

Proof. This follows from Theorem 5.4. □

5.3. Eight intermediate results

Let $\overline{\partial}^{B*}$ be the formal adjoint of the operator $\overline{\partial}^{B}$ acting on $\Omega^\bullet(B, R^\bullet\pi_*\xi)$, with respect to the metrics $h^{TB}, h^{H(X,\xi|_X)}$. Set

$$D^B = \overline{\partial}^B + \overline{\partial}^{B*}, \quad F = \mathrm{Ker}(D^B). \tag{5.10}$$

By the Hodge theory,

$$H^\bullet(B, R^\bullet\pi_*\xi) \simeq F. \tag{5.11}$$

Let Q be the orthogonal projection from $\Omega^\bullet(B, R^\bullet\pi_*\xi)$ on F with respect to the Hermitian product (1.18) attached to the metrics $h^{TB}, h^{H(X,\xi|_X)}$. Set $Q^\perp = 1 - Q$.

Let $a \in]0,1]$ be such that the operator $D^{B,2}$ has no eigenvalues in $]0, 2a]$.

Definition 5.6. For $T > 0$, set

$$E_T = \mathrm{Ker}(D_T^{M,2}). \tag{5.12}$$

Let P_T be the orthogonal projection operator from $\Omega^\bullet(M, \xi)$ on E_T with respect to $\langle\ \rangle_T$.

Let $E_T^{[0,a]}$ (resp. $E_T^{]0,a]}$) be the direct sum of the eigenspaces of $D_T^{M,2}$ associated with eigenvalues $\lambda \in [0,a]$ (resp. $\lambda \in]0,a]$). Let $D_T^{M,2,[0,a]}$(resp. $D_T^{M,2,]0,a]}$) be the restriction of $D_T^{M,2}$ to $E_T^{[0,a]}$ (resp. $E_T^{]0,a]}$). Let $P_T^{[0,a]}$ (resp. $P_T^{]0,a]}$) be the orthogonal projection operator from $\Omega^\bullet(M,\xi)$ on $E_T^{[0,a]}$ (resp. $E_T^{]0,a]}$) with respect to $\langle\ \rangle_T$. Set $P^{]a,+\infty[} = 1 - P_T^{[0,a]}$. Set

$$\chi(\xi) = \sum_k (-1)^k \dim H^k(M,\xi), \quad \chi(R^j\pi_*\xi) = \sum_k (-1)^k \dim H^k(B, R^j\pi_*\xi).$$

We now state eight intermediate results contained in Theorems 5.7–5.14 which play an essential role in the proof of Theorem 5.1. The proof of Theorems 5.7–5.14 are deferred to Sections 5.5–5.8.

Theorem 5.7. *For any $u > 0$,*

$$\lim_{T\to+\infty} \mathrm{Tr}_s\left[N \exp(-u^2 D_T^{M,2})\right] = \mathrm{Tr}_s\left[N \exp(-u^2 D^{B,2})\right]. \tag{5.13}$$

For any $u > 0$, there exists $C > 0$ such that for $T \geq 1$,

$$\left|\mathrm{Tr}_s[N_V \exp(-u^2 D_T^{M,2})] - \sum_{j=0}^{\dim X} (-1)^j \chi(R^j\pi_*\xi)\right| \leq \frac{C}{T}. \tag{5.14}$$

For any $\varepsilon > 0$, there exists $C > 0$ such that for $u \geq \varepsilon$, $T \geq 1$,

$$\left| \operatorname{Tr}[\exp(-u^2 D_T^{M,2})] \right| \leq C. \tag{5.15}$$

Theorem 5.8. *For any $u > 0$,*

$$\lim_{T\to+\infty} \operatorname{Tr}_s \left[N \exp(-u^2 D_T^{M,2}) P^{]a,+\infty[} \right] = \operatorname{Tr}_s \left[N \exp(-u^2 D^{B,2}) Q^{\perp} \right]. \tag{5.16}$$

There exist $c > 0, C > 0$ such that for $u \geq 1, T \geq 1$,

$$\left| \operatorname{Tr}[N \exp(-u D_T^{M,2}) P^{]a,+\infty[}] \right| \leq c \exp(-Cu). \tag{5.17}$$

Theorem 5.9. *The following identity holds,*

$$\lim_{T\to+\infty} \operatorname{Tr} \left[D_T^{M,2,[0,a]} \right] = 0. \tag{5.18}$$

For $T \geq 1$ large enough, for $0 \leq k \leq \dim M$,

$$\dim E_T^{[0,a],k} = \sum_{j=0}^{k} \dim H^j(B, R^{k-j}\pi_*\xi). \tag{5.19}$$

Let (E_r, d_r) $(r \geq 2)$ be the spectral sequence of the Dolbeault complex $(\Omega^\bullet(M,\xi), \overline{\partial}^M)$ filtered as in [2, §1a)]. Then as in [2, §4], for $r \geq 2$, E_r is equipped with a metric h^{E_r} associated to h^{TM}, h^{TB}, h^{ξ}. For $r \geq 2$, let ${}_r| \quad |_{\lambda_M(\xi)}$ be the corresponding metric on $\lambda_M(\xi) \simeq (\det E_r)^{-1}$

For $r \geq 1$, let $N_{|E_r}, N_{H|E_r}, N_{V|E_r}$ be the restrictions of N, N_H, N_V to E_r.

Theorem 5.10. *The following identity holds,*

$$\lim_{T\to+\infty} \Big\{ \operatorname{Tr}_s[N \log(D_T^{M,2,]0,a]})] + 2\sum_{r\geq 2}(r-1)\Big(\operatorname{Tr}_s[N_{|E_r}] - \operatorname{Tr}_s[N_{|E_{r+1}}]\Big)\log(T)\Big\}$$
$$= \log\left(\frac{{}_\infty| \quad |_{\lambda_M(\xi)}}{{}_2| \quad |_{\lambda_M(\xi)}}\right)^2. \tag{5.20}$$

For $T \geq 1$, let $| \quad |_{\lambda_M(\xi),T}$ be the L_2 metric on the line $\lambda_M(\xi)$ associated to the metrics h_T^{TM}, h^{ξ} on TM, ξ.

Theorem 5.11. *The following identity holds,*

$$\lim_{T\to+\infty} \Big\{ \log\left(\frac{| \quad |_{\lambda_M(\xi),T}}{| \quad |_{\lambda_M(\xi)}}\right)^2 + 2\Big(-\dim X\chi(\xi) + \operatorname{Tr}_s[N_{V|E_\infty}]\Big)\log(T)\Big\}$$
$$= \log\left(\frac{{}_\infty| \quad |_{\lambda_M(\xi)}}{| \quad |_{\lambda_M(\xi)}}\right)^2. \tag{5.21}$$

For $u > 0$, let B_u be the Bismut superconnection on $\Omega^\bullet(X, \xi|_X)$ constructed in Section 1.3 which is attached to h^{TM}, h^{ξ} on TM, ξ. Let $\widetilde{N}_u$ be the operator defined in Section 1.3 associated with the metric $\widetilde{h}^{TM}$.

Theorem 5.12. *For any* $T \geq 1$,

$$\begin{aligned}&\lim_{\varepsilon\to 0} \operatorname{Tr}_s\left[*_{T/\varepsilon}^{-1} \frac{\partial}{\partial T}(*_{T/\varepsilon}) \exp(-\varepsilon^2 D_{T/\varepsilon}^{M,2})\right] \\ &= \frac{2}{T}\int_{B\cup\Sigma B} \operatorname{Td}^{\Sigma}(TB, h^{TB})\Phi \operatorname{Tr}_s^{\Sigma}\left[\widetilde{N}_{T^2}\exp(-B_{T^2}^2)\right] - \frac{2}{T}\dim X\chi(\xi).\end{aligned} \tag{5.22}$$

Let $\omega^M, \widetilde{\omega}^M, \omega^B$ be the Kähler forms associated with $h^{TM}, \widetilde{h}^{TM}, h^{TB}$. Let ∇_T^{TM} be the holomorphic Hermitian connection on (TM, h_T^{TM}), and let R_T^{TM} be its curvature.

Theorem 5.13. *There exists* $C > 0$ *such that for* $\varepsilon \in]0,1], \varepsilon \leq T \leq 1$,

$$\begin{aligned}\Big| &\operatorname{Tr}_s\left[*_{T/\varepsilon}^{-1} \frac{\partial}{\partial T}(*_{T/\varepsilon}) \exp\Big(-\varepsilon^2 D_{T/\varepsilon}^{M,2}\Big)\right] \\ &- \frac{2}{T^3}\int_{M\cup\Sigma M} \frac{\widetilde{\omega}^M}{2\pi}\operatorname{Td}^{\Sigma}(TM)\operatorname{ch}^{\Sigma}(\xi) \\ &+ \int_{M\cup\Sigma M} \frac{\partial}{\partial b}\operatorname{Td}^{\Sigma}\Big(\frac{-R_{T/\varepsilon}^{TM}}{2i\pi} - b(h_{T/\varepsilon}^{TM})^{-1}\frac{\partial}{\partial T}(h_{T/\varepsilon}^{TM})\Big)_{b=0}\operatorname{ch}^{\Sigma}(\xi, h^{\xi})\Big| \leq C.\end{aligned} \tag{5.23}$$

Theorem 5.14. *There exist* $\delta \in]0,1], C > 0$ *such that for* $\varepsilon \in]0,1], T \geq 1$,

$$\begin{aligned}\Big| &\operatorname{Tr}_s\left[*_{T/\varepsilon}^{-1} \frac{\partial}{\partial T}(*_{T/\varepsilon}) \exp(-\varepsilon^2 D_{T/\varepsilon}^{M,2})\right] \\ &- \frac{2}{T}\Big(\sum_{j=0}^{\dim X}(-1)^j j\chi(R^j\pi_*\xi) - \dim X\chi(\xi)\Big)\Big| \leq \frac{C}{T^{1+\delta}}.\end{aligned} \tag{5.24}$$

Besides, at a formal level, Theorems 5.7–5.14 can be obtained formally from [2, Theorems 4.8–4.15]. This will permit us to transfer formally the discussion in [2, §4] to our situation.

5.4. A proof of Theorem 5.1

By Theorem 5.5, Theorems 5.7–5.14 and proceeding as in [2, §4c), d)], we get (5.4).

5.5. A proof of Theorems 5.7–5.11

The proof of Theorems 5.7–5.11 is essentially the same as the proof of [2, Theorems 4.8–4.12] given in [2, §5, §6], where the corresponding results were established when M, B are manifolds. Now we use the notation of [2, §5].

By Proposition 1.4, for each $b \in B$, there exists a small neighbourhood $(G_b, \widetilde{V}_b) \to V_b$, an orbifold $\widetilde{M}_b$, such that π is induced by a G_b-equivariant orbifold submersion $\widetilde{\pi}_b : \widetilde{M}_b \to \widetilde{V}_b$ with compact fiber X.

Then $\operatorname{Ker}(D_T^X)$ is a G_b-equivariant vector bundle on $\widetilde{V}_b$. This defines an orbifold Hermitian vector bundle $\operatorname{Ker}(D_T^X)$ on B.

For $T \in [1, +\infty]$, let $E_{1,T}$ be the vector space of the smooth sections on B of $\operatorname{Ker}(D_T^X)$. As in [2, (5.26)], we have

$$E_{1,T} \simeq E_1. \tag{5.25}$$

The proof of Theorems 5.7–5.11 then proceeds as in [2, §5, §6] by using (3.15).

5.6. A proof of Theorem 5.12

Now we use the notation of [33, §7].

By Proposition 1.4, for each $b \in B$, there exists a small neighbourhood $(G_b, \widetilde{V}_b) \to V_b$ ($\widetilde{V}_b$ is a neighbouhood of $0 \in \mathbb{C}^m$ and G_b acts linearly on $\mathbb{C}^m$), an orbifold $\widetilde{M}_b$, such that π is induced by a G_b-equivariant orbifold submersion $\widetilde{\pi}_b : \widetilde{M}_b \to \widetilde{V}_b$ with compact fibre X.

Let $(G_{b_i}, \widetilde{V}_{b_i})_{i\in I}$ be a cover of B such that $(G_{b_i}, \frac{1}{2}\widetilde{V}_{b_i})_{i\in I}$ also is a cover of B. Let $\beta = \inf_{i\in I}\{\text{injectivity radius of } b_i \text{ on } \widetilde{V}_{b_i}\}$. Let $\alpha \in]0, \beta/8]$.

If $b \in B$, let $B^B(b, r)$ be the open ball of center b and radius r in B.

Proposition 5.15. *For $\delta > 0$, there exist $c > 0$, $C > 0$ such that for $0 < \varepsilon \leq \delta$, $T \geq 1$,*

$$\Big|\operatorname{Tr}_s\Big[*_T^{-1}\frac{\partial}{\partial T}(*_T)G_{\frac{\varepsilon}{T}}(\frac{\varepsilon}{T}D_T^M)\Big]\Big| \leq c\exp\Big(-\frac{CT^2}{\varepsilon^2}\Big). \tag{5.26}$$

Proof. The proof of (5.26) is essentially the same as the proof of [2, Proposition 8.3]. □

For $T \geq 1$ fixed, we use (5.26) with $\varepsilon = T$ and T replace by T/ε, we find

$$\Big|\operatorname{Tr}_s\Big[*_{T/\varepsilon}^{-1}\frac{\partial}{\partial T}(*_{T/\varepsilon})G_\varepsilon(\varepsilon D_{T/\varepsilon}^M)\Big]\Big| \leq c\exp\Big(-\frac{C}{\varepsilon^2}\Big). \tag{5.27}$$

Set

$$A'_{\varepsilon,T} = \Big(\frac{T}{\varepsilon}\Big)^{N_V}\varepsilon D_{T/\varepsilon}^M\Big(\frac{T}{\varepsilon}\Big)^{-N_V}. \tag{5.28}$$

Let $F_\varepsilon(\varepsilon D_{T/\varepsilon}^M)(x,x')$, $F_\varepsilon(A'_{\varepsilon,T})(x,x')(x,x' \in M)$ be the smooth kernel associated with $F_\varepsilon(\varepsilon D_{T/\varepsilon}^M)$, $F_\varepsilon(A'_{\varepsilon,T})$ with respect to the volume form $\frac{dv_M(x')}{(2\pi)^{\dim M}}$. Using (2.11), (4.9) and finite propagation speed [20, §7.8], [35, Appendix D. 2], it is clear that for $\varepsilon \in]0,1]$, $T \geq 1$, $x, x' \in M$, if $d^B(\pi(x), \pi(x')) \geq \alpha$, then

$$F_\varepsilon(\varepsilon D_{T/\varepsilon}^M)(x,x') = 0$$

and moreover, given $x \in M$, $F_\varepsilon(\varepsilon D_{T/\varepsilon}^M)(x,\cdot)$ only depends on the restriction of $D_{T/\varepsilon}^M$ to $\pi^{-1}(B^B(\pi(x),\alpha))$.

Let ρ_i be a partition of unity subordinate to the cover $(G_{b_i}, \frac{1}{2}\widetilde{V}_{b_i})_{i\in I}$ of B. Then by (5.28), we get as in [2, (7.8)]

$$\operatorname{Tr}_s\Big[*_{T/\varepsilon}^{-1}\frac{\partial}{\partial T}(*_{T/\varepsilon})F_\varepsilon(\varepsilon D_{T/\varepsilon}^M)\Big] = \operatorname{Tr}_s\Big[*_{T/\varepsilon}^{-1}\frac{\partial}{\partial T}(*_{T/\varepsilon})F_\varepsilon(A'_{\varepsilon,T})\Big]. \tag{5.29}$$

We replace $\widetilde{M}_{b_i}$ by $(\widetilde{TB})_{b_i} \times X_{b_i} = \mathbb{C}^m \times X_{b_i}$ and trivialize the vector bundles as indicated in [33, §7b)].

As in [2, §9b)], for $\alpha > 0$ small enough, there is also a smooth $\mathbb{Z}$-graded vector bundle $K \subset \Omega_{b_i}$ over $(\widetilde{TB})_{b_i} \simeq \mathbb{R}^{2m}$ which coincides with $\operatorname{Ker}(D^X)$ on $B(0, 4\alpha)$, with $\operatorname{Ker}(D^X_{b_i})$ over $(\widetilde{TB})_{b_i}\backslash B(0,6\alpha)$ and such that if $K^\perp$ is the orthogonal bundle to K in Ω_{b_i},

$$K^\perp \cap \operatorname{Ker}(D^X_{b_i}) = \{0\}. \tag{5.30}$$

Let P_b be the orthogonal projection operator from Ω_{b_i} on K_b. Set $P_b^\perp = 1 - P_b$.

Let Δ^{TB} be the standard Laplacian on the vector space $(\widetilde{TB})_{b_i}$ with respect to the metric $h^{TB}_{b_i}$. Let $dv_{T_{b_i}B}$ be the Riemannian volume form on $((\widetilde{TB})_{b_i}, h^{TB}_{b_i})$.

Let $\varphi : \mathbb{R} \to [0,1]$ be a $\mathscr{C}^\infty$ function which is equal 1 if $|t| \le 2\alpha$, equal 0 if $|t| \ge 4\alpha$. Let $L^1_{\varepsilon,T}$ be the operator on $\mathbb{C}^m \times X_{b_0}$

$$L^1_{\varepsilon,T} = \varphi^2(|Y|)A'^2_{\varepsilon,T} + (1-\varphi^2(|Y|))\Big(\frac{-\varepsilon^2\Delta^{TB}}{2} + T^2 P_Y^\perp D^{X,2}_{b_i} P_Y^\perp\Big). \tag{5.31}$$

Let $\widetilde{F}_\varepsilon(L^1_{\varepsilon,T})\Big((Y,x),(Y',x')\Big)$ $((Y,x),(Y',x') \in (\widetilde{TB})_{b_i} \times X_{b_0})$ be the smooth kernels associated with $\widetilde{F}_\varepsilon(L^1_{\varepsilon,T})$ with respect to $\dfrac{dv_{T_{b_i}B}(Y')dv_{X_{b_i}}(x')}{(2\pi)^{\dim M}}$.

For $(Y,x) \in (\widetilde{TB})_{b_i} \times X_{b_i}$, $|Y| < \beta/4$, set

$$dv_M(Y,x) = k(Y,x)dv_{T_{b_i}B}dv_{X_{b_i}}. \tag{5.32}$$

Using finite propagation speed and (2.11), we see that if $(Y,x) \in (\widetilde{TB})_{b_i} \times X_{b_i}$, $|Y| < \alpha$, then

$$F_\varepsilon(A'_{\varepsilon,T})\Big((Y,x),(Y,x)\Big) = \sum_{h\in G_{b_i}} k(Y,x)h\widetilde{F}_\varepsilon(L^1_{\varepsilon,T})\Big(h^{-1}(Y,x),(Y,x)\Big). \tag{5.33}$$

By (5.33), and proceeding as in [33, §7], we have Theorem 5.12.

5.7. A proof of Theorem 5.13

As in [2, §8] or [33, §8], the following theorem implies Theorem 5.13.

Theorem 5.16. *There exists $C>0$ such that for $0 < u \le 1, T \ge 1$,*

$$\begin{aligned}\Big|\operatorname{Tr}_s\Big[*_T^{-1}\frac{\partial *_T}{\partial T}\exp(-\frac{u^2}{T^2}D_T^{M,2})\Big] - \frac{2}{u^2}\int_{M\cup\Sigma M}\frac{\widetilde{\omega}^M}{2\pi T}\operatorname{Td}^\Sigma(TM)\operatorname{ch}^\Sigma(\xi) \\ + \int_{M\cup\Sigma M}\frac{\partial}{\partial b}\operatorname{Td}^\Sigma\Big(\frac{-R_T^{TM}}{2i\pi} - b(h_T^{TM})^{-1}\frac{\partial}{\partial T}(h_T^{TM})\Big)_{b=0}\operatorname{ch}^\Sigma(\xi,h^\xi)\Big| \le \frac{Cu^2}{T}.\end{aligned} \tag{5.34}$$

Proof. By (5.28)

$$A'_{1/T,1} = T^{N_V}\frac{1}{T}D_T^M T^{-N_V}. \tag{5.35}$$

Therefore

$$\operatorname{Tr}_s\Big[*_T^{-1}\frac{\partial *_T}{\partial T}\exp(-\frac{u^2}{T^2}D_T^{M,2})\Big] = \operatorname{Tr}_s\Big[*_T^{-1}\frac{\partial *_T}{\partial T}\exp(-u^2A'^2_{1/T,1})\Big]. \tag{5.36}$$

By (3.15), we can replace M by $(\mathbb{C}^m \times X_{b_i})/K_{b_i}$, and trivialize the vector bundles as indicated in [33, §7b)]. Then we will prove (5.34) in this situation.

Let $P_{u,T}(x,x')$ be the smooth kernel associated with the operator $\exp(-u^2A'^2_{1/T,1})$ with respect to $\frac{dv_M(x')}{(2\pi)^{\dim M}}$. Let

$$P^1_{\varepsilon,T,u}((Y,x),(Y',x'))\ ((Y,x),(Y',x')\in(\widetilde{TB})_{b_i}\times X_{b_i})$$

be the smooth kernel associated with the operator $\exp(-u^2L^1_{\varepsilon,T})$ with respect to $\frac{dv_{T_{b_i}B}(Y')\times dv_{X_{b_i}}(x')}{(2\pi)^{\dim M}}$. By Proposition 5.15, as $u\to 0$, uniformly on $T\geq 1$, the asymptotics of the following three terms is the same

$$\begin{aligned}
&\int_M \rho_i \operatorname{Tr}_s\Big[*_T^{-1}\frac{\partial *_T}{\partial T}F_{u/T}(uA'_{1/T,1})(x,x')\Big]dv_M/(2\pi)^{\dim M},\\
&\quad\int_M \rho_i \operatorname{Tr}_s\Big[*_T^{-1}\frac{\partial *_T}{\partial T}P_{u,T}(x,x')\Big]dv_M/(2\pi)^{\dim M},\\
&\int_{(\widetilde{TB})_{b_i}\times X_{b_i}}\rho_i\sum_{h\in G_{b_i}}\frac{1}{|G_{b_i}|}\operatorname{Tr}_s\Big[h*_T^{-1}\frac{\partial *_T}{\partial T}P^1_{1/T,1,u}(h^{-1}(Y,x),(Y,x))\Big]\\
&\qquad\qquad k(Y,x)dv_{T_{b_i}B}(Y)\times dv_{X_{b_i}}(x)/(2\pi)^{\dim M}.
\end{aligned}\tag{5.37}$$

By [33, §8], (5.36), (5.37), we get Theorem 5.16. □

5.8. A proof of Theorem 5.14

Proposition 5.17. *There exists $C>0$, such that for $0<\varepsilon\leq 1$, $T\geq 1$*

$$\begin{aligned}
&\Big|\operatorname{Tr}_s\Big[*_{T/\varepsilon}^{-1}\frac{\partial}{\partial T}(*_{T/\varepsilon})G_\varepsilon(\varepsilon D^M_{T/\varepsilon})\Big]\\
&\quad-\frac{2}{T}\Big(\sum_{j=0}^{\dim X}(-1)^j j\chi(R^j\pi_*\xi)-\dim X\chi(\xi)\Big)G_\varepsilon(0)\Big|\leq\frac{C}{T^2}.
\end{aligned}\tag{5.38}$$

Proof. By an analogue of the McKean–Singer formula [1, Theorem 3.50], we find that

$$\operatorname{Tr}_s[N_V G_\varepsilon(\varepsilon D^B)]=\sum_{j=0}^{\dim X}(-1)^j j\chi(R^j\pi_*\xi)G_\varepsilon(0).\tag{5.39}$$

Using (5.39) and proceeding as in [2, Proposition 9.1], we have (5.38). □

By (4.10) and (5.38), to establish Theorem 5.14, we only need to establish the following result,

Theorem 5.18. *If $\alpha>0$ is small enough, there exist $\delta>0, C>0$, such that for $0<\varepsilon\leq 1$, $T\geq 1$*

$$\begin{aligned}
&\Big|\operatorname{Tr}_s\Big[*_{T/\varepsilon}^{-1}\frac{\partial}{\partial T}(*_{T/\varepsilon})F_\varepsilon(\varepsilon D^M_{T/\varepsilon})\Big]\\
&\quad-\frac{2}{T}\Big(\sum_{j=0}^{\dim X}(-1)^j j\chi(R^j\pi_*\xi)-\dim X\chi(\xi)\Big)F_\varepsilon(0)\Big|\leq\frac{C}{T^{1+\delta}}.
\end{aligned}\tag{5.40}$$

Proof. Using (5.28), we deduce that

$$\mathrm{Tr}_s\left[*_{T/\varepsilon}^{-1}\frac{\partial}{\partial T}(*_{T/\varepsilon})F_\varepsilon(\varepsilon D^M_{T/\varepsilon})\right]=\mathrm{Tr}_s\left[*_{T/\varepsilon}^{-1}\frac{\partial}{\partial T}(*_{T/\varepsilon})\widetilde{F}_\varepsilon(A'^2_{\varepsilon,T})\right]. \tag{5.41}$$

Let $\widetilde{F}_\varepsilon(A'^2_{\varepsilon,T})(x,x')(x,x'\in M)$ be the smooth kernel associated with $\widetilde{F}_\varepsilon(A'^2_{\varepsilon,T})$ with respect to $dv_M(x')/(2\pi)^{\dim M}$. Using finite propagation speed, it is clear that if $x\in M$, $\widetilde{F}_\varepsilon(A'^2_{\varepsilon,T})(x,\cdot)$ only depends on the restriction of $A'_{\varepsilon,T}$ to $\pi^{-1}(B^B(\pi(x),\alpha))$.

We use the same trivialization and notation as in Section 5.6. If $(Y,x)\in(\widetilde{TB})_{b_i}\times X_{b_i}$, $|Y|<\alpha$, then

$$\begin{aligned}\rho_i(Y,x)&\widetilde{F}_\varepsilon(A'^2_{\varepsilon,T})((Y,x),(Y,x))\\&=\rho_i\sum_{h\in G_{b_i}}k(Y,x)h\widetilde{F}_\varepsilon(L^1_{\varepsilon,T})(h^{-1}(Y,x),(Y,x)).\end{aligned} \tag{5.42}$$

By [33, §9], (5.41), (5.42), we get Theorem 5.18. □

The proof of Theorem 5.14 is completed. □

References

[1] N. Berline, E. Getzler, and M. Vergne, *Heat kernels and Dirac operators*, Grundlehren Text Editions, Springer-Verlag, Berlin, 2004, Corrected reprint of the 1992 original.

[2] A. Berthomieu and J.-M. Bismut, *Quillen metrics and higher analytic torsion forms*, J. Reine Angew. Math. **457** (1994), 85–184.

[3] J.-M. Bismut, *The Atiyah–Singer index theorem for families of Dirac operators: two heat equation proofs*, Invent. Math. **83** (1986), no. 1, 91–151.

[4] J.-M. Bismut, *Equivariant short exact sequences of vector bundles and their analytic torsion forms*, Compositio Math. **93** (1994), no. 3, 291–354.

[5] J.-M. Bismut, *Equivariant immersions and Quillen metrics*, J. Differential Geom. **41** (1995), no. 1, 53–157.

[6] J.-M. Bismut, *Holomorphic families of immersions and higher analytic torsion forms*, Astérisque no. 244, (1997), viii+275 pp.

[7] J.-M. Bismut and J. Cheeger, *η-invariants and their adiabatic limits*, J. Amer. Math. Soc. **2** (1989), 33–70.

[8] J.-M. Bismut and D. Freed, *The analysis of elliptic families. I. Metrics and connections on determinant bundles*, Comm. Math. Phys. **106** (1986), no. 1, 159–176.

[9] J.-M. Bismut and D. Freed, *The analysis of elliptic families. II. Dirac operators, eta invariants, and the holonomy theorem*, Comm. Math. Phys. **107** (1986), no. 1, 103–163.

[10] J.-M. Bismut, H. Gillet, and C. Soulé, *Analytic torsion and holomorphic determinant bundles. I. Bott–Chern forms and analytic torsion*, Comm. Math. Phys. **115** (1988), no. 1, 49–78.

[11] J.-M. Bismut, H. Gillet, and C. Soulé, *Analytic torsion and holomorphic determinant bundles. II. Direct images and Bott–Chern forms*, Comm. Math. Phys. **115** (1988), no. 1, 79–126.

[12] J.-M. Bismut, H. Gillet, and C. Soulé, *Analytic torsion and holomorphic determinant bundles. III. Quillen metrics on holomorphic determinants*, Comm. Math. Phys. **115** (1988), no. 2, 301–351.

[13] J.-M. Bismut and K. Köhler, *Higher analytic torsion forms for direct images and anomaly formulas*, J. Algebraic Geom. **1** (1992), no. 4, 647–684.

[14] J.-M. Bismut, F. Labourie, Symplectic geometry and the Verlinde formulas. Surveys in differential geometry: differential geometry inspired by string theory, 97–311, Surv. Differ. Geom., 5, Int. Press, Boston, MA, 1999.

[15] J.-M. Bismut and G. Lebeau, *Complex immersions and Quillen metrics*, Inst. Hautes Études Sci. Publ. Math. no. 74 (1991), ii+298 pp.

[16] J.-M. Bismut and X. Ma. *Holomorphic immersions and equivariant torsion forms.* J. Reine Angew. Math., **575** (2004),189–235.

[17] J. Burgos Gil, G. Freixas i Montplet, R. Liţcanu, *Generalized holomorphic analytic torsion.* J. Eur. Math. Soc. (JEMS) 16 (2014), no. 3, 463–535.

[18] J. Burgos Gil, G. Freixas i Montplet, R. Liţcanu, *The arithmetic Grothendieck–Riemann–Roch theorem for general projective morphisms.* Ann. Fac. Sci. Toulouse Math. (6), **23** (2014), 513–559.

[19] H. Cartan, *Quotient d'un espace analytique par un groupe d'automorphismes.* Algebraic Geometry and Topology. Fox R.H., Spencer D.C., and Tucker A.W. eds. Princeton Univ. Press, Princeton 1957, 90–102.

[20] J. Chazarain and A. Piriou, *Introduction à la théorie des équations aux dérivées partielles linéaires*, Gauthier-Villars, Paris, 1981.

[21] X. Dai, J. Yu, *Comparison between two analytic torsions on orbifolds.* Math. Z. **285** (2017), 1269–1282.

[22] H. Gillet and C. Soulé, *Analytic torsion and the arithmetic Todd genus*, Topology **30** (1991), no. 1, 21–54, With an appendix by D. Zagier.

[23] H. Gillet and C. Soulé, *An arithmetic Riemann–Roch theorem*, Invent. Math. **110** (1992), 473–543.

[24] H. Gillet, D. Roessler, and C. Soulé. *An arithmetic Riemann–Roch theorem in higher degrees.* Ann. Inst. Fourier, **58** (2008), 2169–2189.

[25] H. Grauert, Ein Theorem der analytischen Garbentheorie und die Modulräume komplexer Strukturen. Inst. Hautes Études Sci. Publ. Math. No. **5** (1960), 64 pp.

[26] T. Kawasaki, *The Signature theorem for V-manifolds.* Topology **17** (1978) 75–83.

[27] T. Kawasaki, *The Riemann–Roch theorem for V-manifolds.* Osaka J. Math **16** (1979) 151–159.

[28] F.F. Knudsen and D. Mumford, *The projectivity of the moduli space of stable curves. I. Preliminaries on "det" and "Div"*, Math. Scand. **39** (1976), no. 1, 19–55.

[29] K. Köhler and D. Roessler. *A fixed point formula of Lefschetz type in Arakelov geometry. I. Statement and proof.* Invent. Math., **145**, (2001), 333–396.

[30] H.B. Lawson and M.-L. Michelsohn, *Spin geometry*, Princeton Mathematical Series, vol. 38, Princeton University Press, Princeton, NJ, 1989.

[31] X. Ma, *Formes de torsion analytique et familles de submersions. I*, Bull. Soc. Math. France **127** (1999), no. 4, 541–621.

[32] X. Ma, *Formes de torsion analytique et familles de submersions. II*, Asian J. Math. 4 (2000), no. 3, 633–667.

[33] X. Ma, *Submersions and equivariant Quillen metrics*, Ann. Inst. Fourier (Grenoble) **50** (2000), no. 5, 1539–1588.

[34] X. Ma, *Orbifolds and analytic torsions*, Trans. Amer. Math. Soc. **357** (2005), 2205–2233.

[35] X. Ma and G. Marinescu, *Holomorphic Morse inequalities and Bergman kernels*, Progress in Mathematics, vol. 254, Birkhäuser Boston Inc., Boston, MA, 2007.

[36] V. Maillot, D. Roessler, Formes automorphes et théorèmes de Riemann–Roch arithmétiques. *From Probability to Geometry. Volume in honor of the 60th birthday of Jean-Michel Bismut*, Astérisque no. **328** (2009), 237–253.

[37] V. Maillot, D. Roessler, On a canonical class of Green currents for the unit sections of abelian schemes. Doc. Math. **20** (2015), 631–668.

[38] D. Quillen, *Determinants of Cauchy–Riemann operators on Riemann surfaces*, Functional Anal. Appl. **19** (1985), no. 1, 31–34.

[39] D.B. Ray and I.M. Singer, *Analytic torsion for complex manifolds*, Ann. of Math. (2) **98** (1973), 154–177.

[40] I. Satake, *The Gauss–Bonnet theorem for V-manifolds*, J. Math. Soc. Japon. **9** (1957), 464–492.

[41] S. Shen, J. Yu, *Flat vector bundles and analytic torsion on orbifolds*, arXiv: 1704.08369, Comm. Anal. Geom. to appear.

[42] C. Soulé, *Lectures on Arakelov geometry*, Cambridge Studies in Advanced Mathematics, vol. 33, Cambridge University Press, Cambridge, 1992, With the collaboration of D. Abramovich, J.-F. Burnol and J. Kramer.

[43] K.-I. Yoshikawa, K3 surfaces with involution, equivariant analytic torsion, and automorphic forms on the moduli space. Invent. Math., **156** (2004), 53–117.

[44] K.-I. Yoshikawa, K3 surfaces with involution, equivariant analytic torsion, and automorphic forms on the moduli space, II: A structure theorem for $r(M) > 10$. J. Reine Angew. Math. **677** (2013), 15–70.

[45] K.-I. Yoshikawa, Analytic torsion for Borcea–Voisin threefolds. *Geometry, Analysis and Probability – in honor of Jean-Michel Bismut*, Progr. Math., **310**, 279–361, Birkhäuser/Springer, Cham, 2017.

Xiaonan Ma
Institut de Mathématiques de Jussieu-Paris Rive Gauche
Université de Paris, CNRS
F-75013 Paris, France
e-mail: xiaonan.ma@imj-prg.fr

Progress in Mathematics, Vol. 338, 179–212

Analytic Torsion, Regulators and Arithmetic Hyperbolic Manifolds

Jean Raimbault

Abstract. This paper is a survey of some topics pertaining to the regulators associated to the homology of Riemannian manifolds. These were originally introduced by Ray–Singer in order to give a definition of Reidemeister torsion for a closed Riemannian manifold. This torsion was conjectured to be equal to the analytic torsion defined by purely analytic means, which was proven shortly afterwards by Cheeger and Müller. We will give a short account of this theorem before turning to our main theme of interest which concerns arithmetic groups and locally symmetric spaces.

In recent work of Bergeron–Venkatesh and others regulators have appeared as an obstruction in the study of torsion in the cohomology of arithmetic groups. In this context regulators also have an arithmetic significance, and this was explored further by Calegari–Venkatesh and Bergeron–Şengün–Venkatesh. Finally, since many interesting arithmetic locally symmetric spaces are not compact it is natural to attempt to extend the definition of regulators, and the Cheeger–Müller theorem, to them. We will survey the work of Calegari–Venkatesh, Pfaff and Pfaff together with the author on this topic.

Mathematics Subject Classification (2010). 11F75, 22E40, 58J52, 57M50.

Keywords. Arithmetic group, hyperbolic manifold, analytic torsion, Reidemeister torsion, homology.

Introduction

This survey paper aims at giving an introduction to the interaction between Riemannian and spectral geometry on one side and number theory on the other, from the point of view of analytic torsion and the Cheeger–Müller Theorem. The former is a spectral invariant and the latter relates it to a topological invariant, the Reidemeister torsion. *Regulators* are one of the elements in the relation. The connection to number theory arises when applying this theory to the locally symmetric spaces associated to arithmetic lattices in semisimple Lie groups. The simplest examples where this gives interesting results is when the Lie group is $SL_2(\mathbb{C})$, and the associated manifolds are then hyperbolic 3-manifolds. The points I want to discuss are more precisely the following:

- Bounding the growth of regulators in sequences of congruence covers of an arithmetic manifold is a step in the program of understanding asymptotic growth of homology in these covers laid in [8];
- Regulators themselves seem to have interesting arithmetic properties;
- Analytic torsion and the Cheeger–Müller Theorem only make sense for closed Riemannian manifolds. It is desirable and possible to extend some parts of the theory to manifolds associated to nonuniform arithmetic lattices such as the Bianchi groups $SL_2(R_D)$, where R_D is a ring of imaginary quadratic integers.

I will concentrate on the first and third points, after giving a short survey of analytic and Reidemeister torsion and hyperbolic manifolds including arithmetic constructions. I will briefly discuss the growth of torsion homology in congruence covers (including some new numerical data); a more complete survey on this topic is given in [6]. The contents of the paper are as follows:

- Section 1 is a short introduction to Reidemeister and analytic torsion, and the Cheeger–Müller Theorem relating the two. This is rather old material, from [15] and [32] essentially. This section is mainly here to provide context for the rest of the survey.
- Section 2 concerns hyperbolic 3-manifolds. This section contains a very short introduction to their geometry and topology and then reviews the relations between regulators and geometric invariants, following [7] and [11].
- Arithmetic lattices in $SL_2(\mathbb{C})$ are introduced in Section 3 and a few results about their regulators and cohomology are explained there. The contents of this section are mainly taken from [8], [14] and [7].
- The last section (Section 4) is concerned with the extension to the noncompact case of the results mentioned above, and how to adapt some parts of the previous section to this setting. The contents are mainly from [14] and [37].

As noted above the study of arithmetic groups involves both number theory and differential geometry. I tried to separate as much as possible those results that involve only the latter (and so would be valid in the more general context of locally symmetric spaces) from those which make essential use of the number-theoretical origin of the spaces we consider (and might be false in a more general context).

This survey grew out of a talk I gave about [37] and [39] at the conference "Regulators IV", the contents of which intersected mainly with the last section.

Acknowledgments

Thanks are due to the organisers of "Regulators IV", especially Gerard Freixas, for asking me to write a survey on this topic. I am grateful to Nathan Dunfield, Michael Lipnowski and Aurel Page for valuable comments on and corrections to a preliminary version of this survey and to Nicolas Bergeron for explanations on base-change as used in his joint work with Haluk Şengün and Akshay Venkatesh. Finally, this work benefited from a careful reading by referees whose remarks helped improve its presentation and remove some imprecisions.

1. Reidemeister torsion and the Cheeger–Müller equality

1.1. Reidemeister torsions of CW-complexes

We will only give a short review of the general theory of Reidemeister torsion, which we will not use much in the sequel. For a complete survey see the book by V. Turaev [46], and some parts in what follows are essentially lifted from [8, Section 2].

If X is a finite CW-complex and $\mathbb{K}$ a commutative ring we let $C_*(X;\mathbb{K})$, be its chain complex, the differential of which we denote by d_*. Suppose that $\mathbb{K}$ is a field and $C_*(X;\mathbb{K})$ is acyclic[1]. Choose graded bases c_*, z_*, b_* for the spaces $C_*, \ker(d_*)$ and $\mathrm{Im}(d_*)$ respectively. Then the *Reidemeister torsion* of X is defined as follows:

$$\tau(X;\mathbb{K}) = \prod_i \det{}_{c_i}\left(d_{i+1}(b_{i+1}) \oplus z_i\right)^{(-1)^i} \tag{1.1}$$

where $\oplus$ denotes concatenation of bases. An elementary verification shows that it does not depend on the various choices, and it is also possible to prove that it is a homeomorphism invariant. If $C_*(X;\mathbb{K})$ is not acyclic then one needs to add a choice of bases h_* for the homology and define the torsion as

$$\tau(X;\mathbb{K}) = \prod_i \det{}_{c_i}\left(d_{i+1}(b_{i+1}) \oplus z_i \oplus \tilde{h}_i\right)^{(-1)^i} \tag{1.2}$$

where $\tilde{h}_*$ is a graded lift of the base h_* of $H_*(X;\mathbb{K})$ to $C_*(X;\mathbb{K})$. The torsion thus defined does depend on the choice of h_*. When $\mathbb{K}$ has characteristic 0 the most natural choice is to take bases coming from the integral structure: since the base change matrices have determinant ± 1 the torsion thus obtained depends on the choice only up to sign. A simple computation shows that in this case we have up to sign:

$$\tau(X;\mathbb{Q}) = \prod_i |H_i(X;\mathbb{Z})_{\mathrm{tors}}|^{(-1)^i} \tag{1.3}$$

where A_{tors} denotes the torsion subgroup of an abelian group A. In general we must add for each i the factor corresponding to the determinant of the chosen base in a $\mathbb{Z}$-base.

When $\mathbb{K}$ is $\mathbb{R}$ or $\mathbb{C}$ another natural choice for the bases h_* is as follows. There is a natural choice of a basis for C_* given by the cells in each degree. This gives each $C_i(X;\mathbb{K})$ the structure of a Euclidean or Hermitian space, and then we can identify $H_i(X;\mathbb{K})$ with a subspace of $C_i(X;\mathbb{K})$, namely the orthogonal of $\mathrm{Im}(d_{i+1})$ in $\ker(d_i)$. Then we may take for h_i any orthonormal basis of this subspace and

[1]Of course if X is a triangulation of a manifold then $C_*(X;\mathbb{K})$ is never acyclic but taking coefficients in a local system may remedy this issue.

the resulting torsion does not depend on this choice up to sign. It can be computed via the formula

$$\tau = \prod_{i=0}^{\dim(X)} \det{}'(d_i)^{(-1)^i} \tag{1.4}$$

where $\det{}'(f)^2$ is the product of all nonzero "singular values" of f, which are the eigenvalues of the associated self-adjoint map f^*f. The relation with the torsion in 1.3 is easily worked out to be (see [8, (2.2.4)]):

$$\prod_{i=0}^{\dim(X)} \det{}'(d_i)^{(-1)^i} = \prod_i \left(\frac{|H_i(X;\mathbb{Z})_{\mathrm{tors}}|}{R_i(X)}\right)^{(-1)^i} \tag{1.5}$$

where the *combinatorial regulator* $R_i(X)$ is the covolume of the lattice

$$H_i(X;\mathbb{Z})_{\mathrm{free}} = H_i(X;\mathbb{Z})/H_i(X;\mathbb{Z})_{\mathrm{tors}}$$

in the Euclidean subspace of $H_i(X;\mathbb{K})$ that it spans.

1.2. Reidemeister torsions of Riemannian manifolds

We saw above (cf. 1.4) how to relate the homological torsion of a CW-complex to the "geometry" of its cellular complex via determinants and regulators. The definition given for the latter depended on identifying the free part of the homology to a lattice in a Euclidean space. In this section we will take a CW-complex coming from a triangulation of a smooth manifold and relate its homological torsion to a quantity defined using a Riemannian metric on the manifold. For this purpose D. Ray and I. Singer defined the Reidemeister torsion of a Riemannian manifold: see [40, Definitions 1.1 and 3.6] and also [15, Section 1]. We will define it ex-nihilo in terms of regulators.

Let M be a compact Riemannian manifold with no boundary and of dimension $\dim(M) = d$. We fix a smooth triangulation X of M. For $0 \leq k \leq d$ let $\mathscr{H}^k(M)$ be the space of harmonic forms on M. By the Hodge-de Rham Theorem the linear map

$$\Phi^k : \mathscr{H}^k(M) \to H^k(X;\mathbb{R})$$

given by

$$\Phi^k(\omega)(c) = \int_c \omega$$

(here c is a k-cycle of X, so in particular it is a smooth k-submanifold away from a codimension 1 subset and the integral makes sense) is an isomorphism. On the space $\mathscr{H}^k(M)$ there is the L^2-inner product $\langle \cdot, \cdot \rangle_{L^2}$ defined by integrating the pointwise inner product induced by the Riemannian metric[2]. The kth regulator of M is then defined to be the covolume of $H^k(X;\mathbb{Z})_{\mathrm{free}}$ with respect to the inner product $(\Phi^k)_* \langle \cdot, \cdot \rangle_{L^2}$ on $H^k(X;\mathbb{R})$. We will denote it by $R_k(M)$; note that in general it depends on the choice of the Riemannian metric (but not on the

[2]Equivalently $\langle u, v \rangle = \int_M u \wedge *v$ where $*$ is the Hodge star operator asocietd to the Riemannian metric.

triangulation) but this will not matter for us. Using the explicit form of Φ^k we see that if $c_1, \dots, c_{b_k}$ is a $\mathbb{Z}$-basis of $H_k(X;\mathbb{Z})_{\text{free}}$ and $\omega_1, \dots, \omega_{b_k}$ is an orthonormal basis of $\mathscr{H}^k(M)$ then we have

$$R_k(M) = \det\left(\int_{c_i} \omega_j\right)_{1\le i,j\le b_k}. \tag{1.6}$$

The Reidemeister torsion of M is then defined to be:

$$\tau(M) = \prod_{i=0}^{d}\left(\frac{|H^k(X;\mathbb{Z})_{\text{tors}}|}{R_k(M)}\right)^{(-1)^k}. \tag{1.7}$$

Note that through (1.5) this can be seen as the particular case of (1.2) where one takes integral bases for the $C^k(X;\mathbb{R})$ and L^2-orthonormal bases for the $H^k(X;\mathbb{R})$.

1.3. Analytic torsion and the Cheeger–Müller Theorem

Now instead of determinants of the differentials of the complex $C^*(X;\mathbb{R})$ we will consider "determinants" of the de Rham differentials. As the spaces $\Omega^k(M)$ are not finite dimensional, and the differentials are not continuous these are not defined as usual. Rather, Ray–Singer used the spectral theory of the Hodge–Laplace operators to define the so-called regularised determinants. We will now describe this without paying attention to the underlying functional analysis.

We will denote by $L^2\Omega^k(M)$ the Hilbert space of square-integrable k-forms. The Hodge–Laplace operator Δ_k on $\Omega^k(M)$ is defined by:

$$\Delta_k = d_k^* d_k + d_{k+1} d_{k+1}^*$$

where d_k^* is the formal adjoint of d_k. The operator Δ_k is unbounded but it has a unique essentially self-adjoint extension in $L^2\Omega^k(M)$. The spectral theorem in Riemannian geometry states that its spectrum is a discrete subset of $[0,+\infty[$. All this means that there exists an Hilbert basis $(\omega_i)_{1\le i}$ of $L^2\Omega^k(M)$ and a nondecreasing sequence sequence $(\lambda_i)_{1\le i}$ such that λ_i goes to infinity and $\Delta_k\omega_i = \lambda_i\omega_i$. Moreover Weyl's law states that

$$|\{i : |\lambda_i| \le \lambda\}| \sim_{\lambda\to+\infty} C(d)\binom{d}{k}\operatorname{vol}(M)\lambda^{d/2}. \tag{1.8}$$

This means that the series

$$\zeta_k(s) = \sum_{\lambda_i>0} \lambda_i^{-s}$$

converges in the half-plane $\operatorname{Re}(s) > d/2$. Its sum actually extends to a meromorphic function on $\mathbb{C}$, which is often called the Minakshisundaram–Pleijel zeta function when $k = 0$. It is regular at $s = 0$ and we define the regularised determinant of Δ_k by the following formula:

$$\det{}'(\Delta_k) = \exp(\zeta_k'(0)) \tag{1.9}$$

(note that formally, the value on the right equals the (nonconvergent) product $\prod_{i\ge b_k+1}\lambda_i$).

The *analytic torsion* of M is defined, in analogy with (1.4), by:

$$T(M) = \prod_{k=0}^{d} (\det{}'(\Delta_k))^{k(-1)^k}. \tag{1.10}$$

The main motivation for this definition, as well as its use for us, resides in the following result which was conjectured by Ray–Singer and proven shortly afterwards by J. Cheeger and W. Müller independently of each other in [15] and [31] respectively.

Theorem 1.1 (Cheeger, Müller). *Let M be a closed Riemannian manifold. Then*

$$\tau(M) = T(M).$$

1.4. Local systems

The work of Ray–Singer as well as that of Cheeger and Müller was done in the slightly more general case where one considers cohomology and differential forms with coefficients in a flat orthogonal vector bundle. For later applications we will need an even larger setting, where the generalisation of the Cheeger–Müller Theorem 1.1 was worked out (independently) by J.-M. Bismut–W. Zhang [9] and W. Müller [32].

Let M, X be as above and let ρ be a representation of $\pi_1(M)$ on a finite-dimensional real vector space V, such that $\rho(\pi_1(M)) \subset \mathrm{SL}(V)$. Then we can associate to ρ a *flat unimodular vector bundle* F on M, obtained by quotienting $\widetilde{M}\times V$ by the left-action $\gamma\cdot(x,v) = (\gamma\cdot x, \rho(\gamma)\cdot v)$. There is also an associated cochain complex of finite-dimensional real vector spaces $C^*(X;V) = C^*(\widetilde{X};\mathbb{R})\otimes_{\mathbb{R}[\pi_1(M)]} V$, with cohomology $H^*(X;V)$. To proceed further one needs a metric on F: in general there is no canonical choice so we will assume that an arbitrary choice has been made. With this metric comes an identification of each $H^k(X;V)$ with a space $\mathscr{H}^k(M;F)$ of harmonic forms on M. It also allows to choose a prefered class of bases of each $C^k(X;V)$ and with these choices we can define a Reidemeister torsion $\tau(M;F)$ as in (1.2). On the analytic side the spectral theory is similar to that described above and the analytic torsion $T(M;F)$ is defined in the same manner. The result of Müller and Bismut–Zhang is then stated as follows.

Theorem 1.2 (Bismut–Zhang, Müller). *Let M be a closed Riemannian manifold and F a flat unimodular bundle over M. Then $\tau(M;F) = T(M;F)$.*

To end this section we will describe a particular case in which the Reidemeister torsion can be expressed with a formula similar to (1.7). Suppose that V contains a lattice L such that $\rho(\pi_1(M))(L) = L$. Then the integral cochain complex $C^*(X;L)$ and its homology $H^*(X;L)$ are well defined. Thus it is possible to take integral bases to compute the Reidemeister torsion $\tau(M;F)$, and to define the regulators $R_k(M;L)$ as the covolume of $H^k(X;L)_{\mathrm{free}}$ with respect to the L^2-inner product. Then we have the formula:

$$\tau(M;F) = \prod_{k=0}^{d} \left(\frac{|H^k(X;L)_{\mathrm{tors}}|}{R_k(M;L)} \right)^{(-1)^k}. \tag{1.11}$$

2. Regulators of hyperbolic 3-manifolds

2.1. Hyperbolic manifolds

A *hyperbolic manifold* is a complete Riemannian manifold all of whose sectional curvatures are equal to -1. Riemann's Theorem states that this determines a unique local isometry class, so that there is a unique such manifold which is simply connected. It is referred to as the *d-dimensional hyperbolic space* and usually denoted by $\mathbb{H}^d$. The d-dimensional hyperbolic space admits an algebraic description as the symmetric space associated to the Lie group $\mathrm{SO}(d,1)$, in other words it is isometric to $\mathrm{SO}(d,1)/\mathrm{O}(d)$ with the (suitably normalised) left-$\mathrm{SO}(d,1)$-invariant Riemannian metric. In the sequel we will mostly restrict ourselves to $d = 3$. In this dimension there is a local isomorphism $\mathrm{SO}(3,1) \cong \mathrm{SL}_2(\mathbb{C})$ and we will use the latter group since it is easier to deal with algebraically.

2.1.1. Thick-thin decomposition. Recall that the injectivity radius $\mathrm{inj}_x(M)$ of a Riemannian manifold M at a point x is the maximal radius of a ball (in the Riemannian distance) around x which is embedded. For negatively curved manifolds, an alternative description is that $\mathrm{inj}_x(M)$ equals half the minimal length of a closed curve passing through x which is nontrivial in $\pi_1(M)$. For a closed manifold the global injectivity radius $\mathrm{inj}(M)$ is defined to be the minimum of $\mathrm{inj}_x(M)$ over all $x \in M$. We have $\mathrm{inj}(M) > 0$.

The most basic tool to describe the global structure of finite-volume hyperbolic manifolds is the Margulis lemma. We will use one of its corollary, namely the *thick-thin decomposition* of hyperbolic manifolds which we will describe in what follows. For an $\varepsilon > 0$ the *ε-thin part* of M is the subset

$$M_{\leq \varepsilon} = \{x \in M : \mathrm{inj}_x(M) \leq \varepsilon\}$$

and the ε-thick part $M_{>\varepsilon}$is its complement in M. There exists a constant μ (depending on d) called the *Margulis constant*, such that for all d-dimensional hyperbolic manifolds M of finite volume and $\varepsilon \leq \mu$ the ε-thin part of M is diffeomorphic to a finite union of *cusps* and *tubes*. Cusps are the noncompact components, and each is diffeomorphic to $N \times [0, +\infty[$ where N is a closed $(d-1)$-dimensional flat manifold, for example $N = \mathbb{T}^{d-1}$, the flat torus (this is the only possibility in dimension $d = 3$). Tubes are compact, and each of them is obtained as a tubular neighbourhood of a closed geodesic, in particular it is diffeomorphic to $\mathbb{B}^{d-1} \times \mathbb{S}^1$. It then follows from the van Kampen Theorem that for $d \geq 3$ the fundamental group of M is generated by that of its subset $M_{\geq \varepsilon}$. If $d \geq 4$ the manifolds M and $M_{\geq \varepsilon}$ have the same fundamental group.

2.1.2. Triangulations of hyperbolic manifolds. Suppose that M is a closed Riemannian manifold such that $\mathrm{inj}(M) \geq \varepsilon > 0$. Let $x_1, \ldots, x_m \in M$ such that the open balls $B(x_i, \varepsilon/6)$ cover M, and the distances $d(x_i, x_j)$ for $i \neq j$ are all at least $\varepsilon/12$. We can look at the pattern of intersections between the balls $B(x_i, \varepsilon/4)$ to define a simplicial complex X, namely X has (x_i) as its 0-skelethon and vertices $x_{i_0}, \ldots, x_{i_k}$ define a k-simplex of X if and only if the intersection $B(x_{i_0}, \varepsilon/4) \cap \cdots \cap B(x_{i_k}, \varepsilon/4)$ is nonempty (this complex is usually called the *nerve* of the covering by the $B(x_i, \varepsilon/4)$). For ε small enough (depending only on the local geometry of the manifold) this simplicial complex will be homotopy equivalent to M (but not necessarily homeomorphic to it: it can happen that it is not a manifold). Moreover it has at most $C(k, \varepsilon)$ simplices in degree k, and every simplex is adjacent to at most $d(\varepsilon)$ other simplices. In particular $b_k(M) \leq C(k, \varepsilon)$.

In the case where M is an hyperbolic manifold and ε is the Margulis constant, and $\mathrm{inj}(M) < \varepsilon$ it is still possible to construct a triangulation from the thick part $M_{\geq\varepsilon}$ which is homotopy equivalent to it (see [12]). Since in dimension 3 the fundamental group of M is a quotient of that of $M_{\geq\varepsilon}$, and in higher dimension the two manifolds are homotopy equivalent, we see that there exists a constant $C(d)$ such that for every d-dimensional hyperbolic manifold we have:

$$b_k(M) \leq C(d) \cdot \mathrm{vol}(M). \tag{2.1}$$

Less obviously (see for example [42]) the homotopy equivalence of $M_{\geq\varepsilon}$ with such a simplicial complex also implies that

$$\log |H_k(M_{\geq\varepsilon}; \mathbb{Z})_{\mathrm{tors}}| \leq C(d) \cdot \mathrm{vol}(M). \tag{2.2}$$

If M has tubes in its thin part then M and $M_{\geq\varepsilon}$ are not necessarily homotopy equivalent. However (2.2) still implies that $\log |H_k(M; \mathbb{Z})_{\mathrm{tors}}| \leq C(d) \cdot \mathrm{vol}(M)$ for $k \neq d-2$, using a long exact sequence to relate $H_*(M_{\geq\varepsilon}; \mathbb{Z})$ to $H_*(M; \mathbb{Z})$. In general there is no reason for this bound to hold for $k = d-2$: for example in dimension 3 Dehn surgeries on a given manifold can have arbitrarily large torsion in their first homology group while keeping the volume bounded. We do not know whether there is such a uniform bound for $H_{d-2}(M; \mathbb{Z})$ for $d \geq 4$.

2.1.3. Reflection groups. Classical constructions of low-dimensional hyperbolic manifolds proceed by using a Coxeter polyhedron and Poincaré's Theorem. For example there exists a compact tetrahedron T_2 (we take the notation from [29, Figure 13.1]) in $\mathbb{H}^3$ with Coxeter symbol

Let Γ_{T_2} be the subgroup of $\mathrm{Isom}(\mathbb{H}^3)$ generated by reflections on the faces of T_2 (we'll denote the reflection in a face by the same letter). Then Γ_{T_2} is a cocompact discrete subgroup in $\mathrm{Isom}(\mathbb{H}^3)$. It has an index-2 subgroup $\Gamma_{T_2}^+$ of orientation-preserving isometries. The latter is generated by $\alpha = ab, \beta = ac$ and $\gamma = ad$ and has the presentation

$$\Gamma_{T_2}^+ = \langle \alpha, \beta, \gamma | \alpha^5, \beta^2, \gamma^3, (\beta\alpha)^3, (\gamma\alpha)^2, (\beta\gamma)^4 \rangle.$$

It is possible to compute an explicit matrix representation of $\Gamma_{T_2}^+$ to $\mathrm{PSL}_2(\mathbb{C})$. More generally formulae for the representations of 3-dimensional hyperbolic tetrahedral groups are given in [24].

Since $\Gamma_{T_2}^+$ is not torsion-free the quotient $\Gamma_{T_2}^+\backslash\mathbb{H}^3$ is not a manifold. There are two ways to find a finite-index subgroup in $\Gamma_{T_2}^+$ which is torsion-free. We will see the algebraic one later, here we describe a beautiful geometric construction due (in greater generality) to A. Vesnin. The subgroup of Γ_{T_2} fixing the vertex $x_0 = a \cap b \cap c$ is isomorphic to S_5. Thus we get a morphism $\pi : \Gamma_{T_2} \to S_5$ by sending the generator d to the identity. The polyhedron in $\mathbb{H}^3$ corresponding to the subgroup $\ker(\pi)$ is a right-angled dodecahedron D with center x_0 and the group $\Gamma_D = \ker(\pi)$ is generated by the reflections in the sides of D. These sides correspond to the vertices of an icosahedron. Choose a colouring of the icosahedral graph with 4 colours and define a morphism $\rho : \Gamma_D \to (\mathbb{Z}/2\mathbb{Z})^3$ by sending a generator to $(1,0,0)$, $(0,1,0)$, $(0,0,1)$ or $(1,1,1)$ according to its color. Then $\Gamma = \ker(\rho)$ is a torsion-free group and the manifold $\Gamma\backslash\mathbb{H}^3$ is obtained by gluing eight copies of D along their faces according to a certain pattern. Its volume is about $.03588 \times 120 \times 8 \cong 34.4448$.

For later use we note that doubling T_2 along the face d yields another Coxeter tetrahedron T_4, whose Coxeter symbol is

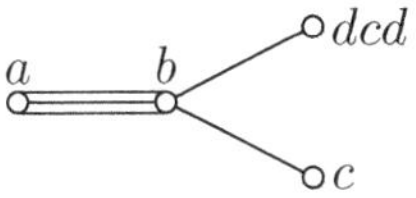

A presentation for its reflection group is

$$\Gamma_{T_4}^+ = \langle \alpha, \beta, \gamma | \alpha^5, \beta^2, \gamma^2, (\beta\alpha)^3, (\gamma\alpha)^3, (\beta\gamma)^2 \rangle . \tag{2.3}$$

Moreover, the following matrices[3] generate a discret cocompact subgroup of $\mathrm{PSL}_2(\mathbb{C})$ isomorphic to $\Gamma_{T_4}^+$:

$$\tilde{\alpha} = \frac{1}{2}\begin{pmatrix} a^2 & -e \\ e & a^2 \end{pmatrix}, \; \tilde{\beta} = \frac{1}{5}\begin{pmatrix} 0 & \left(\left(a^2+2\right)i - 2a^2 + 1\right)e \\ \left(\left(a^2+2\right)i + 2a^2 - 1\right)e & 0 \end{pmatrix},$$
$$\tilde{\gamma} = \frac{1}{5}\begin{pmatrix} 5\left(-a^3+a\right)i & \left(\left(a^2-3\right)i - 2a^2 + 1\right)e \\ \left(\left(a^2-3\right)i + 2a^2 - 1\right)e & 5\left(a^3-a\right)i \end{pmatrix}. \tag{2.4}$$

Here a is an imaginary root of $X^4 - X^2 - 1$ and $e = \sqrt{4 - a^4}$, so the coefficients belong to an algebraic number field of degree 16.

Finally, we will also refer to the tetrahedron T_8 which occurs in [44]. Its Coxeter symbol is

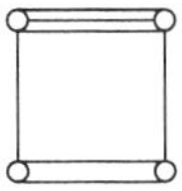

[3]They are different from those in [24], we simply solved the trace equations given by the presentation (2.3) to find them.

and a presentation for its reflection group is

$$\Gamma_{T_8}^+ = \langle \alpha, \beta, \gamma | \alpha^3, \beta^2, \gamma^5, (\gamma\beta)^3, (\alpha\gamma^{-1})^2, (\alpha\beta)^4 \rangle .$$

2.2. Regulators and geometry

Note that by Poincaré duality, for a closed Riemannian 3-manifold M it follows from formula (1.6) that:

$$R_2(M) = R_1(M)^{-1}.$$

So to prove upper and lower bound for either R_1 and R_2 it suffices to prove upper bounds for both. We will describe how to do so in both cases, following [7]: in degree 1 the result is good enough for applications, but degree 2 is more complicated.

2.2.1. Bounding R_1. In degree 1 it is rather easy to estimate regulators. The following is a slightly modified version of [7, Proposition 3.1].

Proposition 2.1 (Bergeron–Şengün–Venkatesh). *There exists a constant $C > 0$ such that for every closed hyperbolic 3-manifold M we have*

$$\log(R_1(M)) \leq Cb_1(M)\log(\mathrm{vol}(M)). \tag{2.5}$$

To get geometric estimates for the regulators here and later one needs to be able to estimate the integral of a harmonic form on a submanifold. In full generality this follows from the Sobolev inequalities, but for hyperbolic manifolds Brock and Dunfield prove the following more precise result (see the proof of [11, Theorem 4.1]).

Theorem 2.2 (Brock–Dunfield). *There exists a constant $C_1 > 0$ such that if M is an hyperbolic manifold and α a harmonic form on M then we have the following bound for the pointwise norms $|\alpha_x|$:*

$$\forall x \in M : |\alpha_x| \leq \frac{C_1}{\sqrt{\mathrm{inj}_x(M)}} \|\alpha\|_{L^2}.$$

Now we can explain the proof of Proposition 2.1. Let μ be the Margulis constant for $\mathbb{H}^3$ and $M_{\geq\mu}$ the μ-thick part of M. Then as we saw it follows from the thick-thin decomposition that $\pi_1(M)$ is generated by $\pi_1(M_{\geq\mu})$. On the other hand, there exists an absolute constant C_2 such that one can find a family of closed curves $\gamma_1, \ldots, \gamma_m$ which are contained in the thick part $M_{\geq\mu}$ and generate $\pi_1(M_{\geq\mu})$, such that each has length at most $C_2 \cdot \mathrm{diam}(M_{\geq\mu})$. We note that $\mathrm{diam}(M_{\geq\mu}) \leq C_0 \,\mathrm{vol}(M)$ where C_0 is the volume of a ball of radius μ in $\mathbb{H}^2$. It follows, using Theorem 2.2, that for ω a harmonic 1-form on M we have for $i = 1, \ldots, m$ the estimate:

$$\left| \int_{\gamma_i} \omega \right| \leq C_3 \,\mathrm{vol}(M) \|\omega\|_{L^2}. \tag{2.6}$$

where C_3 is absolute (we can take $C_3 = C_0C_1C_2/\sqrt{\mu}$). Now there is a subfamily, say $\gamma_1, \ldots, \gamma_b$, such that the singular homology classes $[\gamma_1], \ldots, [\gamma_b]$ generate a subgroup of finite index in $H_1(M;\mathbb{Z})_{\mathrm{free}}$. If $\omega_1,\dots,\omega_b$ is an orthonormal basis of $H^1(M,\mathbb{R})$ for the L^2 norm we get

$$R_1(M) \le \det\left(\int_{\gamma_i}\omega_j\right)_{1\le i,j\le b} \le \prod_{i=1}^{b}\sqrt{\sum_{j=1}^{b}\left|\int_{\gamma_i}\omega_j\right|^2} \le (\sqrt{b})^b(C_3\,\mathrm{vol}(M))^b$$

so that (assuming $b>0$, as otherwise $R_1(M)=1$):

$$\log R_1(M) \ll b\log(b) + b\log\mathrm{vol}(M).$$

The conclusion of Proposition 2.1 follows since we have $b=b_1(M)\ll \mathrm{vol}(M)$.

2.2.2. Cycle complexity and R_2. Let S be a 2-cycle in M. The Poincaré dual of the class $[S]\in H_2(M;\mathbb{Z})$ is by definition the unique harmonic 1-form $\omega_S\in H^1(M;\mathbb{R})$ which satisfies

$$\forall\omega\in H^2(M;\mathbb{R}): \int_S\omega = \int_M \omega_S\wedge\omega.$$

It follows immediately (by the Cauchy–Schwarz inequality) that

$$\left|\int_S\omega\right| \le \|\omega_S\|_{L^2}\cdot\|\omega\|_{L^2}.$$

So, to estimate the regulator R_2 in a manner similar to the case of R_1 we need bounds on the L^2-norm $\|\omega_S\|_{L^2}$. Getting these bounds is much more intricate than for 1-cycles: in fact there is no equivalent to (2.5) in degree 2. Brock–Dunfield [11, Theorem 1.8] construct a sequence M_n of closed hyperbolic 3-manifolds where $R_2(M_n)\ge c^{\mathrm{vol}(M_n)}$ for some explicit $c>1$. However, restricting to specific classes of hyperbolic manifolds (namely arithmetic congruence ones) there is a hope that regulators can be estimated by the volume. We will explain below (see 3.2) some results in this direction due to Bergeron–Şengün–Venkatesh.

In this section we will explain the first step in their argument. For this we need the notion of cycle complexity introduced by them. Let X be a triangulation of M. If $c\in C_2(X;\mathbb{Z})$ then we can represent the homology class $[c]$ by a singular homology class $[S]$ where S is a smooth embedded surface (not necessarily connected). Then a natural notion of complexity for $[c]$ would be the Euler characteristic of S. Note that in a closed negatively-curved manifold any embedded sphere or torus represents the null class in singular homology. The result is then a relation between the cycle complexity and the L^2-norm. The following sharpened version of [7, Proposition 4.1] was proven by Brock–Dunfield [11, Theorem 1.3].

Theorem 2.3 (Bergeron–Şengün–Venkatesh, Brock–Dunfield). *There exists a constant C with the following property. If M is a closed hyperbolic 3-manifold and S is a smooth surface embedded in M then*

$$\|\omega_S\|_{L^2} \le \frac{C}{\sqrt{\mathrm{inj}(M)}}|\chi(S)|.$$

The proof given by Bergeron–Şengün–Venkatesh uses the relation between cycle complexity and the Gromov–Thurston norm (Brock and Dunfield use a more direct differential-geometric argument). This is a result due to D. Gabai, which states that $[S]$ can be represented by a singular chain $\sum_i t_i\sigma_i$ where σ_i are triangles and $\sum_i |t_i| = -\chi(S)$. One can then replace each σ_i by a totally geodesic triangle without changing the homology class of the cycle. Since hyperbolic triangles have area at most π it follows by applying Theorem 2.2 that

$$\|\omega_S\|_{L^2}^2 = \int_S \omega_S \leq C_1\|\omega_S\|_{L^2}\pi \sum_k |t_k| = C|\chi(S)| \cdot \|\omega_S\|_{L^2}$$

which proves the inequality (note that this uses only the "easy" inequality in Gabai's result, as was pointed out to me by N. Dunfield).

2.2.3. Remarks.

1. There is also a lower bound for cycle complexity in terms of the L^2-norm: Brock–Dunfield prove that $\|\omega_S\| \geq \pi\,\mathrm{vol}(M)^{-1/2}|\chi(S)|$ is S has maximal Euler characteristic in its singular homology class, and Bergeron–Şengün–Venkatesh prove a weaker version of this. The proof is more complicated.
2. Brock–Dunfield also give examples showing that it is not possible to remove the dependency on the injectivity radius in Theorem 2.3 (nor that on the volume in the other inequality).

2.3. Small eigenvalues

We end this section by discussing a recent work of M. Lipnowski and M. Stern [26], which is not immediately related to regulators but has a similar flavour and which we will have the occasion of mentioning again later in this survey. They study the problem of giving a bound for the lowest eigenvalue $\lambda_1(M)$ of the Laplace operator on 1-forms on a manifold M depending only on its geometry. They work only in the setting where M is a finite cover of a compact orbifold M_0. The result they obtain is a bound of the form

$$\lambda_1(M)^{-\frac{1}{2}} \leq C(M) \cdot \left(1 + \sup_\gamma \frac{\mathrm{sArea}(\gamma)}{\ell(\gamma)}\right).$$

Here $C(M)$ is a constant which (in the setting above) is bounded by a polynomial in the volume of M. The *stable area* $\mathrm{sArea}(\gamma)$ is the more interesting part of the bound: the supremum is taken over all closed geodesics γ which are null-homologous, and for such the stable area is the infimum of a bounding surface:

$$\mathrm{sArea}(\gamma) = \inf_{m>0} \inf_{\partial S = m[\gamma]} \left(\frac{\mathrm{vol}(S)}{m}\right).$$

The problem of bounding $\mathrm{sArea}(\gamma)$ is to some extent similar to that of bounding the regulator: instead of low-complexity 1-cycles one looks for low-complexity 2-chains bounding a given curve.

3. Regulators and homology of compact arithmetic manifolds

3.1. Arithmetic manifolds

In this section we will recall the construction of the closed hyperbolic arithmetic 3-manifolds: an *arithmetic manifold* is $\Gamma\backslash\mathbb{H}^3$ where $\Gamma \subset \mathrm{SL}_2(\mathbb{C})$ is a torsion-free arithmetic lattice. We now have to explain what arithmetic lattices are: we will first describe a subclass of these, the *congruence lattices*, which are the main point of focus in this survey. Such a lattice is described by the following data:

1. A number field k which has exactly one complex place;
2. A quaternion algebra A over k which ramifies at all real places of k;
3. There is then a simply connected algebraic k-group G given by the kernel of the reduced norm of A. To finish the description we need a compact-open subgroup K_f in $\mathrm{G}(\mathbb{A}_f)$ (where $\mathbb{A}_f$ are the finite adèles of k).

The congruence lattice associated to these items is the group $\mathrm{G}(k) \cap K_f$ which we will denote by Γ_K. It is naturally a subgroup in $\mathrm{G}(\mathbb{C}) \cong \mathrm{SL}_2(\mathbb{C})$ if we consider k as a subfield of $\mathbb{C}$. It is always discrete in $\mathrm{SL}_2(\mathbb{C})$ and it is cocompact there if and only if A is a division algebra, which is equivalent to it not being isomorphic to the matrix algebra $M_2(k)$. The condition that Γ_K be torsion-free does not translate nicely in terms of the data above, but a sufficient condition for this is that K_v be a normal pro-p subgroup at some finite place dividing a prime $p > 2$ and at which A is not ramified.

Now an *arithmetic lattice* is a subgroup of $\mathrm{SL}_2(\mathbb{C})$ which is commensurable to a congruence lattice, that is $\Gamma \le \mathrm{SL}_2(\mathbb{C})$ is an arithmetic lattice if and only if there exists a congruence lattice Γ' (as constructed above) such that $\Gamma \cap \Gamma'$ is of finite index in both Γ and Γ'. Non-congruence arithmetic lattices exist in $\mathrm{SL}_2(\mathbb{C})$ and we give an explicit example below – they are in fact rather more aboundant than congruence ones – but in this text we will mostly not consider them.

The description above is not very useful to do explicit computations. Rather than using the adélic setting we can describe some arithmetic manifolds in the commensurability class defined by A using orders. An *order* $\mathcal{O}$ in A is a finitely generated subring which generates A as a k-vector space. The group $\mathcal{O}^1$ of norm one elements in $\mathcal{O}$ then fits in the family described above. To see this note that if v is a finite place and R_v the ring of v-adic integers then $\mathcal{O}_v = \mathcal{O} \otimes R_v$ is a compact-open subset of $A \otimes k_v$, and the closure of $\mathcal{O}^1$ in there is a compact-open subgroup in $\mathrm{G}(k_v)$. Let K_f be the product of these over all finite places, then it is clear that $\mathcal{O}^1 = \mathrm{G}(k) \cap K_f$. If K'_f is contained in a group K_f constructed in this way we say that $\Gamma_{K'}$ is *derived from a quaternion algebra*.

There are criteria to see whether a group generated by given matrices in $\mathrm{PSL}_2(\mathbb{C})$ is arithmetic or not (assuming one known it is of finite covolume). The simplest test is to look at its *invariant trace field*, which is the field generated by the traces of its elements in the adjoint representation. This is always a number field, and if the lattice is arithmetic then this field equals the field k used in the definition above. Moreover, it is then derived from a quaternion algebra if and only if its trace field (the field generated by the traces of its preimage in $\mathrm{SL}_2(\mathbb{C})$) equals the invariant trace field. This is a necessary but not sufficient condition for a lattice in $\mathrm{PSL}_2(\mathbb{C})$ to be arithmetic; it is also possible to recover the quaternion algebra A in a similar manner, which gives a complete characterisation. We will work out the example of $\Gamma_{T_4}^+$ below, for a more complete description of the contents of this paragraph we refer to [29, 3.5, 3.6, 8.3]. We also note that these methods do noy say anything about whether a group is congruence or not, once it is known that it is arithmetic; for the latter problem one needs a more group-theoretical approach and we will not discuss this here.

3.1.1. Properties of congruence manifolds. There are multiple reasons to focus on congruence manifolds instead of all hyperbolic manifolds of finite volume, or even arithmetic ones. One such reason, which we will not discuss here, is their particular significance to number theory (a point developed in [8] and [14] to cite only works directly relevant to the present survey).

There is also the vague question asked by William Thurston in his influential paper [45], to "find topological and geometrical properties of quotient spaces of arithmetic subgroups of $\mathrm{PSL}_2(\mathbb{C})$" (see loc. cit., Question 19 on page 380), which seems to have particular resonance in the context of congruence manifolds. As a sample of such properties of congruence manifolds (which are not shared by the larger class of all arithmetic manifolds) let us give the following examples:

- Their Cheeger constants (or first eigenvalue of the Laplace operator Δ_0) are uniformly bounded away from zero (see [16]).
- For any $R > 0$ the volume of their R-thin part is a (uniform) $o(\mathrm{vol})$ (this is known as Benjamini–Schramm convergence, see [19]).
- As a consequence of either of the properties above there are only finitely many of them with a given Heegard genus (see [23] for the former and [4] for the latter).

3.1.2. Hecke operators. Arithmetic manifolds also have a distinguished family of operators acting on differential forms, the Hecke operators. We will give a short description since they occur in the arguments of [7]. By strong approximation, there is a natural bijection between the manifold $\Gamma_K\backslash\mathbb{H}^3$ and the quotient $\mathrm{G}(k)\backslash\mathrm{G}(\mathbb{A})/K$ where $K = K_\infty K_f$, $K_\infty = \mathrm{SU}(2)$. For each finite place v of k the space $\mathrm{G}(k_v)/K_v$ is a finite union of trees (one for each coset of K_v in a maximal subgroup) and as such has an averaging operator δ_v which commutes with the left-$\mathrm{G}(k_v)$-action. Thus these operators act on $C^0(\mathrm{G}(k)\backslash\mathrm{G}(\mathbb{A})/K)$, and they actually extend to bounded operators on $L^2(\mathrm{G}(k)\backslash\mathrm{G}(\mathbb{A})/K)$. The action of δ_v on $L^2(M)$ is denoted by T_v. Similar constructions can be made for differential forms.

It is immediate that the Hecke operators T_v commute with each other and with the Laplacians. From this it follows that each eigenspace $\ker(\Delta_k - \lambda_i)$ has a decomposition into summands which are eigenspaces for all T_v simultaneously. This applies in particular to the spaces $\mathscr{H}^k(M) = \ker(\Delta_k)$. In general it is not possible to exactly compute eigenspaces of Δ_k (for example there is no exactly known Maass cusp form for $\mathrm{SL}_2(\mathbb{Z})$) but the cohomology can be computed from a triangulation X of M, which can be obtained for example from a fundamental domain, see [34]. With further refinements it is also possible to compute the action of the Hecke operators on $H^*(X;\mathbb{Z})$, see [21]. (We note that the explicit computations use a combinatorial definition for Hecke operators on cohomology, which is obviously equivalent to the analytic definition given just above but does not generalise to other automorphic forms.)

3.1.3. Examples. The simplest example for k as above is to take an imaginary quadratic field, for example $\mathbb{Q}(\sqrt{-1})$. Then any quaternion algebra A satisfies condition (2) above. In the next section we will discuss the algebraically simplest example, when $A = M_2(k)$ is split. However, for the compact case many examples do not come from quadratic fields.

For example, the cocompact lattice $\Gamma_{T_2}^+$ from 2.1.3 turns out to be arithmetic. In fact its index 2 subgroup $\Gamma_{T_4}^+$ is derived from a maximal order in the quaternion algebra A over the quartic field

$$k = \mathbb{Q}(a),\ a^4 - a^2 - 1 = 0$$

ramified only at real places. This is recorded in [29, 13.1], we will shortly explain how to prove it. (Before getting to the computation let us remark that the commensurable subgroup $\Gamma_{T_2}^+$ is of necessity arithmetic, but because of the maximality of $\Gamma_{T_4}^+$ it cannot be derived from a quaternion algebra. The Vesnin construction gives an example of a non-congruence subgroup.)

We start by computing the traces of all possible products of tuples of pairwise distinct generators of $\Gamma_{T_4}^+$: according to the presentation (2.4) all generators save α and all products of pair of generators have trace 0 or ± 1, and the remaining traces are:

$$\mathrm{tr}(\tilde{\alpha}) = a^2,\ \mathrm{tr}(\tilde{\alpha}\tilde{\beta}\tilde{\gamma}) = -a.$$

It then follows by [29, Lemma 3.5.2] that the invariant trace field is the field k generated by $\mathrm{tr}(\alpha\beta\gamma)$ and also that all traces of elements of $\Gamma_{T_4}^+$ are integral. We see immediately that k has exactly one complex place and two real ones. Now let A be the quaternion algebra generated (as a subalgebra of $M_2(\mathbb{C})$) by $\Gamma_{T_4}^+$. Since $\langle a, b, c\rangle$ generate a group isomorphic to S_5 it is not possible for A to split at a real place (otherwise we would get an embedding of S_5 into $\mathrm{GL}_2(\mathbb{R})$), so it must ramify at all real places. We can finally put all this together and apply Theorem 8.3.2, loc. cit to conclude that Γ_{T4}^+ is arithmetic, and in fact contained with finite index in the unit group of an order in a quaternion algebra. It remains to see that $\Gamma_{T_4}^+$ is the unit group of a maximal order, which can be done by computing the volume $\mathrm{vol}(T_4)$ and comparing it to the covolume of a maximal order given by the volume formula Theorem 11.1.3, loc. cit. We note that we did not check the finite ramification of A, but using a scheme similar to that in section 4.7 of loc. cit. it is possible to get a Hilbert symbol for A and check by hand that it splits at all of them.

There also exists nonarithmetic lattices in $\mathrm{SL}_2(\mathbb{C})$ (in fact "most" of them are nonarithmetic). An example is the relection group $\Gamma_{T_8}^+$ introduced in 2.1.3: indeed, its (invariant) trace field is of degree 8 with two complex places (see [29, 13.1]) and thus cannot be the trace field of an arithmetic manifold.

Finally we describe some congruence subgroups of $\Gamma_{T_4}^+$. Let p be a rational prime and v a place of k dividing p. By weak approximation we have a dense embedding of $\Gamma_{T_4}^+$ in $\mathrm{PSL}_2(R_v)$. Let f_v be the residue field of R_v, then we get a surjective map $\Gamma_{T_4}^+ \to \mathrm{PSL}_2(f_v)$. The kernel $\Gamma(\mathfrak{p}_v)$ of this map is a congruence subgroup ("principal congruence subgroup of level $\mathfrak{p}_v$"), as is the preimage $\Gamma_0(\mathfrak{p}_v)$ of the subgroup of upper triangular matrices ("Hecke congruence subgroup of level $\mathfrak{p}_v$"). The former is torsion-free if $p > 2$ but the latter is not.

Let us be more explicit for some primes. Assume that p is a rational prime such that the polynomial $X^4 - X^2 - 1$ has a root a in $\mathbb{F}_p$. Suppose in addition that $4 - a^4$ and -1 are quadratic residues modulo p (the latter actually implies that $X^4 - X^2 - 1$ splits in $\mathbb{F}_p$). Then the matrices in (2.4) make sense in $\mathbb{F}_p$ and "reduction modulo $\mathfrak{p}$" (for $\mathfrak{p}$ one of the prime ideals of k above p) defines a map $\Gamma_{T_4}^+ \to \mathrm{PSL}_2(\mathbb{F}_p)$. Then the subgroup $\Gamma_0(\mathfrak{p})$ is equal to those matrices whose reduction modulo $\mathfrak{p}$ is upper triangular.

Note that in general, for there to be a surjective map $\Gamma_{T_4}^+ \to \mathrm{PSL}_2(\mathbb{F}_p)$ is equivalent to $X^4 - X^2 - 1$ having a root in $\mathbb{F}_p$ (but not necessarily being split). To define the "reduction modulo p" map is in general not obvious using the matrices given in (2.4). However it is easy to solve the equations corresponding to the relations in (2.3) to define a morphism $\Gamma_{T_4}^+ \to \mathrm{PSL}_2(\mathbb{F}_p)$. It is then automatically conjugated to a "reduction modulo $\mathfrak{p}$" morphism.[4]

3.2. Regulators of closed arithmetic manifolds: estimation by the volume

The following theorem is a slight modification of [7, Theorem 6.1] (in the statement we conflate between a closed submanifold and the cohomology class it represents).

Theorem 3.1 (Bergeron–Şengün–Venkatesh). *Let M be a congruence arithmetic manifold defined over an imaginary quadratic field k. Let $H_2^{\mathrm{bc}}(M;\mathbb{Q})$ be the subspace of $H_2(M;\mathbb{Q})$ spanned by closed imbedded totally geodesic surfaces in M. Then there exist such surfaces $S_1, \dots, S_b$ which span $H_2^{\mathrm{bc}}(M;\mathbb{Q})$ and such that*

$$\forall 1 \le i \le b : |\chi(S_i)| \le \mathrm{vol}(M)^C$$

where the constant C depends only on the field k.

The proof of this is rather involved and we will not describe it but we will introduce its main ingredients.

---

[4]This is because $A_5 = \langle \alpha, \beta \rangle$ admits only two representations in $\mathrm{PSL}_2(\mathbb{F}_p)$, which are conjugated to each other, and there are then exactly two nontrivial choices for the remaining generator γ which are equivalent.

3.2.1. Totally geodesic surfaces in arithmetic manifolds. Let k, A and K describe an arithmetic manifold or orbifold as explained above. Then the orbifold $M_K = \Gamma_K \backslash \mathbb{H}^3$ contains totally geodesic 2-dimensional suborbifolds if and only if:

- k is a quadratic extension of a totally real field l;
- $A = B \otimes_l k$ where B is a quaternion algebra over l which splits at exactly one real place.

Note that there are infinitely many choices for the quaternion algebra B (indeed, we can always add ramification at a finite place which is inert in k/l). However there is a "minimal" choice for B, which ramifies exactly at the places of l which lie below pairs of conjugated places of k. In the sequel we will always assume that B is chosen thus.

As in [7] we will now only consider the case where $l = \mathbb{Q}$, so that B is an anisotropic quaternion algebra over $\mathbb{Q}$ which splits over the reals. Let H be the unit group of B. It is a $\mathbb{Q}$-subgroup of (the Weil restriction to $\mathbb{Q}$ of) G. With respect to $\mathbb{R}$-points this gives an embedding $\mathrm{SL}_2(\mathbb{R}) \to \mathrm{SL}_2(\mathbb{C})$, which in turn induces a totally geodesic embedding $\mathbb{H}^2 \subset \mathbb{H}^3$. In addition, for any $g \in \mathrm{G}(k)$, the subgroup $\Lambda_{g,K} = g\mathrm{H}(k)g^{-1} \cap K$ is a cocompact lattice in $g\mathrm{H}(\mathbb{R})g^{-1}$ and the image $S_{g,K}$ of the imbedding $\Lambda_{g,K} \backslash \mathbb{H}^2 \to \Gamma_K \backslash \mathbb{H}^3$ is a totally geodesic surface in M.

Note that since we fixed B not all totally geodesic surfaces are obtained by the construction above, as there are infinitely many commensurability classes of totally geodesic surfaces in M [29, Theorem 9.5.6]. But at the cohomological level only we will see below that everything is obtained from the single algebra B (that is, every class in $H_2^{\mathrm{bc}}(M_K; \mathbb{Q})$ has a representative of the form $S_{g,K}$ for some g.

3.2.2. Base change. We will now explain brielfy the analytic construction of the cohomology classes dual to totally geodesic surfaces. It uses the *base-change* construction of R. Langlands. The latter associates (in our special case) an automorphic representation of SL_2/k to an automorphic representation of SL_2/l. To use it in the case of a cocompact lattice one needs to use in addition the *Jacquet–Langlands correspondence* which is a bijection between automorphic representations of an anisotropic group G as above and SL_2/l. We may sum up these constructions via the diagram:

$$\begin{array}{ccc}
\mathscr{A}(\mathrm{SL}_2(\mathbb{A}_l)) & \xrightarrow{\text{base change}} & \mathscr{A}(\mathrm{SL}_2(\mathbb{A}_k)) \\
\text{Jacquet–Langlands} \big\uparrow & & \big\uparrow \text{Jacquet–Langlands} \\
\mathscr{A}(\mathrm{H}(\mathbb{A}_l)) & & \mathscr{A}(\mathrm{G}(\mathbb{A}_k))
\end{array}$$

where $\mathscr{A}$ denotes the space of automorphic forms on a group. The Jacquet–Langlands maps are not bijections (see 3.5.2 below for more information on its image) but the one on the right is when restricted to the image by base-change of $\mathscr{A}(\mathrm{H}(\mathbb{A}_l))$ and we get a map $\mathscr{A}(\mathrm{H}(\mathbb{A}_l)) \to \mathscr{A}(\mathrm{G}(\mathbb{A}_k))$ which completes the diagram above. We will call it the base change map associated to $B \subset A$. It respects Laplace eigenvalues and it is Hecke-equivariant, and it follows in particular that they define a Hecke-invariant subspace $H^2_{\mathrm{bc}}(M;\mathbb{C})$. We note that this subspace is defined over $\mathbb{Q}$. The following lemma is a less precise version of [7, Proposition 6.9]

Lemma 3.2. *The spaces $H^2_{\mathrm{bc}}(M;\mathbb{C})$ and $H_2^{\mathrm{bc}}(M;\mathbb{C})$ are dual to each other.*

This proves the claim above: the image of the base change map between $\mathscr{A}(\mathrm{H}(\mathbb{A}_l))$ and $\mathscr{A}(\mathrm{G}(\mathbb{A}_k))$ depends on the choice of B but the image of the map for our "minimal" choice above contains all the others, hence by the lemma any totally geodesic surface is cobordant to a class $S_{g,K}$.

Now the volume of such a surface has an upper bound in terms of the reduced discriminant of B and the volume of K and the denominator d_g of g by the volume formula. Since $\mathrm{disc}(B)^2 = \mathrm{disc}(A)$ we see that $\log \mathrm{disc}(B) \ll \log \mathrm{vol}(M)$ and we get a rough estimate:

$$\chi(S_{g,K}) = 2\pi \,\mathrm{vol}(S_{g,K}) \leq (d_g \,\mathrm{vol}(M))^C \tag{3.1}$$

for some absolute C.

3.2.3. The Hecke operators T_v act on $H^2(M;\mathbb{Q})$ and by duality on $H_2(M;\mathbb{Q})$ as well. As we said above this action preserves $H^2_{\mathrm{bc}}(M;\mathbb{Q})$ and by the lemma its dual action preserves the subspace spanned by the $S_{g,K}$. In fact it follows from the combinatorial definition of the Hecke operators acting on cohomology that $T_v S_{g,K}$ is a linear combination of $S_{g_i,K}$ where $d_{g_i} \leq C d_g q_v$. What Bergeron–Şengün–Venkatesh prove, which implies the statement of Theorem 3.1 in view of (3.1), is the following:

> *There exists $v_1, \dots, v_l$ and $g_1, \dots, g_l$ such that $d_{g_i}, q_{v_i} \leq \mathrm{vol}(M)^C$ and the classes $T_{v_i} S_{g_i,K}$ span $H_2^{\mathrm{bc}}(M;\mathbb{Q})$.*

We will not discuss the proof of this result and instead refer the reader to [7, Section 6.10] where a detailed outline is provided.

3.3. Growth of torsion homology

We saw in (2.2) that for M a closed hyperbolic manifold the torsion subgroup of $H_1(M;\mathbb{Z})$ has its size bounded above by $C^{\mathrm{vol}(M)}$ for some C depending only on the injectivity radius of M. It was recently proven by M. Frączyk [19] that the bound $|H_1(M;\mathbb{Z})_{\mathrm{tors}}| \leq C^{\mathrm{vol}(M)}$ holds with a uniform C for the class of all congruence hyperbolic manifolds (conditionally on a positive answer to Lehmer's question the injectivity radius of all arithmetic manifolds is bounded away from zero but Frączyk's proof holds unconditionally). A much simpler observation is that if we restrict attention to the finite covers of a given manifold then the injectivity radius is bounded below and so the bound applies with an absolute C. A natural question in either context is whether it is sharp. A recent preprint of Y. Liu [27, Theorem 1.2] together with known results (for example [8, Theorem 7.3]) imply that the exponential aspect is. However the covers produced are not congruence and the constant C is not specified independently of M. Closer to our center of interest Bergeron and Venkatesh made the following more precise conjecture.

Conjecture 3.3. *Let Γ be an arithmetic lattice in $\mathrm{SL}_2(\mathbb{C})$ and Γ_n a sequence of pairwise distinct congruence subgroups of Γ. Then*

$$\lim_{n\to+\infty}\left(|H_1(\Gamma_n;\mathbb{Z})_{\mathrm{tors}}|^{\frac{1}{\mathrm{vol}(\Gamma_n\backslash\mathbb{H}^3)}}\right)=e^{\frac{1}{6\pi}}.$$

As a corollary of Theorem 3.1, the Cheeger–Müller Theorem 1.1, and the "limit multiplicity for analytic torsion" from [8] we get the following theorem (see [7, Theorem 1.2], note that the assumption the the cohomology is purely base change implies the hypothesis on the Betti numbers there).

Theorem 3.4 (Bergeron–Şengün–Venkatesh). *Let M_n be a sequence of closed congruence arithmetic hyperbolic manifolds. Assume that the M_n have "few small eigenvalues" (this is an asymptotic condition precised in* (i) *of* [7, Theorem 1.2]) *and that the characteristic zero cohomology of each M_n consists only of totally geodesic classes. Then Conjecture* 3.3 *holds for M_n.*

Unfortunately there is no known example of a sequence satifying the "few small eigenvalues conditions" (the results of Lipnowski–Stern [26] mentioned in 2.3 are still far from providing even a sufficiently good lower bound on the first nonzero eigenvalue). F. Calegari and N. Dunfield [13] construct examples of sequences M_n with $b_1 = 0$ (so the condition on cohomology is vacuously true) but there is no example known of an infinite sequence of congruence subgroups, where there are nontrivial classes in the second homology but they all are totally geodesic.

Regarding the upper limit in Conjecture 3.3, a much more general result due to T. Lê holds.

Theorem 3.5 (Lê). *Let M be a closed hyperbolic 3-manifold and M_n a sequence of finite covers which converges in the Benjamini–Schramm sense to $\mathbb{H}^3$. Then*

$$\limsup_{n\to+\infty}\frac{\log|H_1(M_n;\mathbb{Z})_{\mathrm{tors}}|}{\mathrm{vol}(M_n)}\leq\frac{1}{6\pi}.$$

We note that the proof of this last theorem is mostly topological; see [28] for a survey of its setting.

3.3.1. Numerical computations. The first computational evidence for Conjecture 3.3 was given by Şengün in [43]. He computed the abelianisation of $\Gamma_0(\mathfrak{p})$ for Γ a Bianchi group and a set of prime ideals $\mathfrak{p}$ with norm close to 20000 (note that $\Gamma_0(\mathfrak{p})$ is not torsion-free) and found the size to be in accordance with Bergeron and Venkatesh's prediction. For cocompact lattices N. Dunfield made computations for $\Gamma_0(\mathfrak{n})$ where $\mathfrak{n}$ is prime or a prime power and Γ is an arithmetic orbifold coming from a simple topological construction. The results of these computations are recorded in [10] and they also support the conjecture. An interesting phenomenon in this numerical data is that prime levels exhibit a much faster convergence towards the $1/6\pi$ growth rate.

There are also computations for congruence covers of nonarithmetic manifolds, by Şengün [44] (for example for the tetrahedral group Γ_{T_8}) and Dunfield

[10]. In both cases the data seems to indicate that when $b_1(M_n) > 0$ the size $|H_1(M_n; \mathbb{Z})_{\text{tors}}|^{1/\operatorname{vol}(M_n)}$ can be much smaller than $e^{1/6\pi}$.

We also performed some new computations for subgroups $\Gamma_0(\mathfrak{p})$ in $\Gamma_{T_4}^+$. Note that the tetrahedral group $\Gamma_{T_4}^+$ is not commensurable to the examples in Şengün's computations, but it is commensurable to the fundamental group of the orbifold $T(2,5)$ appearing in Dunfield's computations in [13] and [10].

We performed the calculations using GAP [20], the raw data is available at [38]. The range of computation was $1000 < |\mathfrak{p}| < 26000$. We only recorded data from subgroups $\Gamma_0(\mathfrak{p})$ for ideals $\mathfrak{p}$ of degree 1, since in any case very few ideals of degree 2 and none of degree 4 have norm in this range. Figure 1 is a graphic representation of the ratios of $\log |H_1(M_p; \mathbb{Z})_{\text{tors}}|$ by $\operatorname{vol}(M_p)/(6\pi)$ where p is a rational prime which totally splits in the trace field k and M_p is one of the congruence covers $\Gamma_0(\mathfrak{p})\backslash\mathbb{H}^3$ for $\mathfrak{p}$ a prime factor of p. Blue dots correspond to subgroups with $b_1 = 0$, and red ones to the others.

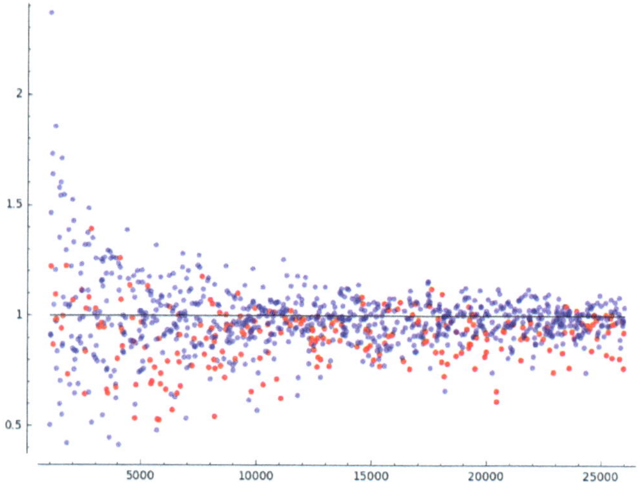

FIGURE 1. Torsion ratio for $\Gamma_0(\mathfrak{p})$, $1000 < |\mathfrak{p}| < 26000$

3.4. Homology with coefficients

3.4.1. Strongly acyclic local systems. Conjecture 3.3 can be formulated in greater generality: instead of looking only at the homology group $H_1(\Gamma_n; \mathbb{Z})$ one can study the groups $H_1(\Gamma_n; L)$ where L is an *arithmetic* Γ*-module*, meaning it is of the type discussed before (1.11): there exists a finite-dimensional representation $\rho : \mathrm{SL}_2(\mathbb{C}) \to \mathrm{SL}(V)$ such that L is a lattice in V preserved by $\rho(\Gamma)$. The lattice Γ being arithmetic means that such modules do exist (their description depends on the algebraic group G – we will give specific ones below). Perhaps counter-intuitively, a huge simplification can occur in this more general setting: as proven in [8], certain

ρ are *strongly acyclic*, meaning that the Hodge–Laplace operators on the spaces of square-integrable forms with coefficients in the flat bundle F associated to ρ have a uniform – not depending on the base manifold – spectral gap.[5] In particular there are no regulator terms in (1.11) and thus Theorem 1.2 establishes a direct relation between the order of H_1 and the analytic torsion. Using the uniformity of the spectral gap (an extremal case of the "few small eigenvalues" condition mentioned above) it is relatively easy to establish the asymptotic behaviour of the analytic torsion (see [8, Section 4]), leading to the following result.

Theorem 3.6 (Bergeron–Venkatesh). *Let Γ be a uniform congruence arithmetic lattice in $\mathrm{SL}_2(\mathbb{C})$. Let ρ be a real linear representation of $G = \mathrm{PSL}_2(\mathbb{C})$ on a vector space V which has no irreducible component fixed by the Cartan involution of G. Suppose that L is a lattice in V which is preserved by $\rho(\Gamma)$. Then there is a constant $c_\rho > 0$ (effectively computable from the decomposition of ρ into irreducible factors) such that for any sequence M_n of torsion-free congruence subgroups of M we have*

$$\lim_{n\to+\infty} \frac{\log|H_1(\Gamma_n; L)|}{[\Gamma : \Gamma_n]} = c_\rho \operatorname{vol}(\Gamma\backslash\mathbb{H}^3).$$

An example of such a representation is obtained by the following construction (see [8, Section 8.2]). Let A be a quaternion algebra (satisfying the conditions in 3.1) and $\mathcal{O}$ and order in A. Let Γ be the group of norm 1 elements in $\mathcal{O}$ and L the sub-$\mathbb{Z}$-module of elements of trace 0 in $\mathcal{O}$. Then L is a Γ-stable lattice in $A \otimes_{\mathbb{Q}} \mathbb{R}$ on which G acts by conjugation. As a representation of $\mathrm{SL}_2(\mathbb{C})$ it is isomorphic to $3^{[k:\mathbb{Q}]-2}$ copies of the adjoint representation and hence satisfies the conditions of the theorem. The constant c_ρ in this case has been computed in [8] to be equal to $13/6\pi$.

3.4.2. Remarks.

1. It is possible to prove a similar statement for a sequence of pairwise noncommensurable arithmetic lattices, using the Benjamini–Schramm convergence of arithmetic lattices proven in [19] and elementary arguments. Note that in this case the lattices L are not fixed, and if the degree of the field of definition goes to infinity then their rank has to go to infinity.
2. Bergeron–Venkatesh prove a result which is valid for all arithmetic locally symmetric spaces. In its general form it does not state that torsion homology in a certain degree witnesses exponential growth but only that in some cases (depending only on the ambient Lie group) at least one does (its parity is the only thing that is determined). They conjecture a precise statement for the general case. Some numerical data for case beyond $\mathrm{SL}_2(\mathbb{C})$ is collected in [3].

[5] Note that this is never the case for trivial coefficients: in addition to the 0 eigenvalue on functions, even when there is a uniform spectral gap on functions the spectral gap on 1-forms always tends to 0 in a sequence of manifolds with volume going to infiinity.

3.4.3. Changing coefficients. Another type of result on torsion growth, which has the same scheme of proof as outlined above, consists in fixing the lattice Γ and changing the coefficients modules. We will only quote the following result from [30].

Theorem 3.7 (Marshall–Müller). *Let Γ be a uniform torsion-free congruence arithmetic lattice in $\mathrm{SL}_2(\mathbb{C})$, defined over an imaginary quadratic field. Let $L_2 \subset \mathfrak{sl}_2(\mathbb{C})$ a lattice preserved by $\mathrm{Ad}(\Gamma)$ and $L_{2m} = \mathrm{Sym}^m L_2$. Then*

$$\lim_{m \to +\infty} \frac{\log H_1(\Gamma; L_{2m})}{m^2} = \frac{2\,\mathrm{vol}(\Gamma\backslash\mathbb{H}^3)}{\pi}.$$

3.5. Rationality properties of regulators of closed arithmetic manifolds

We will say a few short words about some results concerning the fine arithmetic properties of regulators of arithmetic manifolds proven in [14] and [5].

3.5.1. Regulators and L-values. Let M be an arithmetic congruence manifold defined by a compact subgroup K as in 3.1 above. In [14, Theorem 5.2.3] Calegari and Venkatesh rationally relate $R_1(M)$ to a product of (essentially) L-values associated to the cohomological representations occuring in the space of K-invariant automorphic forms. The precise statement of their result needs too much notation for us to quote here. While interesting from a number-theoretical point of view this result is useless if we want to estimate the size of R_2: to do this we would need to "pin down $R_2(M)$ integrally rather than rationally" to paraphrase [14]. We will see below (cf. 4.3) how to do this in a similar but simpler context; in general we are not aware of any results on this for closed manifolds. There are some hints of this (from a number-theoretical point of view) in [14, 5.2.5].

3.5.2. Spectrally related manifolds. Suppose that Γ is a cocompact lattice in $\mathrm{PSL}_2(\mathbb{C})$ and there is a surjection $\pi : \Gamma \to Q$ where Q is a finite group which contain two subgroups H_1, H_2 which are not conjugated by an element of $\mathrm{Aut}(Q)$ but such that $\mathbb{C}[Q/H_i]$ are isomorphic as Q-spaces (H_1 and H_2 are then said to be *almost conjugated*). There are numerous examples of such Γ and Q. Assuming further that $\Gamma_i := \pi^{-1}(H_i)$ are torsion-free it was discovered by T. Sunada that the manifolds $M_i = \Gamma_i\backslash\mathbb{H}^3$ are isospectral to each other (together they are called a *Sunada pair*). By the Cheeger–Müller Theorem 1.1 and (1.7) it follows that

$$\frac{R_1(M_1) \cdot R_2(M_2)}{R_2(M_1) \cdot R_1(M_2)} \in \mathbb{Q}^\times. \tag{3.2}$$

A. Bartel and A. Page give in [5, Theorem 1.7] a more precise description of the quotient. It relies only on the representation theory of the finite group Q, and so in particular it is independent of the Cheeger–Müller Theorem.

The other well-known construction of isospectral hyperbolic 3-manifolds is by M. F. Vignéras, who uses nonconjugated maximal orders in a quaternion algebra. A different but somewhat similar construction is that of *Jacquet–Langlands pairs*. These are not isospectral but the spectra are still related. The construction proceeds as follows: take two quaternion algebras A_1, A_2 and let Γ_i be the lattice associated to a maximal order in A_i. Let S_1, S_2 be the sets of finite places where A_1, A_2 respectively ramify, and assume $S_1 \neq S_2$. Let $S \supset S_1 \cup S_2$ and let $\Gamma_{0,i}(N_i)$ be the Hecke subgroups of level N_i in each $M_i = \Gamma_i$, where $N_i = \prod_{v \in S \setminus S_i} \mathfrak{p}_v$. Jacquet and Langlands proved that there exists a relation between the spaces of automorphic forms on $\Gamma_i \backslash \mathrm{PSL}_2(\mathbb{C})$ and in [14] this is exploited to relate regulators on M_1 and M_2. The relation is more complicated than for Sunada pairs and they only prove partial results. The statements are similar in form to (3.2) but unfortunately they are too involved to be explained here.

In the same work Calegari and Venkatesh also give statements relating the sizes of torsion homology groups of such manifolds, for example in the Introduction to loc. cit., Theorems A, A† and B (whose description unfortunately lies outside our scope here).

4. Regulators of finite-volume arithmetic manifolds

4.1. Non-compact hyperbolic manifolds

In this preliminary subsection we will detail a bit more the structure of noncompact hyperbolic manifolds of finite volume, and the arithmetic examples of such.

4.1.1. Cusps and height functions. For a noncompact manifold of finite volume the injectivity radius is always 0 (as the manifold must scrunch at infinity to have finite volume). Another metric invariant in this case is the *systole* $\mathrm{sys}(M)$, which by definition is the smallest length of a closed geodesic on M (in the case where M is closed it equals twice the injectivity radius). In this section we will work under the assumption (which holds for all noncompact arithmetic hyperbolic manifolds) that $\mathrm{sys}(M) > \varepsilon > 0$ for a fixed $\varepsilon > 0$. Thus, according to the thick-thin decomposition, the ε-thin part $M_{\leq \varepsilon}$ consists only of cusps. The metric description of a cusp is as follows: let T be a 2-dimensional torus with holomorphic coordinate z and flat Riemannian metric $|dz|^2$. Then the warped product

$$C_T = T \times [1, +\infty[, \quad \frac{|dz|^2 + dy^2}{y^2}$$

is locally hyperbolic and of finite volume. For ε smaller than the Margulis constant of $\mathbb{H}^3$ the thin part of M is then made of a finite disjoint union of cusps $C_{T_1}, \dots, C_{T_h}$. We see that this decomposition specifies a *height function* on M, which in a cusp C_{T_i} is given by the y-coordinate and which we take to equal 1 on the ε-thick part. We note that this function is well defined independently of the choice of ε (or more generally of a parametrisation of each cusp) only up to multiplication by a constant in each cusp but this will not affect the objects we will define later in a way that we need to worry about.

In the sequel we will also use the Borel–Serre compactification $\overline{M}$ of a manifold M. This is the disjoint union $M \cup \partial\overline{M}$ where

$$\partial\overline{M} = T_1 \sqcup \cdots \sqcup T_h$$

and the topology in $C_{T_i} \cup T_i$ is defined by $(y_n, x_n) \to x$ if $y_n \to +\infty$ and $x_n \to x$. It is a smooth manifold with boundary, diffeomorphic to the level sets of a height function, and the inclusion $M \subset \overline{M}$ is a homotopy equivalence.

4.1.2. Reflection group examples. It is much easier to come with examples of noncompact manifolds than compact ones. For example if T_r is a regular ideal tetrahedron in $\mathbb{H}^3$ then its dihedral angles are equal to $\pi/3$ and hence the reflection group Γ_{T_r} is a nonuniform lattice in $\mathrm{PSL}_2(\mathbb{C})$. It is also possible to glue the faces of T_r to obtain a nonorientable manifold (the Gieseking manifold), which is the noncompact hyperbolic manifold of smallest volume [1]. Its orientation cover is the figure eight-knot complement. These examples are commensurable to each other and arithmetic (more on this below). Another arithmetic example (which is not commensurable to the previous ones) is the reflection group of the regular ideal octahedron.

There are also nonarithmetic reflection examples. A list is given in [29], the simplest example there is U_1 whose Coxeter symbol is:

4.1.3. Arithmetic examples. As for the geometric constructions, the arithmetic construction of lattices in $\mathrm{SL}_2(\mathbb{C})$ is much simpler. By general theory an arithmetic lattice is non-cocompact if and only if its algebraic group is not anisotropic. In the construction given in 3.1 above this amounts to the quaternion algebra A not being a division algebra, which in turn means that A is isomorphic to the matrix algebra $M_2(k)$ for some imaginary quadratic field k. If R is the ring of integers of k then $M_2(R)$ is an order in A and the associated group of units is $\mathrm{SL}_2(R)$. These groups are called Bianchi groups, and as a consequence of this paragraph every nonuniform arithmetic lattice in $\mathrm{SL}_2(\mathbb{C})$ is commensurable (up to conjugation) to one of these. As we mentioned above the figure-eight complement is arithmetic; the image of its holonomy map is in fact contained in $\mathrm{PSL}_2(\mathbb{Z}[e^{2i\pi/3}])$ as a subgroup of finite index[6].

We also note that it is much easier to describe congruence subgroups for the Bianchi groups than for uniform lattices. For example if $\Gamma = \mathrm{SL}_2(R)$ and $\mathfrak{n}$ is an ideal in R the Hecke congruence subgroup of level $\mathfrak{n}$ is simply

$$\Gamma_0(\mathfrak{n}) = \left\{ \begin{pmatrix} a & b \\ c & d \end{pmatrix} \in \mathrm{SL}_2(R) : c \in \mathfrak{n} \right\}.$$

4.2. Definitions of the regulators

In this section we will use $H^*(M; \mathbb{K})$ (where $\mathbb{K} = \mathbb{R}$ or $\mathbb{C}$) to denote de Rham cohomology, which is when needed identified with the cohomology of a triangulation of M. We will denote the class of a differential form ω in de Rham cohomology by $[\omega]$. When A is a subgring of $\mathbb{C}$ we write $H^*(M; A)$ to denote the A-submodule of those de Rham classes whose integral against integral cycles (i.e., submanifolds) lie in A.

[6]This is actually how its hyperbolic structure was first discovered by R. Riley, see [41].

If Γ is a discrete, torsion-free subgroup of $\mathrm{PSL}_2(\mathbb{C})$ then there is a well-defined maximal essentially self-adjoint extension of the Hodge–Laplace operators from smooth compactly supported forms to square-integrable ones. On the other hand if M is the hyperbolic manifold $\Gamma\backslash\mathbb{H}^3$ it is not true that $H^k(M;\mathbb{R}) \cong \mathscr{H}^k(M)$ when M is noncompact. For related reasons it is not possible to define the analytic torsion using the same definition as in the compact case. In the case where M is of finite volume it is possible to extend both Hodge theory and the Cheeger–Müller Theorem using the theory of Eisenstein series, which gives an explicit description of the orthogonal complement to the subspace where the Laplacian has a discrete spectrum.

4.2.1. Eisenstein cohomology. Let $M = \Gamma\backslash\mathbb{H}^3$ be a non-compact hyperbolic manifold of finite volume. We will first describe the cohomology group $H^1(M;\mathbb{R})$ by analytic means. The first important fact is that the "cuspidal" and "L^2" cohomologies coincide in our setting.

Lemma 4.1. *Let $H^1_{\mathrm{cusp}}(M;\mathbb{R})$ be the kernel of the restriction map $i_1^* : H^1(M;\mathbb{R}) \to H^1(\partial\overline{M},\mathbb{R})$ and $\mathscr{H}^1_{L^2}(M) = \ker(\Delta_1)$. Then the Hodge–de Rham map $\mathscr{H}^1_{L^2}(M) \to H^1_{\mathrm{cusp}}(M;\mathbb{R})$ is an isomorphism.*

This gives an isomorphism

$$H^1(M;\mathbb{R}) \cong \mathscr{H}^1_{L^2}(M) \oplus \mathrm{Im}(i_1^*).$$

Duality and the long exact sequence imply that the image has dimension half that of $H^1(\partial\overline{M};\mathbb{R})$, which is equal to the number h of cusps of M. To proceed further we need to give a description of the second summand by de Rham classes associated to automorphic forms on M. The theory of Eisenstein cohomology developed in [22], of which we will give a very short account here, does this.

To introduce Eisenstein classes we need to extend scalars from $\mathbb{R}$ to $\mathbb{C}$. The cohomology $H^1(\partial\overline{M},\mathbb{C})$ has a Hodge decomposition: if $T_1,\dots,T_h$ are the boundary components of $\overline{M}$ then each of them has a holomorphic coordinate z_i. The forms dz_i and $d\overline{z}_i$ are both harmonic and the cohomology is then decomposed as

$$H^1(\partial\overline{M};\mathbb{C}) = H^{1,0}(\partial\overline{M};\mathbb{C}) \oplus H^{0,1}(\partial\overline{M};\mathbb{C})$$

where

$$H^{1,0}(\partial\overline{M};\mathbb{C}) = \bigoplus_{i=1}^{h} \mathbb{C}[dz_i] \text{ and } H^{0,1}(\partial\overline{M};\mathbb{C}) = \bigoplus_{i=1}^{h} \mathbb{C}[d\overline{z}_i].$$

There is then a distinguished injective map

$$E^1 : H^{1,0}(\partial\overline{M};\mathbb{C}) \to H^1(M;\mathbb{C})$$

such that $H^1(M;\mathbb{C}) = \mathscr{H}^1_{L^2}(M) \oplus \mathrm{Im}(E^1)$. To define it formally Eisenstein series are needed, which we won't introduce. We will assume that everything is defined using the normalisations from [14, Chapter 6]. If $\omega \in H^{1,0}(\partial\overline{M};\mathbb{C})$ the Eisenstein series $E(0,\omega)$ is a smooth harmonic 1-form on M which is not square-integrable. It is closed and the class $E^1([\omega])$ is defined by $E^1([\omega]) = [E(0,\omega)]$. We will not care about the exact definition of Eisenstein series but we will record the following analytic properties: for a differential form f on M let f_P denote the *constant term*. This is defined as the collection of zeroth Fourier coefficient for restriction of f to the cross-sections of the cusps; in general it depends on the height at which the cross section is taken but for the forms $E(0,\omega)$ it does not.

Lemma 4.2. *There is a linear isomorphism $\Phi(0)$ ("intertwining operator") from $H^{1,0}(\partial\overline{M};\mathbb{C})$ to $H^{0,1}(\partial\overline{M};\mathbb{C})$ such that*

$$E(0,\omega)_P = \omega + \Phi(0)(\omega). \tag{4.1}$$

It follows in particular from the lemma that $\mathrm{Im}(E^1)\cap H^1_{\mathrm{cusp}}(M;\mathbb{R}) = 0$, hence we get an actual decomposition

$$H^1(M;\mathbb{C}) = \mathrm{Im}(E^1) \oplus \mathscr{H}^1_{L^2}(M).$$

In degree 2 the construction of Eisenstein classes is slightly different. As above we define $\mathscr{H}^2_{L^2}(M) = \ker(\Delta_2)$ and $H^2_{\mathrm{cusp}}(M;\mathbb{R}) = \ker(i_2^*)$. The same proof as that of Lemma 4.1 yields that $\mathscr{H}^2_{L^2}(M) \cong H^2_{\mathrm{cusp}}(M;\mathbb{R})$. Then we look at the long exact sequence of $\overline{M}, \partial\overline{M}$ to see that $\dim(\mathrm{Im}(i_2^*)) = h-1$, and let

$$Z = \mathrm{Im}(i_2^*) = \left\{ \sum_i [a_i dz_i \wedge d\overline{z}_i] : \sum_i a_i = 0 \right\}.$$

The Eisenstein map is a map $E^2 : Z \to H^2(M;\mathbb{C})$ such that

$$H^2(M;\mathbb{C}) = \mathrm{Im}(E^2) \oplus H^2_{\mathrm{cusp}}(M;\mathbb{C}).$$

We define it following [14, 6.3.1]: if $\eta \in H^2(\partial\overline{M};\mathbb{C})$ we take its Poincaré dual $f = *\eta$ which is just a collection $(a_1,\ldots,a_h)$. Eisenstein series for the constant function have a pole at $s=1$ but if $\sum_i a_i = 0$ we may still construct the Eisenstein series $E(1,f)$. It is then a harmonic function (not square-integrable) and thus the 2-form $*dE(1,f)$ is closed. Finally, we put $E^2([\eta]) = [*dE(1,f)]$.

4.2.2. Reidemeister torsion. According to the above paragraph we can define a Hermitian inner product on $H^1(M;\mathbb{C})$ by putting:

$$\begin{aligned} \langle [f_1],[f_2]\rangle &= \langle f_1,f_2\rangle_{L^2} && \text{if } f_1,f_2 \in \mathscr{H}^1_{L^2}(M); \\ \langle [f], E^1([\omega])\rangle &= 0 && \text{if } f \in \mathscr{H}^1_{L^2}(M), \omega \in H^1(\partial\overline{M};\mathbb{C}); \\ \langle E^1([\omega_1]), E^1([\omega_2])\rangle &= \langle \omega_1,\omega_2\rangle && \text{if } \omega_1,\omega_2 \in H^1(\partial\overline{M};\mathbb{C}). \end{aligned}$$

and similarly for $H^2(M;\mathbb{C})$. We can then define the regulators $R_i(M)$ for $i=1,2$ using the formula (1.6) (0- and 3-cohomology groups are represented by square-integrable function, see [14]), and the Reidemeister torsion by the formula (1.7). It is also possible to define an analytic torsion $T_R(M)$ using the Selberg trace formula (see [14], [33]).

There should be a "Cheeger–Müller" relation between $\tau(M)$ and $T_R(M)$, possibly with additional terms coming from the cusps. A proof for this is almost given in the course of proving Theorem 6.8.3 in [14], which however misses a crucial ingredient (see the remarks there after the statement of the theorem).

4.3. Estimating regulators

Frow now on we will assume M to be a congruence manifold commensurable to the Bianchi orbifold $\mathrm{PSL}_2(R)\backslash\mathbb{H}^3$ with R the ring if integers in the imaginary quadratic field k. For these manifolds the intertwining operator $\Phi(0)$ is quite well understood and we will explain how to use this to separate the terms coming from cusp forms and from Eisenstein series in the regulators defined in the previous subsection.

4.3.1. Rationality questions. Let l be a subfield of $\mathbb{C}$ and V a $\mathbb{C}$-vector space which contains a spanning l-subspace V_l of the same dimension. In this situation we say that a subspace $W \subset V$ is *defined over* l if $W_l := W \cap V_l$ spans W. A linear map Θ between two subspaces W, W' is said to be defined over l if $\Theta(W_l) \subset W'_l$.

The subspace $H^1_{\mathrm{cusp}}(M;\mathbb{C})$ is obviously defined over $\mathbb{Q}$ (with respect to $H^1(M;\mathbb{Q})$), hence also over any subfield of $\mathbb{C}$. With this $\mathbb{Q}$-structure however, the subspaces $H^{1,0}(\partial\overline{M};\mathbb{C})$ and $H^{0,1}(\partial\overline{M};\mathbb{C})$ are not defined over $\mathbb{Q}$ (in fact not over $\mathbb{R}$). This can be seen as follows: let T be a component of $\partial\overline{M}$ and c_1, c_2 be a basis for the homology $H_1(T;\mathbb{Z})$. Then we can find (by appropriately scaling) an holomorphic form ω on T such that $\int_{c_1}\omega = 1$; as $H_1(T;\mathbb{Z})$ is a lattice in $\mathbb{C}$ commensurable to R we get that $\int_{c_2}\omega$ must be a number in $k \setminus \mathbb{Q}$. This also shows that $H^{1,0}(\partial\overline{M};\mathbb{C})$ is defined over k; and the same proof holds for $H^{0,1}(\partial\overline{M};\mathbb{C})$. We then have the following rationality result.

Proposition 4.3. *The map $\Phi(0)$ is defined over k.*

This is proven for example in [22, Corollary 4.2.1][7]. This result implies that the subspace $\mathrm{Im}(E^1)$ spanned by Eisenstein series is defined over k as well, as it is the graph of a map defined over k. In what follows we will denote $H^1_{\mathrm{Eis}}(M;k)$, etc. the subspaces $\mathrm{Im}(E^1) \cap H^1(M;k)$, etc.

4.3.2. Cohomology spaces have a natural $\mathbb{Z}$-structure, hence a natural A-structure for any subring $A \subset \overline{\mathbb{Z}}$. If V is an Hermitian vector space with a l-structure and L an A-lattice in V (that is a spanning free A-submodule of rank equal to $\dim(V)$) then we define the covolume of L in V as the $[l:\mathbb{Q}]$th root of the absolute value of the absolute norm[8] of the determinant of any linear map sending an orthonormal basis of V_l to a R-basis of L (this is independant of the choices made as two such matrices differ by a matrix whose determinant is a unit of R, hence of norm 1). With this definition the covolume of a lattice is stable under extension of scalars. In particular, the regulator $R_1(M)$ is equal to the covolume of $H^1(M;R)$ in $H^1(M;\mathbb{C})$.

---

[7]The statement in loc. cit. is slightly different in form, essentially because Harder defines the $\mathbb{Q}$-structure on cohomology by taking restriction of scalars from the k-structure we use.
[8]Recall that for an algebraic number $a \in \overline{\mathbb{Q}}$ this is defined as $\prod_{a^\sigma \in \mathrm{Gal}(\overline{\mathbb{Q}}/Q)} a^\sigma \in \mathbb{Q}$.

Over k we have the decomposition

$$H^1(M;k) = H^1_{\mathrm{cusp}}(M;k) \oplus H^1_{\mathrm{Eis}}(M;k).$$

Note that this decomposition is in general not true over R, that is $H^1_{\mathrm{cusp}}(M;R) \oplus H^1_{\mathrm{Eis}}(M;R)$ is a finite index subgroup in $H^1(M;R)$ but they are not equal up to modding out torsion subgroups. It follows that we cannot estimate directly $R_1(M)$ by estimating factors coming from cuspidal and Eisenstein homology. In the rest of this subsection we will mainly explain how one should deal with this problem. Estimating the Eisenstein factor is easy as we will mention below. The cuspidal factor in degree 1 can be dealt with as in the compact case, and in degree 2 it is likely to be much harder to estimate, we present a partial result due to Bergeron–Şengün–Venkatesh about this in the last subsection (Subsection 4.3.6).

4.3.3. Denominators and estimates. Let us now define formally the different parts of the regulator that we will be dealing with.

- The "cuspidal regulator" $R_{\mathrm{cusp},1}$ is defined to be the covolume of $H^1_{\mathrm{cusp}}(M;\mathbb{Z})$ in $\mathscr{H}^1_{L^2}(M;\mathbb{R})$;
- The "Eisenstein regulator" $R_{\mathrm{Eis},1}$ is the covolume of $H^1(\partial\overline{M};R) \cap \mathrm{im}(i_1^*)$ in its $\mathbb{C}$-span in $\mathscr{H}^1(\partial\overline{M};\mathbb{C})$.

It is easy to see that $\log R_{\mathrm{Eis},1}(M) \ll \log \mathrm{vol}(M)$ (see [39, Lemma 6.4]). To relate the product $R_{\mathrm{cusp},1}(M)R_{\mathrm{Eis},1}(M)$ to $R_1(M)$ we need to introduce the *denominator* a_{E^1} of the map E^1: by definition this is the smallest integer $a \geq 1$ such that

$$\forall[\omega] \in H^{1,0}(\partial\overline{M};R) : aE^1([\omega]) \in H^1(M;R).$$

It is then easy to see that

$$a_{E^1} \cdot H^1(M;R) \subset H^1_{\mathrm{cusp}}(M;R) \oplus H^1_{\mathrm{Eis}}(M;R)$$

up to torsion, and this implies that

$$R_1(M) \leq a_{E^1}^{b_1(M)} \cdot R_{\mathrm{cusp},1}(M) \cdot R_{\mathrm{Eis},1}(M). \tag{4.2}$$

4.3.4. Arithmetic structure of the intertwining map. In view of (4.2), to estimate the regulator it is necessary to estimate the denominator a_{E^1} of E^1. In order to do this we need first to do it "on the boundary": we let a_Φ be the smallest integer such that $a_\Phi \Phi(0)([\omega]) \in H^{0,1}(\partial\overline{M};R)$ for all $[\omega] \in H^{0,1}(\partial\overline{M};R)$.

To compute the denominator a_Φ via intertwining integrals we need a finer decomposition of the Eisenstein cohomology. For this we interpret the boundary as

$$\partial\overline{M} \cong \mathrm{P}(k)\backslash\mathrm{G}(\mathbb{A})/K$$

where P is the parabolic k-subgroup of upper triangular matrices, and $\cong$ signifies homotopy retraction. The group $\mathbb{A}^\times/k^\times$ acts on the right-hand side (by multiplication by diagonal matrices). Thus we have a decomposition of $H^1(\partial\overline{M};\overline{\mathbb{Q}})$ according to Hecke characters, we will denote by $H^1(\partial\overline{M};\overline{\mathbb{Q}})_\chi$ eigenspace with eigencharacter χ. Note that $H^1(\partial\overline{M};\overline{\mathbb{Q}})_\chi$ is contained in the $(0,1)$ or $(1,0)$ part of the cohomology according to whether $\chi_\infty = z^2/|z|^2$ or $\overline{z}^2/|z|^2$. The following result follows for example from the proof of [37, (5.17)].

Proposition 4.4. *If χ is a Hecke character with infinite part $z^2/|z|^2$ and $\omega \in H^1(\partial\overline{M};\mathbb{Z})_\chi$ we have the expression:*

$$\Phi(0)(\omega) = \frac{M}{N} \cdot \frac{L(\chi,0)}{L(\chi,1)}\overline{\omega} \tag{4.3}$$

where $\overline{\omega} \in H^1(\partial\overline{M};\overline{\mathbb{Z}})_{\overline{\chi}}$ and M, N are integers with N polynomially bounded in the level of Γ.

4.3.5. Modular symbols. The expression (4.3) deals with a_Φ, in other words it estimates the integrality of classes $E^1([\omega])$ against chains in the image of $H_1(\partial\overline{M};\mathbb{Z})$ inside $H_1(M;\mathbb{Z})$. We will say a few words about the next steps necessary to estimate a_{E^1}.

For this we need to estimate integrality of $E^1([\omega])$ against lifts of chains in $H_1(\overline{M}, \partial\overline{M};\mathbb{Z})$. The latter are represented by *modular symbols*, which are bi-infinite geodesics between two (not necessarily distinct) cusps. We will present a simplified version of the exposition from [14]. For $\alpha, \beta \in \mathbb{P}^1(k)$ let $c_{\alpha,\beta}$ be the image in M of the geodesic line from α to β in $\mathbb{H}^3$. By the exponential decay of cuspidal forms it is possible to integrate any element of $\mathscr{H}^1_{L^2}(M)$ against $c_{\alpha,\beta}$; it is also true that the harmonic forms $E^1(\omega)$ are integrable against the $c_{\alpha,\beta}$, because their constant term is orthogonal to the geodesics going to a cusp. Since the $c_{\alpha,\beta}$ generate $H_1(\overline{M}, \partial\overline{M};\mathbb{Z})$ we see that the class $[E^1(\omega)]$ is integral if and only if $\int_{c_{\alpha,\beta}} E^1(\omega) \in \mathbb{Z}$ for all α, β (see also [14, 6.7.5]). These integrals are computed in [14, 6.7.6] where an expression similar to (4.3) is obtained:

$$\int_{c_{\alpha,\beta}} E(0,\omega) = \frac{M}{N}\sum_{\zeta} \frac{L(1/2,\chi\zeta)L(1/2,\overline{\chi}\zeta)}{L(1,\chi^2)} \tag{4.4}$$

(the sum runs over all Hecke characters obtained from characters of the class group of k).

4.3.6. Degree 2. In degree 2 it is actually much simpler to separate the cuspidal and Eisenstein part in the regulator. This is done in [14, 6.3.3], we will shortly explain the argument. Let i^2_* be the inclusion map $H_2(\partial\overline{M};\mathbb{Z}) \to H_2(M;\mathbb{Z})$, then as $H_2(\overline{M};\partial\overline{M})$ is torsion-free we have

$$H_2(M;\mathbb{Z}) = H_{2,\mathrm{cusp}}(M;\mathbb{Z}) \oplus \mathrm{Im}(i^2_*).$$

In addition, if $f \in \mathscr{H}^2_{L^2}(M)$ then f is a cuspidal form and as such $\int_c f = 0$ for any cycle $c \in \mathrm{Im}(i^*_2)$. Thus the matrix of periods appearing in (1.6) is block-diagonal and it follows that

$$R_2(M) = R_{2,\mathrm{cusp}}(M) \cdot R_{2,\mathrm{Eis}}(M). \tag{4.5}$$

In addition the Eisenstein regulator is immediately computable: per [14] its value is

$$R_{2,\mathrm{Eis}}(M) = \left(\frac{\prod_{i=1}^h \mathrm{vol}(T_i)}{\sum_{i=1}^h \mathrm{vol}(T_i)}\right)^{-1/2}.$$

As in the closed case the cuspidal part remains hard to evaluate because we do not know whether it is possible to generate $H_{2,\mathrm{cusp}}$ with cycles of low complexity (polynomial in the volume). A partial result is given in [7, Theorem 7.2], which we will only informally describe. The authors prove that for the orbifold $M = \Gamma_0(\mathfrak{n})$ (where Γ is a Bianchi group and $\mathfrak{n}$ an ideal in its trace ring), if $\dim(H^1_{\mathrm{cusp}}(M;\mathbb{C})) = 1$, and under additional assumptions related to the number-theoretical side of the Langlands programme, then $H^1_{\mathrm{cusp}}(M;\mathbb{Z})$ can be generated by a harmonic form of L^2-norm at most $\mathrm{vol}(M)^C$ (for some C which a priori depends on the base field). Of course the dual cycle must have small area, hence small complexity.

4.4. Nontrivial coefficients

As above let k, R be an imaginary quadratic field and its ring of integers. If Γ is a subgroup of the Bianchi group $\mathrm{SL}_2(R)$ there are natural Γ-modules which fit in the setup of 1.4[9]. These are the symmetric powers $\mathrm{Sym}^m(R^2)$, which are lattices in the space $\mathrm{Sym}^m(\mathbb{C}^2)$ on which $\mathrm{SL}_2(\mathbb{C})$ acts. For each m there are Eisenstein series with coefficients in the associated fiber bundle F_m and there are also Eisenstein maps

$$E^i : H^i(\partial\overline{M}; F_m) \to H^i(M; F_m)$$

for $i = 1, 2$ which are defined respectively by $[\omega] \mapsto [E(m,\omega)]$ and $[*f] \mapsto [*dE(m+1, f)]$. The major difference with respect to trivial coefficients is that for $m \geq 1$ the cuspidal homology vanishes, hence $H^i(M; F_m) = \mathrm{im}(E^i)$. Thus the only difficulty in estimating regulators lies in the denominator a_{E^1} of E^1. Moreover there is an equality due to J. Pfaff [36] (see also [2]) between the Reidemeister torsion $\tau_{\mathrm{Eis}}(M; F_m)$ (as defined above) and a suitably defined analytic torsion $T_R(M; F_m)$ which states that

$$\tau_{\mathrm{Eis}}(M; F_m) = T_R(M; F_m) + B \tag{4.6}$$

where B depends only on the conformal structure of the boundary $\partial\overline{M}$ and the choices made to define both the Reidemeister and analytic torsions.

4.5. Application to homology growth

It is possible to generalise in part Theorem 3.6 to the case of congruence subgroups of the Bianchi groups. The upper limit in the following theorem is proven to hold in [39] (we note that the main result of [25] includes finite volume manifolds, and the proof is actually simpler in this case). The lower bound is due to J. Pfaff and given in [35].

[9]There exists congruence lattices commensurable to $\mathrm{SL}_2(R)$ which are not contained in $\mathrm{SL}_2(k)$, for which not all of these modules are defined.

Theorem 4.5. *Let Γ be a torsion-free nonuniform lattice in $\mathrm{PSL}_2(\mathbb{C})$ and Γ_n a sequence of congruence subgroups. Fix notation as in the statement of Theorem 3.6. Assume in addition that the sequence Γ_n is "cusp-uniform"*[10]. *Then*

$$\left(c_\rho - \frac{12}{\pi}\right) \mathrm{vol}(\Gamma\backslash\mathbb{H}^3) \le \liminf_{n\to+\infty} \frac{\log|H_1(\Gamma_n; L)_{\mathrm{tors}}|}{[\Gamma:\Gamma_n]}$$

and

$$\limsup_{n\to+\infty} \frac{\log|H_1(\Gamma_n; L)_{\mathrm{tors}}|}{[\Gamma:\Gamma_n]} \le c_\rho \,\mathrm{vol}(\Gamma\backslash\mathbb{H}^3).$$

The proof uses an argument similar to that of Theorem 3.6; the hypothesis on the cusps (which can be slightly weakened) is used to control the additional terms in the trace formula. The reason why we do not get an equality as in Theorem 3.6 is because we lack a good upper bound for the norm of the algebraic part of the L-values appearing in (4.3) and (4.4). The lower bound obtained by Pfaff is based on a trick using duality and the long exact sequence to show that the growth of R_1 must be compensated by an equivalent growth of the torsion in H_1, and which avoids directly dealing with intertwining operators.

We won't give more detail about this argument here, rather we will explain the proof of the following theorem from [37] which uses similar arguments but also an estimate for a_Φ. The result itself is a partial generalisation of Marshall and Müller's theorem 3.7. We note that we give here a slighty less precise statement that the one in loc. cit.; R is as above a ring of quadratic integers.

Theorem 4.6. *There exists $C > 0$ such that for Γ is a principal congruence subgroup of large enough level in the Bianchi group $\mathrm{SL}_2(R)$ and $L(m) = \mathrm{Sym}^m(R_D^2)$ then we have*

$$\frac{1}{C}\mathrm{vol}(\Gamma\backslash\mathbb{H}^3) \le \liminf_{m\to+\infty} \frac{\log|H_1(\Gamma; L(m))_{\mathrm{tors}}|}{m^2}$$

and

$$\limsup_{m\to+\infty} \frac{\log|H_1(\Gamma; L(m))_{\mathrm{tors}}|}{m^2} \le C\,\mathrm{vol}(\Gamma\backslash\mathbb{H}^3).$$

The interesting part is the proof of the lower bound; the upper bound follows from the generic arguments explained before 2.2. The first step towards proving it is to obtain upper and lower bounds for the growth of the Reidemeister torsions $\tau_{\mathrm{Eis}}(M; F_m)$ appearing in Theorem [36]. This follows from earlier work of Müller and Pfaff together with (4.6), using a trick comparing growth between two different Γs at once (this explains the indeterminacy of the constant $1/C$ in the lower bound). Then, instead of estimating the regulator directly we use the following lemma [37, Lemma 4.1].

Lemma 4.7. *We have:*

$$R_1(M; \mathcal{L}(m)) \le |H_1(\Gamma; L(m))_{\mathrm{tors}}| \cdot a_\Phi \cdot R_{\mathrm{Eis},1}(M; \mathcal{L}(M)). \tag{4.7}$$

[10] This means that the subgroups $\Gamma_n \cap P$ for P a parabolic subgroup stay within a compact subset of the moduli space of lattices in $\mathbb{C}$.

Here $\mathcal{L}(m)$ is the local system on M coming from the Γ-module L_m: while the Reidemeister torsion itself does not, the regulators depend on an integral structure on $H^*(M; F_m)$. The proof is elementary. Assuming that we know that $\log(a_\Phi) = o(m^2)$ (and $R_2(M; \mathcal{L}(m)) = o(m^2)$ which is simpler to prove) the lemma then basically gives

$$\liminf_{m\to+\infty} \frac{\log |H_1(\Gamma; L(m))_{\mathrm{tors}}|}{m^2} \geq \frac{1}{2} \liminf \frac{\log \tau_{\mathrm{Eis}}(M; \mathcal{L}(m))}{m^2}$$

which finishes the proof. To estimate the denominator a_Φ we use the formula (4.3) and two papers of R. Damerell [17, 18]. The first one almost states that there exists $\Omega \in \mathbb{C}^\times$ such that $L(\chi, m)/\Omega$ and $L(\chi, m+1)/\Omega$ are both algebraic numbers. We need to refine that statement to get that in addition the absolute norm of both is bounded above by $(m!)^c$ for some c depending only on k, which is done by examining carefully Damerell's argument. The second paper proves that $L(\chi, m)/\Omega, L(\chi, m+1)/\Omega \in 1/c\overline{\mathbb{Z}}$ where $c \in \mathbb{Z}$ depends only on Γ. These two theorems together immediately imply that $\log(a_\Phi) = O(m \log(m))$, which concludes the proof.

References

[1] Colin C. Adams. The noncompact hyperbolic 3-manifold of minimal volume. *Proc. Amer. Math. Soc.*, 100(4):601–606, 1987.

[2] Pierre Albin, Frédéric Rochon, and David Sher. Analytic torsion and R-torsion of Witt representations on manifolds with cusps. *Duke Math. J.*, 167(10):1883–1950, 2018.

[3] Avner Ash, Paul E. Gunnells, Mark McConnell, and Dan Yasaki. On the growth of torsion in the cohomology of arithmetic groups, 2016.

[4] David Bachman, Daryl Cooper, and Matthew E. White. Large embedded balls and Heegaard genus in negative curvature. *Algebr. Geom. Topol.*, 4:31–47 (electronic), 2004.

[5] Alex Bartel and Aurel Page. Torsion homology and regulators of isospectral manifolds. *J. Topol.*, 9(4):1237–1256, 2016.

[6] Nicolas Bergeron. Torsion homology growth in arithmetic groups. to appear in the proceedings of the 7th ECM.

[7] Nicolas Bergeron, Mehmet Haluk Şengün, and Akshay Venkatesh. Torsion homology growth and cycle complexity of arithmetic manifolds. *Duke Math. J.*, 165(9):1629–1693, 2016.

[8] Nicolas Bergeron and Akshay Venkatesh. The asymptotic growth of torsion homology for arithmetic groups. *J. Inst. Math. Jussieu*, 12(2):391–447, 2013.

[9] Jean-Michel Bismut and Weiping Zhang. An extension of a theorem by Cheeger and Müller. *Astérisque*, (205):235, 1992. With an appendix by François Laudenbach.

[10] Jeffrey F. Brock and Nathan M. Dunfield. Injectivity radii of hyperbolic integer homology 3-spheres. *Geom. Topol.*, 19(1):497–523, 2015.

[11] Jeffrey F. Brock and Nathan M. Dunfield. Norms on the cohomology of hyperbolic 3-manifolds. *Invent. Math.*, 210(2):531–558, 2017.

[12] M. Burger, T. Gelander, A. Lubotzky, and S. Mozes. Counting hyperbolic manifolds. *Geom. Funct. Anal.*, 12(6):1161–1173, 2002.

[13] Frank Calegari and Nathan M. Dunfield. Automorphic forms and rational homology 3-spheres. *Geom. Topol.*, 10:295–329, 2006.

[14] Frank Calegari and Akshay Venkatesh. A torsion Jacquet–Langlands correspondence, 2012.

[15] Jeff Cheeger. Analytic torsion and the heat equation. *Ann. of Math. (2)*, 109(2):259–322, 1979.

[16] Laurent Clozel. Démonstration de la conjecture τ. *Invent. Math.*, 151(2):297–328, 2003.

[17] R. M. Damerell. L-functions of elliptic curves with complex multiplication. I. *Acta Arith.*, 17:287–301, 1970.

[18] R. M. Damerell. L-functions of elliptic curves with complex multiplication. II. *Acta Arith.*, 19:311–317, 1971.

[19] M. Fraczyk. Strong Limit Multiplicity for arithmetic hyperbolic surfaces and 3-manifolds. *ArXiv e-prints*, December 2016.

[20] The GAP Group. *GAP – Groups, Algorithms, and Programming, Version* 4.8.9, 2017.

[21] Paul E. Gunnells. Lectures on computing cohomology of arithmetic groups. In *Computations with modular forms*, volume 6 of *Contrib. Math. Comput. Sci.*, pages 3–45. Springer, Cham, 2014.

[22] G. Harder. Eisenstein cohomology of arithmetic groups. The case GL_2. *Invent. Math.*, 89(1):37–118, 1987.

[23] Marc Lackenby. Heegaard splittings, the virtually Haken conjecture and property (τ). *Invent. Math.*, 164(2):317–359, 2006.

[24] Grant Lakeland. Matrix realizations of the hyperbolic tetrahedral groups in $\mathrm{PSL}_2((\mathbb{C})$, 2010. Available at `http://www.ux1.eiu.edu/~gslakeland/tetrahedral_realizations.pdf`.

[25] Thang T. Q. Lê. Growth of homology torsion in finite coverings and hyperbolic volume. *Ann. Inst. Fourier (Grenoble)*, 68(2):611–645, 2018.

[26] Michael Lipnowski and Mark Stern. Geometry of the smallest 1-form Laplacian eigenvalue on hyperbolic manifolds, 2016. To appear in GAFA.

[27] Yi Liu. Virtual homological spectral radii for automorphisms of surfaces, 2017.

[28] Wolfgang Lück. Approximating L^2-invariants by their classical counterparts. *EMS Surv. Math. Sci.*, 3(2):269–344, 2016.

[29] Colin Maclachlan and Alan W. Reid. *The arithmetic of hyperbolic 3-manifolds*, volume 219 of *Graduate Texts in Mathematics*. Springer-Verlag, New York, 2003.

[30] Simon Marshall and Werner Müller. On the torsion in the cohomology of arithmetic hyperbolic 3-manifolds. *Duke Math. J.*, 162(5):863–888, 2013.

[31] Werner Müller. Analytic torsion and R-torsion of Riemannian manifolds. *Adv. in Math.*, 28(3):233–305, 1978.

[32] Werner Müller. Analytic torsion and R-torsion for unimodular representations. *J. Amer. Math. Soc.*, 6(3):721–753, 1993.

[33] Werner Müller and Jonathan Pfaff. Analytic torsion of complete hyperbolic manifolds of finite volume. *J. Funct. Anal.*, 263(9):2615–2675, 2012.

[34] Aurel Page. Computing arithmetic Kleinian groups. *Math. Comp.*, 84(295):2361–2390, 2015.

[35] Jonathan Pfaff. Exponential growth of homological torsion for towers of congruence subgroups of Bianchi groups. *Ann. Global Anal. Geom.*, 45(4):267–285, 2014.

[36] Jonathan Pfaff. A gluing formula for the analytic torsion on hyperbolic manifolds with cusps. *J. Inst. Math. Jussieu*, 16(4):673–743, 2017.

[37] Jonathan Pfaff and Jean Raimbault. On the torsion in symmetric powers on congruence subgroups of bianchi groups, 2015.

[38] Jean Raimbault. Data for the homology of congruence subgroups of $\Gamma_{T_4}^+$. `https://www.math.univ-toulouse.fr/~jraimbau/data_totally_split.csv`.

[39] Jean Raimbault. Analytic, Reidemeister and homological torsion for congruence three–manifolds, 2013.

[40] D. B. Ray and I. M. Singer. R-torsion and the Laplacian on Riemannian manifolds. *Advances in Math.*, 7:145–210, 1971.

[41] Robert Riley. A personal account of the discovery of hyperbolic structures on some knot complements. *Expo. Math.*, 31(2):104–115, 2013.

[42] R. Sauer. Volume and homology growth of aspherical manifolds. *Geom. Topol.*, 20:1035–1059, 2016.

[43] Mehmet Haluk Şengün. On the integral cohomology of Bianchi groups. *Exp. Math.*, 20(4):487–505, 2011.

[44] Mehmet Haluk Şengün. On the torsion homology of non-arithmetic hyperbolic tetrahedral groups. *Int. J. Number Theory*, 8(2):311–320, 2012.

[45] William P. Thurston. Three-dimensional manifolds, Kleinian groups and hyperbolic geometry. *Bull. Amer. Math. Soc. (N.S.)*, 6(3):357–381, 1982.

[46] Vladimir Turaev. *Introduction to combinatorial torsions*. Lectures in Mathematics ETH Zürich. Birkhäuser Verlag, Basel, 2001. Notes taken by Felix Schlenk.

Jean Raimbault
Institut de Mathématiques de Toulouse; UMR5219
Université de Toulouse; CNRS
UPS IMT, F-31062 Toulouse Cedex 9, France
e-mail: `Jean.Raimbault@math.univ-toulouse.fr`

Progress in Mathematics, Vol. 338, 213–246

A Local Refinement of the Adams–Riemann–Roch Theorem in Degree One

Damian Rössler

Abstract. We prove that the Adams–Riemann–Roch theorem in degree one (i.e., at the level of the Picard group) can be lifted to an isomorphism of line bundles, compatibly with base change.

Mathematics Subject Classification (2010). 14C40, 19L10.

Keywords. Grothendieck–Riemann–Roch, vector bundles, equivariant geometry, fibration, fixed point formula.

1. Introduction

The aim of this text is to provide a proof of the following theorem.

Let B be a scheme.

Let $\mathcal{S}_{\mathrm{line,B}}$ be the category whose objects are pairs (S, M), where S is a locally Noetherian B-scheme and where M is a line bundle (i.e., a locally free sheaf of rank one) on S. An arrow $(S', M') \to (S, M)$ in $\mathcal{S}_{\mathrm{line,B}}$ is a morphism of B-schemes $\phi : S' \to S$, together with an isomorphism $\phi^*(M) \cong M'$.

Let $\mathcal{S}_{\mathrm{rel,line},B}$ be the category, whose objects are pairs $(Y \to S, L)$, where $Y \to S$ is a smooth and locally projective morphism of B-schemes with geometrically connected fibres and constant relative dimension, S is a locally Noetherian B-scheme and L is a line bundle on Y. An arrow $(Y' \to S', L') \to (Y \to S, L)$ in $\mathcal{S}_{\mathrm{rel,line},B}$ is a Cartesian diagram of B-schemes

$$\begin{array}{ccc} Y' & \xrightarrow{\rho} & Y \\ \downarrow & & \downarrow \\ S' & \longrightarrow & S \end{array}$$

together with an isomorphism $\rho^*(L) \cong L'$.

If $(Y \to S, L)$ is an object of $\mathcal{S}_{\mathrm{rel,line},B}$, we shall write $\dim(Y/S)$ for the dimension of some (and hence any) geometric fibre of the morphism $Y \to S$.

Recall that to say that $Y \to S$ is locally projective means that every point in S has an open neighbourhood U, such that there is a factorisation of $\pi|_U$ into a closed U-immersion $Y_U \to \mathbb{P}^N_U$ followed by projection to U, for some $N \geq 0$ which depends on U.

We let $\mathcal{S}_{\mathrm{rel,line,cf},B}$ be the full subcategory of $\mathcal{S}_{\mathrm{rel,line},B}$, which consists of those pairs

$$(\pi : Y \to S, L),$$

where L is cohomologically flat over S. Recall that to say that L is cohomologically flat over S means that $\mathrm{R}^i\pi_*(L)$ is a locally free sheaf for all $i \geq 0$.

If $\pi : Y \to S$ is a proper and flat morphism of locally Noetherian schemes and F is a vector bundle (i.e., a coherent locally free sheaf) on Y, we shall write $\lambda(F) := \det(\mathrm{R}^\bullet\pi_*(F))$. Here $\det(\cdot)$ is the Knudsen–Mumford determinant of a perfect complex (note that $\mathrm{R}^\bullet\pi_*(F)$ is a perfect complex by the semicontinuity theorem because π is proper and flat). We shall denote by $\mathrm{Sym}^k(F)$ the kth symmetric power of F and we shall write $F^\vee := \underline{\mathrm{Hom}}(F, \mathcal{O}_X)$ for the dual of F. If M is a line bundle on Y and $k \in \mathbb{Z}$, we define $M^{\otimes k} := \otimes_{i=1}^k M$ if $k \geq 0$ and $M^{\otimes k} := \otimes_{i=1}^{-k} M^\vee$ if $k < 0$. As is costumary, we shall write $\Omega_{Y/S} = \Omega_\pi$ for the sheaf of differentials of π.

Note that the rule, which associates the line bundle

$$\lambda(L)^{\otimes 2^{2\dim(Y/S)+2}}$$

with the object $(Y \to S, L)$ of $\mathcal{S}_{\mathrm{rel,line},B}$, naturally defines a functor from $\mathcal{S}_{\mathrm{rel,line},B}$ to $\mathcal{S}_{\mathrm{line},B}$. We shall denote this functor LRR.

Similarly, the rule, which associates the line bundle

$$\bigotimes_{j=0}^{2\dim(Y/S)} \lambda(L^{\otimes 2} \otimes \mathrm{Sym}^j(\Omega_{Y/S}))^{\otimes (-1)^j \sum_{i=0}^{2\dim(Y/S)-j} \binom{2\dim(Y/S)+1}{i}}$$

with the object $(Y \to S, L)$ of $\mathcal{S}_{\mathrm{rel,line},B}$, naturally defines a functor from $\mathcal{S}_{\mathrm{rel,line},B}$ to $\mathcal{S}_{\mathrm{line},B}$. We shall denote this functor RRR.

Theorem 1.1. *Suppose that $B = \mathrm{Spec}\,\mathbb{Z}[\frac{1}{2}]$. Then the restrictions of the functors* LRR *and* RRR *to $\mathcal{S}_{\mathrm{rel,line,cf},B}$ are isomorphic.*

In other words, it is possible to associate with any locally projective and smooth morphism of locally Noetherian $\mathbb{Z}[\frac{1}{2}]$-schemes $Y \to S$ and any line bundle L on Y, which is cohomologically flat over S, an isomorphism

$$\lambda(L)^{\otimes 2^{2\dim(Y/S)+2}} \cong \bigotimes_{j=0}^{2\dim(Y/S)} \lambda(L^{\otimes 2} \otimes \mathrm{Sym}^j(\Omega_{Y/S}))^{\otimes (-1)^j \sum_{i=0}^{2\dim(Y/S)-j} \binom{2\dim(Y/S)+1}{i}} \tag{1}$$

compatibly with base change to any locally Noetherian scheme.

Remark 1.2. (1) We conjecture that the assumption that L is cohomologically flat over S is unnecessary. In other words, we conjecture that the functors LRR and RRR are isomorphic if $B = \mathrm{Spec}\,\mathbb{Z}[\frac{1}{2}]$ (and not only their restrictions to

$\mathcal{S}_{\text{rel,line,cf},B}$). Proving this boils down to a problem in the linear algebra of perfect complexes. See Remark 7.4 below for details.

(2) Note if S is a scheme of characteristic 0 then the trivial line bundle $\mathcal{O}_Y$ is cohomologically flat over S by a theorem of Deligne (see [4, Th. 5.5]).

(3) It is actually plausible that LRR and RRR are isomorphic if $B = \operatorname{Spec}\mathbb{Z}$ (this would generalize conjecture (1) above in this remark). This is suggested by Proposition 1.3 below and Deligne's theorem [3, Th. 9.9 (3)]. See the discussion after Proposition 1.3.

(4) Our construction of the isomorphism I between the restrictions of the functors LRR and RRR to $\mathcal{S}_{\text{rel,line,cf},B}$ depends on a slew of arbitrary combinatorial choices. These choices are all contained in the proof of Lemma 4.1 below. One might conjecture that, up to sign, the isomorphism I does not depend on these choices but proving this seems to be a formidable task. Presumably it is possible to show that there is only one isomorphism I, up to sign, provided it satisfies some axiomatic conditions. It would be very interesting to determine such conditions.

For example, suppose that $\dim(Y/S) = 1$. We then get an isomorphism

$$\lambda(L)^{\otimes 16} \cong \lambda(L^{\otimes 2})^{\otimes 7} \otimes \lambda(L^{\otimes 2} \otimes \Omega_{Y/S})^{\otimes(-4)} \otimes \lambda(L^{\otimes 2} \otimes \Omega_{Y/S}^{\otimes 2}) . \qquad (2)$$

In particular, writing $\lambda_k := \lambda(\Omega_{Y/S}^{\otimes k})$ for any $k \geq 0$, (2) gives

$$\lambda_k^{\otimes 16} \cong \lambda_{2k}^{\otimes 7} \otimes \lambda_{2k+1}^{\otimes(-4)} \otimes \lambda_{2k+2}.$$

By the Grothendieck duality, there is a canonical isomorphism $\lambda_0 \cong \lambda_1$. Thus, setting $k = 0$ we obtain an isomorphism

$$\lambda_1^{\otimes 13} \cong \lambda_2. \qquad (3)$$

In [26] Mumford also constructs such an isomorphism and also proves that it is invariant under base change (and he does not need the assumption that 2 is invertible on S). Our isomorphism presumably coincides with his up to a universal constant of the form $\pm 2^k$ ($k \in \mathbb{Z}$) but we did not verify this.

Suppose that $\pi : Y \to S$ is an elliptic scheme (i.e., an abelian scheme of relative dimension 1) over S. We then have a canonical isomorphism $\Omega_{Y/S}^{\otimes k} \cong \pi^*(\pi_*(\Omega_{Y/S}^{\otimes k}))$ for any $k \in \mathbb{Z}$. Furthermore, we have

$$\mathrm{R}^1\pi_*(\mathcal{O}_{Y/S}) \cong \pi_*(\Omega_{Y/S})^\vee$$

by Grothendieck duality. Using the projection formula, we can thus compute

$$\begin{aligned}\lambda_k &= \det((\mathcal{O}_S - \mathrm{R}^1\pi_*(\mathcal{O}_{Y/S})) \otimes \pi_*(\Omega_{Y/S})^{\otimes k}) \\ &= \det(\pi_*(\Omega_{Y/S})^{\otimes k} - \pi_*(\Omega_{Y/S})^{\otimes(k-1)}) \cong \pi_*(\Omega_{Y/S})\end{aligned}$$

for all $k \geq 0$. In particular, we have an isomorphism $(\pi_*(\Omega_{Y/S}))^{\otimes 12} \cong \mathcal{O}_S$. Again, possibly up to multiplication by a term of the form $\pm 2^k$ ($k \in \mathbb{Z}$), this is presumably the classical discriminant modular form (but we did not verify this). This suggests that the isomorphism in Theorem 1.1 is in some sense optimal.

When Y is an elliptic scheme over S and L is a non-trivial torsion line bundle, whose order is prime to the characteristic of all the residue fields of S, then $\mathrm{R}^\bullet\pi_*(L) = 0$. In that case, both sides of (1) are canonically isomorphic to the trivial line bundle. Thus the isomorphism (1) provides an element of $\Gamma(S, \mathcal{O}_S^*)$, in other words an elliptic unit. It seems likely that one can construct all the Siegel units in this way but to prove this, one will probably have to wait for a metric version of Theorem 1.1. See below for a discussion.

Returning to the general situation, recall that if S is of characteristic 0, the trivial sheaf $\mathcal{O}_Y$ is cohomologically flat over S by a result of Deligne. Let us suppose that S is of characteristic 0 and $\dim(Y/S) = 2$. We then get the isomorphism

$$\lambda(\mathcal{O}_Y)^{\otimes 64} \cong \lambda(\mathcal{O}_Y)^{\otimes 31} \otimes \lambda(\Omega_{Y/S})^{\otimes(-26)} \otimes \lambda(\mathrm{Sym}^2(\Omega_{Y/S}))^{\otimes 16} \\ \otimes \lambda(\mathrm{Sym}^3(\Omega_{Y/S}))^{\otimes(-6)} \otimes \lambda(\mathrm{Sym}^4(\Omega_{Y/S}))$$

from Theorem 1.1. This is equivalent to

$$\lambda(\mathcal{O}_Y)^{\otimes 33} \otimes \lambda(\Omega_{Y/S})^{\otimes 26} \otimes \lambda(\mathrm{Sym}^3(\Omega_{Y/S}))^{\otimes 6} \\ \cong \lambda(\mathrm{Sym}^2(\Omega_{Y/S}))^{\otimes 16} \otimes \lambda(\mathrm{Sym}^4(\Omega_{Y/S})).$$

and there are similar identities in any relative dimension.

Here is our method of proof. We first give a proof of the geometric fixed formula for an involution, which avoids any reference to K-theory and uses only the geometric properties of quotients. This is Theorem 6.1, which is of independent interest. The idea to use quotients to prove the fixed point formula is due to Thomason (see [30]) and most probably many earlier authors but our proof relies on the crucial fact that when the fixed point scheme is a Cartier divisor then the quotient morphism is flat. This seems to be a well-known fact (J. Oesterlé kindly explained the proof to me many years ago) but we could find no proof of it in the literature in the required generality and we provide one in Proposition 2.5 (1). Our proof of the geometric fixed point formula is sufficiently explicit to provide isomorphisms at every step (rather than equalities in the Picard group) but ends with an error term, which turns out to be a line bundle arising from a higher-dimensional version of the Deligne pairing. This pairing was studied by Ducrot in [7] and we use his results to show that this line bundle is canonically trivial, compatibly with any base change to a locally Noetherian scheme. We then apply this formula to the space $Y \times_S Y$ with the involution swapping the factors. Nori (see [27]) was apparently the first one to notice that the fixed point formula applied to this situation recovers the Adams–Riemann–Roch for the Adams operation ψ^2 and using our method we thus recover a refinement of this formula (in degree one). This is formula (1).

In [10] Eriksson gives a proof of a functorial refinement of the Adams–Riemann–Roch formula (see also [9] for an announcement), which can also be used to prove a weaker version of Theorem 1.1. It is weaker in the sense that the provided isomorphism, although invariant under base change, will include a

2^∞-torsion line bundle, which is undetermined and also because the resulting linear combination in the symmetric powers of $\Omega_{Y/S}$ will a priori depend on the dimension of the total space.

Similarly, using Franke's work in [11], it is possible to prove a weak version of Theorem 1.1, where an undetermined (not necessarily 2^∞) torsion line bundle will be included (but on the other hand the linear combination in the symmetric powers of $\Omega_{Y/S}$ should be the same as ours and should thus not depend on the dimension of the total space).

One interesting aspect of our result is thus that it removes this indeterminacy. However, the main interest of the present text is the method of proof, which is elementary (whereas Franke's and Eriksson's approaches require a vast categorical apparatus and use higher K-theory, resp. the homotopy theory of schemes). Our isomorphism is constructed very explicitly, making it in principle possible to compute its norm, when both sides are endowed with metrics (e.g., Quillen metrics). We hope to return to this question in a later article.

Note that other constructions of the higher-dimensional Deligne pairing were given in [31] and [8] but they cannot be used in our context, because they are not described in terms of determinants of cohomology and therefore cannot easily be compared with our error term. In [1], a canonical isomorphism between Ducrot's pairing and Zhang's pairing is announced (in a restricted setting), which could be used to bypass the use of Ducrot's pairing in certain situations. However, the details of the proof of Theorem 1 of [1] have not appeared yet (thank you to one of the referees for pointing this out). In [5] Ducrot's pairing is also considered.

Finally, note that in the situation where $\dim(Y/S) = 1$, Deligne also constructed an isomorphism similar to (1) (see [3]). Deligne's work was in fact the initial motivation for the work of Franke and Eriksson. Under the assumptions of Theorem 1.1 and when $\dim(Y/S) = 1$, Deligne's theorem [3, Th. 9.9 (3)] provides in particular an isomorphism

$$\lambda(L)^{\otimes 18} \cong \lambda(\mathcal{O}_Y)^{18} \otimes \lambda(L^{\otimes 2} \otimes \Omega_{Y/S}^\vee)^{\otimes 6} \otimes \lambda(L \otimes \Omega_{Y/S}^\vee)^{\otimes(-6)}, \tag{4}$$

which is invariant under any base change to a locally Noetherian scheme (note that Deligne's theorem is expressed in terms of the Deligne pairing; Deligne's pairing can be expressed using the determinant of cohomology – see Section 4 below – and (4) is the expression one obtains when using only the determinant of cohomology). This can be seen as a variant of the isomorphism (1) when $\dim(Y/S) = 1$ and Deligne shows that it holds even if 2 is not invertible on S and L is not cohomologically flat over S.

Using Theorem 1.1 for $\dim(Y/S) = 1$, we prove

Proposition 1.3. *Under the assumptions of Theorem* 1.1 *and when* $\dim(Y/S) = 1$, *there is an isomorphism*

$$\left(\lambda(L)^{\otimes 18}\right)^{\otimes 8} \cong \left(\lambda(\mathcal{O}_Y)^{18} \otimes \lambda(L^{\otimes 2} \otimes \Omega_{Y/S}^\vee)^{\otimes 6} \otimes \lambda(L \otimes \Omega_{Y/S}^\vee)^{\otimes(-6)}\right)^{\otimes 8} \tag{5}$$

which is invariant under any base change to a locally Noetherian scheme.

In other words, we give a new proof of Deligne's theorem, up to a torsion line bundle of order 8 (and under the running assumption that 2 is invertible on S and that L is cohomologically flat over S). The proof of Proposition 1.3 actually also shows that one can deduce Theorem 1.1 for $\dim(Y/S) = 1$ from Deligne's theorem, up to a torsion line bundle of order 9. Thus, when $\dim(Y/S) = 1$ and under the running assumption that 2 is invertible on S and that L is cohomologically flat over S, Theorem 1.1 and Deligne's theorem are equivalent up to torsion.

The structure of the article is as follows. In Section 2 we recall various facts about quotients of schemes by finite groups and we prove various supplementary properties of these in the situation where the group is isomorphic to a diagonalisable group scheme, whose order is prime and invertible in the base scheme and the fixed point scheme is a Cartier divisor. In Section 4 we recall the part of Ducrot's work that is relevant to this text. In Section 6, we give a proof of a local refinement of the fixed formula for an involution, in the situation where the fixed scheme is regularly immersed. In Section 7, we apply this formula to the fibre product of a relative scheme by itself and we prove Theorem 1.1. In the final Section 8 we give the proof of Proposition 1.3. Note that the core of the proof of Theorem 1.1 amounts to a detailed analysis of the geometry of the blow-up along the diagonal of the relative fibre product of X with itself. This is intriguing, since this particular space was believed to be relevant to a possible solution of the standard conjectures in the early days of scheme theory. It would be interesting to relate our construction to statements about algebraic cycles.

Notation. We shall say that a morphism $h : Z \to T$ of schemes is strongly projective if there is a factorisation of h into a closed T-immersion $Z \to \mathbb{P}^N_T$ followed by projection to T, for some $N \geq 0$. The notion of a locally projective morphism is defined at the beginning of the introduction. If Z is a locally Noetherian scheme, we write $\mathrm{Coh}(Z)$ for the category of coherent sheaves on Z. If F is an $\mathcal{O}_Z$-module on a scheme Z and $l \geq 0$, we shall write $F^{\otimes l} := \bigotimes_{k=1}^{l} F$. If Z is a scheme, we write $D(Z)$ (resp. $D^b(Z)$) for the derived category of complexes of $\mathcal{O}_Z$-modules (resp. the derived category of bounded complexes of $\mathcal{O}_Z$-modules) on Z.

Acknowledgments. We are grateful to Jean-Michel Bismut and Vincent Maillot for interesting discussions around this article. Warm thanks to the referees, whose very detailed reading (to say the least!) led to many improvements. This text would be much less clear without their input.

2. The geometry of quotients by finite groups

Let G be a finite group.

A scheme T together with a group homomorphism $G \to \mathrm{Aut}(T)$ will be called a G-equivariant scheme, or an equivariant scheme for short (if there is no ambiguity). A G-equivariant morphism of G-equivariant schemes is a morphism commuting with the action of G on source and target. We shall say that the action of G on the G-equivariant scheme T is trivial if the image of $G \to \mathrm{Aut}(T)$ is the identity morphism.

A G-equivariant sheaf (or equivariant sheaf for short) F on a G-equivariant scheme is a quasi-coherent sheaf F together with a morphism of sheaves $\alpha_g = \alpha_{F,g} : F \to g_*(F)$ for every $g \in G$, such that $g_*(\alpha_h) \circ \alpha_g = \alpha_{g \circ h}$ for any $g, h \in G$ and $\alpha_{\mathrm{Id}_G} = \mathrm{Id}_F$.

Suppose that T is a G-equivariant scheme with trivial action and that F is a G-equivariant sheaf on T. The G-equivariant structure on F then amounts to a homomorphism of groups $G \to \mathrm{Aut}(F)$. We then write F^G for the quasi-coherent sheaf on T such that

$$F^G(U) = F(U)^G$$

for every open set $U \subseteq T$. Here $F(U)^G$ is the subgroup of elements of $F(U)$, which are fixed under the action of G.

Suppose that $\phi : T \to Z$ is a morphism of schemes, where T is locally Noetherian. Assume also that T carries G-equivariant structure and that $\phi \circ g = \phi$ for all $g \in G$. Let F be a G-equivariant sheaf. Then the sheaf $\phi_*(F)$ is also quasi-coherent. Furthermore, if Z is viewed as a G-equivariant scheme carrying the trivial G-equivariant structure, then $\phi_*(F)$ carries the G-equivariant structure given for any $g \in G$ by the composition of arrows

$$\phi_*(F) \xrightarrow{\sim} \phi_*(g_*(F)) \xrightarrow{\sim} \phi_*(F)$$

arising from the equivariant structure on F and the identity $\phi \circ g = \phi$.

Suppose that $\phi : T \to Z$ is a morphism of schemes, that T carries a G-equivariant structure and that $\phi \circ g = \phi$ for all $g \in G$. View Z as a G-equivariant scheme endowed with the trivial G-equivariant structure. Let F be a G-equivariant sheaf on Z. Then the quasi-coherent sheaf $\phi^*(F)$ carries a natural G-equivariant structure, given for any $g \in G$ by the composition of arrows

$$\phi^*(F) \overset{\phi^*(g_*)}{\xrightarrow{\sim}} \phi^*(F) \xrightarrow{\sim} g^{-1,*}(\phi^*(F)) = g_*(\phi^*(F))$$

where the first arrow comes by functoriality from the arrow $g_*(F) \to g_*(F)$, the second arrow from the identity $\phi \circ g = \phi$ and the third arrow from the identification of functors $g^{-1,*} = g_*$.

If $x \in X$, then we define $G_d(x)$ to be the stabiliser in G of x viewed as a subset of X. This group is called the decomposition group of x. The group $G_d(x)$ naturally acts on the residue field $\kappa(x)$ of x. The kernel of the homomorphism $G_d(x) \to \mathrm{Aut}(\kappa(x))$ is called the inertia group $G_i(x)$ of x.

Suppose that X is a G-equivariant scheme. A (categorical) quotient X/G of X by G (if it exists) is a G-equivariant scheme X/G together with an G-equivariant morphism $q : X \to X/G$, with the following properties:

- X/G carries the trivial action;
- if X' is a scheme with a trivial G-action and $q' : X \to X'$ is a morphism then there is a unique morphism $h : X/G \to X'$, such that $h \circ q = q'$.

These properties clearly determine X/G up to unique isomorphism.

We recall the following

Proposition 2.1. *Let X be a G-equivariant scheme. Suppose that the orbit of every point in X is contained in an affine open subscheme. Then the quotient X/G of X by G exists and*

(1) *The canonical morphism $q : X \to X/G$ is integral and surjective.*
(2) *The natural morphism of sheaves $\mathcal{O}_{X/G} \to q_*(\mathcal{O}_X)$ factors through $(q_*(\mathcal{O}_X))^G$ and induces an isomorphism $\mathcal{O}_{X/G} \to (q_*(\mathcal{O}_X))^G$.*
(3) *The underlying set of X/G is the quotient of the set X by the action of G and the topology of X/G is the quotient topology.*
(4) *if $Z \to X/G$ is a flat morphism then the natural one $(Z \times_{X/G} X)/G \to Z$ is an isomorphism.*
(5) *Consider the X/G-morphism $\phi : G \times X \to X \times_{X/G} X$ given in set-theoretic notation by the formula $(g, x) \mapsto (g(x), x)$. Suppose that ϕ is an isomorphism. Then*
   - *q is étale;*
   - *if M is a G-equivariant locally free sheaf of finite rank on X then the natural morphism $q^*(q_*M)^G \to M$ is an isomorphism.*
(6) *If $G_i(x) = 0$ then $\mathcal{O}_{X,x}$ is étale over $\mathcal{O}_{X/G,q(x)}$.*

Proof. See [15, Chap. V, 1 and 2]. □

Corollary 2.2. *Suppose that there is a morphism of finite type $f : X \to S$, where S is a locally Noetherian scheme. Assume that the action of G on X factors through* $\mathrm{Aut}_S(X)$ *and that the orbit of every point in X is contained in an affine open subscheme. Then the quotient X/G of X by G exists. Moreover, the morphism $q : X \to X/G$ is finite.*

Corollary 2.2 follows from the fact that under the listed assumptions, the quotient morphism is integral and of finite type and hence finite.

Suppose that X is a G-equivariant scheme. Suppose given a morphism $X \to S$ and assume that the action of G on X factors through $\mathrm{Aut}_S(X)$. We say that X is a G-equivariant S-scheme. The fixed scheme X_G (if it exists) is a closed subscheme of X, which represents the functor on S-schemes

$$T \mapsto X(T)^G.$$

Note the following link with decomposition and inertia groups: if $x \in X$ and

$$G_d(x) = G_i(x) = G$$

then $x \in X_G$. This simply follows from the fact that the morphism $\mathrm{Spec}\, \kappa(x) \to X$ then lies in $X(\mathrm{Spec}\, \kappa(x))^G$.

Proposition 2.3. *Suppose that X is separated over S. Then X_G exists.*

Proof. For each $g \in G$, let Γ_g be the graph of g in $X \times_S X$. Let $\Delta = \Gamma_{\mathrm{Id}_X} \cong X$ be the diagonal of X over S. From the separatedness assumption, each Γ_g is a closed subscheme of $X \times_S X$. The closed subscheme $X_G = \cap_{g \in G} \Gamma_g$ is naturally a closed subscheme of Δ and can thus be viewed as a closed subscheme of X. It follows from the definitions that X_G is the fixed scheme of G. $\square$

If X_G exists, we shall write $N_{X_G/X}$ for the conormal sheaf of X_G in X. Recall that if $\mathcal{I}$ is the ideal sheaf of X_G in X, we have by definition $N_{X_G/X} = \mathcal{I}/\mathcal{I}^2$. The sheaf $\mathcal{I}/\mathcal{I}^2$ has a natural structure of $\mathcal{O}_{X_G}$-module. The conormal sheaf $N_{X_G/X}$ is thus a quasi-coherent sheaf on X_G and it carries a natural action of G.

Suppose now that G is a finite cyclic group of order n. Let us write $\widetilde{G}$ for the group scheme over $\operatorname{Spec}\mathbb{Z}$ corresponding to G. Note that we then have a canonical identification $\widetilde{G}(\operatorname{Spec}\mathbb{Z}) \cong G$. Suppose now that $\widetilde{G}_S \cong \mu_{n,S}$, where $\mu_n = \operatorname{Spec}\mathbb{Z}[t]/(1-t^n)$ is the diagonalisable group scheme associated with the cyclic group $\mathbb{Z}/n\mathbb{Z}$. Note that there exists an isomorphism $\widetilde{G}_S \cong \mu_{n,S}$ is iff n is invertible in S and the polynomial $x^n - 1$ splits into linear factors in $\Gamma(S, \mathcal{O}_S)$. We fix an isomorphism $G_S \cong \mu_{n,S}$.

Note the following two facts.

Suppose in this paragraph only that $X = \operatorname{Spec} R$ is affine. Then the action of G on X is given by a ring grading $R \cong \oplus_{k\in\mathbb{Z}/n\mathbb{Z}} R_k$, such that the morphism $X \to S$ factors through $\operatorname{Spec} R_0$. Furthermore, the ideal of X_G is then $R \cdot R_{\neq 0}$, where

$$R_{\neq 0} := \oplus_{k\in\mathbb{Z}/n\mathbb{Z},\, k\neq 0} R_k.$$

See [30, proof of Prop. 3.1] (this is also a good exercise for the reader).

Suppose that the action of G on X is trivial. Let F be a G-equivariant sheaf on X. The G-equivariant structure on F is then given by a $\mathbb{Z}/n\mathbb{Z}$-grading of $\mathcal{O}_X$-modules

$$F \cong \oplus_{k\in\mathbb{Z}/n\mathbb{Z}} F_k.$$

Let $g \in G$. By the above, the element g gives an element of $\widetilde{G}(\operatorname{Spec}\mathbb{Z})$ and hence after base change an element $z \in G(S)$. Applying the isomorphism $G_S \cong \mu_{n,S}$ we obtain an element $z \in \mu_n(S)$. The action of g on F is then by construction given by the formula

$$g((f_0, f_1, \ldots, f_{n-1})) = (1 \cdot f_0, z \cdot f_1, \ldots, z^{n-1} \cdot f_{n-1}),$$

where f_k is a local section of F_k. In particular, we have $F_0 = F^G$.

We record the following

Lemma 2.4. *Let X be an G-equivariant S-scheme. Suppose that the orbit of every point in X is contained in an affine open subscheme. Assume that G is a finite cyclic group of order n and that $G_S \cong \mu_{n,S}$. If $Z \to X/G$ is a morphism then the natural morphism $(Z \times_{X/G} X)/G \to Z$ is an isomorphism.*

In other words, when $G_S \cong \mu_{n,S}$, the quotient construction commutes with any base change on X/G (not only flat base changes as in Proposition 2.1 (4)).

Proof. By Proposition 2.1 (4), we may assume that Z and X are affine, say $Z = \operatorname{Spec} B$ and $X = \operatorname{Spec} A$. In this case, we have to prove that the morphism of

A_0-modules

$$B \to (B \otimes_{A_0} A)_0$$

given by the formula $b \mapsto b \otimes 1$ is an isomorphism. We have

$$B \otimes_{A_0} A = B \otimes_{A_0} \bigoplus_{k \in \mathbb{Z}/n\mathbb{Z}} A_k = \bigoplus_{k \in \mathbb{Z}/n\mathbb{Z}} B \otimes_{A_0} A_k$$

so that $(B \otimes_{A_0} A)_0 = B \otimes_{A_0} A_0 = B$, proving the assertion. □

The next proposition collects the main results of this section.

Proposition 2.5. *Suppose that X is a G-equivariant S-scheme such that S is locally noetherian and the morphism $X \to S$ is separated and of finite type. Assume that the orbit of every point in X is contained in an affine open subscheme. Finally, suppose that G is a finite cyclic group of order n and that $G_S \cong \mu_{n,S}$. Let $\iota : X_G \to X$ be the fixed point scheme of X. Then:*

1. *Suppose that n is prime and that X_G is a (possibly empty) Cartier divisor. Then q is flat.*
2. *Suppose that X_G is a Cartier divisor. Then $(N_{X_G/X})_0 = 0$.*
3. *The morphism $q \circ \iota : X_G \to X/G$ is a closed immersion and we have the set-theoretic equality $q^{-1}(q(X_G)) = X_G$. Thus we have a natural isomorphism $(X/G)\backslash q(X_G) \cong (X\backslash X_G)/G$.*
4. *Let $U = X\backslash X_G$ (so that $U/G = (X/G)\backslash q(X_G)$ by (3)). Consider the U/G-morphism*
$$\phi : G \times U \to U \times_{U/G} U$$
given in set-theoretic notation by the formula $(g, u) \mapsto (g(u), u)$. If n is prime then ϕ is an isomorphism.
5. *Let M be a G-equivariant locally free sheaf of finite rank on X. Suppose that $\iota^* M$ carries the trivial action, that q is flat and that n is prime. Then the natural morphism $q^*(q_* M)_0 \to M$ is an isomorphism.*
6. *If $X \to S$ is smooth and $X_G \to S$ is flat then $X_G \to S$ is smooth.*
7. *If $X \to S$ is smooth, X_G is a Cartier divisor in X and $X_G \to S$ is flat then $X/G \to S$ is also smooth.*

Remark 2.6. A variant (for algebraic varieties) of (5) is proven in [6, Th. 2.3]. See also [20, Lemma 4.8] and [21, Proposition (3.3.4.i)], where most of the above proposition is proven in the restricted context of algebraic varieties.

Proof. We begin with (1). We may suppose that $X = \mathrm{Spec}(R)$ is affine. Then $X/G = \mathrm{Spec}(R_0)$ by Proposition 2.1 (2). To show that R is flat over R_0, it is sufficient to show that for all $\mathfrak{p} \in \mathrm{Spec}(R)$, the ring $R_\mathfrak{p}$ is flat over the ring $R_{0,\mathfrak{p}\cap R_0}$. If $\mathfrak{p} \not\supseteq R \cdot R_{\neq 0}$, then $\mathfrak{p} \notin X_G$ by the previous discussion. Thus $G_i(x) \neq G$ and thus $G_i(x) = 0$ since n is prime; thus $R_\mathfrak{p}$ is flat over the ring $R_{0,\mathfrak{p}\cap R_0}$ by Proposition 2.1 (6). Thus we may assume that $\mathfrak{p} \supseteq R \cdot R_{\neq 0}$. The prime ideal $\mathfrak{p}$ is then graded by construction (if $r \in \mathfrak{p}$, write $r = r_0 + \cdots + r_{n-1}$, where the r_i are homogenous for the grading; by assumption $r_1, \ldots, r_{n-1} \in \mathfrak{p}$; thus $r_0 \in \mathfrak{p}$ as well). The ring

$R_{\mathfrak{p}}$ is thus naturally a $\mathbb{Z}/n\mathbb{Z}$-graded local ring. Now notice that we have a natural identification

$$R_{0,\mathfrak{p}\cap R_0} = (R_{\mathfrak{p}})_0$$

(use the fact that $R\backslash\mathfrak{p} \subseteq R_0$). Also by construction the ideal generated by the image of the ideal $R \cdot R_{\neq 0}$ in $R_{\mathfrak{p}}$ is $R_{\mathfrak{p}} \cdot R_{\mathfrak{p},\neq 0}$. Thus the assumption that $R \cdot R_{\neq 0}$ is a Cartier divisor implies that there exists $t \in R_{\mathfrak{p}}$, which is not a zero divisor, such that $(t) = R_{\mathfrak{p}} \cdot R_{\mathfrak{p},\neq 0}$.

Thus we may assume without restriction of generality that R is a local ring and that $R \cdot R_{\neq 0}$ is generated by an element t, which is not a zero divisor.

We claim that t can be taken to be homogeneous of degree $\neq 0 \pmod n$. To verify the claim, let

$$R \cdot R_{\neq 0} = (a_1, \ldots, a_k)$$

where the $a_i \in R_{\neq 0}$ are homogeneous and of degree $\neq 0 \pmod n$ (recall that R is Noetherian). We take k minimal. We may assume that $k > 1$, otherwise there is nothing to prove. Then for some family of $x_i \neq 0$, we have

$$x_1 a_1 + \cdots + x_k a_k = t\,.$$

Let $b_1 \in R$ be such that $a_1 = t \cdot b_1$. If b_1 is a unit then $R \cdot R_{\neq 0} = (a_1)$ contradicting the assumption that $k > 1$. Thus b_1 is not a unit and thus $1 - x_1 b_1$ is a unit since R is local. We compute

$$t = \frac{x_2}{1 - x_1 b_1} a_2 + \cdots + \frac{x_k}{1 - x_1 b_1} a_k$$

contradicting minimality again. Thus $k = 1$ and the claim is verified.

So we may suppose that $(t) = R \cdot R_{\neq 0}$ where t is homogenous of degree $\neq 0 \pmod n$.

I am grateful to one of the referees for suggesting the argument below.

Sub-lemma 2.7. *For any $i \in \mathbb{Z}/n\mathbb{Z}$, we have $t^i \cdot R_0 = R_{i\deg(t)}$.*

Proof of the sublemma. The proof is by induction on i, where i is viewed as an element of the ordered set $\{0, \ldots, n-1\}$. The identity of course holds if $i = 0$. We suppose that $t^j \cdot R_0 = R_{j\deg(t)}$ for all $j < i$. Note first that we certainly have $t^i \cdot R_0 \subseteq R_{i\cdot\deg(t)}$. To conclude the proof, we need to show that $R_{i\cdot\deg(t)} \subseteq t^i \cdot R_0$. To show this, let $e \in R_{i\cdot\deg(t)}$. By assumption e can be written in the form $e = t \cdot r$, with $r \in R$. For $k \in \{0, \ldots, n-1\}$, let r_k be the homogenous component of degree k of r. We have

$$e = t \cdot r = t \cdot r_0 + \cdots + t \cdot r_{n-1}$$

so that $t \cdot r_{(i-1)\cdot\deg(t)} = t \cdot r = e$. By induction, we have $r_{(i-1)\cdot\deg(t)} \in t^{i-1} R_0$ so that $e \in t \cdot (t^{i-1} R_0) = t^i \cdot R_0$, as required. □

Now since n is prime and $\deg(t) \neq 0 \pmod n$, every element of $\mathbb{Z}/n\mathbb{Z}$ is a multiple of $\deg(t)$. We can thus conclude from the sub-lemma that R is a direct sum of copies of R_0 so in particular R is flat over R_0.

To prove (2), localising at points of X_G, we may still assume that $X = \mathrm{Spec}(R)$, where R is a local ring and $R \cdot R_{\neq 0}$ is generated by a single element t, which is not a zero divisor. In the proof of (1), it was shown that we may suppose that t is homogenous of degree $\neq 0$. The sheaf $N_{X_G/X}$ corresponds to the R-module $(t)/(t^2)$ and thus $(N_{X_G/X})_0 = 0$, since t is of degree $\neq 0 \pmod n$.

Proof of (3). We may suppose that $X = \operatorname{Spec} R$ is affine. The first statement now corresponds to the statement that $R_0 \to R/(R{\cdot}R_{\neq 0})$ is surjective. This follows from the definitions. The fact that $q^{-1}(q(X_G)) = X_G$ follows from Proposition 2.1 (3). The third assertion follows from Proposition 2.1 (4).

Proof of (4). Note that for all $x \in X\backslash X_G$, we have $G_i(x) \neq G$ and thus $G_i(x) = 0$, since n is prime. By Proposition 2.1 (6) this implies that q is étale, in particular flat. Hence the morphism $U \to U/G$ is finite and flat.

We first compute its degree. For this, let $u_0 \in U/G$ and let H be the spectrum of the strict henselisation of $\mathcal{O}_{U/G,u_0}$. Then $H \cong (U \times_{U/G} H)/G$ by Proposition 2.1 (4) and the fact that H is flat over $\mathcal{O}_{U/G,u_0}$ (see [12, I, 1, 1.20] for this). We only have to compute the degree of $U \times_{U/G} H$ over H. Now note that $U \times_{U/G} H$ is a disjoint union $\coprod_{i \in I} H_i$ of copies of H, since H is strictly henselian and $U \times_{U/G} H \to H$ is étale. Furthermore, the group G permutes the H_i and also the closed points of the H_i. Hence the degree is the cardinality of the orbit of a closed point $P \in H_{i_0}$ (i_0 arbitrary). Since $G_i(P) = G_d(P)$, we must have $G_d(P) = 0$, since n is prime and $(U \times_{U/G} H)_G$ is empty. Hence the orbit of P has n elements and thus the degree of $U \to U/G$ is n.

Now consider the morphism $\phi : G \times U \to U \times_{U/G} U$. Let T be a connected scheme. The map $G(T) \times U(T) \to U(T) \times_{(U/G)(T)} U(T)$ is injective. To see this note that otherwise there is $e \in U(T)$ and $g \in G(T)$ such that $g \neq 0$ and $g(e) = e$; since $G(T)$ is of prime order this means that $e \in U(T)^G$ and thus $e \in U_G(T)$, which is not possible, since U_G is empty. Since T was arbitrary, the morphism ϕ is a monomorphism of schemes. Since it is also proper (because $G \times U$ and $U \times_{U/G} U$ are proper over U/G), it is a closed immersion (see [14, IV.3, 8.11.5] for this). Since both $G \times U$ and $U \times_{U/G} U$ are flat and finite of the same rank over U by the previous paragraph, this implies that ϕ is an isomorphism.

Proof of (5). Consider the natural morphism

$$\alpha : q^*(q_*M)_0 \to M\,.$$

The restriction $\alpha_{X\backslash X_G}$ is an isomorphism by (4) and Proposition 2.1 (5). Now both sides are locally free of finite rank by (1). Thus, by Nakayama's lemma, it is sufficient to show that $\alpha_{\kappa(x)}$ is surjective for $x \in X_G$. In particular, it is sufficient to show that the restriction $\iota^*(\alpha)$ of α to X_G is an isomorphism. Now note that since q is an affine morphism, the natural adjunction morphism $\alpha : q^*(q_*M) \to M$ is a surjection and thus we have a surjection

$$\iota^*(q^*(q_*M)) \to \iota^*(M)$$

restricting α. Hence we have a surjection

$$\iota^*(q^*((q_*M)_0)) \to \iota^*(M)_0$$

and since $\iota^*(M)_0 = \iota^*(M)$ by assumption we get a surjection

$$\iota^*(q^*((q_*M)_0)) \to \iota^*(M)$$

which must be an isomorphism, since both sides are locally free of the same rank.

Proof of (6). We need to check that the geometric fibres $X/G \to S$ are regular. So let $\operatorname{Spec} k \to S$ be a geometric point. By assumption, X_k is regular and by [30, Prop. 3.1], $(X_k)_G = (X_G)_k$ is then also regular.

Proof of (7). Since q is faithfully flat, we see that $X/G \to S$ is also flat. To see that $X/G \to S$ is smooth, we need to check that the geometric fibres $X/G \to S$ are regular. Now since X_G is flat over S and a Cartier divisor, we see that for any base change $T \to S$, $(X_T)_G \to T$ is also flat and a Cartier divisor. Furthermore, by Lemma 2.4, for any base change $T \to S$, we have $(X/G)_T \cong (X_T)/G$. So let $\operatorname{Spec} k \to S$ be a geometric point. By assumption X_k is regular and since $(X_k)_G$ is a Cartier divisor, we see that $(X_k)/G = (X/G)_k$ is regular, since q_k is faithfully flat by (1) and Proposition 2.1 (1). □

3. Free algebras

This section is mainly here to fix some notation that will be needed in Section 4. Let I be a finite set. We shall write $\langle I\rangle$ for the free monoid generated by the set I. See for instance [24, Chap. I] for this. Recall that the set $\langle I\rangle$ consists of all the finite words written in the alphabet I. A finite word is a map $\{1,\dots,n\} \to I$, where n is a positive integer. The integer n is called the length of the word. If

$$w_1 : \{1,\dots,n_1\} \to I$$

and

$$w_2 : \{1,\dots,n_2\} \to I$$

are two finite words, their concatenation w_1w_2 is by definition the map

$$w_1w_2 : \{1,\dots,n_1+n_2\} \to I,$$

such that $w_1w_2(k) = w_1(k)$ if $k \le n_1$ and $w_1w_2(k) = w_2(k-n_1)$ if $k > n_1$. The monoid structure of $\langle I\rangle$ is given by the concatenation of finite words. We shall write $\mathbb{Z}\langle I\rangle$ for the free $\mathbb{Z}$-module with basis the elements of $\langle I\rangle$. If $I = \{X_1,\dots,X_n\}$ then we shall use the shorthand

$$\mathbb{Z}\{X_1,\dots,X_n\} := \mathbb{Z}\langle\{X_1,\dots,X_n\}\rangle\,.$$

The set $\mathbb{Z}\langle I\rangle$ has the structure of a unital ring, where the addition is given by the addition on $\mathbb{Z}\langle I\rangle$ provided by its structure of $\mathbb{Z}$-module and the multiplication $\cdot$ is given by the formula

$$\Big(\sum_{w\in\langle I\rangle} n_w\cdot w\Big)\cdot\Big(\sum_{v\in\langle I\rangle} m_v\cdot v\Big) := \sum_{h\in\langle I\rangle}\left(\sum_{w,v\in\langle I\rangle,\ wv=h} n_w\cdot m_v\right)\cdot h.$$

Note that there is an equivalence relation $\sim_{\mathrm{ro}}$ on $\langle I\rangle$, defined as follows. If w_1 and w_2 are two finite words as above, then $w_1 \sim_{\mathrm{ro}} w_2$ iff $n_1 = n_2$ and there is a bijection $\sigma : \{1, \ldots, n_1\} \to \{1, \ldots, n_1\}$ such that $w_1 = w_2 \circ \sigma$. We shall write $[w]_{\mathrm{ro}}$ for the equivalence class of a finite word w in $\langle I \rangle$.

We shall write $\mathbb{Z}[I]$ for the polynomial ring over the set I (i.e., the polynomial ring with coefficients in $\mathbb{Z}$ where each element of I is a variable). This is by definition the free $\mathbb{Z}$-module with basis the free commutative monoid generated by I.

Note that there is an obvious surjective map of rings

$$\mathbb{Z}\langle I \rangle \to \mathbb{Z}[I] .$$

Lemma 3.1. *An element $\sum_{w \in \langle I \rangle} n_w \cdot w$ is in the kernel of $\mathbb{Z}\langle I \rangle \to \mathbb{Z}[I]$ iff for all $v \in \langle I \rangle$, we have*

$$\sum\nolimits_{w \in [v]_{\mathrm{ro}}} n_w = 0 .$$

Proof. Left to the reader. □

Abusing language, we shall say that $\mathbb{Z}\langle I \rangle$ is the ring of non-commutative polynomials with variables I and with coefficients in $\mathbb{Z}$. In particular $\mathbb{Z}\{X_1, \ldots, X_n\}$ is the ring of non-commutative polynomials in the variables $X_1, \ldots, X_n$ and coefficients in $\mathbb{Z}$.

If $P \in \mathbb{Z}\langle I \rangle$, then for each $w \in \langle I \rangle$, the integer P_w is defined by the equality

$$P = \sum_{w \in \langle I \rangle} P_w \cdot w .$$

4. The determinant of cohomology and Ducrot's generalisation of the Deligne pairing

Let $f : X \to S$ be a flat and strongly projective morphism. Suppose that S is locally Noetherian. Let I be a finite set. Let $\{F_i\}_{i \in I}$ be a collection of vector bundles on X indexed by I. If $w = i_1 i_2 \ldots i_k$ is a non-empty word in the alphabet I, then we shall write

$$\lambda(w) := \lambda\Big(\bigotimes_{t=1}^{k} F_{i_t} \Big)$$

where $\lambda(F_{i_t}) := \det(\mathrm{R}^\bullet f_*(F_{i_t}))$ is the determinant of cohomology of the vector bundle F_{i_t}, relatively to f (see beginning of the introduction). If w is the empty word then by convention $\lambda(w) := \lambda(\mathcal{O}_X)$.

If we are given a non-commutative polynomial $P = P((F_i)_{i \in I})$ with variables in I and integral coefficients (see Section 3), we shall write

$$\lambda(P) := \bigotimes_{w \in \langle I \rangle} \lambda(w)^{\otimes P_w} .$$

Note that with this definition, if P and Q are two non-commutative polynomials with variables in I and integral coefficients, then in view of the distributivity of the tensor product, there is a canonical isomorphism

$$\lambda(P + Q) \cong \lambda(P) \otimes \lambda(Q).$$

Abusing language, we shall mostly write non-commutative polynomials $P = P((F_i)_{i\in I})$ with variables in I using the F_i as variable symbols instead of the elements of the index set I. Also we shall mostly use the tensor product symbol $\otimes$ instead of the symbol $\cdot$. So, e.g., if $I = \{X_1, X_2\}$ we would write

$$F_1 \otimes F_2 + F_2 \otimes F_2$$

instead of $X_1 \cdot X_2 + X_2 \cdot X_2$.

Lemma 4.1. *Let $\{F_i\}_{i\in I}$ be a finite collection of vector bundles on X. Let $P = P((F_i)_{i\in I}) \in \mathbb{Z}\langle I\rangle$ be a non-commutative polynomial with integral coefficients in the F_i and suppose that P lies in the kernel of the natural map of rings $\mathbb{Z}\langle I\rangle \to \mathbb{Z}[I]$. Then there is an isomorphism $\lambda(P) \cong \mathcal{O}_S$, which can be chosen compatibly with any base change to a locally Noetherian scheme.*

Proof. Let us write O for the set of equivalence classes of the relation $\sim_{\mathrm{ro}}$ in $\langle I\rangle$ (see Section 3 for the definition). Choose a representative $w'(o) \in o$ (arbitrary but fixed) for each $o \in O$. Furthermore, for each $o \in O$ and each $w \in o$, choose an automorphism σ_w of $\{1, \ldots, \mathrm{length}(w'(o))\}$ such that $w'(o) = \sigma_w \circ w$. Finally, choose an isomorphism $\alpha_o : o \cong \{1, \ldots, \#o\}$ for each $o \in O$.

According to Lemma 3.1, if we write

$$P = \sum_{o\in O} \sum_{w\in o} P_w \cdot w$$

then $\sum_{w\in o} P_w = 0$ for each $o \in O$. On the other hand, by the definition of $\lambda(P)$ and the distributivity of the tensor product, we have a canonical isomorphism

$$\lambda(P) \cong \bigotimes_{o\in O} \bigotimes_{w\in o} \lambda(w)^{P_w}$$

and by the commutativity of the tensor product, there is a canonical isomorphism

$$\lambda(w) \cong \lambda(w'(o))$$

for each $w \in o$, which depends of the choice of the automorphism σ_w. Hence there is a canonical isomorphism

$$\lambda(P) \cong \bigotimes_{o\in O} \lambda(w'(o))^{\sum_{w\in o} P_w},$$

which depends on the isomorphism α_o. The conclusion follows. Note that this isomorphism depends a priori on the choices of the representatives $w'(o) \in o$ and of the automorphisms σ_w and α_o. One might conjecture that different choices of representatives and automorphisms will lead to the same isomorphism $\lambda(P) \cong \mathcal{O}_S$, up to sign (but this is irrelevant to the conclusion of the lemma, which contains no unicity statement). $\square$

Let $L_1, \ldots, L_{d+1}$ be line bundles on X. Suppose that X is of constant relative dimension d over S. We shall write

$$I_{X/S}(L_1, \ldots, L_{d+1}) := \lambda((\mathcal{O}_X - L_1) \otimes (\mathcal{O}_X - L_2) \otimes \cdots \otimes (\mathcal{O}_X - L_{d+1}))^{\otimes(-1)^d}$$

where the expression defining $I_{X/S}(L_1,\dots,L_{d+1})$ is to be read with the above notational conventions in mind. In particular, the expression

$$(\mathcal{O}_X - L_1)\otimes(\mathcal{O}_X - L_2)\otimes\cdots\otimes(\mathcal{O}_X - L_{d+1})$$

should be understood as a non-commutative polynomial in the line bundles

$$\mathcal{O}_X, L_1,\dots,L_{d+1} \quad \text{and} \quad \lambda((\mathcal{O}_X - L_1)\otimes(\mathcal{O}_X - L_2)\otimes\cdots\otimes(\mathcal{O}_X - L_{d+1}))$$

is to be computed according to the conventions described above.

So for example, if $d = 1$,

$$\begin{aligned} & I_{X/S}(L_1, L_2)^\vee = \lambda((\mathcal{O}_X - L_1)\otimes(\mathcal{O}_X - L_2)) \\ = \; & \lambda(\mathcal{O}_X\otimes\mathcal{O}_X - \mathcal{O}_X\otimes L_2 - L_1\otimes\mathcal{O}_X + L_1\otimes L_2) \\ = \; & \lambda(\mathcal{O}_X\otimes\mathcal{O}_X)\otimes\lambda(\mathcal{O}_X\otimes L_2)^\vee\otimes\lambda(L_1\otimes\mathcal{O}_X)^\vee\otimes\lambda(L_1\otimes L_2) \\ \cong \; & \lambda(\mathcal{O}_X)\otimes\lambda(L_2)^\vee\otimes\lambda(L_1)^\vee\otimes\lambda(L_1\otimes L_2). \end{aligned} \tag{6}$$

Ducrot showed in [7, 5] that the line bundle $I_{X/S}(L_1,\dots,L_{d+1})$ is multiadditive in the line bundles $L_1,\dots,L_{d+1}$. In particular, he shows that if Q is a line bundle on X, then there is a canonical isomorphism

$$I_{X/S}(L_1\otimes Q,\dots,L_{d+1}) \cong I_{X/S}(L_1,\dots,L_{d+1})\otimes I_{X/S}(Q,\dots,L_{d+1}). \tag{7}$$

The canonical isomorphism (7) is compatible with any base change to a locally Noetherian scheme. See [7, Th. 4.2 (BC)].

We may thus compute

$$\begin{aligned} & \lambda((\mathcal{O}_X - Q)\otimes(\mathcal{O}_X - L_1)\otimes(\mathcal{O}_X - L_2)\otimes\cdots\otimes(\mathcal{O}_X - L_{d+1})) \\ \overset{(1)}{\cong} \; & \lambda((\mathcal{O}_X - L_1)\otimes(\mathcal{O}_X - L_2)\otimes\cdots\otimes(\mathcal{O}_X - L_{d+1})) \\ \otimes \; & \lambda((Q - Q\otimes L_1)\otimes(\mathcal{O}_X - L_2)\otimes(\mathcal{O}_X - L_3)\otimes\cdots\otimes(\mathcal{O}_X - L_{d+1}))^\vee \\ \overset{(2)}{\cong} \; & \lambda((\mathcal{O}_X - L_1)\otimes(\mathcal{O}_X - L_2)\otimes\cdots\otimes(\mathcal{O}_X - L_{d+1})) \\ \otimes \; & \lambda\big((\mathcal{O}_X - L_1\otimes Q - (\mathcal{O}_X - Q))\otimes(\mathcal{O}_X - L_2)\otimes(\mathcal{O}_X - L_3)\otimes\cdots \\ & \qquad\qquad\qquad\qquad\qquad\qquad \cdots\otimes(\mathcal{O}_X - L_{d+1})\big)^\vee \\ \overset{(3)}{\cong} \; & I_{X/S}(L_1,\dots,L_{d+1})^{\otimes(-1)^d} \\ \otimes \; & I_{X/S}(L_1\otimes Q, L_2,\dots,L_{d+1})^{\otimes(-1)^{d+1}}\otimes I_{X/S}(Q, L_2,\dots,L_{d+1})^{\otimes(-1)^d} \\ \overset{(4)}{\cong} \; & I_{X/S}(L_1, L_2,\dots,L_{d+1})^{\otimes(-1)^d}\otimes I_{X/S}(L_1, L_2,\dots,L_{d+1})^{\otimes(-1)^{d+1}} \\ \otimes \; & I_{X/S}(Q, L_2,\dots,L_{d+1})^{\otimes(-1)^{d+1}}\otimes I_{X/S}(Q, L_2,\dots,L_{d+1})^{\otimes(-1)^d} \overset{(5)}{\cong} \mathcal{O}_X \end{aligned}$$

and this trivialisation is invariant under any base change to a locally Noetherian scheme. The isomorphisms (1), (2), (3) are formal consequences of Lemma 4.1 and of the polynomial equalities

$$\begin{aligned}
&(1-y)(1-x_1)\cdots(1-x_{d+1})\\
&\quad=(1-x_1)\cdots(1-x_{d+1})-(y-yx_1)(1-x_2)\cdots(1-x_{d+1})\\
&\quad=(1-x_1)\cdots(1-x_{d+1})-((1-x_1y)-(1-y))(1-x_2)\cdots(1-x_{d+1})\\
&\quad=(1-x_1)\cdots(1-x_{d+1})-(1-x_1y)(1-x_2)\cdots\\
&\qquad\qquad\cdots(1-x_{d+1})+(1-y)(1-x_2)\cdots(1-x_{d+1})
\end{aligned}$$

(in the same order). The isomorphism (4) comes from the multiadditivity of the symbol $I_{X/S}$ described above. Isomorphism (5) is just a cancellation.

The following theorem summarises the discussion and it is one of the main consequences of the theory developed in [7].

Theorem 4.2. *Suppose that $X \to S$ is flat, strongly projective and of relative dimension d. Suppose that S is locally Noetherian. Let $L_1, \ldots, L_{d+2}$ be line bundles on X. Then the line bundle*

$$\lambda((\mathcal{O}_X - L_1) \otimes (\mathcal{O}_X - L_2) \otimes \cdots \otimes (\mathcal{O}_X - L_{d+2}))$$

is canonically trivial and the trivialisation is invariant under base change to any locally Noetherian scheme.

See also [2, Th. A.21, Appendix], where it is verified that some Noetherian assumptions in Theorem 4.2 can be removed (we do not exploit this because Noetherian assumptions are needed elsewhere in this text).

Corollary 4.3. *Let $\mathcal{F} := n_1M_1 + \cdots + n_kM_k$, where M_i is a line bundle on X (resp. $n_i \in \mathbb{Z}$) for all $i \in \{1, \ldots, k\}$. Let $L_1, \ldots, L_{d+1}$ be line bundles on X. Suppose that $\sum_i n_i = 0$. Then the line bundle*

$$\lambda(\mathcal{F} \otimes (\mathcal{O}_X - L_1) \otimes (\mathcal{O}_X - L_2) \otimes \cdots \otimes (\mathcal{O}_X - L_{d+1}))$$

is canonically trivial and the trivialisation is invariant under base change to any Noetherian scheme.

Proof of Corollary 4.3. By Theorem 4.2, there is a canonical isomorphism

$$\begin{aligned}
&\lambda((n_iM_i) \otimes (\mathcal{O}_X - L_1) \otimes (\mathcal{O}_X - L_2) \otimes \cdots \otimes (\mathcal{O}_X - L_{d+1}))\\
&\quad\cong \lambda((\mathcal{O}_X - L_1) \otimes (\mathcal{O}_X - L_2) \otimes \cdots \otimes (\mathcal{O}_X - L_{d+1}))^{\otimes n_i}
\end{aligned}$$

for any n_i. The corollary follows from this. □

5. Equivariant derived functors

We first recall the definition of a perfect complex. Let Z be a locally Noetherian scheme. We shall as usual write $D^b(Z)$ for the derived category of bounded complexes of $\mathcal{O}_Z$-modules. A complex $J^\bullet$ of $\mathcal{O}_Z$-modules is said to be of finite tor-dimension if there are integers $a < b$ such that for all $\mathcal{O}_Z$-modules M, we have $\underline{\mathrm{Tor}}^k(J^\bullet, M) = 0$ if $k < a$ or $k > b$. A *bounded* complex $J^\bullet$ is said to be perfect if

- the homology sheaves $\mathcal{H}^k(J^\bullet)$ are coherent for all $k \in \mathbb{Z}$;
- there is a covering (U_i) of Z by open subschemes, such that $J^\bullet|_{U_i}$ is of finite tor-dimension.

In view of this definition, we see that the property of being perfect depends only on the image of $J^\bullet$ in the category $D^b(Z)$.

Let G be a finite group. If Z is a G-equivariant locally Noetherian scheme, we let $\mathrm{Coh}^{\mathrm{eq}}(Z)$ be the category of coherent G-equivariant sheaves on Z. Recall also that $\mathrm{Coh}(Z)$ refers to the category of coherent sheaves on Z. Note that the category $\mathrm{Coh}^{\mathrm{eq}}(Z)$ has a natural structure of abelian category. We shall write $D^b(\mathrm{Coh}^{\mathrm{eq}}(Z))$ for the derived category of bounded complexes in $\mathrm{Coh}^{\mathrm{eq}}(Z)$. Note that there is a natural forgetful functor from $D^b(\mathrm{Coh}^{\mathrm{eq}}(Z))$ to $D^b(\mathrm{Coh}(Z))$ and thus also to $D^b(Z)$ via the forgetful functor $D^b(\mathrm{Coh}(Z)) \to D^b(Z)$.

Lemma 5.1. *Suppose that $f : X \to Y$ is a strongly projective morphism of G-equivariant Noetherian schemes. Let $J^\bullet$ be a bounded complex of G-equivariant coherent sheaves on X. Then there is a bounded complex $H^\bullet$ of G-equivariant f-acyclic coherent sheaves on X and a G-equivariant quasi-isomorphism $J^\bullet \to H^\bullet$.*

Recall that if F is a quasi-coherent sheaf on X, one says that F is f-acyclic if $\mathrm{R}^k f_*(F) = 0$ when $k > 0$.

Proof. When the action of G on X and Y is trivial, this is standard. The proof in the equivariant situation is completely similar and we skip it. □

In view of Lemma 5.1 and [18, Th. I.5.1], in the situation of Lemma 5.1 the functor f_*^{eq} has a right derived functor

$$\mathrm{R}^\bullet f_*^{\mathrm{eq}} : D^b(\mathrm{Coh}^{\mathrm{eq}}(X)) \to D^b(\mathrm{Coh}^{\mathrm{eq}}(S)).$$

The functor $\mathrm{R}^\bullet f_*^{\mathrm{eq}}$ is compatible with the usual right derived functor

$$\mathrm{R}^\bullet f_* : D^b(\mathrm{Coh}(X)) \to D^b(\mathrm{Coh}(Y))$$

via the forgetful functors $D^b(\mathrm{Coh}^{\mathrm{eq}}(X)) \to D^b(\mathrm{Coh}(X))$ and $D^b(\mathrm{Coh}^{\mathrm{eq}}(Y)) \to D^b(Y)$. If $f : X \to Y$ and $h : X \to Y$ are strongly projective morphism of G-equivariant Noetherian schemes then we have a natural isomorphism of functors $\mathrm{R}^\bullet(h \circ f)_*^{\mathrm{eq}} \cong \mathrm{R}^\bullet h_*^{\mathrm{eq}} \circ \mathrm{R}^\bullet f_*^{\mathrm{eq}}$. This follows from [18, Prop. 5.4 and following remark]. We leave the details to the reader. The point is that for any bounded complex of G-equivariant coherent sheaves $J^\bullet$ on X, there is a bounded complex $H^\bullet$ of G-equivariant coherent sheaves on X, which is both $f-$ and $h \circ f$-acyclic, and a G-equivariant quasi-isomorphism $J^\bullet \to H^\bullet$.

If F is a G-equivariant locally free sheaf on a G-equivariant locally Noetherian scheme Z, we have a functor $F \otimes (\cdot) : D^b(\mathrm{Coh}^{\mathrm{eq}}(Z)) \to D^b(\mathrm{Coh}^{\mathrm{eq}}(Z))$ (resp. a functor $(\cdot) \otimes F : D^b(\mathrm{Coh}^{\mathrm{eq}}(Z)) \to D^b(\mathrm{Coh}^{\mathrm{eq}}(Z))$. This functor simply sends a complex $J^\bullet \in D^b(\mathrm{Coh}(Z))$ on the complex $J^\bullet \otimes F$ (resp. the complex $F \otimes J^\bullet$).

We have a projection formula:

Proposition 5.2. *Suppose that $f : X \to Y$ is a strongly projective morphism of G-equivariant Noetherian schemes. Let F be a G-equivariant locally free sheaf on Y. Then there is a natural isomorphism of functors*

$$\mathrm{R}f_*^{\mathrm{eq}}(f^*(F) \otimes (\cdot)) \cong \mathrm{R}f_*^{\mathrm{eq}}(\cdot) \otimes F .$$

Proof. Left to the reader. Apply the usual projection formula to the definition of $\mathrm{R}f_*^{\mathrm{eq}}(\cdot)$. □

Recall that a morphism $h : T \to S$ of locally Noetherian schemes is called lci (local complete intersection), if locally on S, there is a factorisation of h into a regular closed immersion $T \to T_1$ followed by a smooth morphism $T_1 \to S$. We recall the

Proposition 5.3. *If a morphism $h : T \to S$ of Noetherian schemes is lci and strongly projective and $F^\bullet$ is an object of $D^b(\mathrm{Coh}(T))$, which is a perfect complex then $\mathrm{R}^\bullet f_*(F^\bullet)$ is also a perfect complex.*

Proof. See [17, Cor. 4.8.1, Exp. III]. □

Suppose now that Z is a locally Noetherian scheme, that $G = \mathbb{Z}/2\mathbb{Z}$ and that 2 is invertible on Z. Suppose also that the scheme Z is endowed with a trivial G-equivariant structure. If F is an equivariant coherent sheaf on Z, we shall write

$$F_+ := F_0 \qquad \text{and} \qquad F_- := F_1.$$

The functors

$$(\cdot)_- = (\cdot)_1 : \mathrm{Coh}^{\mathrm{eq}}(Z) \to \mathrm{Coh}(Z)$$

and

$$(\cdot)_+ = (\cdot)_0 : \mathrm{Coh}^{\mathrm{eq}}(Z) \to \mathrm{Coh}(Z)$$

are exact functors and so they uniquely extend to functors from $D^b(\mathrm{Coh}^{\mathrm{eq}}(Z))$ to $D^b(\mathrm{Coh}(Z))$, which are their right and left derived functors simultaneously. We shall also call these extensions $(\cdot)_+$ and $(\cdot)_-$.

If $F^\bullet$ is an object in $D^b(\mathrm{Coh}^{\mathrm{eq}}(Z))$ then we have by construction a canonical direct sum decomposition $F^\bullet \cong (F^\bullet)_+ \oplus (F^\bullet)_-$ in $D^b(\mathrm{Coh}(Z))$. In particular, if the image of $F^\bullet$ in $D^b(\mathrm{Coh}(Z))$ is a perfect complex, so are $(F^\bullet)_+$ and $(F^\bullet)_-$. We shall say that an object $F^\bullet$ of $D^b(\mathrm{Coh}^{\mathrm{eq}}(Z))$ is a perfect complex if its image in $D^b(\mathrm{Coh}(Z))$ (or $D^b(Z)$) is a perfect complex. If $F^\bullet$ is an object of $D^b(\mathrm{Coh}^{\mathrm{eq}}(Z))$, which is a perfect complex, we can thus write

$$\det{}^{\mathrm{eq}}(F^\bullet) := \det((F^\bullet)_+) \otimes \det((F^\bullet)_-)^\vee .$$

If

$$F^\bullet \to H^\bullet \to J^\bullet \to F^\bullet[1]$$

is a triangle of perfect complexes in $D^b(\mathrm{Coh}^{\mathrm{eq}}(Z))$, we then have a canonical isomorphism

$$\det{}^{\mathrm{eq}}(F^\bullet) \otimes \det{}^{\mathrm{eq}}(J^\bullet) \cong \det{}^{\mathrm{eq}}(H^\bullet)$$

by the standard properties of determinants (see [22]) and the fact that the functors

$$(\cdot)_{\pm} : D^b(\mathrm{Coh}^{\mathrm{eq}}(Z)) \to D^b(\mathrm{Coh}(Z))$$

respect triangulations (because they are derived functors).

Let $f : X \to Y$ be a locally projective and lci morphism of G-equivariant locally Noetherian schemes, where the G-action on Y is trivial. Suppose that $G = \mathbb{Z}/2\mathbb{Z}$ and that 2 is invertible on Y (and thus on X). Let $F^\bullet$ be an object of $D^b(\mathrm{Coh}^{\mathrm{eq}}(X))$, which is a perfect complex. Let $U \subseteq Y$ be an open subset, such that $f|_U : f^{-1}(U) \to U$ is strongly projective. By Corollary 5.3 and the above discussion, we may define

$$\lambda^{\mathrm{eq}}(F^\bullet|_{f^{-1}(U)}) := \det\big((\mathrm{R}^\bullet(f^{\mathrm{eq}}|_U)_*(F^\bullet|_{f^{-1}(U)}))_+\big) \\ \otimes \det\big((\mathrm{R}^\bullet(f^{\mathrm{eq}}|_U)_*(F^\bullet|_{f^{-1}(U)}))_-\big)^\vee$$

which is a line bundle on U. Since this line bundle is defined locally on Y, by varying U, we obtain a line bundle on all of Y, which we denote by $\lambda^{\mathrm{eq}}(F^\bullet)$. If the equivariant structure on X and F is trivial, then we of course have a canonical identification

$$\lambda^{\mathrm{eq}}(F^\bullet) \cong \lambda(F^\bullet).$$

Note that if

$$F^\bullet \to H^\bullet \to J^\bullet \to F^\bullet[1]$$

is a triangle of perfect complexes in $D^b(\mathrm{Coh}^{\mathrm{eq}}(X))$, then we have canonically

$$\lambda^{\mathrm{eq}}(F^\bullet) \otimes \lambda^{\mathrm{eq}}(J^\bullet) \cong \lambda^{\mathrm{eq}}(H^\bullet) \tag{8}$$

(because $\mathrm{R}^\bullet f_*^{\mathrm{eq}}(\cdot)$ respects triangles, locally in Y).

Let I be a finite set. Let $\{F_i\}_{i \in I}$ be a collection of equivariant vector bundles on X indexed by I. For any non-commutative polynomial $P = ((F_i)_{i \in I})$ with integral coefficients and variables in I, we may now define $\lambda^{\mathrm{eq}}(P)$ in a manner entirely similar to the non-equivariant case (see beginning of Section 4). The evident equivariant analog of Lemma 4.1 then also holds.

Finally, we shall write $\{-1\}$ for the trivial sheaf $\mathcal{O}_X$, endowed with the G-equivariant structure such that for any $g \in G$ the isomorphism

$$\alpha_{g,\{-1\}} : \{-1\} \to g_*(\{-1\})$$

composed with the canonical non-equivariant identification $g_*(\{-1\}) \cong \{-1\}$ is given by multiplication by $(-1)^g$. If F is a G-equivariant sheaf on X, we shall write $F\{-1\}$ for $F \otimes \{-1\}$. Note that if F is an equivariant coherent locally free sheaf on X and $l \in \mathbb{Z}$, we have canonical isomorphisms

$$\lambda^{\mathrm{eq}}((F\{-1\})^{\otimes l}) \cong \lambda^{\mathrm{eq}}(F^{\otimes l})^{\otimes(-1)^l} \cong \lambda^{\mathrm{eq}}((-F)^{\otimes l}). \tag{9}$$

6. Local refinement of the fixed point formula for an involution

Let S be a locally Noetherian scheme and let $f : X \to S$ be a separated morphism of finite type. Suppose that 2 is invertible in S. Let $G = \mathbb{Z}/2$, so that we have a canonical isomorphism $G_S \cong \mu_{2S}$. Suppose that we have a G-equivariant structure on X over S. Suppose finally that the orbit of every point in X is contained in an open affine subscheme. Let $\iota : X_G \hookrightarrow X$ be the fixed scheme of X and let $q : X \to X/G$ be the quotient morphism. These morphisms exist by Proposition 2.3 and Theorem 2.1. Note that if q is flat then it is faithfully flat (since it is surjective) and thus if q and f are flat the natural morphism $X/G \to S$ is also flat. Similarly, if f is locally projective then so is the natural morphism $X/G \to S$.

In this section, we shall prove a version of the relative geometric fixed point formula for the G-action of G on X, which avoids K-theory entirely, replacing all the equalities in a Grothendieck group or a Picard group by explicit isomorphisms. This is the following Theorem.

Theorem 6.1. *Suppose in addition that f is smooth, locally projective and that f has constant relative dimension d. Suppose also that the morphism $X_G \to S$ is flat. Then $X_G \to S$ is smooth and thus X_G is regularly immersed in X. Let $N = N_{X_G/X}$ be the conormal bundle of $\iota : X_G \hookrightarrow X$, endowed with its canonical G-equivariant structure. Let M be a G-equivariant line bundle on X. We have a canonical isomorphism*

$$\lambda^{\mathrm{eq}}(M)^{\otimes 2^{d+1}} \cong \bigotimes_{j=0}^{d} \lambda^{\mathrm{eq}}(\iota^*(M) \otimes \mathrm{Sym}^j(N))^{\otimes \sum_{i=0}^{d-j} \binom{d+1}{i}}$$

which is compatible with any base change $h : S' \to S$ such that S' is locally Noetherian.

For the proof, we shall need the following

Lemma 6.2. *Let $Z \to T$ be a morphism of locally Noetherian schemes. Let $C \hookrightarrow Z$ be a regular closed immersion. Suppose that C and Z are flat over T. Let $h : T' \to T$ be a morphism of schemes, where T' is locally Noetherian. Then*

(a) *the natural morphism $\mathrm{Bl}_{C_{T'}}(Z_{T'}) \to \mathrm{Bl}_C(Z)_{T'}$ is an isomorphism;*

(b) *$\mathrm{Bl}_C(Z)$ is flat over T.*

Proof. Let I be the sheaf of ideals of C in Z. By definition, we have

$$\mathrm{Bl}_C(Z) := \mathrm{Proj}\Big(\bigoplus_{i\geq 0} I^i\Big)$$

so that

$$\mathrm{Bl}_C(Z)_{T'} := \mathrm{Proj}\Big(\bigoplus_{i\geq 0} h_Z^*(I^i)\Big)$$

where $h_Z : Z_{T'} \to Z$ is the base change of h to Z and $h_Z^*(I^i)$ is the pull-back to $Z_{T'}$ of I^i as a coherent sheaf on Z. On the other hand, we have again by definition

$$\mathrm{Bl}_{C_{T'}}(Z_{T'}) := \mathrm{Proj}\Big(\bigoplus_{i\geq 0} h_Z^{-1}(I)^i\Big) = \mathrm{Proj}\Big(\bigoplus_{i\geq 0} h_Z^{-1}(I^i)\Big)$$

where $h_Z^{-1}(I^i)$ is the ideal sheaf on $Z_{T'}$, which is the image of $h_Z^*(I^i)$ in $\mathcal{O}_{Z_{T'}}$. The surjection of sheaves $h_Z^*(I^i) \to h_Z^{-1}(I^i)$ provide a natural $Z_{T'}$-morphism from $\mathrm{Bl}_{C_{T'}}(Z_{T'})$ to $\mathrm{Bl}_C(Z)_{T'}$, which is the natural map mentioned in the lemma. To prove (a), we need to show that this morphism is an isomorphism. For this, it is sufficient to show that the surjection $h_Z^*(I^i) \to h_Z^{-1}(I^i)$ is an isomorphism for all $i \geq 0$. We will show that the closed subscheme of Z defined by I^i is flat over T, from which this immediately follows. Now note that because C is regularly immersed in Z we have $I^k/I^{k+1} \cong \mathrm{Sym}^k(N_{C/Z})$ for all $k \geq 0$. Here $N_{C/Z}$ is the conormal sheaf of C in Z. See, e.g., [13, IV, par. 2, Cor. 2.4] for this. Since $N_{C/Z}$ is locally free over C and C is flat over T, we see that I^k/I^{k+1} is flat over T for all $k \geq 0$. Since $\mathcal{O}_Z/I^i$ has a natural filtration, whose quotients are of the form I^k/I^{k+1}, we conclude that $\mathcal{O}_Z/I^i$ is also flat over T. In other words, the closed subscheme of Z defined by I^i is flat over T. This concludes the proof of (a). For (b), note that since Z is flat over T and $\mathcal{O}_Z/I^i$ is flat over T (see the proof of (a)), the sheaf I^i is also flat over T (for all $i \geq 0$). Thus the graded $\mathcal{O}_Z$-algebra $\bigoplus_{i\geq 0} I^i$ is flat over T, which implies that $\mathrm{Bl}_C(Z)$ is flat over T. □

Proof. (of Theorem 6.1). First note that since the advertised isomorphism of line bundles is local on S, we may assume that S is affine. In particular, we may assume that f is a strongly projective morphism.

We start with an identity in $\mathbb{Z}[t]$. Define

$$P_k(t) := 2^k + 2^{k-1}(2-t) + 2^{k-2}(2-t)^2 + \cdots + (2-t)^k \in \mathbb{Z}[t].$$

Setting $q := 1 - \frac{t}{2}$, we have

$$\begin{aligned} tP_k(t) &= 2(1-q)2^k(1 + q + \cdots + q^k) = -2^{k+1}(q^{k+1} - 1) \\ &= 2^{k+1} - (2q)^{k+1} = 2^{k+1} - (2-t)^{k+1}. \end{aligned}$$

(I am grateful to one of the referees for providing a simplification of earlier calculations.)

Now suppose first that X_G is a Cartier divisor. Let $L := \mathcal{O}(-X_G)$.

We have an exact sequence

$$0 \to L \otimes M \to M \to \iota_*(\iota^*(M)) \to 0\,. \tag{10}$$

The existence of this sequence, unspectacular as it may seem, is the linchpin of the proof.

Note that by the adjunction formula (or by definition, according to taste) we have a canonical equivariant isomorphism $\iota^*(L) \cong N$. Note also that by Proposition 2.5 (2), G acts by -1 on N. Let $J := q_*(L\{-1\})_0$. Proposition 2.5 (5) implies that this is a line bundle on X/G such that $q^*(J) \simeq L\{-1\}$.

Now we compute

$$\begin{aligned}
&\lambda^{\mathrm{eq}}(\iota^*(M)\otimes P_k(\mathcal{O}_{X_G}-N)))\\
&\overset{(b)}{\cong}\lambda^{\mathrm{eq}}(\iota^*(M)\otimes P_k(\mathcal{O}_{X_G}-\iota^*(L))) \overset{(c)}{\cong} \lambda^{\mathrm{eq}}(M\otimes(\mathcal{O}_X-L)\otimes P_k(\mathcal{O}_X-L))\\
&\overset{(d)}{\cong}\lambda^{\mathrm{eq}}(M\otimes(\mathcal{O}_X^{\oplus 2^{k+1}}-(\mathcal{O}_X^{\oplus 2}-(\mathcal{O}_X-L))^{\otimes(k+1)}))\\
&\overset{(e)}{\cong}\lambda^{\mathrm{eq}}(M\otimes(\mathcal{O}_X^{\oplus 2^{k+1}}-(\mathcal{O}_X^{\oplus 2}-(\mathcal{O}_X+L\{-1\}))^{\otimes(k+1)}))\\
&\overset{(f)}{\cong}\lambda^{\mathrm{eq}}(M\otimes(\mathcal{O}_X^{\oplus 2^{k+1}}-(\mathcal{O}_X-L\{-1\})^{\otimes(k+1)}))\\
&\overset{(g)}{\cong}\lambda^{\mathrm{eq}}(M)^{\otimes 2^{k+1}}\otimes\lambda^{\mathrm{eq}}(M\otimes(\mathcal{O}_X-L\{-1\})^{\otimes(k+1)})^{\vee}\\
&\overset{(h)}{\cong}\lambda^{\mathrm{eq}}(M)^{\otimes 2^{k+1}}\otimes\lambda^{\mathrm{eq}}(q_*(M)\otimes(\mathcal{O}_{X/G}-J)^{\otimes(k+1)})^{\vee}\\
&\overset{(i)}{\cong}\lambda^{\mathrm{eq}}(M)^{\otimes 2^{k+1}}\otimes\lambda((q_*(M)_+-q_*(M)_-)\otimes(\mathcal{O}_{X/G}-J)^{\otimes(k+1)})^{\vee}\\
&\overset{(j)}{\cong}\lambda^{\mathrm{eq}}(M)^{\otimes 2^{k+1}}\otimes\lambda(((\mathcal{O}_{X/G}-q_*(M)_-)-(\mathcal{O}_{X/G}-q_*(M)_+))\\
&\qquad\qquad\otimes(\mathcal{O}_{X/G}-J)^{\otimes(k+1)})^{\vee}\\
&\overset{(k)}{\cong}\lambda^{\mathrm{eq}}(M)^{\otimes 2^{k+1}}\otimes\lambda((\mathcal{O}_{X/G}-q_*(M)_-)\otimes(\mathcal{O}_{X/G}-J)^{\otimes(k+1)})^{\vee}\\
&\otimes\lambda((\mathcal{O}_{X/G}-q_*(M)_+)\otimes(\mathcal{O}_{X/G}-J)^{\otimes(k+1)}).
\end{aligned}$$

Equality (b) is justified by the adjunction formula. Equality (c) follows from the existence of the exact sequence (10) and the compatibility of $\lambda^{\mathrm{eq}}(\cdot)$ with triangles. Equality (d) follows from the equality $t\cdot P_k(t)=2^{k+1}-(2-t)^{k+1}$ and the equivariant analogue of Lemma 4.1. Equality (e) follows from (9). Equality (f) is a simple cancellation and so is equality (g). Equality (h) follows from the projection formula 5.2, the compatibility of equivariant derived functors with compositions of morphisms (see before Proposition 5.2) and the fact that we have $q^*(J)\cong L\{-1\}$. Equality (i) follows from the definition of $\lambda^{\mathrm{eq}}(\cdot)$. Equality (j) is a simple cancellation and so is equality (k).

Now if we let $k=d$, we obtain by Theorem 4.2 canonical trivialisations

$$\begin{aligned}
&\lambda^{\mathrm{eq}}((\mathcal{O}_{X/G}-q_*(M)_-)\otimes(\mathcal{O}_{X/G}-J)^{\otimes(k+1)})\\
&\quad\cong\lambda((\mathcal{O}_{X/G}-q_*(M)_-)\otimes(\mathcal{O}_{X/G}-J)^{\otimes(k+1)})\cong\mathcal{O}_S
\end{aligned}$$

and

$$\begin{aligned}
&\lambda^{\mathrm{eq}}((\mathcal{O}_{X/G}-q_*(M)_+)\otimes(\mathcal{O}_{X/G}-J)^{\otimes(k+1)})\\
&\quad\cong\lambda((\mathcal{O}_{X/G}-q_*(M)_+)\otimes(\mathcal{O}_{X/G}-J)^{\otimes(k+1)})\cong\mathcal{O}_S
\end{aligned}$$

and thus a canonical isomorphism

$$\lambda^{\mathrm{eq}}(\iota^*(M)\otimes P_d(\mathcal{O}_{X_G}-N))\cong\lambda^{\mathrm{eq}}(M)^{\otimes 2^{d+1}}. \tag{11}$$

Note that all the isomorphisms (b),..., (k) are compatible with any base change to a locally Noetherian scheme. This follows from that fact that $X\to S$ and $X_G\to S$ are flat, from Lemma 2.4 and from Theorem 4.2.

We repeat the calculation for $M=\mathcal{O}_X$ and $d=1$ (i.e., when $X\to S$ is a fibration in curves) to make the calculation completely explicit in a simple situation. In the case $d=1$, we may choose $k=d=1$ (see above). We then have $P_k(t)=P_1(t)=4-t$. We shall write $F:=q_*(\mathcal{O}_X)_-$. We compute

$$\begin{aligned}
&\lambda^{\mathrm{eq}}(\mathcal{O}_{X_G})^{\otimes 3}\otimes\lambda^{\mathrm{eq}}(N)\\
&\overset{\alpha}{\cong}\lambda^{\mathrm{eq}}(\mathcal{O}_X)^{\otimes 3}\otimes\lambda^{\mathrm{eq}}(L)^{\otimes(-3)}\otimes\lambda^{\mathrm{eq}}(L)\otimes\lambda^{\mathrm{eq}}(L^{\otimes 2})^{\otimes(-1)}\\
&\overset{\beta}{\cong}\lambda^{\mathrm{eq}}(\mathcal{O}_X)^{\otimes 3}\otimes\lambda^{\mathrm{eq}}(L\{-1\})^{\otimes 2}\otimes\lambda^{\mathrm{eq}}(L\{-1\}^{\otimes 2})^{\otimes(-1)}\\
&\overset{\gamma}{\cong}\lambda^{\mathrm{eq}}(\mathcal{O}_X)^{\otimes 3}\otimes\lambda(J)^{\otimes 2}\otimes\lambda(J\otimes F)^{\otimes(-2)}\otimes\lambda(J^{\otimes 2})^{\otimes(-1)}\otimes\lambda(J^{\otimes 2}\otimes F)\\
&\overset{\delta}{\cong}\lambda^{\mathrm{eq}}(\mathcal{O}_X)^{\otimes 4}\otimes\lambda(\mathcal{O}_{X/G})^{\otimes(-1)}\otimes\lambda(F)\otimes\lambda(J)^{\otimes 2}\otimes\lambda(J\otimes F)^{\otimes(-2)}\\
&\qquad\qquad\otimes\lambda(J^{\otimes 2})^{\otimes(-1)}\otimes\lambda(J^2\otimes F)\\
&\overset{\epsilon}{\cong}\lambda^{\mathrm{eq}}(\mathcal{O}_X)^{\otimes 4}\otimes\lambda((1-F)\otimes(1-J)\otimes(1-J))^{\otimes(-1)}\\
&\overset{\zeta}{\cong}\lambda^{\mathrm{eq}}(\mathcal{O}_X)^{\otimes 4}.
\end{aligned}$$

The isomorphism α comes from the adjunction formula, the exact sequence (10) and the identity (8). The isomorphism β is a consequence of the identities (9). The isomorphisms γ and δ come from the equivariant projection formula (Proposition 5.2) and the fact that equivariant derived functors are compatible with compositions of morphisms (see before Proposition 5.2). Isomorphism ϵ is just a reshuffling of terms, taking into account the commutativity of the tensor product. Isomorphism ζ comes from Theorem 4.2.

We now go back to the general situation. If X_G is not a Cartier divisor let $\widetilde{X}$ be the blow-up of X along X_G and let $b:\widetilde{X}\to X$ be the canonical morphism. Note that since S is affine, the scheme X carries an ample line bundle. In particular the morphism b is strongly projective. Also, the scheme $\widetilde{X}$ is flat over S by Lemma 6.2 (b) and it has geometrically regular fibres over S by Lemma 6.2 (a) and the fact that $X_G\to S$ is smooth. Thus $\widetilde{X}$ is smooth over S and this implies that b is lci. The scheme $\widetilde{X}$ is canonically G-equivariant since the sheaf of ideals of X_G is equivariant. The exceptional divisor E of $\widetilde{X}$ is isomorphic to the projectivised bundle $\mathbb{P}(N)$. Since G acts by multiplication by -1 on N, we see that the action of G is trivial on E. Hence $E=\widetilde{X}_G$ and $\widetilde{X}_G$ is a Cartier divisor, which is clearly smooth over S.

Let $\mu : \widetilde{X}_G \hookrightarrow \widetilde{X}$ and $p : \widetilde{X}_G \to X_G$ be the canonical morphisms. From equality (11), we obtain

$$\begin{aligned}
&\lambda^{\mathrm{eq}}(M)^{\otimes 2^{d+1}} \overset{(p)}{\cong} \lambda^{\mathrm{eq}}(b^*(M))^{\otimes 2^{d+1}} \\
&\overset{(o)}{\cong} \lambda^{\mathrm{eq}}(\mu^*(b^*(M)) \otimes P_d(\mathcal{O}_{\widetilde{X}_G} - N_{\widetilde{X}_G/\widetilde{X}})) \\
&\overset{(l)}{\cong} \lambda^{\mathrm{eq}}(\iota^*(M) \\
&\quad \otimes \mathrm{R}^\bullet p_*^{\mathrm{eq}}\Big(\mathcal{O}_{\widetilde{X}_G}^{\oplus 2^d} + 2^{d-1}(\mathcal{O}_{\widetilde{X}_G}^{\oplus 2} - (\mathcal{O}_{\widetilde{X}_G} - N_{\widetilde{X}_G/X})) \\
&\qquad + 2^{d-2}(\mathcal{O}_{\widetilde{X}_G}^{\oplus 2} - (\mathcal{O}_{\widetilde{X}_G} - N_{\widetilde{X}_G/\widetilde{X}}))^{\otimes 2} + \cdots + (\mathcal{O}_{\widetilde{X}_G}^{\oplus 2} - (\mathcal{O}_{\widetilde{X}_G} - N_{\widetilde{X}_G/\widetilde{X}}))^{\otimes d}\Big) \\
&\overset{(m)}{\cong} \lambda^{\mathrm{eq}}(\iota^*(M) \\
&\quad \otimes \mathrm{R}^\bullet p_*^{\mathrm{eq}}\Big(\mathcal{O}_{\widetilde{X}_G}^{\oplus 2^d} + 2^{d-1}(\mathcal{O}_{\widetilde{X}_G} + N_{\widetilde{X}_G/\widetilde{X}}) \\
&\qquad + 2^{d-2}(\mathcal{O}_{\widetilde{X}_G} + N_{\widetilde{X}_G/\widetilde{X}})^{\otimes 2} + \cdots + (\mathcal{O}_{\widetilde{X}_G} + N_{\widetilde{X}_G/\widetilde{X}})^{\otimes d}\Big) \\
&\overset{(n)}{\cong} \lambda^{\mathrm{eq}}(\iota^*(M) \otimes \mathrm{R}^\bullet p_*^{\mathrm{eq}}\Bigg(\sum_{i=0}^{d}\sum_{j=0}^{i} 2^{d-i}\binom{i}{j}(N_{\widetilde{X}_G/\widetilde{X}})^{\otimes j}\Bigg).
\end{aligned}$$

For equality (l), use the projection formula (Proposition 5.2) and the fact that the functors $\mathrm{R}f_*^{\mathrm{eq}} \cdot \mathrm{R}b_*^{\mathrm{eq}}$ and $\mathrm{R}(f \circ b)_*^{\mathrm{eq}}$ are naturally isomorphic (see discussion after Lemma 5.1). Equality (m) is a simple cancellation. Equality (n) follows from the equivariant analogue of Lemma 4.1 and from the polynomial identity

$$P_d(1-t) = \sum_{i=0}^{d}\sum_{j=0}^{i} 2^{d-i}\binom{i}{j} t^j,$$

which itself follows from the binomial formula. Equality (o) follows from (11). Equality (p) follows from the projection formula and the fact that $\mathrm{R}^\bullet b_*(\mathcal{O}_{\widetilde{X}}) = \mathcal{O}_X$ (see [13, VI, 4, proof of Prop. 4.1] for lack of a better reference).

Now since $\widetilde{X}_G = \mathbb{P}(N)$ we have

$$\mathrm{R}^\bullet p_*(N_{\widetilde{X}_G/\widetilde{X}}^{\otimes j}) \cong \mathrm{Sym}^j(N)$$

(see [19, Lemma 3.1]) and we obtain

$$\lambda^{\mathrm{eq}}(M)^{\otimes 2^{d+1}} \cong \lambda^{\mathrm{eq}}\Big(\iota^*(M) \otimes \sum_{i=0}^{d}\sum_{j=0}^{i} 2^{d-i}\binom{i}{j}\mathrm{Sym}^j(N)\Big).$$

Now note that we have the formal equality

$$\sum_{i=0}^{d}\sum_{j=0}^{i} 2^{d-i}\binom{i}{j}\mathrm{Sym}^j(N) = \sum_{j=0}^{d}\Big[\sum_{i=0}^{d-j} 2^{d-j-i}\binom{i+j}{j}\Big]\mathrm{Sym}^j(N).$$

To simplify this expression, we shall make use of the following combinatorial lemma, that was kindly communicated to us by E. Gomezllata Marmolejo.

Lemma 6.3 (E. Gomezllata Marmolejo). *For $0 \leq j \leq d$, we have*

$$\sum_{i=0}^{d-j} 2^{d-j-i}\binom{i+j}{j} = \sum_{i=0}^{d-j}\binom{d+1}{i}.$$

Proof of Lemma 6.3. The equality clearly holds if $d = j$. We prove it by induction on d, starting at $d = j$:

$$\begin{aligned}\sum_{i=0}^{d-j}\binom{d+1}{i} &= \binom{d}{0} + \sum_{i=1}^{d-j}\left[\binom{d}{i} + \binom{d}{i-1}\right] = 2\left[\sum_{i=0}^{d-j-1}\binom{d}{i}\right] + \binom{d}{d-j} \\ &= 2\left[\sum_{i=0}^{(d-1)-j}\binom{(d-1)+1}{i}\right] + \binom{d}{j} = 2\left[\sum_{i=0}^{(d-1)-j} 2^{(d-1)-j-i}\binom{i+j}{j}\right] + \binom{d}{j} \\ &= \sum_{i=0}^{d-j} 2^{d-j-i}\binom{i+j}{j}. \end{aligned} \tag{12}$$

The first and third equality in (12) follow from standard properties of binomial coefficients, the second and last one are just simplifications and the fourth one relies on the inductive hypothesis. □

Using Lemma 6.3, we finally get the advertised canonical isomorphism

$$\begin{aligned}\lambda^{\mathrm{eq}}(M)^{\otimes 2^{d+1}} &\cong \lambda^{\mathrm{eq}}\Big(\iota^*(M) \otimes \sum_{j=0}^{d}\Big[\sum_{i=0}^{d-j}\binom{d+1}{i}\Big]\mathrm{Sym}^j(N)\Big) \\ &\cong \bigotimes_{j=0}^{d} \lambda^{\mathrm{eq}}(\iota^*(M) \otimes \mathrm{Sym}^j(N))^{\otimes \sum_{i=0}^{d-j}\binom{d+1}{i}}.\end{aligned}$$

Note again that this isomorphism is invariant under any base change to a locally Noetherian scheme by Lemma 6.2 and by the fact that it is invariant under any base change to a locally Noetherian scheme when X_G is a Cartier divisor. □

7. Local refinement of the Adams–Riemann–Roch formula

We shall now prove Theorem 1.1. We recall the terminology. We let $\pi : Y \to S$ be a smooth and locally projective morphism of locally Noetherian schemes. We suppose that the fibres of π are geometrically connected and that π has constant relative dimension $d > 0$. We suppose that 2 is invertible on S. We want to prove that there is a canonical isomorphism

$$\lambda(L)^{\otimes 2^{2d+2}} \cong \bigotimes_{j=0}^{2d} \lambda(L^{\otimes 2} \otimes \mathrm{Sym}^j(\Omega_{Y/S}))^{\otimes (-1)^j \sum_{i=0}^{2d-j}\binom{2d+1}{i}}$$

(this is (1) in Theorem 1.1) which is invariant under any base change to a locally Noetherian scheme.

We shall write

$$X := Y \times_S Y$$

and we shall write $\pi_1 : X \to Y$ and $\pi_2 : X \to Y$ for the two projections. The group scheme $G = \mathbb{Z}/2\mathbb{Z}$ acts on X by swapping the coordinates, with fixed point scheme the relative diagonal Δ. The diagonal Δ is then regularly immersed.

Note that we used the fact that the fibres of π are smooth and geometrically connected here. If π is only supposed to be smooth, the diagonal Δ might not be regularly immersed. This can be seen on the example of a finite and étale morphism. In that case, the immersion of the diagonal is open and closed and thus Δ is not a Cartier divisor.

Let L be a line bundle on Y and suppose that L is cohomologically flat over S (see the beginning of the introduction for the definition of cohomological flatness). The line bundle $M = \pi_1^*(L) \otimes \pi_2^*(L)$ is naturally G-equivariant and $M|_\Delta \cong L^{\otimes 2}$ carries the trivial action. Furthermore $N_{\Delta/X} \cong \Omega_{Y/S}$ by definition.

Proposition 7.1. *We have a canonical isomorphism*

$$\lambda^{\mathrm{eq}}(M) \cong \lambda(L)^{\otimes 2}$$

where $\lambda^{\mathrm{eq}}(M)$ is computed using the above equivariant structure on M. This isomorphism is invariant under any base change to a locally Noetherian scheme.

Lemma 7.2. *Let W be a vector bundle on a locally Noetherian scheme T. Suppose that 2 is invertible on T. Endow $W \otimes W$ with the G-action which swaps the factors. There is a canonical isomorphism*

$$\det{}^{\mathrm{eq}}(W \otimes W) := \det((W \otimes W)_+) \otimes \det((W \otimes W)_-)^\vee \cong \det(W)^{\otimes 2} \tag{13}$$

which is compatible with any base change to a locally Noetherian scheme.

Proof. (of Lemma 7.2) Note that we have by definition

$$\mathrm{Sym}^2(W) := (W \otimes W)_+$$

and

$$\Lambda^2(W) := (W \otimes W)_-.$$

The identity (13) can be proven "by pure thought". We sketch the argument, leaving some of the details to the reader. Let $r := \mathrm{rk}(W)$. Recall that there is an additive and exact functor A from the additive category of the GL_r-comodules (i.e., representations of the group scheme GL_r), which are finitely generated and free $\mathbb{Z}$-modules, to the additive category of vector bundles over T. This functor can be described as follows. Choose an open covering (U_i) of S, such that $W|_{U_i} \cong \mathcal{O}_{U_i}^{\oplus r}$ for all indices i. This leads to transition functions $\tau_{ij} : U_i \cap U_j \to \mathrm{GL}_r(U_i \cap U_j)$. Now let $h \geq 0$ and choose a GL_r-comodule structure on $\mathbb{Z}^h$. This corresponds to a homomorphism of group schemes $\rho : \mathrm{GL}_r \to \mathrm{GL}_h$. We then define the vector bundle $A(W)$ as the vector bundle described by the transition functions $\rho(\tau_{ij})$. See [17, Exp. VI, after Th. 3.3] for more details on this. The functor A is compatible by construction with all the usual tensor constructions (tensor powers, exterior powers, etc.) and the construction of A is naturally compatible with any base change of W. Let now V be the standard representation of GL_r (so that $V = \mathbb{Z}^r$ as a $\mathbb{Z}$-module). Consider the two GL_r-comodules $\det(\Lambda^2(V))$ and $\det(V)$. These are both one-dimensional GL_r-representations. Since the one-dimensional GL_r-comodules are all of the form $(\det(V))^{\otimes n}$ for some $n \in \mathbb{Z}$ (see, e.g., [29, par. 3.8] for this), we see that there exists a uniquely determined integer m and an isomorphism of comodules

$$\det(\Lambda^2(V)) \cong \det(V)^{\otimes m}.$$

We fix one such isomorphism (it is actually fixed up to sign, since $\det(W)$ is a one-dimensional $\mathbb{Z}$-module). In view of the definition of the functor $A(\cdot)$, we see that this isomorphism induces an isomorphism of vector bundles

$$\det(\Lambda^2(W)) \cong \det(W)^{\otimes m}.$$

To compute m, it is sufficient to find a locally Noetherian scheme Z and a vector bundle J of rank r on Z, such that $\det(\Lambda^2(J))$ is isomorphic to at most one tensor power of $\det(J)$. The scheme $\mathbb{P}^1$ has this property, since $\mathrm{Pic}(\mathbb{P}^1)) = \mathbb{Z}$, provided $\det(J) \not\cong \mathcal{O}_{\mathbb{P}^1}$. So supposing that $Z = \mathbb{P}^1$ and $J = \mathcal{O}(1)^{\oplus r}$, we compute

$$\det(\Lambda^2(J)) \cong \det\left(\bigoplus_{1\leq i<j\leq r} \mathcal{O}(1)\otimes\mathcal{O}(1)\right) \cong \mathcal{O}\left(2\sum_{1\leq i<j\leq r} 1\right) = \mathcal{O}\left(2\binom{r}{2}\right).$$

We can repeat this reasoning for $\mathrm{Sym}^2(\cdot)$ in place of $\Lambda^2(\cdot)$ and we obtain

$$\det(\mathrm{Sym}^2(J)) = \mathcal{O}(2\binom{r+1}{r-1}).$$

We conclude that for any T and W, we have

$$\det(\Lambda^2(W)) \cong \det(W)^{\otimes\frac{(r-1)!}{(r-2)!}} \cong \det(W)^{\otimes(r-1)}$$

and

$$\det(\mathrm{Sym}^2(W)) \cong \det(W)^{\otimes\frac{(r+1)!}{r(r-1)!}} \cong \det(W)^{\otimes(r+1)}$$

and the lemma follows from these two equations. □

Lemma 7.3. *Let W be a vector bundle on a locally Noetherian scheme T. Suppose that 2 is invertible on T. Let G be the action of G on $W \oplus W$, which swaps the summands. Then there is a canonical isomorphism*

$$\det{}^{\mathrm{eq}}(W\oplus W) := \det((W\oplus W)_+)\otimes\det((W\oplus W)_-)^\vee \cong \mathcal{O}_T, \tag{14}$$

which is compatible with any base change to a locally Noetherian scheme.

Proof of Lemma 7.3. Note that the diagonal morphism of sheaves $W \to W\oplus W$ identifies $(W\oplus W)_+$ with W. Similarly, the antidiagonal morphism $W \to W\oplus W$ (given by the formula $w \mapsto (w, -w)$) identifies $(W\oplus W)_-$ with W. The lemma follows from this. □

Proof of Proposition 7.1. Let $f : X \to S$ be the canonical morphism. By the Künneth formula (see [14, III, par. 6, Th. 6.7.3]), we have a canonical isomorphism

$$\mathrm{R}^i f_*(M) \cong \bigoplus_t \mathrm{R}^t\pi_*(L) \otimes \mathrm{R}^{i-t}\pi_*(L). \tag{15}$$

Note that we used the fact that L is cohomologically flat here. The vector bundle

$$\bigoplus_t \mathrm{R}^t\pi_*(L) \otimes \mathrm{R}^{i-t}\pi_*(L)$$

carries a natural G-action by permutation, namely the action such that the non-trivial element of G sends $\bigoplus_t w_t \otimes w_{i-t}$ to $\bigoplus_t (-1)^{t(i-t)} w_{i-t} \otimes w_t$. By the Koszul rule of signs, the isomorphism (15) becomes G-equivariant with this choice of G-action on the righthand side. Let $\mathrm{sgn} : G \to \{0, 1\}$ be the non-trivial character of G. Let us first suppose that i is odd. We compute

$$\begin{aligned}&\det{}^{\mathrm{eq}}(\mathrm{R}^i f_*(M))\\ &\cong \bigotimes_{0\leq t\leq \lfloor i/2\rfloor} \det{}^{\mathrm{eq}}(\mathrm{R}^t\pi_*(L) \otimes \mathrm{R}^{i-t}\pi_*(L) \oplus \mathrm{R}^{i-t}\pi_*(L) \otimes \mathrm{R}^t\pi_*(L)).\end{aligned} \tag{16}$$

In the right-hand side of the isomorphism (16), the terms

$$\mathrm{R}^t\pi_*(L) \otimes \mathrm{R}^{i-t}\pi_*(L) \oplus \mathrm{R}^{i-t}\pi_*(L) \otimes \mathrm{R}^t\pi_*(L)$$

carry a G-equivariant structure of the form considered in Lemma 7.3. We thus see that we have a canonical isomorphism

$$\det{}^{\mathrm{eq}}(\mathrm{R}^i f_*(M)) \cong \mathcal{O}_S\,.$$

Now suppose that i is even. We then have

$$\begin{aligned}\det{}^{\mathrm{eq}}(\mathrm{R}^i f_*(M)) &\cong \det{}^{\mathrm{eq}}(\mathrm{R}^{i/2}\pi_*(L) \otimes \mathrm{R}^{i/2}\pi_*(L))\\ &\otimes \bigotimes_{0\leq t< i/2} \det{}^{\mathrm{eq}}(\mathrm{R}^t\pi_*(L) \otimes \mathrm{R}^{i-t}\pi_*(L) \oplus \mathrm{R}^{i-t}\pi_*(L) \otimes \mathrm{R}^t\pi_*(L)).\end{aligned}$$

Here the summands $\mathrm{R}^t\pi_*(L) \otimes \mathrm{R}^{i-t}\pi_*(L) \oplus \mathrm{R}^{i-t}\pi_*(L) \otimes \mathrm{R}^t\pi_*(L)$ carry a G-equivariant structure of the type considered in Lemma 7.3 multiplied by sgn^t and the summand $\mathrm{R}^{i/2}\pi_*(L) \otimes \mathrm{R}^{i/2}\pi_*(L)$ carries the equivariant structure considered in Lemma 7.2 multiplied by $\mathrm{sgn}^{i/2}$. As before, we conclude that

$$\det{}^{\mathrm{eq}}(\mathrm{R}^i f_*(M)) \cong \det{}^{\mathrm{eq}}(\mathrm{R}^{i/2}\pi_*(L) \otimes \mathrm{R}^{i/2}\pi_*(L)).$$

On the other hand, by Lemma 7.2, we have

$$\det{}^{\mathrm{eq}}(\mathrm{R}^{i/2}\pi_*(L) \otimes \mathrm{R}^{i/2}\pi_*(L)) \cong \det(\mathrm{R}^{i/2}\pi_*(L))^{\otimes 2(-1)^{i/2}}\,.$$

Summarising, we have

$$\det{}^{\mathrm{eq}}(\mathrm{R}^i f_*(M)) \cong \det(\mathrm{R}^{i/2}\pi_*(L))^{\otimes 2(-1)^{i/2}}$$

if i is even and

$$\det{}^{\mathrm{eq}}(\mathrm{R}^i f_*(M)) \cong \mathcal{O}_S$$

if i is odd. We conclude that we have

$$\lambda^{\mathrm{eq}}(M) = \bigotimes_{i\geq 0} \det{}^{\mathrm{eq}}(\mathrm{R}^i f_*(M))^{\otimes(-1)^i} \cong \bigotimes_{i\geq 0,\, i \text{ even}} \det{}^{\mathrm{eq}}(\mathrm{R}^i f_*(M))$$
$$\cong \bigotimes_{j\geq 0} \det(\mathrm{R}^j \pi_*(L))^{\otimes 2(-1)^j} = \lambda(L)^{\otimes 2}$$

which is what we wanted to prove. □

Remark 7.4. Lemma 7.1 is the only place in the proof of Theorem 1.1 where we use the assumption that L is cohomologically flat over S. We conjecture that Lemma 7.1 holds without that assumption. If this is true then Theorem 1.1 holds without the assumption that L is cohomologically flat over S. If one tries to prove Lemma 7.1 without the assumption of cohomological flatness, one is faced with a difficult problem in the linear algebra of perfect complexes that to date we have not been able to solve. See also [28] about this.

Finally, combining Proposition 7.1 and Theorem 6.1 we get an isomorphism

$$\lambda(L)^{\otimes 2^{2d+2}} \cong \bigotimes_{j=0}^{2d} \lambda(L^{\otimes 2} \otimes \mathrm{Sym}^j(\Omega_{Y/S}))^{\otimes(-1)^j \sum_{i=0}^{2d-j} \binom{2d+1}{i}}. \tag{17}$$

and this completes the proof of Theorem 1.1.

8. Proof of Proposition 1.3

We work with the assumptions and terminology of Theorem 1.1 and we suppose that $d = 1$. Consider the formal linear combinations of line bundles

$$\mathrm{MT}(L) := 7L^{\otimes 2} - 4\Omega_{Y/S} \otimes L^{\otimes 2} + L^{\otimes 2} \otimes \Omega_{Y/S}^{\otimes 2}$$

and

$$\mathrm{DT}(L) := 18 + 6L^{\otimes 2} \otimes \Omega_{Y/S}^{\vee} - 6L \otimes \Omega_{Y/S}^{\vee}\,.$$

Theorem 1.1 for $\dim(Y/S) = 1$ says that we have a canonical isomorphism

$$\lambda(\mathrm{MT}(L)) \cong \lambda(L)^{\otimes 16}.$$

Similarly, Deligne's theorem (4) implies that there is a canonical isomorphism

$$\lambda(\mathrm{DT}(L)) \cong \lambda(L)^{\otimes 18}.$$

We shall prove that the line bundle $\lambda\Big(9\mathrm{MT}(L) - 8\mathrm{DT}(L)\Big)$ is canonically trivial, even without the assumption that L is cohomologically flat over S. Assuming Theorem 1.1 for $\dim(Y/S) = 1$, this will prove that $\lambda(8\mathrm{DT}(L))$ is canonically trivial, which is the conclusion of Proposition 1.3 (note that $9 \cdot 16 = 8 \cdot 18 = 144$).

Now since L is arbitrary, it is sufficient to prove that

$$\lambda\Big(9\mathrm{MT}(L \otimes \Omega_{Y/S}) - 8\mathrm{DT}(L \otimes \Omega_{Y/S})\Big)$$

is canonically trivial.

We first compute

$$\begin{aligned}&9\mathrm{MT}(L\otimes\Omega_{Y/S})-8\mathrm{DT}(L\otimes\Omega_{Y/S})\\&\quad=(9\Omega_{Y/S}^{\otimes 4}-36\Omega_{Y/S}^{\otimes 3}+63\Omega_{Y/S}^{\otimes 2}-48\Omega_{Y/S})\otimes L^{\otimes 2}+48L-144.\end{aligned}$$

Let

$$P(x,y):=(9y^4-36y^3+63y^2-48y)x^2+48x-144\in\mathbb{Z}[x,y].$$

We compute

$$\begin{aligned}P(x,y)&=P(1-(1-x),1-(1-y))\\&=(9(1-y)^4+9(1-y)^2-6(1-y)-12)(1-x)^2\\&\quad+(-18(1-y)^4-18(1-y)^2+12(1-y)-24)(1-x)\\&\quad+(9(1-y)^4+9(1-y)^2-6(1-y)-108)\\&=-12(1-x)^2+(12(1-y)-24)(1-x)+(9(1-y)^2-6(1-y)-108)\\&\qquad\mathrm{mod}\,((1-y)^3,(1-y)(1-x)^2,(1-y)^2(1-x),(1-x)^3)\end{aligned}$$

(where $((1-y)^3,(1-y)(1-x)^2,(1-y)^2(1-x),(1-x)^3)$ refers to the ideal of $\mathbb{Z}[x,y]$ generated by $(1-y)^3$, $(1-y)(1-x)^2$, $(1-y)^2(1-x)$ and $(1-x)^3$).

We deduce from this identity, Lemma 4.1 and Corollary 4.3 that we have a canonical isomorphism

$$\begin{aligned}&\lambda\Big(9\mathrm{MT}(L\otimes\Omega_{Y/S})-8\mathrm{DT}(L\otimes\Omega_{Y/S})\Big)\\&\quad\cong\lambda\Big(-12(1-L)^2+(12(1-\Omega_{Y/S})-24)(1-L)\\&\qquad+(9(1-\Omega_{Y/S})^2-6(1-\Omega_{Y/S})-108)\Big)\\&\quad\cong\lambda\Big(-12L^{\otimes 2}+(12\Omega_{Y/S}+36)\otimes L+(9\Omega_{Y/S}^{\otimes 2}-24\Omega_{Y/S}-129)\Big).\end{aligned}$$

Note that by Grothendieck duality, we have a canonical isomorphism

$$\lambda(\Omega_{Y/S}\otimes L)\cong\lambda(L^{\vee}).$$

We deduce that we have

$$\begin{aligned}&\lambda\Big(9\mathrm{MT}(L\otimes\Omega_{Y/S})-8\mathrm{DT}(L\otimes\Omega_{Y/S})\Big)\\&\quad\cong\lambda\Big(-12L^{\otimes 2}+12L^{\vee}+36L+9\Omega_{Y/S}^{\otimes 2}-24\Omega_{Y/S}-129\Big).\end{aligned}\tag{18}$$

Now by Corollary 4.3, we have a canonical trivialisation

$$\lambda(L^{\vee}\otimes(1-L)^{\otimes 3})\cong\mathcal{O}_S$$

or in other words a canonical isomorphism

$$\lambda(L^{\vee})\cong\lambda(L^{\otimes 2}-3L+3).$$

Merging this with (18), we obtain a canonical isomorphism

$$\begin{aligned}\lambda\Big(9\mathrm{MT}(L)-8\mathrm{DT}(L)\Big) &\cong \lambda(9\Omega_{Y/S}^{\otimes 2}-117)\\ &\cong (\lambda(\Omega_{Y/S}^{\otimes 2})\otimes\lambda(\mathcal{O}_X)^{\otimes -13})^{\otimes 9}.\end{aligned} \tag{19}$$

To conclude, notice that by (3), we have canonically

$$\lambda(\Omega_{Y/S}^{\otimes 2})\cong\lambda(\mathcal{O}_X)^{\otimes 13}.$$

References

[1] Indranil Biswas, Georg Schumacher, and Lin Weng, *Deligne pairing and determinant bundle*, Electron. Res. Announc. Math. Sci. **18** (2011), 91–96, DOI 10.3934/era.2011.18.91.

[2] Sébastien Boucksom and Dennis Eriksson, *Spaces of norms, determinant of cohomology and Fekete points in non-archimedean geometry.* ArXiv 1805.01016.

[3] P. Deligne, *Le déterminant de la cohomologie*, Current trends in arithmetical algebraic geometry (Arcata, Calif., 1985), Contemp. Math., vol. 67, Amer. Math. Soc., Providence, RI, 1987, pp. 93–177, DOI 10.1090/conm/067/902592 (French).

[4] ———, *Théorème de Lefschetz et critères de dégénérescence de suites spectrales*, Inst. Hautes Études Sci. Publ. Math. **35** (1968), 259–278 (French).

[5] Paolo Dolce, *Explicit Deligne pairing.* arXiv:1911.05367.

[6] J.-M. Drezet and M.S. Narasimhan, *Groupe de Picard des variétés de modules de fibrés semi-stables sur les courbes algébriques*, Invent. Math. **97** (1989), no. 1, 53–94 (French).

[7] François Ducrot, *Cube structures and intersection bundles*, J. Pure Appl. Algebra **195** (2005), no. 1, 33–73, DOI 10.1016/j.jpaa.2004.06.002.

[8] R. Elkik, *Fibrés d'intersections et intégrales de classes de Chern*, Ann. Sci. École Norm. Sup. (4) **22** (1989), no. 2, 195–226 (French).

[9] Dennis Eriksson, *Un isomorphisme de type Deligne–Riemann–Roch*, C. R. Math. Acad. Sci. Paris **347** (2009), no. 19-20, 1115–1118, DOI 10.1016/j.crma.2009.09.003 (French, with English and French summaries).

[10] ———, *Un isomorphisme de Deligne–Riemann–Roch.* Thesis, Université Paris 6, 2008, see `https://www.theses.fr/2008PA112190`.

[11] Jens Franke, *Riemann–Roch in functorial form.* Preprint, IAS, early nineties.

[12] Eberhard Freitag and Reinhardt Kiehl, *Étale cohomology and the Weil conjecture*, Ergebnisse der Mathematik und ihrer Grenzgebiete (3) [Results in Mathematics and Related Areas (3)], vol. 13, Springer-Verlag, Berlin, 1988. Translated from the German by Betty S. Waterhouse and William C. Waterhouse; With an historical introduction by J.A. Dieudonné.

[13] William Fulton and Serge Lang, *Riemann–Roch algebra*, Grundlehren der Mathematischen Wissenschaften [Fundamental Principles of Mathematical Sciences], vol. 277, Springer-Verlag, New York, 1985.

[14] A. Grothendieck and J. Dieudonné, *Éléments de géométrie algébrique.* Inst. Hautes Études Sci. Publ. Math. **4, 8, 11, 17, 20, 24, 28, 32** (1960–1967).

[15] *Revêtements étales et groupe fondamental* (*SGA* 1), Documents Mathématiques (Paris) [Mathematical Documents (Paris)], vol. 3, Société Mathématique de France, Paris, 2003 (French). Séminaire de géométrie algébrique du Bois Marie 1960–61. [Algebraic Geometry Seminar of Bois Marie 1960–61]; Directed by A. Grothendieck; With two papers by M. Raynaud; Updated and annotated reprint of the 1971 original [Lecture Notes in Math., 224, Springer, Berlin.

[16] Philippe Gille and Patrick Polo (eds.), *Schémas en groupes* (*SGA* 3). *Tome III. Structure des schémas en groupes réductifs*, Documents Mathématiques (Paris) [Mathematical Documents (Paris)], vol. 8, Société Mathématique de France, Paris, 2011 (French). Séminaire de Géométrie Algébrique du Bois Marie 1962–64. [Algebraic Geometry Seminar of Bois Marie 1962–64]; A seminar directed by M. Demazure and A. Grothendieck with the collaboration of M. Artin, J.-E. Bertin, P. Gabriel, M. Raynaud and J.-P. Serre; Revised and annotated edition of the 1970 French original.

[17] *Théorie des intersections et théorème de Riemann–Roch*, Lecture Notes in Mathematics, Vol. 225, Springer-Verlag, Berlin-New York, 1971 (French). Séminaire de Géométrie Algébrique du Bois-Marie 1966–1967 (SGA 6); Dirigé par P. Berthelot, A. Grothendieck et L. Illusie. Avec la collaboration de D. Ferrand, J.P. Jouanolou, O. Jussila, S. Kleiman, M. Raynaud et J.-P. Serre.

[18] Robin Hartshorne, *Residues and duality*, Lecture notes of a seminar on the work of A. Grothendieck, given at Harvard 1963/64. With an appendix by P. Deligne. Lecture Notes in Mathematics, No. 20, Springer-Verlag, Berlin-New York, 1966.

[19] ______, *Ample vector bundles*, Inst. Hautes Études Sci. Publ. Math. **29** (1966), 63–94.

[20] Olivier Haution, *Involutions and Chern numbers of varieties*. arXiv:1903.07304.

[21] ______, *Involutions of varieties and Rost's degree formula*. arXiv:1403.3604.

[22] Finn Faye Knudsen and David Mumford, *The projectivity of the moduli space of stable curves. I. Preliminaries on "det" and "Div"*, Math. Scand. **39** (1976), no. 1, 19–55, DOI 10.7146/math.scand.a-11642.

[23] Bernhard Köck, *Computing the homology of Koszul complexes*, Trans. Amer. Math. Soc. **353** (2001), no. 8, 3115–3147, DOI 10.1090/S0002-9947-01-02723-4.

[24] Serge Lang, *Algebra*, 3rd ed., Graduate Texts in Mathematics, vol. 211, Springer-Verlag, New York, 2002.

[25] Hideyuki Matsumura, *Commutative ring theory*, 2nd ed., Cambridge Studies in Advanced Mathematics, vol. 8, Cambridge University Press, Cambridge, 1989. Translated from the Japanese by M. Reid.

[26] David Mumford, *Stability of projective varieties*, Enseignement Math. (2) **23** (1977), no. 1-2, 39–110.

[27] Madhav V. Nori, *The Hirzebruch–Riemann–Roch theorem*, Michigan Math. J. **48** (2000), 473–482, DOI 10.1307/mmj/1030132729. Dedicated to William Fulton on the occasion of his 60th birthday.

[28] Damian Rössler, *Determinantal identities for perfect complexes*. Question asked on MathOverflow. See `https://mathoverflow.net/questions/354214/determinantal-identities-for-perfect-complexes`.

[29] Jean-Pierre Serre, *Groupes de Grothendieck des schémas en groupes réductifs déployés*, Inst. Hautes Études Sci. Publ. Math. **34** (1968), 37–52 (French).

[30] R.W. Thomason, *Une formule de Lefschetz en K-théorie équivariante algébrique*, Duke Math. J. **68** (1992), no. 3, 447–462, DOI 10.1215/S0012-7094-92-06817-7 (French).

[31] Shouwu Zhang, *Heights and reductions of semi-stable varieties*, Compositio Math. **104** (1996), no. 1, 77–105.

Damian Rössler
Mathematical Institute
University of Oxford
Andrew Wiles Building
Radcliffe Observatory Quarter
Woodstock Road
Oxford OX2 6GG, United Kingdom

Progress in Mathematics, Vol. 338, 247–299

Analytic Torsion and Dynamical Flow: A Survey on the Fried Conjecture

Shu Shen

Abstract. Given an acyclic and unitarily flat vector bundle on a closed manifold, Fried conjectured an equality between the analytic torsion and the value at zero of the Ruelle zeta function associated to a dynamical flow. In this survey, we review the Fried conjecture for different flows, including the suspension flow, the Morse–Smale flow, the geodesic flow, and the Anosov flow.

Mathematics Subject Classification (2010). 58J20, 58J52, 11F72, 11M36, 37C30.

Keywords. Index theory and related fixed point theorems, analytic torsion, Selberg trace formula, Ruelle dynamical zeta function.

Contents

The author is indebted to Xiaolong Han, Jianqing Yu, and referees for reading this paper very carefully.

Introduction

The purpose of this survey is to review some recent progress of the Fried conjecture, which affirms an equality between the analytic torsion and the value at zero of the Ruelle dynamical zeta function.

In the first two sections of the paper, we describe the Ray–Singer analytic torsion of a flat vector bundle and the Ruelle zeta function of a dynamical flow. The next three sections are devoted to the study on the Fried conjecture for certain flows, including the suspension flow, the Morse–Smale flow, the geodesic flow, and the Anosov flow.

The paper is written in an informal way. We only sketch the proofs, and refer to the original papers when necessary.

We now describe in more details the content of this paper, and give the proper historical perspective to the results described in the paper.

0.1. The combinatorial and analytic torsions

Let Z be a smooth closed manifold. Let F be a complex unitarily flat vector bundle on Z, which amounts to specifying a unitary finite-dimensional representation ρ of the fundamental group $\pi_1(Z)$. Denote by $H^\cdot(Z, F)$ the cohomology of the sheaf of locally constant sections of F. Assume that F is acyclic, i.e., $H^\cdot(Z, F) = 0$.

The Reidemeister combinatorial torsion [Re35, Fr35, dR50] is a positive real number defined with the help of a triangulation on Z. However, it does not depend on the triangulation and becomes a topological invariant. It is the first invariant that can distinguish closed manifolds such as lens spaces which are homotopy equivalent but not homeomorphic.

The analytic torsion was introduced by Ray and Singer [RS71] as an analytic counterpart of the Reidemeister torsion. In order to define the analytic torsion one has to choose a Riemannian metric on Z and a Hermitian metric on F. The analytic torsion is a certain weighted alternating product of regularized determinants of the Hodge Laplacians acting on the space of differential forms with values in F.

The celebrated Cheeger–Müller Theorem [C79, M78] tells us that the Ray–Singer analytic torsion coincides with the Reidemeister combinatorial torsion. Bismut–Zhang [BZ92] and Müller [M93] simultaneously considered generalizations of this result. Müller [M93] extended this result to the case where F is unimodular, i.e., $|\det \rho(\gamma)| = 1$ for all $\gamma \in \pi_1(Z)$. Bismut and Zhang [BZ92, Theorem 0.2] generalised the original Cheeger–Müller Theorem to arbitrary flat vector bundles with arbitrary Hermitian metrics. There are also various extensions to the equivariant case by Lott–Rothenberg [LoRo91], Lück [Lü93], and Bismut–Zhang [BZ94], to the family case by Bismut–Goette [BG01] under the assumption of the existence of the fibrewise Morse function, and to manifolds with boundaries by Brüning–Ma [BrMa13].

0.2. Dynamical zeta function

The grandmother of all zeta functions is the Riemann zeta function defined for $\mathrm{Re}\,(s) > 1$ by

$$\zeta(s) = \sum_{n=1}^{\infty} \frac{1}{n^s}. \tag{0.1}$$

Riemann showed that $\zeta(s)$ extends meromorphically to $\mathbf{C}$ with a single pole at $s = 1$ and that there is a functional equation relating $\zeta(s)$ and $\zeta(1-s)$. The Euler product formula asserts that for $\mathrm{Re}\,(s) > 1$,

$$\zeta(s) = \prod_{p:\text{prime}} \left(1 - p^{-s}\right)^{-1}. \tag{0.2}$$

It tells us that $\zeta(s)$ encodes the distribution of the prime numbers. Some statistical properties, like the prime number theorem, can be deduced from the information on the poles and zeros of $\zeta(s)$.

After Riemann's work, several zeta functions with similar properties have been introduced. In particular, Weil [W49] constructed a zeta function using the Frobenius map T defined on an algebraic variety Z over a finite field. It counts the closed orbits of the discrete dynamical system (Z, T). To a smooth closed manifold with a diffeomorphism, similar zeta function was also introduced by Artin and Mazur [ArMaz65].

To a flat vector bundle on the underlying manifold, we can associate a weight to the dynamical systems. In [Ru76a], Ruelle introduced his zeta function for a weighted dynamical system of a diffeomorphism or a flow.

For the geodesic flow on the unit tangent bundle of a Riemann surface of genus $\geqslant 2$, an application of the Selberg trace formula [Sel56] shows that the Ruelle dynamical zeta function has a meromorphic extension to $\mathbf{C}$. Similar methods can be generalized to hyperbolic manifolds by Fried [F86a], to the locally symmetric space of rank 1 by Bunke and Olbrich [BuO95], to the locally symmetric space of higher ranks by Moscovich–Stanton [MoSt91], Shen [Sh18], and Shen–Yu [ShY17].

Independent of the above Selberg theory, which is based on the spectral theory of the Laplacian, a thermodynamical formalism, which is based on the spectral theory of the transfer operator, is used to study the Anosov flow (or more generally Axiom A flow[1]) by Ruelle [Ru76b]. He showed that if the flow itself and the stable and unstable foliations are all analytic, then his zeta function has a meromorphic extension to **C**. Fried [F86c] generalized Ruelle's result by requiring only the flow and the stable foliation to be analytic. An important extension of the above results was given by Rugh [Rug96], for three-dimensional manifolds, and then by Fried [F95], in arbitrary dimensions, but still assuming the analyticity of the flow. The extension of such results to the C^∞ setting was only given very recently by Giulietti, Liverani, and Pollicott [GiLiPo13] and by Dyatlov and Zworski [DyZ16] (See also [FaT17, DyGu16, DyGu18, BWSh20] for related works). This recent progress on the dynamical zeta function is based on the introduction of the anisotropic space. We refer the reader to the book of Baladi [Ba18] for an introduction of these techniques.

0.3. The Fried conjecture

It was Milnor [Mi68a] who observed a remarkable similarity between the Reidemeiter torsion and the Weil zeta function. A precise and quantitative description of their relation was obtained by Fried [F86a, F86b] for the geodesic flow on the unit tangent bundle of a hyperbolic manifold. He showed that the analytic/combinatorial torsion of an acyclic unitarily flat vector bundle on a unit tangent bundle of a closed oriented hyperbolic manifold is equal to the value at zero of the Ruelle dynamical zeta function[2] associated to the geodesic flow. He conjectured later [F87, F95] that similar results hold true for more general flows.

Four kinds of flows will be examined in this survey. As a warm up, we will begin with two simple flows: the suspension flow and Morse–Smale flow. As we will see, the Fried conjecture for the suspension flow is just the Lefschetz fixed point formula, and the Fried conjecture for the Morse–Smale flow is a consequence of the Cheeger–Müller/Bismut–Zhang theorem (see [F87, Theorem 3.1] and [ShY18, Theorem 0.2]). Next, we will consider the geodesic flow on the unit tangent bundle of the locally symmetric spaces. In this case, the Fried conjecture can be deduced formally via the V-invariant of Bismut–Goette [BG04]. Following previous contributions by Moscovici–Stanton [MoSt91], a rigorous proof is given by the author [Sh18] using Bismut's orbital integral formula [B11]. In [ShY17], Shen and Yu made a further generalization to closed locally symmetric orbifolds. Finally, we will study the Anosov flow. If the underlying manifold has dimension 3, it was known by Sánchez-Morgado [SM93, SM96a] that the Fried conjecture holds true

---

[1] In this case, the meromorphic extension problem is called the Smale conjecture [Sm67].

[2] Here the Ruelle dynamical zeta function should be twisted by the holonomy of the flat vector bundle.

if the flow is transitive and analytic, and if the flat vector bundle satisfies certain holonomy conditions. Using a variation formula, Dang, Guillarmou, Rivière, and Shen [DaGuRiSh20] removed the analyticity assumption in Sánchez-Morgado's result. For general Anosov flow, e.g., the geodesic flow on the unit tangent bundle of a negatively curved manifold, the Fried conjecture is still open. The only known result is that under certain spectral condition on the transfer operator, in [DaGuRiSh20], the authors have shown that the value at zero of the Ruelle dynamical zeta function does not depend on a small perturbation of the Anosov flow.

Due to the limitation of the length of this survey, there are several interesting topics which have not been included. For example, the Fried conjecture for hyperbolic manifolds with cusps [Pa09] or for non-unitarily flat vector bundles [Wo08, M12, M20, Sh20a, Sh20b, Sp20], the relation between the dynamical zeta function with the eta invariant [Mil78, MoSt89, B19] and with the holomorphic torsion [F88, MoSt18]. We refer the reader to the cited references for more information.

0.4. Organization of the paper

This paper is organized as follows. In Section 1, we give the main properties of the Ray–Singer analytic torsion.

In Section 2, we introduce the dynamical zeta function and state the Fried conjecture.

In Sections 3–6, we discuss the Fried conjecture for the suspension flow, the Morse–Smale flow, the geodesic flow, and the Anosov flow.

Throughout the paper, we use the supersymmetric convention. For a matrix A acting on a $\mathbf{Z}$-graded space $E^\cdot$, the supertrace of A is defined by

$$\mathrm{Tr}_{\mathrm{s}}^{E^\cdot}[A] = \mathrm{Tr}^{E^\cdot}[(-1)^N A], \tag{0.3}$$

where N is the number operator on $E^\cdot$ which sends $e \in E^q$ to $qe \in E^q$. We use the notation

$$\mathbf{R}_+ = [0,\infty), \quad \mathbf{R}_+^* = (0,\infty), \quad \mathbf{N} = \{0,1,2,\ldots\}, \quad \mathbf{N}^* = \{1,2,3,\ldots\}. \tag{0.4}$$

For a finite set A, we denote by $|A|$ the cardinality of A. By a closed orbit we mean a non-trivial closed orbit, i.e., its period is positive.

1. The Ray–Singer analytic torsion

The purpose of this section is to recall the definition of the Ray–Singer analytic torsion [RS71] of a Hermitian flat vector bundle on a closed Riemannian manifold.

This section is organized as follows. In Section 1.1, we introduce the flat vector bundle.

In Section 1.2, we recall the definition of the Ray–Singer analytic torsion.

In Section 1.3, we consider a fibration $M \to Z$ of closed Riemannian manifolds. We give a formula relating the analytic torsions of a Hermitian flat vector bundle on M and of its direct image on Z, which is equipped with an L^2 Hermitian metric.

1.1. The flat vector bundle

Let Z be a smooth closed manifold. Let F be a complex flat vector bundle on Z with flat connection ∇^F. Equivalently, F can be obtained via a finite-dimensional complex representation[3] $\rho : \pi_1(Z) \to \mathrm{GL}_r(\mathbf{C})$ of the fundamental group $\pi_1(Z)$ of Z. The flat vector bundle F is called unitarily flat if ρ is unitary.

Let $\left(\Omega^\cdot(Z,F), d^Z\right)$ be the de Rham complex of smooth sections of $\Lambda^\cdot(T^*Z)\otimes_{\mathbf{R}} F$ on Z. Let $H^\cdot(Z,F)$ be the cohomology of the above complex. We say that F is acyclic if $H^\cdot(Z,F)=0$. The Euler characteristic number of F is then given by

$$\chi(Z,F) = \sum_{q=0}^{\dim Z} (-1)^q \dim H^q(Z,F). \tag{1.1}$$

When F is the real trivial line bundle $\mathbf{R}$, we use the notation $(\Omega^\cdot(Z), d^Z), H^\cdot(Z)$ and $\chi(Z)$.

Example 1.1. Take $Z = \mathbb{S}^1 = \mathbf{R}/\mathbf{Z}$. Set $\alpha \in \mathbf{C}$. Let F be the trivial complex line bundle $\mathbf{C}$ equipped with the connection $\nabla^F = d + \alpha dt$. The holonomy ρ is then given by $\rho : n \in \mathbf{Z} \to e^{n\alpha} \in \mathbf{C}^*$. It is easy to see that $H^\cdot(Z,F)=0$ if and only if $\alpha \notin 2i\pi\mathbf{Z}$, and that F is unitarily flat if and only if $\alpha \in i\mathbf{R}$.

Example 1.2. Let $0 \to F_0 \to F_1 \to F_2 \to 0$ be an exact sequence of flat vector bundles on Z. Using the associated long exact sequence

$$\cdots \to H^p(Z,F_0) \to H^p(Z,F_1) \to H^p(Z,F_2) \to \cdots, \tag{1.2}$$

we see that if two of F_i are acyclic, then the third one is also acyclic.

1.2. Hodge Laplacian and analytic torsion

Let g^{TZ} be a Riemannian metric on TZ, and let g^F be a Hermitian metric on F. For $s_1, s_2 \in \Omega^\cdot(Z,F)$, put

$$\langle s_1, s_2\rangle_{\Omega^\cdot(Z,F)} = \int_{z\in Z} \langle s_1(z), s_2(z)\rangle_{\Lambda^\cdot(T^*Z)\otimes_{\mathbf{R}}F} dv_Z, \tag{1.3}$$

where $\langle\cdot,\cdot\rangle_{\Lambda^\cdot(T^*Z)\otimes_{\mathbf{R}}F}$ is the metric on $\Lambda^\cdot(T^*Z)\otimes_{\mathbf{R}} F$ induced by g^{TZ}, g^F, and dv_Z is the Riemannian volume of (Z, g^{TZ}). Then, $\langle\cdot,\cdot\rangle_{\Omega^\cdot(Z,F)}$ defines an L^2-metric on $\Omega^\cdot(Z,F)$.

Let $d^{Z,*}$ be the formal adjoint of d^Z with respect to $\langle\,,\rangle_{\Omega^\cdot(Z,F)}$. Put

$$\Box^Z = d^Z d^{Z,*} + d^{Z,*} d^Z. \tag{1.4}$$

Then, $\Box^Z$ acts on $\Omega^\cdot(Z,F)$ and preserves its degree. It is a formally self-adjoint non-negative second-order elliptic differential operator. Since Z is compact, $\Box^Z$ has a unique self-adjoint extension, which we still denote by $\Box^Z$.

[3] We use the convention that $F = \pi_1(Z)\backslash(X\times\mathbf{C}^r)$, where $\pi_1(Z)$ acts on the left on the universal covering X of Z by the deck transformation, and acts on the left on $\mathbf{C}^r$ by ρ.

Take $0 \leqslant q \leqslant \dim Z$. Let $\square_q^Z$ be the restriction of $\square^Z$ on $\Omega^q(Z,F)$. By the Hodge theory, we have a canonical isomorphism of vector spaces,

$$\ker \square_q^Z \simeq H^q(Z,F). \tag{1.5}$$

Denote by $\left(\square_q^Z\right)^{-1}$ the inverse of $\square_q^Z$ acting on the orthogonal space of $\ker \square_q^Z$ in $\Omega^q(Z,F)$.

Definition 1.3. For $s \in \mathbf{C}$ such that $\mathrm{Re}\,(s) > \dim Z/2$, set

$$\theta_q(s) = \mathrm{Tr}\left[\left(\square_q^Z\right)^{-s}\right]. \tag{1.6}$$

By [Se67] or [BeGeVe04, Proposition 9.35], $\theta_q(s)$ has a meromorphic extension to $\mathbf{C}$ which is holomorphic at $s=0$. The regularized determinant of $\square_q^Z$ is a positive number defined by $\exp(-\theta_q'(0))$. We write

$$\det\left(\square_q^Z\right) = \exp\left(-\theta_q'(0)\right). \tag{1.7}$$

Definition 1.4. The Ray–Singer analytic torsion [RS71] of F is defined by

$$T_F(Z) = \prod_{q=1}^{\dim Z} \det\left(\square_q^Z\right)^{(-1)^q q/2} = \exp\left(\frac{1}{2}\sum_{q=1}^{\dim Z}(-1)^{q-1} q\,\theta_q'(0)\right) \in \mathbf{R}_+^*. \tag{1.8}$$

Let $o(TZ)$ be the orientation line bundle of Z. Let $\overline{F}^*$ be the antidual bundle of F. The following proposition is a consequence of the Poincaré duality.

Proposition 1.5. *The following identity holds,*

$$T_F(Z) = \left(T_{\overline{F}^*\otimes o(TZ)}(Z)\right)^{(-1)^{\dim Z - 1}}. \tag{1.9}$$

In particular, if Z has even dimension and is orientable, and if F is unitarily flat, then

$$T_F(Z) = 1. \tag{1.10}$$

The next theorem is a special case of the anomaly formula of Bismut–Zhang [BZ92, Theorem 4.7], which generalizes a result of Ray–Singer [RS71, Theorem 2.1]. Its proof is based on the local index techniques, and consists in calculating the constant term in the asymptotic expansion of $\mathrm{Tr}[A\exp(-t\square^Z)]$ as $t \to 0$, where A is a certain smooth section of $\mathrm{End}(\Lambda^\cdot(T^*Z)\otimes_{\mathbf{R}} F)$.

Theorem 1.6. *Assume $H^\cdot(Z,F)=0$. The following statements hold.*

- *If Z has odd dimension, then $T_F(Z)$ is independent of g^{TZ}, g^F.*
- *If F is unimodular (i.e., $|\det\rho(\gamma)| = 1$ for any $\gamma \in \pi_1(X)$), then $T_F(Z)$ is independent of g^{TZ} and the unimodular metric g^F (i.e., g^F indues a flat metric on the determinant line bundle* $\det F = \Lambda^{\max} F$).

In either of the above situations, the analytic torsion becomes a topological invariant. The celebrated Cheeger–Müller/Bismut–Zhang Theorem [C79, M78, M93, BZ92] compares the Ray–Singer analytic torsion with the Reidemeister combinatorial torsion [Re35, Fr35, dR50]. We refer the reader to the above references for more details.

Example 1.7. We use the notation in Example 1.1. We equip TZ with the standard metric $(dt)^2$ and equip F with the trivial metric. Then,

$$d^Z = dt\frac{\partial}{\partial t} + \alpha dt. \quad d^{Z,*} = -i_{\frac{\partial}{\partial t}}\frac{\partial}{\partial t} + \overline{\alpha} i_{\frac{\partial}{\partial t}}, \quad \square^Z = -\left(\frac{\partial}{\partial t} - \overline{\alpha}\right)\left(\frac{\partial}{\partial t} + \alpha\right). \tag{1.11}$$

Using the Fourier transform, we get the spectrum

$$\operatorname{Sp}\square_0^Z = \operatorname{Sp}\square_1^Z = \left\{\left|\alpha + 2ik\pi\right|^2 : k \in \mathbf{Z}\right\}, \tag{1.12}$$

and

$$T_F(Z) = \begin{cases} \left|2\sinh\left(\frac{\alpha}{2}\right)\right|^{-1} & , \quad \alpha \notin 2i\pi\mathbf{Z}, \\ 1 & , \quad \alpha \in 2i\pi\mathbf{Z}. \end{cases} \tag{1.13}$$

We also have [BZ92, Theorem 0.3].

Theorem 1.8. *We use the notation in Example* 1.2. *Assume that two of F_i are acyclic. If Z has odd dimension, or if g^{F_0}, g^{F_1}, and g^{F_2} induce a flat metric on* $\det F_0 \otimes (\det F_1)^{-1} \otimes \det F_2$, *then*

$$T_{F_1}(Z) = T_{F_0}(Z) T_{F_2}(Z). \tag{1.14}$$

Here, for a line bundle L, the notion L^{-1} denotes the dual bundle of L.

1.3. A formula for fibrations

Let $\pi : M \to Z$ be a fibration of closed manifolds with closed fibre Y. Let $TY \subset TM$ be the relative tangent bundle on M of the fibre Y. Let F be a flat vector bundle on M.

Let g^{TM}, g^{TY}, g^{TZ} be the Riemannian metrics on TM, TY, TZ. Let g^F be a Hermitian metric on F. Let $T_F(M)$ be the Ray–Singer analytic torsion with respect to g^{TM} and g^F.

Moreover, the restriction $F|_Y$ of F to the fibre Y is still a flat vector bundle. So we can consider the fibrewise Ray–Singer analytic torsion $T_{F|_Y}(Y)$ with respect to g^{TY} and $g^F|_Y$. It is a smooth function on Z.

Also, for $0 \leqslant q \leqslant \dim Y$, the fibrewise cohomology $H^q(Y, F|_Y)$ of Y forms a flat vector bundle on Z (see [BLo95, Section III.f]). Using the Hodge theory (1.5), the L^2 metric on $\Omega^q(Y, F|_Y)$ induces a Hermitian metric $g^{H^q(Y,F|_Y)}$ on $H^q(Y, F|_Y)$. Let $T_{H^q(Y,F|_Y)}(Z)$ be the Ray–Singer analytic torsion with respect to g^{TZ} and $g^{H^p(Y,F|_Y)}$.

The following theorem is a special case of a very general result of Ma [Ma02, Theorem 0.1 and (0.6)], which is a generalization of Dai–Melrose [DMe12] and Lück–Schick–Thielmann [LüScTh98, Theorem 0.2].

Theorem 1.9. *Assume $H^\cdot(M,F)=0$ and $H^\cdot(Z,H^\cdot(Y,F|_Y))=0$. If M has odd dimension or if F is unimodular, then*

$$T_F(M)=\exp\left(\int_Z e\left(TZ,\nabla^{TZ}\right)\log T_{F|_Y}(Y)\right)\prod_{q=0}^{\dim Y}\left(T_{H^q(Y,F|_Y)}(Z)\right)^{(-1)^q}, \quad (1.15)$$

where $e\left(TZ,\nabla^{TZ}\right)\in\Omega^{\dim Z}(Z,o(TZ))$ is the Euler characteristic form on Z with respect to the Levi-Civita connection ∇^{TZ}. In addition, if Z has odd dimension, then

$$T_F(M)=\prod_{q=0}^{\dim Y}\left(T_{H^q(Y,F|_Y)}(Z)\right)^{(-1)^q}. \quad (1.16)$$

2. The Fried conjecture

It was known dating back to Poincaré, Hopf, Lefschetz, etc., that certain topological invariants, like Euler characteristic number, can be expressed with the help of some dynamical objects, like the fixed point set. Fried considered a similar problem relating the Ray–Singer analytic torsion with the closed orbits of certain dynamical systems, which he called the Lefschetz formula for flows.

More precisely, given a diffeomorphism of a manifold, the Lefschetz fixed point formula asserts that the Lefschetz number can be written as a sum of the Lefschetz indices of each fixed point of the diffeomorphism. Formally, the Lefschetz index is a way to count cohomologically the cardinality of the fixed point set.

For a dynamical flow, Fried uses the Fuller index [Fu67] to count cohomologically the cardinality of the set of closed orbits of the flow. However, due to the infinite numbers of closed orbits, the Fuller index of the set of all closed orbits is not well defined and should be regularized. For this Fried relies on the meromorphic extension and the regularity at 0 of the Ruelle dynamical zeta function. The purpose of this section is to formulate the above as a general conjecture, known as the Fried conjecture.

This section is organized as follows. In Section 2.1, we consider a diffeomorphism on a manifold. We recall the Lefschetz fixed point formula.

In Section 2.2, we consider a flow on a manifold. We introduce the Fuller index.

In Section 2.3, we construct the Ruelle dynamical zeta function, and state the Fried conjecture.

2.1. Lefschetz index

Let Z be a closed manifold. Let $T\in\mathrm{Diffeo}(Z)$ be a diffeomorphism of Z. The pull back T^* defines a morphism of the complex $(\Omega^\cdot(Z),d)$. It induces a morphism on the cohomology $H^{\cdot}(Z)$. The Lefschetz number of T is defined by

$$L_T = \mathrm{Tr}_{\mathrm{s}}^{H^{\cdot}(Z)}[T^*] = \sum_{q=0}^{\dim Z} (-1)^q \,\mathrm{Tr}\left[T^*|_{H^q(Z)}\right]. \tag{2.1}$$

Clearly, if T is the identity map, then the Lefschetz number L_T is equal to the Euler characteristic number $\chi(Z)$.

Let

$$Z^T = \{z \in Z : Tz = z\} \tag{2.2}$$

be the set of the fixed points of T. Recall that a fixed point $z \in Z^T$ is called non-degenerate if $\det(1 - D_zT)|_{T_zZ} \neq 0$. For such a fixed point $z \in Z^T$, the Lefschetz index of T at z is defined by

$$\mathrm{ind}_L(T, z) = \mathrm{sgn}\big(\det(1 - D_zT)|_{T_zZ}\big). \tag{2.3}$$

If T has only non-degenerate fixed points, then Z^T is finite, and the Lefschetz fixed point theorem [BeGeVe04, Corollary 6.7] states that

$$L_T = \sum_{z \in Z^T} \mathrm{ind}_L(T, z). \tag{2.4}$$

More generally, if Z^T is not finite, we need the following assumption.

Assumption 2.1. *We assume that*

(1) *the fixed point set Z^T is a disjoint union of finitely many closed connected submanifolds $Z_i \subset Z$, so that*

$$Z^T = \coprod_{i \in I} Z_i; \tag{2.5}$$

(2) *for $z \in Z^T$, the eigenspace of $D_zT : T_zZ \to T_zZ$ associated to the eigenvalue 1 coincides with its characteristic space, and is equal to T_zZ^T.*

Let $N_{Z_i/Z} = (TZ|_{Z_i})/TZ_i$ be the normal bundle of Z_i in Z. Our assumption (2) is equivalent to

$$\det(1 - D_zT)|_{N_{Z_i/Z,z}} \neq 0, \qquad \text{for all } i \in I, z \in Z_i. \tag{2.6}$$

Note that for each $i \in I$, the sign of the above determinant does not depend on $z \in Z_i$.

Definition 2.2. The Lefschetz index of the diffeomorphism T at Z_i is defined by

$$\mathrm{ind}_L(T, Z_i) = \mathrm{sgn}\left(\det(1 - D_zT)|_{N_{Z_i/Z,z}}\right) \cdot \chi(Z_i) \in \mathbf{Z}, \tag{2.7}$$

where $z \in Z_i$.

Under Assumption 2.1, we have the Lefschetz fixed point theorem[4]

$$L_T = \sum_{i \in I} \mathrm{ind}_L(T, Z_i). \tag{2.8}$$

Formally, Lefschetz index $\mathrm{ind}_L(T, Z_i)$ counts cohomologically the cardinality of Z_i. Equation (2.8) says that the Lefschetz number can be obtained by counting cohomologically the cardinality of the fixed point set Z^T.

2.2. Fuller index

Let M be a closed manifold. Let $V \in C^\infty(M, TM)$ be a smooth vector field on M, and let $(\phi_t)_{t \in \mathbf{R}}$ be the flow generated by V.

Definition 2.3. We define the periodic set

$$\wp(\phi.) = \{(x,t) \in M \times \mathbf{R}_+^* : \phi_t(x) = x\}, \tag{2.9}$$

and the length spectrum

$$\ell(\phi.) = \{t \in \mathbf{R}_+^* : \text{ there is } x \in M \text{ such that } \phi_t(x) = x\}. \tag{2.10}$$

A point $(x,t) \in \wp(\phi.)$ corresponds to a closed orbit $\{\phi_s(x)\}_{0 \leqslant s \leqslant t}$ starting from x of period t. And $\ell(\phi.)$ represents all the possible periods. Clearly, if $\ell \in \ell(\phi.)$, then $k\ell \in \ell(\phi.)$ for all $k \in \mathbf{N}^*$.

Since our ultimate object "closed orbit" disregards the starting points, we will consider certain quotient space of $\wp(\phi.)$. Define an $\mathbf{R}$-action on $\wp(\phi.)$ by

$$s \cdot (x,t) = (\phi_{ts} x, t), \qquad \text{for } s \in \mathbf{R}, (x,t) \in \wp(\phi.). \tag{2.11}$$

By (2.11), the group $\mathbb{S}^1 = \mathbf{R}/\mathbf{Z}$ acts on $\wp(\phi.)$. Set

$$\overline{\wp}(\phi.) = \wp(\phi.)/\mathbb{S}^1. \tag{2.12}$$

We identify $\overline{\wp}(\phi.)$ with the space of closed orbits of the flow $\phi.$.

To count cohomologically the points in the $\overline{\wp}(\phi.)$, as in Section 2.1, we need an analogue of Assumption 2.1 for flows.

Assumption 2.4. *We assume that*

(1) *the length spectrum* $\ell(\phi.) \subset (0, \infty)$ *is discrete;*
(2) *for any* $\ell \in \ell(\phi.)$, *the diffeomorphism* $\phi_\ell \in \mathrm{Diffeo}(M)$ *satisfies Assumption* 2.1.

Recall that $M^{\phi_\ell} \subset M$ is the fixed point set of ϕ_ℓ in M. By (1) of Assumption 2.4, we can write

$$\wp(\phi.) = \coprod_{\ell \in \ell(\phi.)} M^{\phi_\ell} \times \{\ell\}. \tag{2.13}$$

[4]When T is an isometry of Z, a proof can be found in [BeGeVe04, Section 6.4]. For general T, the Lefschetz fixed point formula can be proved by adapting the argument given in [BZ92, Theorem 4.20].

To avoid the problem of disconnectedness of M^{ϕ_ℓ}, we write instead

$$\wp(\phi_\cdot) = \coprod_{i \in I} M_i \times \{\ell_i\}, \tag{2.14}$$

where M_i is a connected component of $M^{\phi_{\ell_i}}$. Note that not all the ℓ_i are necessarily distinct. Moreover, (1) of Assumption 2.4 implies that the $\mathbb{S}^1$-action on $\wp(\phi_\cdot)$ is locally free. Thus, $M_i/\mathbb{S}^1$ is a closed connected orbifold. By (2.12), we have

$$\overline{\wp}(\phi_\cdot) = \coprod_{i \in I} M_i/\mathbb{S}^1 \times \{\ell_i\}. \tag{2.15}$$

Denote by $\chi_{\mathrm{orb}}(M_i/\mathbb{S}^1) \in \mathbf{Q}$ the orbifold Euler characteristic number [Sa57] of $M_i/\mathbb{S}^1$. Set

$$m_i = \left|\ker\left(\mathbb{S}^1 \to \mathrm{Diffeo}(M_i)\right)\right| \in \mathbf{N}^* \tag{2.16}$$

to be the generic multiplicity of the closed orbits in $M_i/\mathbb{S}^1 \times \{\ell_i\}$.

By (2) of Assumption 2.4, as in Section 2.1, the sign

$$\mathrm{sgn}\left(\det(1 - D\phi_{\ell_i})|_{N_{M_i/M}}\right) \in \{\pm 1\} \tag{2.17}$$

is well defined for all $i \in I$.

Definition 2.5. The Fuller index of the flow $\phi_\cdot$ at $M_i \times \{\ell_i\}$ is defined by

$$\mathrm{ind}_F(\phi_\cdot, M_i) = \mathrm{sgn}\left(\det(1 - D\phi_{\ell_i})|_{N_{M_i/M}}\right) \cdot \frac{\chi_{\mathrm{orb}}(M_i/\mathbb{S}^1)}{m_i} \in \mathbf{Q}. \tag{2.18}$$

Remark 2.6. The pair $(M_i, \mathbb{S}^1)$ defines a non-effective orbifold. Its "non-effective" Euler characteristic number is given by $\frac{\chi_{\mathrm{orb}}(M_i/\mathbb{S}^1)}{m_i}$. In this way, the Fuller index $\mathrm{ind}_F(\phi_\cdot, M_i)$ is a strict analogue of the Lefschetz index (2.7).

More generally, let F be a flat vector bundle on M with holonomy ρ. Since M_i is connected, for any $x \in M_i$, the closed orbit $\{\phi_t(x)\}_{0 \leqslant t \leqslant \ell_i}$ lies in the same freely homotopy class of the loops on M. Up to conjugation, the holonomy[5] ρ_i of F associated to the closed orbit $\{\phi_t(x)\}_{0 \leqslant t \leqslant \ell_i}$ does not depend on the choice of $x \in M_i$. We can define the twisted Fuller index by

$$\mathrm{ind}_F(\phi_\cdot, M_i, F) = \mathrm{ind}_F(\phi_\cdot, M_i)\,\mathrm{Tr}\left[\rho_i\right]. \tag{2.19}$$

2.3. The Ruelle dynamical zeta function

We assume that the flow satisfies Assumption 2.4. As an analogue of (2.4), Fried raised the question: for what kind of flows the infinite sum

$$\sum_{i \in I} \mathrm{ind}_F(\phi_\cdot, M_i, F) \tag{2.20}$$

[5]The morphism ρ_i can be obtained by the parallel transport with respect to the flat connection along the closed orbit $\{\phi_t(x)\}_{0 \leqslant t \leqslant \ell_i}$ from $t = \ell_i$ to $t = 0$.

can be regularized, and is equal to some topological invariant. More precisely, Fried conjectured that if F is unitarily flat and acyclic, then the regularized sum (2.20) is equal to $\log T_F(M)$.

To regularize the sum (2.20), Fried used the Ruelle dynamical zeta function [Ru76b, F87].

Definition 2.7. For $\sigma \in \mathbf{C}$, we define formally

$$R_{\phi,\rho}(\sigma) = \exp\left(\sum_{i\in I} \mathrm{ind}_F(\phi_\cdot, M_i, F)e^{-\ell_i\sigma}\right). \tag{2.21}$$

We call $R_{\phi,\rho}$ is well defined if the following properties hold.

1. There is $\sigma_0 > 0$ such that for $\sigma \in \mathbf{C}$ with $\mathrm{Re}(\sigma) > \sigma_0$, the sum on the right-hand side of (2.21) converges to a holomorphic function;
2. The holomorphic function $R_{\phi,\rho}(\sigma)$, defined for $\mathrm{Re}(\sigma) > \sigma_0$, has a meromorphic extension to $\sigma \in \mathbf{C}$.

Remark 2.8. The above definition of the Ruelle dynamical zeta function is the reciprocal of the one introduced by Fried [F87, Section 5].

The Fried conjecture now can be formulated as follows.

Conjecture 2.9. *For a wide class of flows on a closed manifold M, if F is a unitarily flat vector bundle on M and $H^\cdot(M,F) = 0$, then the Ruelle dynamical zeta function $R_{\phi,\rho}(\sigma)$ is well defined, and is regular at $\sigma = 0$ such that*

$$|R_{\phi,\rho}(0)| = T_F(M). \tag{2.22}$$

Example 2.10. Consider the rotation flow $\phi_\cdot$ on $\mathbb{S}^1$, i.e.,

$$\phi_t(x) = x + t \mod \mathbf{Z}, \qquad \text{for } t \in \mathbf{R}, x \in \mathbf{R}/\mathbf{Z}. \tag{2.23}$$

For the flat vector bundle defined in Example 1.1, the Ruelle dynamical zeta function is given by

$$R_{\phi,\rho}(\sigma) = (1 - e^{-\sigma+\alpha})^{-1}. \tag{2.24}$$

If F is acyclic and unitarily flat (i.e., $\alpha \in i\mathbf{R}$ and $\alpha \notin 2i\pi\mathbf{Z}$), by (1.13) and (2.24), we have

$$|R_{\phi,\rho}(0)| = T_F\left(\mathbb{S}^1\right). \tag{2.25}$$

Example 2.11. Consider now the geodesic flow of $\mathbb{S}^1$. That is a flow $\widetilde{\phi}$ defined on $\mathbb{S}^1 \times \{\pm 1\}$. On $\mathbb{S}^1 \times \{1\}$, $\widetilde{\phi}$ is just the rotation flow considered in Example 2.10, and on the other copy $\mathbb{S}^1 \times \{-1\}$, $\widetilde{\phi}$ is the inverse of the rotation flow. Let $\pi : \mathbb{S}^1 \times \{\pm 1\} \to \mathbb{S}^1$ be the natural projection.

Recall that F is defined in Example 1.1. Let π^*F be the pull back of the flat vector bundle on $\mathbb{S}^1 \times \{\pm 1\}$. The Ruelle dynamical zeta function is then given by

$$R_{\widetilde{\phi},\rho}(\sigma) = (1 - e^{-\sigma+\alpha})^{-1}(1 - e^{-\sigma-\alpha})^{-1}. \tag{2.26}$$

If $\alpha \in \mathbf{C}$ and $\alpha \notin 2i\pi\mathbf{Z}$, we have

$$|R_{\widetilde{\phi},\rho}(0)| = T_{\pi^*F}\left(\mathbb{S}^1 \times \{\pm 1\}\right). \tag{2.27}$$

Example 2.10 shows that the Fried conjecture fails for non-unitarily flat vector bundles, while Example 2.11 shows that for some special flow the Fried conjecture holds even for certain non-unitarily flat vector bundles.

However, even for unitarily flat vector bundles, we can not expect that the Fried conjecture holds for general flows. In fact, a fixed point or a closed orbit is a limit set of dimensions 0 or 1. Wilson [Wi66, Corollary 2] has constructed a smooth flow on any closed manifold M with vanishing Euler characteristic number, whose only limit sets are a finite collection of torus of dimension $\dim M - 2$. In particular, if $\dim M \geqslant 4$ with $\chi(M) = 0$, there exists a flow without fixed points or closed orbits. If $\dim M = 3$, Schweitzer [Sch74, Theorem A] has constructed a C^1-flow without fixed points or closed orbits.

In the following sections, we will discuss several different classes of flows where the Fried conjecture has or might have a positive solution. This includes the suspension flow, the Morse–Smale flow, the geodesic flow, and the Anosov flow.

3. Suspension flow of a diffeomorphism

In this section, we consider the suspension flow of a diffeomorphism. We show that in this case the Fried conjecture is a consequence of the Lefschetz fixed point formula (2.8).

This section is organized as follows. In Section 3.1, we introduce the suspension flow.

In Section 3.2, we describe the flat vector bundle on a suspension, and evaluate its analytic torsion.

In Section 3.3, we give an explicit formula for the Ruelle dynamical zeta function associated to the suspension flow. Then, we show the Fried conjecture.

3.1. The definition of the suspension flow

Let Y be a connected closed manifold. Let $T \in \mathrm{Diffeo}(Y)$ be a diffeomorphism of Y. Let M be the suspension of T. Then $M = \mathbf{R} \times_{\mathbf{Z}} Y$ is the quotient space of $\mathbf{R} \times Y$ by the $\mathbf{Z}$-action given by

$$n \cdot (t, y) = (t - n, T^n y), \qquad \text{for } n \in \mathbf{Z}, (t, y) \in \mathbf{R} \times Y. \tag{3.1}$$

Clearly, $M \to \mathbf{R}/\mathbf{Z}$ is a fibration with fibre Y.

The suspension flow $\phi_\cdot$ on M is defined by

$$\phi_s([t, y]) = [t + s, y], \qquad \text{for } s \in \mathbf{R}, [t, y] \in \mathbf{R} \times_{\mathbf{Z}} Y. \tag{3.2}$$

3.2. The flat vector bundle on the suspension

Let E be a flat vector bundle on Y. Assume that the diffeomorphism T lifts to E. This means that there is a bundle morphism T_E of E such that the diagram

$$\begin{array}{ccc} E & \xrightarrow{T_E} & E \\ \downarrow & & \downarrow \\ Y & \xrightarrow{T} & Y \end{array} \tag{3.3}$$

commutes. Let

$$F = \mathbf{R} \times_{\mathbf{Z}} E \tag{3.4}$$

be the suspension of E with respect to T_E. Then F is a flat vector bundle on M.

Proposition 3.1. *All the flat vector bundles on M can be obtained in this way.*

Proof. Using the long exact sequence of the homotopy groups associated to the fibration $M \to \mathbf{R}/\mathbf{Z}$, we get an exact sequence of groups

$$1 \to \pi_1(Y) \to \pi_1(M) \to \mathbf{Z} \to 1. \tag{3.5}$$

The diffeomorphism T defines an isomorphism $T_* : \pi_1(Y) \to \pi_1(Y)$ of groups such that $\pi_1(M)$ is the semidirect product of $\pi_1(Y)$ and $\mathbf{Z}$ with respect to T_*.

Any r-dimensional representation of $\pi_1(M)$ can be constructed by an r-dimensional representation ρ of $\pi_1(Y)$ and an element $A \in \mathrm{GL}_r(\mathbf{C})$ such that

$$A\rho(\gamma) = \rho(T_*\gamma)A, \qquad \text{for all } \gamma \in \pi_1(Y). \tag{3.6}$$

It is easy to check that ρ and A induce our E and T_E. □

As in Section 2.1, the pull back T_E^* defines a morphism on $H^\cdot(Y, E)$. Let $H_T^\cdot(Y, E)$ be a flat vector bundle on $\mathbf{R}/\mathbf{Z}$ defined by the holonomy

$$\rho : n \in \mathbf{Z} \to (T_E^*)^n. \tag{3.7}$$

This is just the direct image $H^\cdot(Y, F|_Y)$ of F described in Section 1.3.

Proposition 3.2. *The flat vector bundle F on M is acyclic if and only if 1 is not an eigenvalue of T_E^* on $H^\cdot(Y, E)$.*

Proof. Let $(E_r^{p,q}, d_r)_{r\geqslant 0}$ be the Leray spectral sequence [BoTu82, p. 169] associated to the fibration $M \to \mathbf{R}/\mathbf{Z}$. By [BoTu82, Theorem 14.18], $(E_r^{p,q}, d_r)_{r\geqslant 0}$ converges to $H^\cdot(M, F)$, and

$$E_2^{p,q} = H^p\big(\mathbf{R}/\mathbf{Z}, H_T^q(Y, E)\big). \tag{3.8}$$

By (3.8), for $r \geqslant 2$, $p \neq 0, 1$, we have $E_r^{p,q} = 0$. Since d_r sends $E_r^{p,q}$ to $E_r^{p+r,q-r+1}$, we see that $d_2 = d_3 = \cdots = 0$. So the spectral sequence degenerates at $r = 2$. It implies

$$\dim H^p(M, F) = \dim H^0\big(\mathbf{R}/\mathbf{Z}, H_T^p(Y, E)\big) + \dim H^1\big(\mathbf{R}/\mathbf{Z}, H_T^{p-1}(Y, E)\big). \tag{3.9}$$

By (3.9), $H^{\cdot}(M,F)=0$ if and only if $H^{\cdot}(\mathbf{R}/\mathbf{Z},H_T^{\cdot}(Y,E))=0$. By Example 1.1, we see that our proposition holds true if T_E^* is diagonalizable. Thanks to Example 1.2, we can deduce our proposition in full generality. □

If A is a matrix acting on a $\mathbf{Z}$-graded space $E^{\cdot}$ which preserves the degree, then write

$$\det_{\mathrm{s}}(A)|_{E^{\cdot}} = \prod_{q\in\mathbf{Z}} \left(\det(A)|_{E^q}\right)^{(-1)^q}. \tag{3.10}$$

Proposition 3.3. *Assume $H^{\cdot}(M,F)=0$. If M has odd dimension or if F is unimodular, then*

$$T_F(M) = \left|\det_{\mathrm{s}}\left(T_E^*\right)|_{H^{\cdot}(Y,E)}\right|^{1/2} \left|\det_{\mathrm{s}}\left(1-T_E^*\right)|_{H^{\cdot}(Y,E)}\right|^{-1}. \tag{3.11}$$

Proof. By Theorem 1.9, we have

$$T_F(M) = \prod_{q=0}^{\dim Y} \left(T_{H_T^q(Y,E)}(\mathbf{R}/\mathbf{Z})\right)^{(-1)^q}. \tag{3.12}$$

Now (3.11) is a consequence of (1.13), (1.14), and (3.12). □

3.3. The Ruelle dynamical zeta function of the suspension flow

Note that if for all $n\in\mathbf{N}^*$, the diffeomorphism T^n satisfies Assumption 2.1, then the suspension flow $\phi_{\cdot}$ associated to T satisfies Assumption 2.4.

Theorem 3.4. *Assume that for all $n\in\mathbf{N}^*$, the diffeomorphism $T^n\in\mathrm{Diffeo}(Y)$ satisfies Assumption 2.1. The Ruelle dynamical zeta function is given by*

$$R_{\phi,\rho}(\sigma) = \left(\det_{\mathrm{s}}\left(1-T_E^*e^{-\sigma}\right)|_{H^{\cdot}(Y,E)}\right)^{-1}. \tag{3.13}$$

If $H^{\cdot}(M,F)=0$, then $R_{\phi,\rho}$ is regular at $\sigma=0$, so that

$$R_{\phi,\rho}(0) = \left(\det_{\mathrm{s}}\left(1-T_E^*\right)|_{H^{\cdot}(Y,E)}\right)^{-1}. \tag{3.14}$$

Proof. For the suspension flow, we have $\ell(\phi_{\cdot})\subset\mathbf{N}^*$. The periodic set is given by

$$\wp(\phi_{\cdot}) = \coprod_{n=1}^{\infty} M^{\phi_n}\times\{n\}. \tag{3.15}$$

Let $\pi:\mathbf{R}\times Y\to M$ be the natural projection. We have

$$M^{\phi_n} = \pi\left([0,1]\times Y^{T^n}\right). \tag{3.16}$$

We claim that Fuller index (twisted by F) of $\phi_{\cdot}$ at M^{ϕ_n} is given by

$$\mathrm{ind}_F(\phi_{\cdot},M^{\phi_n},F) = \frac{1}{n}\mathrm{Tr}_{\mathrm{s}}^{H^{\cdot}(Y,E)}\left[(T_E^*)^n\right]. \tag{3.17}$$

Let us show (3.17) in the case where Y^{T^n} is discrete. The proof for the general case is similar, and we omit the details. For $p\in\mathbf{N}^*$ and $p|n$, let $Y_p^{T^n}\subset Y^{T^n}$ be the subset of Y^{T^n} formed by the points of Y^{T^n} whose prime period is p. Take $y \in Y_p^{T^n}$. By (2.18), the contribution of the closed orbit $\{[t,y]\}_{0\leqslant t\leqslant n}$ to the Fuller index (twisted by F) is given by

$$\frac{\operatorname{sgn}\left(\det\left(1-DT_y^n\right)|_{T_yY}\right)}{n/p}\operatorname{Tr}\left[\tau_{y,n}\right], \tag{3.18}$$

where $\tau_{y,n}$ is the parallel transport of F with respect to the flat connection along the curve $\{[t,y]\}_{0\leqslant t\leqslant n}$ from $t=n$ to $t=0$. We have

$$\operatorname{Tr}\left[\tau_{y,n}\right]=\operatorname{Tr}^{E_y}\left[T_E^{-n}\right]. \tag{3.19}$$

Since $y, Ty, \ldots, T^{p-1}y$ are all in $Y_p^{T^n}$ and correspond to the same closed orbit, we have

$$\begin{aligned}\operatorname{ind}_F(\phi_\cdot, M^{\phi_n}, F) &= \sum_{p|n}\frac{1}{p}\sum_{y\in Y_p^{T^n}}\frac{\operatorname{sgn}\left(\det\left(1-DT_y^n\right)|_{T_yY}\right)}{n/p}\operatorname{Tr}^{E_y}\left[T_E^{-n}\right]\\ &= \frac{1}{n}\sum_{y\in Y^{T^n}}\operatorname{sgn}\left(\det\left(1-DT_y^n\right)|_{T_yY}\right)\operatorname{Tr}^{E_y}\left[T_E^{-n}\right]. \end{aligned}\tag{3.20}$$

By a twisted version of (2.4) (see [BeGeVe04, Theorem 6.6]) and (3.20), we get (3.17).

By (3.17), using

$$\log(1-z) = -\sum_{n=1}^{\infty}\frac{z^n}{n}, \tag{3.21}$$

we get (3.13). The rest of our proposition is a consequence of Proposition 3.2 and (3.13). □

Corollary 3.5. *Suppose that for all $n \in \mathbf{N}^*$ the diffeomorphism $T^n \in \operatorname{Diffeo}(Y)$ satisfies Assumption 2.1. Assume that $H^\cdot(M,F)=0$ and that F is unimodular. Then, we have*

$$|R_{\phi,\rho}(0)| = T_F(M). \tag{3.22}$$

Proof. By (3.11) and (3.14), it is enough to show

$$\left|\det{}_s(T_E^*)|_{H^\cdot(Y,E)}\right| = 1. \tag{3.23}$$

Recall that $H_T^\cdot(Y,E)$ is a flat vector bundle on $\mathbf{R}/\mathbf{Z}$. Equation (3.23) is equivalent to say that the line bundle $\otimes_{q=0}^{\dim Y}\left(\det H_T^q(Y,E)\right)^{(-1)^q}$ is unitarily flat.

Remark that (see [BLo95, Section I.g]) if (W,∇^W) is a flat vector bundle on a manifold S, for any Hermitian metric g^W on W, the 1-form $\frac{1}{2}\operatorname{Tr}\left[g^{W,-1}\nabla^W g^W\right]$ is closed. Moreover, its class in $H^1(S)$ does not depend on the choice of g^W, and is called the first odd Chern class $c_1^{\mathrm{odd}}(W)$ of W. Clearly, $\det(W)$ is unitarily flat if and only if $c_1^{\mathrm{odd}}(W)=0$.

By above, it is equivalent to show the first odd Chern class of the fibrewise cohomology vanishes,

$$\sum_{q=0}^{\dim Y} (-1)^q c_1^{\mathrm{odd}}(H_T^q(Y,E)) = 0. \tag{3.24}$$

Since F is unimodular, (3.24) is a consequence of [BLo95, Theorem 0.1]. □

4. Morse–Smale flow

The Morse–Smale flow is the simplest structurally stable dynamical system which has only two types of recurrent behaviors: closed orbits and fixed points. In this section, we study the Fried conjecture for the Morse–Smale flow. We show that the Fried conjecture in this case is a consequence of Cheeger–Müller/Bismut–Zhang theorem. The result in this section is originally due to Fried [F87, Section 3] and is extended by Shen–Yu [ShY18].

This section is organized as follows. In Section 4.1, we introduce the Morse–Smale flow.

In Section 4.2, we give an explicit formula for the Ruelle dynamical zeta function, and show the Fried conjecture.

4.1. The definition of Morse–Smale flow

Let M be a closed manifold with a smooth vector field $V \in C^\infty(M, TM)$. Let $(\phi_t)_{t\in\mathbf{R}}$ be the flow on M generated by V.

Definition 4.1. The nonwandering set of $\phi_\cdot$ is defined by

$$\Big\{x \in M : \forall \text{ open neighborhood } U \text{ of } x,\ \forall T > 0, \text{ we have } U \cap \bigcup_{t\geqslant T} \phi_t(U) \neq \varnothing\Big\}. \tag{4.1}$$

Definition 4.2. A fixed point $x \in M$ of the flow $\phi_\cdot$ is called hyperbolic if there is a ϕ_t-invariant splitting

$$T_xM = E_x^u \oplus E_x^s, \tag{4.2}$$

and there exist $C > 0, \theta > 0$ and a Riemannian metric on M such that for $v \in E_x^u$, $v' \in E_x^s$, and $t > 0$, we have

$$|\phi_{-t,*}v| \leqslant Ce^{-\theta|t|}\,|v|, \qquad |\phi_{t,*}v'| \leqslant Ce^{-\theta|t|}\,|v'|. \tag{4.3}$$

The index $\mathrm{ind}(x) \in \mathbf{N}$ of x is defined by

$$\mathrm{ind}(x) = \dim E_x^u. \tag{4.4}$$

The unstable and stable manifolds of x are defined by

$$\begin{aligned} W_x^u &= \big\{y \in M : \lim_{t\to-\infty} d_M(\phi_t(y), x) = 0\big\},\\ W_x^s &= \big\{y \in M : \lim_{t\to+\infty} d_M(\phi_t(y), x) = 0\big\}, \end{aligned} \tag{4.5}$$

where d_M is the Riemannian distance on M.

Definition 4.3. A closed orbit γ of the flow $\phi_\cdot$ is called hyperbolic, if there is a ϕ_t-invariant continuous splitting

$$TM|_\gamma = \mathbf{R}V \oplus E^u_\gamma \oplus E^s_\gamma \tag{4.6}$$

of vector bundles over γ such that (4.3) holds. The index $\mathrm{ind}(\gamma) \in \mathbf{N}$ of γ is defined by

$$\mathrm{ind}(\gamma) = \dim E^u_\gamma. \tag{4.7}$$

The unstable and stable manifolds of γ are defined by

$$\begin{aligned} W^u_\gamma &= \bigcup_{x\in\gamma} \Big\{ y \in M : \lim_{t\to-\infty} d_M(\phi_t(y), \phi_t(x)) = 0 \Big\}, \\ W^s_\gamma &= \bigcup_{x\in\gamma} \Big\{ y \in M : \lim_{t\to+\infty} d_M(\phi_t(y), \phi_t(x)) = 0 \Big\}. \end{aligned} \tag{4.8}$$

Denote by A the set of fixed points and by B the set of prime closed orbits.

Definition 4.4. A vector field V or a flow $\phi_\cdot$ is called Morse–Smale (see [PalMe82, Definition, p.118]) if

- the sets A and B are finite and contain only hyperbolic elements;
- the nonwandering set of $\phi_\cdot$ is equal to $A \cup \bigcup_{\gamma\in B} \gamma$;
- the stable manifold of any element in $A \coprod B$ intersects transversally with the unstable manifolds of any element in $A \coprod B$.

In the rest of this section, we assume that V is a Morse–Smale vector field.

4.2. The Ruelle dynamical zeta function of the Morse–Smale flow

For $\gamma \in B$, denote by $\ell_\gamma \in \mathbf{R}^*_+$ its period. A closed orbit $\gamma \in B$ is called untwist if E^u_γ is orientable along γ, and is called twist otherwise. Put

$$\Delta(\gamma) = \begin{cases} 1, & \text{if } \gamma \text{ is untwist,} \\ -1, & \text{if } \gamma \text{ is twist.} \end{cases} \tag{4.9}$$

Recall that F is a flat vector bundle on M with holonomy ρ. For $\gamma \in B$, denote by $\rho(\gamma)$ the holonomy along γ, which is also the parallel transport with respect to the flat connection along γ^{-1} (cf. footnote 3). Clearly, $\rho(\gamma)$ is well defined up to a conjugation.

Proposition 4.5. *The Ruelle dynamical zeta function is given by*

$$R_{\phi,\rho}(\sigma) = \prod_{\gamma\in B} \det\left(1 - \Delta(\gamma)\rho(\gamma)e^{-\sigma\ell_\gamma}\right)^{(-1)^{1+\mathrm{ind}(\gamma)}}. \tag{4.10}$$

Proof. Without loss of generality, we assume that there is only one prime closed orbit γ. For $x \in \gamma$, $k \in \mathbf{N}^*$, let us calculate the sign of $\det(1 - D\phi_{k\ell_\gamma})|_{T_xM/\mathbf{R}V(x)}$. By (4.6), we have

$$\det(1 - D\phi_{k\ell_\gamma})|_{T_xM/\mathbf{R}V(x)} = \det\left(1 - (D\phi_{\ell_\gamma})^k\right)|_{E^s_{\gamma,x}} \det\left(1 - (D\phi_{\ell_\gamma})^k\right)|_{E^u_{\gamma,x}}. \tag{4.11}$$

Note that the non-real eigenvalues α and $\overline{\alpha}$ of $(D\phi_{\ell_\gamma})^k$ come in pair. Note also that by property of E^s, all the absolute value of the real eigenvalues of $(D\phi_{\ell_\gamma})^k$ on $E^s_{\gamma,x}$ is bounded by 1. So the sign of (4.11) comes from the positive eigenvalues of $D\phi_{\ell_\gamma}|_{E^u_{\gamma,x}}$. More precisely, we have

$$\begin{aligned}\operatorname{sgn}\left(\det\left(1-(D\phi_{\ell_\gamma})^k\right)|_{T_xM/\mathbf{R}V(x)}\right) &= \operatorname{sgn}\det\left(-\left(D\phi_{\ell_\gamma}\right)^k|_{E^u_{\gamma,x}}\right)\\ &= (-1)^{\operatorname{ind}(\gamma)}\Delta(\gamma)^k.\end{aligned} \tag{4.12}$$

Equation (4.10) is a consequence of (2.21), (3.21), and (4.12). □

The following theorem is due to Fried [F87, Theorem 3.1] if F is unitarily flat, and can be extended to unimodular vector bundle [ShY18, Proposition 2.12].

Theorem 4.6. *Assume that F is a unimodular flat vector bundle, and that $\phi.$ does not have fixed points. If for all $\gamma \in B$, $\Delta(\gamma)$ is not an eigenvalue of $\rho(\gamma)$, then we have $H^\cdot(M,F)=0$, and $R_{\phi,\rho}$ is regular at $\sigma=0$, so that*

$$|R_{\phi,\rho}(0)| = \prod_{\gamma\in B}\left|\det\left(1-\Delta(\gamma)\rho(\gamma)\right)\right|^{(-1)^{1+\operatorname{ind}(\gamma)}} = T_F(M). \tag{4.13}$$

Proof. Following [Fra82, Definition 9.10], let

$$\varnothing = M^0 \subset M^1 \subset \cdots \subset M^N = M \tag{4.14}$$

be a Smale filtration on M associated to the flow $\phi.$. Note that each $M^p \subset M$ is a submanifold with boundary, and can be constructed by the sublevel set of a smooth Lyapunov function. Also, we have

- on each ∂M^p, V points toward the inside of M^p;
- there is only one prime closed orbit γ in $M^{p+1}\backslash M^p$ and

$$\gamma = \bigcap_{t\in\mathbf{R}} \phi_t\left(M^{p+1}\backslash M^p\right). \tag{4.15}$$

Note that M^{p+1} can be obtained from M^p by attaching a round $\operatorname{ind}(\gamma)$-handle,

$$M^{p+1} = M^p \cup_{\mathbb{S}^1\times\mathbb{S}^{\operatorname{ind}(\gamma)}\times\mathbb{D}^{m-\operatorname{ind}(\gamma)-1}} \mathbb{S}^1\times\mathbb{D}^{\operatorname{ind}(\gamma)}\times\mathbb{D}^{m-\operatorname{ind}(\gamma)-1}, \tag{4.16}$$

where $m=\dim M$. By [Fra82, Theorem 9.11] (see also [SM96b, Section 2]),

$$H^q\left(M^{p+1},M^p,F\right) = H^{q-\operatorname{ind}(\gamma)}\left(\gamma, o(E^u_\gamma)\otimes_{\mathbf{R}} F|_\gamma\right), \tag{4.17}$$

where $o(E^u_\gamma)$ is the orientation line bundle of E^u_γ along the closed orbit γ.

By Examples 1.1 and 1.2, we see that $\Delta(\gamma)$ is not an eigenvalue of $\rho(\gamma)$ if and only if

$$H^\cdot\left(M^{p+1},M^p,F\right)=0. \tag{4.18}$$

Using the long exact sequence

$$\cdots \to H^q(M^p,M^{p-1},F)\to H^q(M^p,F)\to H^q(M^{p-1},F)\to\cdots, \tag{4.19}$$

we see that

$$H^{\cdot}(M,F)=H^{\cdot}(M^N,F)\simeq H^{\cdot}\left(M^{N-1},F\right)\simeq\cdots\simeq H^{\cdot}\left(M^0,F\right)=0. \tag{4.20}$$

The first equality of (4.13) is a trivial consequence of (4.10). Let us show the second equality of (4.13).

We claim that

$$T_F(M)=\prod_{\gamma\in B}\left(T_{o(E^u_\gamma)\otimes_{\mathbf{R}}F|_\gamma}(\gamma)\right)^{(-1)^{\mathrm{ind}(\gamma)}}. \tag{4.21}$$

Indeed, if F is unitary, using (4.16), (4.19), the Cheeger–Müller theorem [C79, M78], we can deduce (4.21) from the excision property of the Reidemeister torsion. If F is only unimodular, using Bismut–Zhang's theorem [BZ92, Theorem 0.2] and a property of Milnor torsion, (4.21) still holds true (see [ShY18, Proposition 2.13]).

By (1.13), (1.14), and (4.21), we get the second equality of (4.13). □

It is natural to ask if a similar result holds when the Morse–Smale flow has both fixed points and closed orbits. However, Sánchez-Morgado [SM96b] has shown that the heteroclinic orbits have a non-trivial contribution in the torsion invariant, and in this way he constructed a counterexample to the Fried conjecture on Seifert manifolds.

In [ShY18], Shen and Yu constructed a Milnor metric, which indeed contains the heteroclinic contributions, and obtained a proof for a modified version of the Fried conjecture for generally Morse–Smale flow with or without fixed points and for arbitrary flat vector bundle with arbitrary Hermitian metric. The formulation and the proof are based on [BZ92], using the determinant line of the cohomology. We refer the reader to [ShY18] for more details.

Let us mention that there is another interpretation of the Ruelle dynamical zeta function provided by Dang–Rivière [DaRi20c]. See also [DaRi19, DaRi20a, DaRi20b, DaRi21] for related works.

5. Geodesic flow

In this section, we discuss the Fried conjecture in the most interesting case the geodesic flow on the unit tangent bundle of a non-positively curved Riemannian manifold of dimension $\geqslant 3$. (The case of surface will be treated in Section 6.4.) In this case, the Fried conjecture can be proved formally via Bismut–Goette's V-invariant [BG04]. However, this argument remains non-rigorous due to the involved infinite-dimensional loop space. For locally symmetric spaces, we give the rigorous argument [F86a, MoSt91, Sh18] which is based on the Selberg trace formula and the Bismut orbital integral formula [B11].

This section is organized as follows. In Section 5.1, we study the flat vector bundle defined on the unit tangent bundle.

In Section 5.2, we study the geodesic flow on the unit tangent bundle of a non-positively curved Riemannian manifold.

In Section 5.3, we sketch the formal proof of the Fried conjecture via Bismut–Goette's V-invariant.

In Section 5.4, we introduce the reductive group, symmetric space, and the locally symmetric space. We state the main Theorem 5.9 of this section.

Sections 5.5–5.8 are devoted to sketching the proof of Theorem 5.9.

In Section 5.5, we recall the Selberg trace formula and the Bismut orbital integral formula, which are our main analytic tools.

In Section 5.6, we give a new proof of Moscovici–Stanton's vanishing theorem [MoSt91], which is due to Bismut [B11].

In Section 5.7, we sketch the proof of the Fried conjecture for the real reductive group of **R**-rank one, which is originally due to Fried [F86a].

In Section 5.8, we sketch the proof of the Fried conjecture for general reductive groups [Sh18].

5.1. Flat vector bundle on the unit tangent bundle

Let (Z, g^{TZ}) be a compact Riemannian manifold of dimension m. Let

$$M = SZ = \{(x, v) \in TZ : |v| = 1\} \tag{5.1}$$

be the unit tangent bundle. Then $\pi : M \to Z$ is a fibration of sphere $\mathbb{S}^{m-1}$.

Proposition 5.1. *Assume* $\dim Z \geqslant 3$. *Then all the flat vector bundles on M are the pull back of a flat vector bundle F on Z. Moreover, F is unitary (resp. unimodular) if and only if $\pi^* F$ is unitary (resp. unimodular).*

Proof. Consider the long exact sequence of homotopy groups

$$\cdots \to \pi_1(\mathbb{S}^{m-1}) \to \pi_1(M) \to \pi_1(Z) \to \pi_0(\mathbb{S}^{m-1}) \to \cdots \tag{5.2}$$

associated to the fibration $M \to Z$. Since $m \geqslant 3$, $\pi_1(\mathbb{S}^{m-1}) = \pi_0(\mathbb{S}^{m-1}) = 1$. By (5.2), we have the isomorphism of the groups

$$\pi_1(M) \simeq \pi_1(Z), \tag{5.3}$$

from which we deduce our proposition. □

Proposition 5.2. *Assume that Z is orientable and that* $\dim Z \geqslant 3$. *Then $\pi^* F$ is acyclic on M if and only if F is acyclic on Z. In additional, if* $\dim Z$ *is odd, then*[6]

$$T_{\pi^* F}(M) = T_F(Z)^2. \tag{5.4}$$

Proof. Let

$$\begin{aligned} \cdots \to H^{p-m}(Z, F) \to H^p(Z, F) \to H^p(M, \pi^* F) \to H^{p-m+1}(Z, F) \\ \to H^{p+1}(Z, F) \to \cdots \end{aligned} \tag{5.5}$$

[6]Since both of M and Z have odd dimensions, by Theorem 1.6, we do not need to specify the metric data to define the analytic torsion.

be the Gysin exact sequence [BoTu82, Proposition 14.33] associated to the fibration $M \to Z$ of orientable $(m-1)$-spheres. Comparing the degrees in (5.5), for $0 \leqslant p \leqslant m-2$ and $m+1 \leqslant q \leqslant 2m-1$, we have

$$H^p(M, \pi^* F) = H^p(Z, F), \qquad H^q(M, \pi^* F) = H^{q-m+1}(Z, F). \tag{5.6}$$

By (5.6), using $m \geqslant 3$, we get the first statement of our proposition. By (1.16), we get (5.4). □

Remark 5.3. Assume that $\dim Z \geqslant 3$ is odd. If we do not assume that Z is orientable, a detailed analysis on the Leray spectral sequence tells us that $\pi^* F$ is acyclic on M if and only if F and $o(TZ) \otimes_{\mathbf{R}} F$ are acyclic on Z. In this case, we have

$$T_{\pi^* F}(M) = T_F(Z) \cdot T_{o(TZ)\otimes_{\mathbf{R}} F}(Z). \tag{5.7}$$

Let us consider the geodesic flow on a negatively curved orientable surface (Z, g^{TZ}) of genus $g \geqslant 2$, which will be used in Section 6.4. The long exact sequence of the homotopy groups (5.2) is simply

$$1 \to \mathbf{Z} \to \pi_1(M) \to \pi_1(Z) \to 1. \tag{5.8}$$

Let $a_0 \in \pi_1(M)$ be a generator of $\mathbf{Z} \subset \pi_1(M)$. Let F be a flat vector bundle of rank r on M with holonomy ρ. Then $\rho(a_0) \in \mathrm{GL}_r(\mathbf{C})$.

Corollary 5.4. *Let us assume that Z is a negatively curved orientable surface. Then $H^\cdot(M, F) = 0$ if and only if 1 is not an eigenvalue of $\rho(a_0)$. In this case, we have*

$$T_F(M) = \left|\det\left(\rho(a_0)\right)\right|^{(1-g)} \cdot \left|\det\left(1-\rho(a_0)\right)\right|^{2(g-1)}. \tag{5.9}$$

Proof. We use the Leray spectral sequence associated to the fibration $M \to Z$ as in the proof of Propositions 3.2. By [BoTu82, Theorem 14.18], $(E_r^{p,q}, d_r)_{r\geqslant 0}$ converges to $H^\cdot(M, F)$, and

$$E_2^{p,q} = H^p\big(Z, H^q(Y, F|_Y)\big). \tag{5.10}$$

Since the fibre Y is a circle, for $r \geqslant 2$ and $q \neq 0$ or 1, we see that $E_r^{p,q} = 0$. Since d_r sends $E_r^{p,q}$ to $E_r^{p+r,q-r+1}$. We get $d_3 = d_4 = \cdots = 0$, and d_2 vanishes except

$$d_2 : E_2^{0,1} \to E_2^{2,0}. \tag{5.11}$$

Indeed, by [BoTu82, p. 178], up to a sign, $d_2|_{E_2^{0,1}}$ is the multiplication by the Euler characteristic class of Z, which is an isomorphism since $g \geqslant 2$. So the spectral sequence degenerates at $r = 3$ such that

$$\begin{aligned} &H^0(M,F) = H^0(Z, H^0(Y, F|_Y)), \quad H^1(M,F) = H^1(Z, H^0(Y, F|_Y)),\\ &H^2(M,F) = H^0(Z, H^1(Y, F|_Y)), \quad H^3(M,F) = H^2(Z, H^1(Y, F|_Y)). \end{aligned} \tag{5.12}$$

Since Y is a circle and since Z is orientable, we have an isomorphism of vector bundles on Z,

$$H^1(Y, F|_Y) \simeq H^0(Y, F|_Y). \tag{5.13}$$

By (5.12) and (5.13), we see that F is acyclic on M if and only if $H^0(Y, F|_Y)$ is acyclic on Z. By the Gauss–Bonnet theorem,

$$\chi(Z, H^0(Y, F|_Y)) = 2\,\mathrm{rk}[H^0(Y, F|_Y)](1-g), \tag{5.14}$$

where $\mathrm{rk}[H^0(Y, F|_Y)]$ denotes the rank of the vector bundle $H^0(Y, F|_Y)$. By (5.14), we see that $H^0(Y, F|_Y)$ is acyclic on Z if and only if $H^0(Y, F|_Y) = 0$. The latter means exactly that 1 is not an eigenvalue of $\rho(a_0)$. By (1.13), (1.14), and (1.15), we get (5.9). □

Remark 5.5. Equation (5.9) is obtained in [F86b] by a purely topological method under the assumption that F is unitarily flat.

5.2. Geodesic flow of the non-positively curved manifolds

Let (Z, g^{TZ}) be a Riemannian manifold with non-positive sectional curvature. Let $(\phi_t)_{t\in\mathbf{R}}$ be the geodesic flow on the unit tangent bundle $M = SZ$. Let $V \in C^\infty(M, TM)$ be the generator of $\phi_\cdot$.

Let $LZ = C^\infty(\mathbb{S}^1, Z)$ be the free loop space of Z. It is equipped with a canonical $\mathbb{S}^1$-action given by

$$s \cdot x_\cdot = x_{\cdot+s}, \qquad \text{for } s \in \mathbb{S}^1, x_\cdot \in LZ. \tag{5.15}$$

Denote by Γ the fundamental group of Z. Let $[\Gamma]$ be the set of conjugacy classes of Γ. We identify $[\Gamma]$ with the set of freely homotopy classes of loops on Z. For $[\gamma] \in [\Gamma]$, let $(LZ)_{[\gamma]} \subset LZ$ be the subset of LZ formed by all the loops on Z with the freely homotopy class $[\gamma]$. Write

$$LZ = \coprod_{[\gamma]\in[\Gamma]} (LZ)_{[\gamma]}. \tag{5.16}$$

Let E be the $\mathbb{S}^1$-invariant energy functional on LZ defined by

$$E : x_\cdot \in LZ \to \frac{1}{2}\int_0^1 |\dot{x}_s|^2 ds. \tag{5.17}$$

Then E is a convex function whose critical points are closed geodesics. Let $B_{[\gamma]}$ be the critical points set of E in $(LZ)_{[\gamma]}$. Since (Z, g^{TZ}) has non-positive sectional curvature,

$$B_{[1]} \simeq Z. \tag{5.18}$$

Let us give a more explicit description of $B_{[\gamma]}$. Take an element γ in $[\gamma]$. Up to evident isomorphisms, our objects constructed below do not depend on the choice of γ in $[\gamma]$. Let X be the universal cover of Z. Let g^{TX} be the induced Riemannian metric on X by g^{TZ}. Let $d_X(\cdot,\cdot)$ be the Riemannian distance on X. Since Z and also X have non-positive sectional curvature, the displacement function $x \in X \to d_X(x, \gamma x)$ is convex. Let $\ell_{[\gamma]} \in \mathbf{R}_+$ be the minimal value[7] of the displacement function on X and let $X(\gamma) \subset X$ be its minimal set. Since Z is compact and since the displacement function is convex, $X(\gamma)$ is a non-empty

[7] As the notation indicates, $\ell_{[\gamma]}$ does not depend on the choice of $\gamma \in [\gamma]$.

convex set (see [BaGrSch85, Section 6], [Ma17, Proposition 3.9]). Let $\Gamma(\gamma)$ be the centralizer of γ in Γ. Then $\Gamma(\gamma)$ acts on $X(\gamma)$. We have the identification

$$x. \in B_{[\gamma]} \simeq [\widetilde{x}_0] \in \Gamma(\gamma)\backslash X(\gamma), \tag{5.19}$$

where $\widetilde{x}.$ is the lifting of $x.$ in X. By the convexity of $X(\gamma)$, we see that $B_{[\gamma]}$ is connected. Recall that $M = SZ$. We can also identify $B_{[\gamma]}$ with a connected component of $M^{\phi_{\ell_{[\gamma]}}}$ by

$$x. \in B_{[\gamma]} \to (x_0, \dot{x}_0/\ell_{[\gamma]}) \in M^{\phi_{\ell_{[\gamma]}}}. \tag{5.20}$$

Since $M^{\phi_{\ell_{[\gamma]}}}$ is compact, we see that $B_{[\gamma]}$ is compact. By (5.20), equations (2.14) and (2.15) become

$$\wp(\phi.) = \coprod_{[\gamma]\in[\Gamma]-\{1\}} B_{[\gamma]} \times \{\ell_{[\gamma]}\}, \quad \overline{\wp}(\phi.) = \coprod_{[\gamma]\in[\Gamma]-\{1\}} B_{[\gamma]}/\mathbb{S}^1 \times \{\ell_{[\gamma]}\}. \tag{5.21}$$

Proposition 5.6. *Assume that (Z, g^{TZ}) has non-positive sectional curvature. The following statements hold.*

(1) *Condition* (1) *of Assumption* 2.4 *holds.*
(2) *Condition* (2) *of Assumption* 2.4 *holds if Z is a negatively curved manifold or if Z is a locally symmetric space.*

Proof. For Condition (1), it is enough to show that for $r \geqslant 0$ the set

$$\{[\gamma] \in [\Gamma] : \ell_{[\gamma]} \leqslant r\} \tag{5.22}$$

is finite. Fix a point $x \in X$ in the universal cover X. Fix a fundamental domain $F_Z \subset X$ of Z in X. There is $z_0 \in F_Z$ and $\gamma_1 \in \Gamma$ such that

$$\ell_{[\gamma]} = d_X(\gamma_1 z_0, \gamma\gamma_1 z_0). \tag{5.23}$$

By (5.23) and by the triangular inequality, we have

$$d_X(\gamma_1 x, \gamma\gamma_1 x) \leqslant \ell_{[\gamma]} + 2 \max_{z\in F_Z} d(x, z). \tag{5.24}$$

Using (5.24), we get

$$|\{[\gamma] \in [\Gamma] : \ell_{[\gamma]} \leqslant r\}| \leqslant |\{\gamma \in \Gamma : d_X(x, \gamma x) \leqslant r + 2 \max_{z\in F_Z} d(x, z)\}|. \tag{5.25}$$

It is known by [Mi68b, Remark p.1, Lemma 2] that the set on the right-hand side of (5.25) is finite[8].

The statement (2) is clear if Z is a negatively curved manifold. When Z is a locally symmetric space, the statement (2) is a consequence of [B11, (3.3.17), (3.5.10)]. □

5.3. The V-invariant and a formal proof of the Fried conjecture

Let us give a formal proof of (2.22) using the V-invariant of Bismut–Goette [BG04].

---

[8]Indeed, it is smaller than Ce^{Cr} for certain $C > 0$.

5.3.1. The V-invariant. Let S be a closed manifold equipped with an action of a compact Lie group L, with Lie algebra $\mathfrak{l}$. If $a \in \mathfrak{l}$, let a^S be the corresponding vector field on S. Bismut–Goette [BG04] introduced the V-invariant $V_a(S) \in \mathbf{R}$.

Let f be an a^S-invariant Morse–Bott function on S. Let $B_f \subset S$ be the critical submanifold. Since $a^S|_{B_f} \in TB_f$, $V_a(B_f)$ is also well defined. By [BG04, Theorem 4.10], $V_a(S)$ and $V_a(B_f)$ are related by a simple formula.

5.3.2. A formal proof of the Fried conjecture. Let us argue formally as in [Sh18, Section 1E]. Let a be the generator of the Lie algebra of $\mathbb{S}^1$ such that $\exp(a) = 1$.

As explained in [B05, Equation (0.3)], if F is a unitarily flat vector bundle on Z such that $H^{\cdot}(Z, F) = 0$, at least formally, we have

$$\log T_F(Z) = - \sum_{[\gamma]\in[\Gamma]} \mathrm{Tr}[\rho(\gamma)] V_a\big((LZ)_{[\gamma]}\big). \tag{5.26}$$

Let (Z, g^{TZ}) be an odd-dimensional oriented Riemannian manifold with non-positive sectional curvature. Assume that the energy functional is Morse–Bott[9]. Applying [BG04, Theorem 4.10] to the infinite-dimensional manifold $(LZ)_{[\gamma]}$ with the $\mathbb{S}^1$-invariant Morse–Bott functional E, we have the formal identity

$$V_a\big((LZ)_{[\gamma]}\big) = V_a\big(B_{[\gamma]}\big). \tag{5.27}$$

Since $B_{[1]} \simeq Z$ is formed by the trivial geodesics, by the definition of the V-invariant,

$$V_a(B_{[1]}) = 0. \tag{5.28}$$

By [BG04, Proposition 4.26], if $[\gamma] \in [\Gamma] - \{1\}$, then

$$V_a(B_{[\gamma]}) = -\frac{\chi_{\mathrm{orb}}(B_{[\gamma]}/\mathbb{S}^1)}{2m_{[\gamma]}}. \tag{5.29}$$

By Proposition 5.2 and by (5.26)–(5.29), we get a formal identity

$$\log T_{\pi^*F}(M) = 2\log T_F(Z) = \sum_{[\gamma]\in[\Gamma]-\{1\}} \mathrm{Tr}[\rho(\gamma)] \frac{\chi_{\mathrm{orb}}(B_{[\gamma]}/\mathbb{S}^1)}{m_{[\gamma]}}, \tag{5.30}$$

which is formally just (2.22)[10].

5.4. Reductive group, globally and locally symmetric space

We give a rigorous argument in the case of locally symmetric space of reductive type. Let us recall some preliminaries that are needed.

---

[9]Since this part is completely formal, we will not give the precise definition of the Morse–Bott function on the loop space LZ. Here we can understand it by the fact that the energy functional is convex and its critical points form a smooth manifold.

[10]We need also to show that the sign that appeared in the Fuller index is positive when $\chi_{\mathrm{orb}}(B_{[\gamma]}/\mathbb{S}^1) \neq 0$. Indeed, it follows from the fact that Z has an odd dimension and is orientable, and the fact that $\chi_{\mathrm{orb}}(B_{[\gamma]}/\mathbb{S}^1) = 0$ if $\dim B_{[\gamma]}$ is even. We omit the details. For negatively curved manifolds, a proof can be found in [GiLiPo13, Appendix B].

5.4.1. Reductive group. Let G be a linear connected real reductive group [K86, p. 3], and let $\theta \subset \mathrm{Aut}(G)$ be the Cartan involution. This means that G is a closed connected group of real matrices that is stable under transpose, and θ is the composition of transpose and inverse of matrices. Let K be the maximal compact subgroup of G which is the fixed point set of θ.

Let $\mathfrak{g}$ be the Lie algebra of G, and let $\mathfrak{k} \subset \mathfrak{g}$ be the Lie algebra of K. The Cartan involution θ acts naturally as a Lie algebra automorphism of $\mathfrak{g}$. Then $\mathfrak{k}$ is the eigenspace of θ associated with the eigenvalue 1. Let $\mathfrak{p}$ be the eigenspace with the eigenvalue -1, so that

$$\mathfrak{g} = \mathfrak{p} \oplus \mathfrak{k}. \tag{5.31}$$

We have

$$[\mathfrak{k}, \mathfrak{k}] \subset \mathfrak{k}, \qquad [\mathfrak{p}, \mathfrak{p}] \subset \mathfrak{k}, \qquad [\mathfrak{k}, \mathfrak{p}] \subset \mathfrak{p}. \tag{5.32}$$

By [K86, Proposition 1.2], we have the diffeomorphism

$$(Y, k) \in \mathfrak{p} \times K \to e^{Y} k \in G. \tag{5.33}$$

Let B be a real-valued non-degenerate bilinear symmetric form on $\mathfrak{g}$ which is invariant under the adjoint action Ad of G on $\mathfrak{g}$, and also under θ. Then (5.31) is an orthogonal splitting of $\mathfrak{g}$ with respect to B. We assume B to be positive on $\mathfrak{p}$, and negative on $\mathfrak{k}$. The form $\langle \cdot, , \cdot \rangle = -B(\cdot, \theta \cdot)$ defines an $\mathrm{Ad}(K)$-invariant scalar product on $\mathfrak{g}$ such that the splitting (5.31) is still orthogonal. We denote by $|\cdot|$ the corresponding norm.

The real (resp. complex) rank of G, denoted by $\mathrm{rk}_{\mathbf{R}}\, G$ (resp. $\mathrm{rk}_{\mathbf{C}}\, G$), is defined by the dimension of the maximal abelian subspace of $\mathfrak{p}$ (resp. the Cartan subalgebra of $\mathfrak{g}$). The fundamental rank of G is defined by

$$\delta(G) = \mathrm{rk}_{\mathbf{C}}\, G - \mathrm{rk}_{\mathbf{C}}\, K \in \mathbf{N}. \tag{5.34}$$

Clearly, $\delta(\mathfrak{g}) \in \mathbf{N}$ is also well defined. The following proposition is a consequence of the classification theory of real simple Lie algebras (see [B11, Remark 7.9.2]).

Theorem 5.7. *The only simple Lie algebras*[11] $\mathfrak{g}$ *with* $\delta(\mathfrak{g}) = 1$ *are given by*

$$\mathfrak{sl}_3(\mathbf{R}) \text{ or } \mathfrak{so}(p, q) \text{ with } pq > 1 \text{ odd}. \tag{5.35}$$

In Table 1, we list some reductive groups G, the maximal compact subgroups K, the dimensions $\mathfrak{p}$, and the fundamental ranks $\delta(G)$.

5.4.2. Semisimple elements. If $\gamma \in G$, we denote by $Z(\gamma) \subset G$ the centralizer of γ in G, and by $\mathfrak{z}(\gamma) \subset \mathfrak{g}$ its Lie algebra. If $a \in \mathfrak{g}$, let $Z(a) \subset G$ be the stabilizer of a in G, and let $\mathfrak{z}(a) \subset \mathfrak{g}$ be its Lie algebra.

An element $\gamma \in G$ is said to be semisimple if γ can be conjugate to $e^{a}k^{-1}$ such that

$$a \in \mathfrak{p}, \qquad k \in K, \qquad \mathrm{Ad}(k)a = a. \tag{5.36}$$

---

[11] We have also $\delta(\mathbf{R}) = 1$. But the abelian Lie algebra $\mathbf{R}$ is not considered as simple Lie algebra.

TABLE 1. Examples of reductive Lie group.

G	$\mathbf{R}$	$\mathrm{GL}_n^+(\mathbf{R})$	$\mathrm{SL}_n(\mathbf{R})$	$\mathrm{SO}^0(p,q)$
K	$\{0\}$	$\mathrm{SO}(n)$	$\mathrm{SO}(n)$	$\mathrm{SO}(p)\times\mathrm{SO}(q)$
$\dim\mathfrak{p}$	1	$\frac{n(1+n)}{2}$	$\frac{n(1+n)}{2}-1$	pq
$\delta(G)$	1	$n-\left[\frac{n}{2}\right]$	$n-1-\left[\frac{n}{2}\right]$	$\left[\frac{p+q}{2}\right]-\left[\frac{p}{2}\right]-\left[\frac{q}{2}\right]$

Let $\gamma = e^a k^{-1}$ be such that (5.36) holds. By [B11, (3.3.4), (3.3.6)], we have

$$Z(\gamma) = Z(a) \cap Z(k), \qquad \mathfrak{z}(\gamma) = \mathfrak{z}(a) \cap \mathfrak{z}(k). \tag{5.37}$$

Set

$$\mathfrak{p}(\gamma) = \mathfrak{z}(\gamma) \cap \mathfrak{p}, \qquad \mathfrak{k}(\gamma) = \mathfrak{z}(\gamma) \cap \mathfrak{k}, \qquad K(\gamma) = Z(\gamma) \cap K. \tag{5.38}$$

From (5.37) and (5.38), we get

$$\mathfrak{z}(\gamma) = \mathfrak{p}(\gamma) \oplus \mathfrak{k}(\gamma). \tag{5.39}$$

By [K02, Proposition 7.25], $Z(\gamma)$ is a reductive subgroup (not necessarily connected) of G with maximal compact subgroup $K(\gamma)$, and with Cartan decomposition (5.39).

5.4.3. The symmetric space. Set $X = G/K$ to be the associated symmetric space. Then

$$p : G \to X = G/K \tag{5.40}$$

is a K-principle bundle.

Let τ be a finite-dimensional orthogonal representation of K on the real Euclidean space E_τ. Then $\mathcal{E}_\tau = G \times_K E_\tau$ is a real Euclidean vector bundle on X. The space of smooth sections $C^\infty(X, \mathcal{E}_\tau)$ on X can be identified with the subspace $C^\infty(G, E_\tau)^K$ of K-invariant smooth E_τ-valued functions on G.

By adjoint action, K acts isometrically on $\mathfrak{p}$. Using the above construction, the tangent bundle of X is given by

$$TX = G \times_K \mathfrak{p}. \tag{5.41}$$

The bilinear form $B|_{\mathfrak{p}}$ induces a G-invariant Riemannian metric g^{TX} on X. It is well known that (X, g^{TX}) is a Riemannian manifold of non-positive sectional curvature. For $x, y \in X$, we denote by $d_X(x, y)$ the Riemannian distance on X.

5.4.4. The locally symmetric space. Let $\Gamma \subset G$ be a discrete torsion-free cocompact subgroup of G. Take $Z = \Gamma\backslash X = \Gamma\backslash G/K$. Then Z is a connected closed orientable Riemannian locally symmetric manifold with non-positive sectional curvature. Since X is contractible, $\pi_1(Z) = \Gamma$ and X is the universal cover of Z.

By [Sel60, Lemmas 1], Γ contains the identity element and non-elliptic semisimple elements. Recall that $\Gamma(\gamma)$ is the centralizer of γ in Γ defined before (5.19). By [Sel60, Lemma 2], $\Gamma(\gamma)$ is cocompact in $Z(\gamma)$.

The following proposition is [DuKV79, Proposition 5.15], relating the geometric object considered in Section 5.2 and the group theoretic object considered in this section.

Proposition 5.8. *For $[\gamma] \in [\Gamma] - \{1\}$, we have the following identifications,*

$$\ell_{[\gamma]} = |a|, \qquad X(\gamma) \simeq Z(\gamma)/K(\gamma), \qquad B_{[\gamma]} \simeq \Gamma(\gamma)\backslash Z(\gamma)/K(\gamma). \tag{5.42}$$

Let us state the main theorem of [Sh18, Theorem 1.1], which generalizes [F86a, Theorem 3] in the case where $\mathrm{rk}_{\mathbf{R}}[G] = 1$ and [MoSt91, Corollary 2.2] in the case where $\delta(G) \neq 1$.

Theorem 5.9. *Assume that $Z = \Gamma\backslash G/K$ and $\dim Z \geqslant 3$ is odd. Let F be a unitarily flat vector bundle on Z. The following statements hold.*

(1) *The dynamical zeta function $R_{\phi,\rho}(\sigma)$ is well defined in the sense of Definition 2.7.*

(2) *There exist explicit constants $C_\rho \in \mathbf{R}^*$ and $r_\rho \in \mathbf{Z}$ (see (5.99)) such that, when $\sigma \to 0$,*

$$R_{\phi,\rho}(\sigma) = C_\rho T_F(Z)^2 \sigma^{r_\rho} + \mathcal{O}\left(\sigma^{r_\rho+1}\right). \tag{5.43}$$

(3) *If $H^\cdot(Z,F) = 0$, then*

$$C_\rho = 1, \qquad r_\rho = 0, \tag{5.44}$$

so that

$$R_{\phi,\rho}(0) = T_F(Z)^2 = T_{\pi^*(F)}(M). \tag{5.45}$$

We sketch the proof of Theorem 5.9 in Sections 5.5–5.8.

Remark 5.10. If we do not require Γ to be torsion free, then $Z = \Gamma\backslash G/K$ is an orbifold. In [ShY17], we show that the above theorem still holds true.

Remark 5.11. Using Hirzebruch proportionality [H66, Theorem 22.3.1] and Bott's formula [Bo65, p. 175], we can show that if $\delta(G) \geqslant 2$, then for all $[\gamma] \in [\Gamma] - \{1\}$,

$$\chi(B_{[\gamma]}/\mathbb{S}^1) = 0. \tag{5.46}$$

So $R_{\phi,\rho}(\sigma) \equiv 1$.

Remark 5.12. We do not need to study the case where $Z = \Gamma\backslash G/K$ with $\dim Z \geqslant 4$ is even. Recall that $\delta(G)$ and $\dim Z$ have the same parity. If $\delta(G) \geqslant 2$, by (1.10) and Remark 5.11, we have $R_{\phi,\rho}(0) = T_F(Z) = 1$. If $\delta(G) = 0$, there are no acyclic flat vector bundles on Z. Indeed, as in (5.46), by Hirzebruch proportionality [H66, Theorem 22.3.1] and Bott's formula [Bo65, p. 175], we can deduce that

$$(-1)^{\frac{1}{2}\dim Z}\chi(Z) > 0. \tag{5.47}$$

Using the Gauss–Bonnet–Chern Theorem, by (5.47), we see that

$$\chi(Z,F) = \mathrm{rk}[F]\chi(Z) \neq 0, \tag{5.48}$$

which implies $H^\cdot(Z,F) \neq 0$.

Remark 5.13. By Remark 5.11, if $\dim Z$ is odd and $\delta(G) \geqslant 3$, to show the Fried conjecture, it is enough to show $T_F(Z) = 1$, which is known as the Moscovici–Stanton vanishing theorem (see Section 5.6).

5.5. The Selberg trace formula

In this section, we recall the Selberg trace formula. We introduce the Casimir operator and its heat kernel, orbital integral, and an explicit orbital integral formula [B11, Theorem 6.1.1] for the heat operator of the Casimir.

5.5.1. Casimir operator. Let $\mathscr{U}(\mathfrak{g})$ be the enveloping algebra of $\mathfrak{g}$. We identify $\mathscr{U}(\mathfrak{g})$ with the algebra of left-invariant differential operators on G. Let $C^{\mathfrak{g}} \in \mathscr{U}(\mathfrak{g})$ be the Casimir element of $(\mathfrak{g}, B)$. If $\{e_1, \ldots, e_{\dim \mathfrak{p}}\}$ is an orthonormal basis of $\mathfrak{p}$, and if $\{f_1, \ldots, f_{\dim \mathfrak{k}}\}$ is an orthonormal basis of $\mathfrak{k}$, then

$$C^{\mathfrak{g}} = -\sum_{i=1}^{\dim \mathfrak{p}} e_i^2 + \sum_{i=1}^{\dim \mathfrak{k}} f_i^2. \tag{5.49}$$

It is well known that $C^{\mathfrak{g}}$ is in the center of $\mathscr{U}(\mathfrak{g})$.

We define $C^{\mathfrak{k}}$ in the same way. Let τ be a finite-dimensional representation of K on E_τ. We denote by $C^{\mathfrak{k}, E_\tau} \in \operatorname{End}(E_\tau)$ the corresponding Casimir operator acting on E_τ, so that

$$C^{\mathfrak{k}, E_\tau} = \sum_{i=1}^{\dim \mathfrak{k}} \tau(f_i)^2. \tag{5.50}$$

Let $C^{\mathfrak{g}, X, \tau}$ be the Casimir element of G acting on $C^\infty(X, \mathcal{E}_\tau)$. Then $C^{\mathfrak{g}, X, \tau}$ is a formally self-adjoint second-order elliptic differential operator which is bounded from below. If $E_\tau = \Lambda^\cdot(\mathfrak{p}^*)$, then $C^\infty(X, \mathcal{E}_\tau) = \Omega^\cdot(X)$. In this case, we write $C^{\mathfrak{g}, X} = C^{\mathfrak{g}, X, \tau}$. By [B11, Proposition 7.8.1], $C^{\mathfrak{g}, X}$ coincides with the Hodge Laplacian acting on $\Omega^\cdot(X)$.

We denote by $\widehat{p} : \Gamma \backslash G \to Z$ and $\pi : X \to Z$ the natural projections, so that the diagram

$$\begin{array}{ccc} G & \longrightarrow & \Gamma\backslash G \\ \downarrow{\scriptstyle p} & & \downarrow{\scriptstyle \widehat{p}} \\ X & \xrightarrow{\ \pi\ } & Z \end{array} \tag{5.51}$$

commutes. Recall that the group Γ acts isometrically on the left on X. This action lifts to all the homogeneous Euclidean vector bundles $\mathcal{E}_\tau$ constructed in Subsection 5.4.3. It descends to a Euclidean vector bundle $\mathcal{F}_\tau = \Gamma \backslash \mathcal{E}_\tau$ on Z. Let F be the unitarily flat vector bundle of rank r on Z. Let $\rho : \Gamma \to \mathrm{U}(r)$ be the holonomy of F. Let $C^{\mathfrak{g}, Z, \tau, \rho}$ be the Casimir element of G acting on $C^\infty(Z, \mathcal{F}_\tau \otimes_{\mathbf{R}} F)$. As before, when $E_\tau = \Lambda^\cdot(\mathfrak{p}^*)$, we write $C^{\mathfrak{g}, Z, \rho} = C^{\mathfrak{g}, Z, \tau, \rho}$. Then,

$$\square^Z = C^{\mathfrak{g}, Z, \rho}. \tag{5.52}$$

5.5.2. The heat kernel of $C^{\mathfrak{g},X,\tau}$. We fix a finite-dimensional real orthogonal representation (τ, E_τ) of K.

Let $p_t^{X,\tau}(x,x')$ be the smooth kernel of $\exp(-tC^{\mathfrak{g},X,\tau}/2)$ with respect to the Riemannian volume dv_X on X. The Gaussian estimate on the heat kernel tells us that for $t>0$, there exist $c>0$ and $C>0$ such that for $x,x'\in X$,

$$\left|p_t^{X,\tau}(x,x')\right| \leqslant C\exp\left(-c\, d_X^2(x,x')\right). \tag{5.53}$$

Set

$$p_t^{X,\tau}(g) = p_t^{X,\tau}(p1,pg). \tag{5.54}$$

It is a smooth $\mathrm{End}(E_\tau)$-valued function on G. For $g\in G$ and $k,k'\in K$, we have

$$p_t^{X,\tau}(kgk') = \tau(k)p_t^{X,\tau}(g)\tau(k'). \tag{5.55}$$

We can recover $p_t^{X,\tau}(x,x')$ by

$$p_t^{X,\tau}(x,x') = p_t^{X,\tau}(g^{-1}g'), \tag{5.56}$$

where $g,g'\in G$ are such that $pg=x, pg'=x'$.

In the sequel, we do not distinguish $p_t^{X,\tau}(x,x')$ and $p_t^{X,\tau}(g)$. We refer to both of them as the smooth kernel of $\exp(-tC^{\mathfrak{g},X,\tau}/2)$.

5.5.3. The Selberg trace formula. We will write the trace of $\exp(-tC^{\mathfrak{g},Z,\tau,\rho}/2)$ with the help of the heat kernel $p_t^{X,\tau}(g)$ on X. Let dv_K be the Haar measure on K such that the volume of K is equal to 1. By (5.33), dv_K and $B|_{\mathfrak{p}}$ induce a Haar measure dv_G on G. Let $dv_{Z(\gamma)}$ be the Haar measure on $Z(\gamma)$ constructed in the same way. Let $dv_{Z(\gamma)\backslash G}$ be the right invariant measure on $Z(\gamma)\backslash G$ such that

$$dv_G = dv_{Z(\gamma)}dv_{Z(\gamma)\backslash G}. \tag{5.57}$$

It is known (see [B11, (3.4.36)]) that for $s\gg 1$,

$$\int_{g\in Z(\gamma)\backslash G} e^{-sd_X(p\gamma g,pg)}dv_{Z(\gamma)\backslash G} < \infty. \tag{5.58}$$

Definition 5.14. Let $\gamma\in G$ be semisimple. We define the orbital integral of $\exp(-tC^{\mathfrak{g},X,\tau}/2)$ by

$$\mathrm{Tr}^{[\gamma]}\left[\exp\left(-tC^{\mathfrak{g},X,\tau}/2\right)\right] = \int_{g\in Z(\gamma)\backslash G}\mathrm{Tr}^{E_\tau}\left[p_t^{X,\tau}(g^{-1}\gamma g)\right]dv_{Z(\gamma)\backslash G}. \tag{5.59}$$

By (5.53) and (5.58), we see that the integration in (5.59) is well defined.

Theorem 5.15. *There exist $c>0$, $C>0$ such that for $t>0$, we have*

$$\sum_{[\gamma]\in[\Gamma]-\{1\}} \mathrm{vol}\left(B_{[\gamma]}\right)\left|\mathrm{Tr}^{[\gamma]}\left[\exp\left(-tC^{\mathfrak{g},X,\tau}/2\right)\right]\right| \leqslant C\exp\left(-\frac{c}{t}+Ct\right). \tag{5.60}$$

For $t > 0$, the following identity holds,

$$\operatorname{Tr}\left[\exp\left(-tC^{\mathfrak{g},Z,\tau,\rho}/2\right)\right] = \sum_{[\gamma]\in[\Gamma]} \operatorname{vol}\left(B_{[\gamma]}\right) \operatorname{Tr}[\rho(\gamma)] \operatorname{Tr}^{[\gamma]}\left[\exp(-tC^{\mathfrak{g},X,\tau}/2)\right]. \tag{5.61}$$

Proof. The estimate in (5.60) and the convergence of (5.61) are consequences of the following fact: there exist $c > 0$, $C > 0$ such that for $t > 0$ and $x \in X$, we have

$$\sum_{\gamma\in\Gamma-\{1\}} \left|p_t^{X,\tau}(x,\gamma x)\right| \leqslant C\exp\left(-\frac{c}{t} + Ct\right). \tag{5.62}$$

The proof of (5.62) can be found in [Sh18, Proposition 4.8] (see also [MaMari15, Theorem 4]), where we use an estimate similar to the one in (5.25).

Let us prove (5.61). We disregard the convergence problem. Let $p_t^{Z,\tau,\rho}(z,z')$ be the smooth kernel of $\exp(-tC^{\mathfrak{g},Z,\tau,\rho}/2)$ with respect to the Riemannian volume on Z. For $x \in X$ and $g \in G$ such that $pg = x$, we have

$$p_t^{Z,\tau,\rho}(\pi x, \pi x) = \sum_{\gamma\in\Gamma} \rho(\gamma) p_t^{X,\tau}\left(g^{-1}\gamma g\right). \tag{5.63}$$

Recall that $F_Z \subset X$ is the fundamental domain of Z in X. Then, $p^{-1}(F_Z) \subset G$ is a fundamental domain of $\Gamma\backslash G$ in G. Since $\operatorname{vol}(K) = 1$, by (5.63), we have

$$\operatorname{Tr}\left[\exp\left(-tC^{\mathfrak{g},Z,\tau,\rho}/2\right)\right] = \int_{g\in p^{-1}(F_Z)} \sum_{\gamma\in\Gamma} \operatorname{Tr}\left[\rho(\gamma)\right] \operatorname{Tr}\left[p_t^{X,\tau}\left(g^{-1}\gamma g\right)\right] dv_G. \tag{5.64}$$

Take $\gamma \in \Gamma$. Recall that $[\gamma] \in [\Gamma]$ is the conjugacy class of γ in Γ. We claim that

$$\int_{g\in p^{-1}(F_Z)} \sum_{\gamma'\in[\gamma]} \operatorname{Tr}\left[p_t^{X,\tau}\left(g^{-1}\gamma g\right)\right] dv_G = \operatorname{vol}(\Gamma(\gamma)\backslash Z(\gamma)) \operatorname{Tr}^{[\gamma]}\left[\exp(-tC^{\mathfrak{g},X,\tau}/2)\right] \tag{5.65}$$

Indeed, since $\gamma' \to (\gamma')^{-1}\gamma\gamma'$ induces an identification of $\Gamma(\gamma)\backslash\Gamma \simeq [\gamma]$, we have

$$\sum_{\gamma'\in[\gamma]} \operatorname{Tr}\left[p_t^{X,\tau}\left(g^{-1}\gamma' g\right)\right] = \sum_{\gamma'\in\Gamma(\gamma)\backslash\Gamma} \operatorname{Tr}\left[p_t^{X,\tau}\left(g^{-1}(\gamma')^{-1}\gamma\gamma' g\right)\right]. \tag{5.66}$$

By (5.66), and by changing variable, we have

$$\int_{g\in p^{-1}(F_Z)} \sum_{\gamma'\in[\gamma]} \operatorname{Tr}\left[p_t^{X,\tau}\left(g^{-1}\gamma' g\right)\right] dv_G \tag{5.67}$$

$$= \int_{\bigcup_{\gamma'\in\Gamma(\gamma)\backslash\Gamma}\gamma' p^{-1}(F_Z)} \operatorname{Tr}\left[p_t^{X,\tau}\left(g^{-1}\gamma g\right)\right] dv_G = \int_{\Gamma(\gamma)\backslash G} \operatorname{Tr}\left[p_t^{X,\tau}\left(g^{-1}\gamma g\right)\right] dv_{\Gamma(\gamma)\backslash G},$$

where in the last equality we use the fact that $\bigcup_{\gamma'\in\Gamma(\gamma)\backslash\Gamma}\gamma' p^{-1}(F_Z)$ is a fundamental domain of $\Gamma(\gamma)\backslash G$ in G. Using the fibration $\Gamma(\gamma)\backslash G \to Z(\gamma)\backslash G$, and using the fact that the integrand $\operatorname{Tr}\left[p_t^{X,\tau}\left(g^{-1}\gamma g\right)\right]$ is constant on the fibre, by (5.59) and (5.67), we get (5.65).

By Proposition 5.8, (5.64), and (5.65), using $\mathrm{vol}(K(\gamma)) = 1$, we get (5.61). □

5.5.4. Bismut orbital integral formula. Let us recall an explicit orbital integral formula for $\mathrm{Tr}^{[\gamma]}\left[\exp(-tC^{\mathfrak{g},X,\tau}/2)\right]$ obtained by Bismut [B11, Theorem 6.1.1].

Let $\gamma = e^a k^{-1} \in G$ be semisimple as in (5.36). Set

$$\mathfrak{p}_0 = \ker(\mathrm{ad}(a)) \cap \mathfrak{p}, \qquad \mathfrak{k}_0 = \ker(\mathrm{ad}(a)) \cap \mathfrak{k}, \qquad \mathfrak{z}_0 = \ker(\mathrm{ad}(a)) \cap \mathfrak{z}. \tag{5.68}$$

Let $\mathfrak{p}_0^\perp \subset \mathfrak{p}$, $\mathfrak{k}_0^\perp \subset \mathfrak{k}$, $\mathfrak{z}_0^\perp \subset \mathfrak{z}$ be the orthogonal spaces of $\mathfrak{p}_0, \mathfrak{k}_0, \mathfrak{z}_0$ with respect to B. Let $\mathfrak{p}_0^\perp(\gamma) \subset \mathfrak{p}_0$, $\mathfrak{k}_0^\perp(\gamma) \subset \mathfrak{k}_0$, $\mathfrak{z}_0^\perp(\gamma) \subset \mathfrak{z}_0$ be the orthogonal spaces of $\mathfrak{p}(\gamma), \mathfrak{k}(\gamma), \mathfrak{z}(\gamma)$ with respect to B. Then

$$\mathfrak{z}_0^\perp = \mathfrak{p}_0^\perp \oplus \mathfrak{k}_0^\perp, \qquad \mathfrak{z}_0^\perp(\gamma) = \mathfrak{p}_0^\perp(\gamma) \oplus \mathfrak{k}_0^\perp(\gamma). \tag{5.69}$$

In particular, we have the orthogonal decompositions with respect to B,

$$\mathfrak{p} = \underbrace{\mathfrak{p}(\gamma) \oplus \mathfrak{p}_0^\perp(\gamma)}_{\mathfrak{p}_0} \oplus \mathfrak{p}_0^\perp, \quad \mathfrak{k} = \underbrace{\mathfrak{k}(\gamma) \oplus \mathfrak{k}_0^\perp(\gamma)}_{\mathfrak{k}_0} \oplus \mathfrak{k}_0^\perp, \quad \mathfrak{g} = \underbrace{\mathfrak{z}(\gamma) \oplus \mathfrak{z}_0^\perp(\gamma)}_{\mathfrak{z}_0} \oplus \mathfrak{z}_0^\perp. \tag{5.70}$$

Let us introduce the J-function [B11, Section 5.5] of Bismut. In [B11, Section 5.5], it has been shown that the analytic function defined for $Y \in \mathfrak{k}(\gamma)$ by

$$\frac{1}{\det\left(1 - \mathrm{Ad}(k^{-1})\right)|_{\mathfrak{z}_0^\perp(\gamma)}} \frac{\det\left(1 - \exp(-i\,\mathrm{ad}(Y))\,\mathrm{Ad}(k^{-1})\right)|_{\mathfrak{k}_0^\perp(\gamma)}}{\det\left(1 - \exp(-i\,\mathrm{ad}(Y))\,\mathrm{Ad}(k^{-1})\right)|_{\mathfrak{p}_0^\perp(\gamma)}} \tag{5.71}$$

has a natural square root, which is still analytic in $Y \in \mathfrak{k}(\gamma)$. For a Hermitian matrix H, define

$$\widehat{A}(H) = \det{}^{1/2}\left(\frac{H/2}{\sinh(H/2)}\right). \tag{5.72}$$

The square root in (5.72) is the positive square root of a positive real number.

Definition 5.16. Let J_γ be the analytic function on $\mathfrak{k}(\gamma)$ defined for $Y \in \mathfrak{k}(\gamma)$ by

$$\begin{aligned} J_\gamma(Y) = {} & \frac{1}{\left|\det\left(1 - \mathrm{Ad}(\gamma)\right)|_{\mathfrak{z}_0^\perp}\right|^{1/2}} \frac{\widehat{A}\left(i\,\mathrm{ad}(Y)|_{\mathfrak{p}(\gamma)}\right)}{\widehat{A}\left(i\,\mathrm{ad}(Y)|_{\mathfrak{k}(\gamma)}\right)} \\ & \left[\frac{1}{\det\left(1 - \mathrm{Ad}(k^{-1})\right)|_{\mathfrak{z}_0^\perp(\gamma)}} \frac{\det\left(1 - \exp(-i\,\mathrm{ad}(Y))\,\mathrm{Ad}(k^{-1})\right)|_{\mathfrak{k}_0^\perp(\gamma)}}{\det\left(1 - \exp(-i\,\mathrm{ad}(Y))\,\mathrm{Ad}(k^{-1})\right)|_{\mathfrak{p}_0^\perp(\gamma)}}\right]^{1/2}. \end{aligned} \tag{5.73}$$

Remark 5.17. There is $C_\gamma > 0$ such that for all $Y \in \mathfrak{k}(\gamma)$, we have

$$|J_\gamma(Y)| \leqslant C_\gamma e^{C_\gamma |Y|}. \tag{5.74}$$

Recall that $C^{\mathfrak{k},\mathfrak{p}}$ and $C^{\mathfrak{k},\mathfrak{k}}$ are defined as in (5.50) associated with the adjoint actions of K on $\mathfrak{p}$ and $\mathfrak{k}$. Write

$$c_{\mathfrak{g}} = -\frac{1}{8}\mathrm{Tr}^{\mathfrak{p}}[C^{\mathfrak{k},\mathfrak{p}}] - \frac{1}{24}\mathrm{Tr}^{\mathfrak{k}}\left[C^{\mathfrak{k},\mathfrak{k}}\right] \in \mathbf{R}_+. \tag{5.75}$$

For $Y \in \mathfrak{k}(\gamma)$, let dY be the Lebesgue measure on $\mathfrak{k}(\gamma)$ induced by $-B$. The main result of [B11, Theorem 6.1.1] is the following.

Theorem 5.18. *For $t > 0$, we have*

$$\operatorname{Tr}^{[\gamma]}\left[\exp\left(-tC^{\mathfrak{g},X,\tau}/2\right)\right] = \frac{1}{(2\pi t)^{\dim \mathfrak{z}(\gamma)/2}} \exp\left(-\frac{|a|^2}{2t} - \frac{c_{\mathfrak{g}}}{2}t\right)$$
$$\int_{Y\in\mathfrak{k}(\gamma)} J_\gamma(Y) \operatorname{Tr}^{E_\tau}\left[\tau\left(k^{-1}\right)\exp(-i\tau(Y))\right]\exp\left(-|Y|^2/2t\right)dY. \quad (5.76)$$

5.6. Moscovici–Stanton's vanishing theorem

In this section, we show Theorem 5.9 in the case of $\delta(G) \neq 1$. By Remark 5.13, it is enough to show $T_F(Z) = 1$.

Let $\mathfrak{t} \subset \mathfrak{k}$ be a Cartan subalgebra of $\mathfrak{k}$. Set

$$\mathfrak{b} = \{Y \in \mathfrak{p} : [Y, \mathfrak{t}] = 0\}. \quad (5.77)$$

Then $\mathfrak{h} = \mathfrak{b} \oplus \mathfrak{t}$ is a fundamental Cartan subalgebra of $\mathfrak{g}$.

Let $\gamma = e^a k^{-1} \in G$ be a semisimple element such that (5.36) holds. Let $\mathfrak{t}(\gamma) \subset \mathfrak{k}(\gamma)$ be a Cartan subalgebra of $\mathfrak{k}(\gamma)$. Set

$$\mathfrak{b}(\gamma) = \{Y \in \mathfrak{p} : [Y, \mathfrak{t}(\gamma)] = 0, \operatorname{Ad}(k)Y = Y\}. \quad (5.78)$$

By (5.77) and (5.78), we have

$$\dim \mathfrak{b}(\gamma) \geqslant \dim \mathfrak{b} = \delta(G). \quad (5.79)$$

The following theorem is [B11, Theorem 7.9.1] (see also [Sh18, Theorem 4.12]).

Theorem 5.19. *Let $\gamma \in G$ be semisimple such that* $\dim \mathfrak{b}(\gamma) \geqslant 2$. *For $Y \in \mathfrak{k}(\gamma)$, we have*

$$\operatorname{Tr}_{\mathrm{s}}^{\Lambda^\cdot(\mathfrak{p}^*)}\left[N^{\Lambda^\cdot(\mathfrak{p}^*)}\operatorname{Ad}(k^{-1})\exp(-i\operatorname{ad}(Y))\right] = 0. \quad (5.80)$$

In particular, for $t > 0$, we have

$$\operatorname{Tr}_{\mathrm{s}}^{[\gamma]}\left[N^{\Lambda^\cdot(T^*X)}\exp\left(-tC^{\mathfrak{g},X}/2\right)\right] = 0. \quad (5.81)$$

Proof. Since the term on the left-hand side of (5.80) is $\operatorname{Ad}\big(K(\gamma)\big)$-invariant, it is enough to show (5.80) for $Y \in \mathfrak{t}(\gamma)$. If $Y \in \mathfrak{t}(\gamma)$, we have

$$\begin{aligned} &\operatorname{Tr}_{\mathrm{s}}^{\Lambda^\cdot(\mathfrak{p}^*)}\left[N^{\Lambda^\cdot(\mathfrak{p}^*)}\operatorname{Ad}(k^{-1})\exp(-i\operatorname{ad}(Y))\right] \\ &= \frac{\partial}{\partial b}\Big|_{b=0}\det\left(1 - e^b\operatorname{Ad}(k)\exp(i\operatorname{ad}(Y))\right)|_{\mathfrak{p}}. \end{aligned} \quad (5.82)$$

Since $\dim \mathfrak{b}(\gamma) \geqslant 2$, by (5.82), we get (5.80) for $Y \in \mathfrak{t}(\gamma)$. By (5.76) and (5.80), we get (5.81). □

In this way, Bismut [B11, Theorem 7.9.3] recovered [MoSt91, Corollary 2.2].

Corollary 5.20. *Let F be a unitarily flat vector bundle on Z. Assume that* $\dim Z$ *is odd and* $\delta(G) \neq 1$. *Then for any $t > 0$, we have*

$$\operatorname{Tr_s}\left[N^{\Lambda^\cdot(T^*Z)} \exp\left(-t\Box^Z/2\right)\right] = 0. \tag{5.83}$$

In particular,

$$T_F(Z) = 1. \tag{5.84}$$

Proof. Since $\dim Z$ and $\delta(G)$ have the same parity, and since $\dim Z$ is odd and $\delta(G) \neq 1$, we get $\delta(G) \geqslant 3$. By (5.79), $\dim \mathfrak{b}(\gamma) \geqslant 3$. Thus, (5.83) is a consequence of (5.52), (5.61), and (5.81). □

5.7. The case $G = \mathrm{SO}^0(p, 1)$ with p odd

In this section, we assume $G = \mathrm{SO}^0(p, 1)$ with p odd. Then G is a semisimple Lie group of $\mathbf{R}$-rank 1, and $X = G/K$ is hyperbolic space of dimension p. The result of this section is due to Fried [F86a].

It is easy to show that up to a sign there exists $\alpha \in \mathfrak{b}^*$ such that for $a \in \mathfrak{b}$, $\mathrm{ad}(a)$ acting on $\mathfrak{g}$ has three eigenvalues $0, \pm\langle\alpha, a\rangle$. We have an orthogonal splitting with respect to B,

$$\mathfrak{g} = \mathfrak{b} \oplus \mathfrak{m} \oplus \mathfrak{n} \oplus \bar{\mathfrak{n}}, \tag{5.85}$$

where $\mathfrak{b} \oplus \mathfrak{m}$ (resp. $\mathfrak{n}$, resp. $\bar{\mathfrak{n}}$) is the eigenspace of $\mathrm{ad}(a)$ for the eigenvalue 0 (resp. $\langle\alpha, a\rangle$, resp. $-\langle\alpha, a\rangle$). Let $\mathcal{M} \subset G$ be the connected subgroup of G associated to $\mathfrak{m}$. Then $\mathcal{M}$ is compact and isomorphic to $\mathrm{SO}(p-1)$. Moreover, the action of $\mathcal{M}$ on $\mathfrak{n}$ is just the $\mathrm{SO}(p-1)$-action on $\mathbf{R}^{p-1}$.

One feature of the group $\mathrm{SO}^0(p, 1)$ with p odd is that any non-elliptic semisimple element of γ can be conjugate to $e^a k^{-1}$ such that $a \in \mathfrak{b}, a \neq 0$ and $k \in \mathcal{M}$. Moreover, the subspaces $\mathfrak{z}_0$ and $\mathfrak{z}_0^\perp$ introduced in (5.68) and (5.69) do not depend on the choice of non-elliptic semisimple element of γ, and is given by

$$\mathfrak{z}_0 = \mathfrak{b} \oplus \mathfrak{m}, \qquad \mathfrak{z}_0^\perp = \mathfrak{n} \oplus \bar{\mathfrak{n}}. \tag{5.86}$$

Definition 5.21. For $\sigma \in \mathbf{C}$, we define a formal Selberg zeta function associated to a representation η of $\mathcal{M}$ and to a representation ρ of Γ by

$$Z_{\eta,\rho}(\sigma) = \exp\left(-\sum_{\substack{[\gamma]\in[\Gamma]-\{1\}\\ \gamma\sim e^a k^{-1}}} \operatorname{Tr}[\rho(\gamma)] \frac{\operatorname{Tr}^{E_\eta}\left[\eta(k^{-1})\right]}{\left|\det\left(1-\mathrm{Ad}(\gamma)\right)|_{\mathfrak{z}_0^\perp}\right|^{1/2}} \frac{e^{-\sigma \ell_{[\gamma]}}}{m_{[\gamma]}}\right). \tag{5.87}$$

The formal Selberg zeta function is said to be well defined if the same conditions as in Definition 2.7 hold.

For $\gamma = e^a k^{-1}$ such that $a \in \mathfrak{b}, a \neq 0$ and $k \in \mathcal{M}$, we have

$$\left|\det(1-\mathrm{Ad}(\gamma))|_{\mathfrak{z}_0^\perp}\right|^{1/2} = \sum_{j=0}^{\dim \mathfrak{n}} (-1)^j \operatorname{Tr}^{\Lambda^j(\mathfrak{n}^*)}\left[\mathrm{Ad}\left(k^{-1}\right)\right] e^{\left(\frac{1}{2}\dim \mathfrak{n} - j\right)|\alpha||a|}. \tag{5.88}$$

If we consider the Selberg zeta function associated to $\eta_j = \Lambda^j(\mathfrak{n}^*)$, using (5.88) and $\ell_{[\gamma]} = |a|$, we have

$$R_{\phi,\rho}(\sigma) = \prod_{j=0}^{\dim \mathfrak{n}} Z_{\eta_j,\rho}\left(\sigma + \left(j - \frac{\dim \mathfrak{n}}{2}\right)|\alpha|\right)^{(-1)^{j-1}}. \tag{5.89}$$

To show the meromorphic extension of $R_{\phi,\rho}$, it is enough to consider the meromorphic extension of $Z_{\eta_j,\rho}$.

There are two remarkable properties of η_j. First, since η_j is just the representation of $\mathrm{SO}(p-1)$ on $\Lambda^j(\mathbf{R}^{p-1})$, the Casimir of $\mathfrak{m}$ acts on η_j is a scalar. Second, η_j has a lift as a virtual representation of K. More precisely, let $\mathrm{RO}(K), \mathrm{RO}(\mathcal{M})$ be the real representation rings of K and $\mathcal{M}$. The restriction induces a morphism of rings

$$\iota : \mathrm{RO}(K) \to \mathrm{RO}(\mathcal{M}). \tag{5.90}$$

Since K and $\mathcal{M}$ have the same complex rank, ι is an injection. The second property is that η_j has a lift $\widehat{\eta}_j = \widehat{\eta}_j^+ - \widehat{\eta}_j^- \in \mathrm{RO}(K)$, where

$$\widehat{\eta}_j^+ = \Lambda^j(\mathfrak{p}^*) + \Lambda^{j-2}(\mathfrak{p}^*) + \cdots, \qquad \widehat{\eta}_j^- = \Lambda^{j-1}(\mathfrak{p}^*) + \Lambda^{j-3}(\mathfrak{p}^*) + \cdots. \tag{5.91}$$

For a virtual representation $\widehat{\eta} = \widehat{\eta}^+ - \widehat{\eta}^- \in \mathrm{RO}(K)$, where $\widehat{\eta}^+$, $\widehat{\eta}^-$ are two representations of K, we use the notation

$$\det{}_{\mathrm{s}}\left(\sigma + C^{\mathfrak{g},Z,\widehat{\eta},\rho}\right) = \frac{\det\left(\sigma + C^{\mathfrak{g},Z,\widehat{\eta}^+,\rho}\right)}{\det\left(\sigma + C^{\mathfrak{g},Z,\widehat{\eta}^-,\rho}\right)}. \tag{5.92}$$

Recall that $c_{\mathfrak{g}} \in \mathbf{R}_+$ is defined in (5.75). We define $c_{\mathfrak{m}} \in \mathbf{R}_+$ in the same way.

Proposition 5.22. *Assume that η has a lift $\widehat{\eta} \in \mathrm{RO}(K)$ and that the Casimir of $\mathfrak{m}$ on E_η is a scalar $C^{\mathfrak{m},\eta} \in \mathbf{R}_-$. Then $Z_{\eta,\rho}(\sigma)$ is well defined such that*

$$Z_{\eta,\rho}(\sigma) = e^{P_\eta(\sigma)}\det{}_{\mathrm{s}}\left(\sigma^2 - c_{\mathfrak{g}} + c_{\mathfrak{m}} - C^{\mathfrak{m},\eta} + C^{\mathfrak{g},Z,\widehat{\eta},\rho}\right). \tag{5.93}$$

where $P_\eta(\sigma)$ is an odd polynomial of σ.

Proof. By a general theorem of elliptic operators [Vo87], we know that the right-hand side of (5.93) has a meromorphic extension to $\mathbf{C}$. So it is enough to show (5.93) for $\sigma \in \mathbf{R}$ and $\sigma \gg 1$. In this case, using the Mellin transform, we have

$$\begin{aligned}&\det{}_{\mathrm{s}}\left(\sigma^2 - c_{\mathfrak{g}} + c_{\mathfrak{m}} - C^{\mathfrak{m},\eta} + C^{\mathfrak{g},Z,\widehat{\eta},\rho}\right) \\ &= \exp\left(-\frac{\partial}{\partial s}\Big|_{s=0}\frac{1}{\Gamma(s)}\int_0^\infty \mathrm{Tr_s}\left[\exp\left(-t\left(\sigma^2 - c_{\mathfrak{g}} + c_{\mathfrak{m}} - C^{\mathfrak{m},\eta} + C^{\mathfrak{g},Z,\widehat{\eta},\rho}\right)\right)\right]t^{s-1}dt\right).\end{aligned} \tag{5.94}$$

We use the Selberg trace formula to evaluate the second line of (5.94). For the group of $\mathbf{R}$-rank 1, Bismut's orbital integral formula has a very simple form

[B11, Theorem 8.2.1]. For $\gamma = e^a k^{-1}$ with $a \in \mathfrak{b}$, $a \neq 0$, and $k \in \mathcal{M}$, we have

$$\begin{aligned}\operatorname{Tr}^{[\gamma]}\left[\exp\left(-tC^{\mathfrak{g},X,\widehat{\eta}}/2\right)\right] \\ = \frac{1}{\left|\det\left(1-\operatorname{Ad}(\gamma)\right)|_{\mathfrak{z}_0^\perp}\right|^{1/2}} \frac{1}{\sqrt{2\pi t}} \exp\left(-\frac{|a|^2}{2t} + \frac{-c_\mathfrak{g}+c_\mathfrak{m}-C^{\mathfrak{m},\eta}}{2}t\right) \operatorname{Tr}\left[\eta(k^{-1})\right],\end{aligned} \tag{5.95}$$

and if $\gamma = 1$, we have

$$\operatorname{Tr}^{[\gamma]}\left[\exp\left(-tC^{\mathfrak{g},X,\widehat{\eta}}/2\right)\right] = \frac{1}{\sqrt{t}} Q_\eta\left(\frac{1}{t}\right)\exp\left(\frac{-c_\mathfrak{g}+c_\mathfrak{m}-C^{\mathfrak{m},\eta}}{2}t\right). \tag{5.96}$$

where Q_η is some explicit polynomial. Using Theorem 5.15, (5.94)–(5.96), we deduce (5.93) for $\sigma \in \mathbf{R}$ and $\sigma \gg 1$. □

Let us give an explicit formula for the constant appearing in Theorem 5.9. For $0 \leqslant j \leqslant \dim \mathfrak{n}$, set

$$r_j = \dim\ker C^{\mathfrak{g},Z,\widehat{\eta}_j^+,\rho} - \dim\ker C^{\mathfrak{g},Z,\widehat{\eta}_j^-,\rho}. \tag{5.97}$$

By (5.52) and (5.91), we have

$$r_j = b_j(Z,F) - b_{j-1}(Z,F) + b_{j-2}(Z,F) - \cdots, \tag{5.98}$$

where $b_j(Z,F) = \dim H^j(Z,F)$ is the Betti number. Clearly, $r_j = r_{\dim\mathfrak{n}-j}$.

Definition 5.23. Set

$$C_\rho = \prod_{j=0}^{\frac{\dim\mathfrak{n}}{2}-1}\left(-4\left(\frac{\dim\mathfrak{n}}{2}-j\right)^2|\alpha|^2\right)^{(-1)^{j-1}r_j}, \quad r_\rho = 2\sum_{j=0}^{\frac{\dim\mathfrak{n}}{2}}(-1)^{j-1}r_j. \tag{5.99}$$

Let us complete the proof of Theorem 5.9 in the case $G = \mathrm{SO}^0(\mathrm{p},1)$ with p odd.

Proof of (5.43) *and* (5.44). Simple calculation shows that

$$c_\mathfrak{g} - c_\mathfrak{m} + C^{\mathfrak{m},\eta_j} = \left|j - \frac{\dim\mathfrak{n}}{2}\right|^2|\alpha|^2. \tag{5.100}$$

By (5.89), (5.93), and (5.100), using the fact that P_η is odd, we see that as $\sigma \to 0$,

$$R_{\phi,\rho}(\sigma) = \left(1+\mathcal{O}(\sigma)\right)\prod_{j=0}^{\dim\mathfrak{n}}\operatorname{det}_{\mathrm{s}}\left(\sigma^2 + 2\sigma\Big(j-\frac{\dim\mathfrak{n}}{2}\Big)|\alpha| + C^{\mathfrak{g},Z,\widehat{\eta}_j,\rho}\right)^{(-1)^{j-1}}. \tag{5.101}$$

On the other hand, by (1.8), (5.52), and (5.91), we have

$$\begin{aligned}\prod_{j=0}^{\dim\mathfrak{n}}\operatorname{det}_{\mathrm{s}}\left(\sigma + C^{\mathfrak{g},Z,\widehat{\eta}_j,\rho}\right)^{(-1)^{j-1}} &= \prod_{q=1}^{\dim Z}\det\left(\sigma + \Box_q^Z\right)^{(-1)^q q} \\ &= \left(1+\mathcal{O}(\sigma)\right)\sigma^{\sum_{q=1}^{\dim Z}(-1)^q q b_q(Z,F)}T_F(Z)^2.\end{aligned} \tag{5.102}$$

Comparing (5.101) and (5.102), by (5.98), we get (5.43).

By (5.98), if $H^\cdot(Z,F) = 0$, then (5.44) follows. □

5.8. The reductive group G with $\delta(G) = 1$

In this section, we assume that G is a real reductive group with $\delta(G) = 1$. From Table 1, we see three examples $\mathbf{R}$, $\mathrm{SL}_3(\mathbf{R})$, and $\mathrm{SO}^0(p,q)$ with $pq > 1$ odd. The result in this section is from [Sh18].

5.8.1. Meromorphic extension. We generalize the construction in Section 5.7 to the case in this section. By [Sh18, Proposition 6.2], we can define $\mathfrak{m}, \mathfrak{n}, \bar{\mathfrak{n}}$ is the same way. So the splitting (5.85) still holds true, and $\dim \mathfrak{n}$ is still even. However the group $\mathcal{M}$ is not necessarily compact. It is a reductive group with maximal compact subgroup $K \cap \mathcal{M}$ and with Cartan decomposition

$$\mathfrak{m} = \mathfrak{p}_{\mathfrak{m}} \oplus \mathfrak{k}_{\mathfrak{m}}. \tag{5.103}$$

One of difference with the $\mathbf{R}$-rank 1 case is that a general non-elliptic semisimple element can not always be conjugate to $e^a k^{-1}$ with $a \in \mathfrak{b}$, $a \neq 0$, and $k \in \mathcal{M} \cap K$. In [Sh18, Proposition 4.1], we observe that this happens if and only if

$$\dim \mathfrak{b}(\gamma) = 1. \tag{5.104}$$

For such an element γ, we still have (5.86). Moreover, if $\dim \mathfrak{b}(\gamma) \geqslant 2$, as in Remark 5.13, we have

$$\chi_{\mathrm{orb}}(B_{[\gamma]}/\mathbb{S}^1) = 0. \tag{5.105}$$

This gives the motivation to introduce the following Selberg zeta function.

Definition 5.24. For $\sigma \in \mathbf{C}$, we define a formal Selberg zeta function associated to a representation η of $\mathcal{M}$ and to a representation ρ of Γ by

$$Z_{\eta,\rho}(\sigma) = \exp\left(- \sum_{\substack{[\gamma]\in[\Gamma]-\{1\} \\ \gamma\sim e^a k^{-1}}} \mathrm{Tr}[\rho(\gamma)] \frac{\mathrm{Tr}^{E_\eta}\left[\eta(k^{-1})\right]}{\left|\det\left(1 - \mathrm{Ad}(\gamma)\right)|_{\mathfrak{z}_0^\perp}\right|^{1/2}} \frac{\chi_{\mathrm{orb}}(B_{[\gamma]}/\mathbb{S}^1)}{m_{[\gamma]}} e^{-\sigma \ell_{[\gamma]}} \right). \tag{5.106}$$

Thanks to (5.105), the semisimple element γ satisfying $\dim \mathfrak{b}(\gamma) \geqslant 2$ does not have a contribution to the Selberg zeta function. Take $\eta_j = \Lambda^j(\mathfrak{n}^*)$ as before. Equation (5.89) still holds true. So we can reduce our problem to showing the meromorphic extension of $Z_{\eta_j,\rho}$.

Note that in this case, the Casimir operator of $\mathfrak{m}$ still acts as a constant on η_j. For technical reasons, we note also that the compact dual of $\mathcal{M}$ acts on $\eta_j \otimes_{\mathbf{R}} \mathbf{C}$. Also, the lifting property still holds but in a more complicated form. Consider the diagram

$$\begin{array}{ccc} & & \mathrm{RO}(\mathcal{M}) \\ & & \big\downarrow \iota_{\mathcal{M}} \\ \mathrm{RO}(K) & \xrightarrow{\iota_K} & \mathrm{RO}(K \cap \mathcal{M}), \end{array} \tag{5.107}$$

where the maps are induced by restriction. In [Sh18, Theorem 6.11], we show that $\iota_{\mathcal{M}}(\eta_j)$, $\Lambda^j(\mathfrak{p}_{\mathfrak{m}}^*) \in \mathrm{RO}(K \cap \mathcal{M})$ have unique lifts in $\mathrm{RO}(K)$.

Proposition 5.25. *Suppose that (η, E_η) is a finite-dimensional real representation of $\mathcal{M}$ such that*

- *the Casimir of $\mathfrak{m}$ acting on E_η is a scalar $C^{\mathfrak{m},\eta} \in \mathbf{R}_-$;*
- *the compact dual of $\mathcal{M}$ acts on $E_\eta \otimes_{\mathbf{R}} \mathbf{C}$;*
- *the restriction $\iota_{\mathcal{M}}(\eta)$ to $K \cap \mathcal{M}$ has a lift in $\mathrm{RO}(K)$.*

Then $Z_{\eta,\rho}(\sigma)$ is well defined such that

$$Z_{\eta,\rho}(\sigma) = e^{P_\eta(\sigma)} \det{}_s \left(\sigma^2 - c_{\mathfrak{g}} + c_{\mathfrak{m}} - C^{\mathfrak{m},\eta} + C^{\mathfrak{g},Z,\widehat{\eta},\rho}\right), \tag{5.108}$$

where $P_\eta(\sigma)$ is an odd polynomial of σ, and $\widehat{\eta} \in \mathrm{RO}(K)$ is a virtual representation of K such that

$$\iota_K(\widehat{\eta}) = \Lambda^{\cdot}(\mathfrak{p}_{\mathfrak{m}}^*) \otimes \iota_{\mathcal{M}}(\eta) \in \mathrm{RO}(K \cap \mathcal{M}). \tag{5.109}$$

Proof. The proof of our proposition is similar to the one of Proposition 5.22, except that the evaluation of the orbital integral is more complicated. Note that when $\dim \mathfrak{b}(\gamma) \geqslant 2$, we have

$$\mathrm{Tr}^{[\gamma]}\left[\exp\left(-tC^{\mathfrak{g},X,\widehat{\eta}}/2\right)\right] = 0. \tag{5.110}$$

This is a refinement of Theorem 5.19. □

Now a method similar to the one given in Section 5.7 shows (5.43) with the constants C_ρ and r_ρ defined by the same formula as in (5.99).

5.8.2. Regularity at $\sigma = 0$. The proof of (5.44) is much more difficult since we do not have a relation between r_j and $b_j(Z,F)$ as in (5.98). We rely on some deep results from the representation theory. Here we only sketch the main steps.

Let $\widehat{G}_u$ be the unitary dual of G. For a unitary representation $V_\pi \in \widehat{G}_u$, denote by $V_{\pi,K}$ the associated Harish-Chandra $(\mathfrak{g}, K)$-module, which is formed by K-finite elements. The center of the complexified enveloping algebra $\mathscr{U}(\mathfrak{g}_{\mathbf{C}})$ acts on $V_{\pi,K}$ as scalars, which is called the infinitesimal character and is denoted by χ_π. Clearly, for $a \in \mathbf{C}$, we have

$$\chi_\pi(a) = a. \tag{5.111}$$

We call χ_π is trivial if χ_π coincides with the infinitesimal character of the trivial representation of G.

Recall that $\widehat{p} : \Gamma\backslash G \to Z$ is the natural projection. The group G acts unitarily on the right on $L^2(\Gamma\backslash G, \widehat{p}^*F)$. By [GeGraPS69, p.23, Theorem], $L^2(\Gamma\backslash G, \widehat{p}^*F)$ decomposes into a discrete Hilbert direct sum with finite multiplicity of unitary representations of G. We can write

$$L^2\left(\Gamma\backslash G, \widehat{p}^*F\right) = \bigoplus_{\pi \in \widehat{G}_u}^{\mathrm{Hil}} n_\rho(\pi) V_\pi, \tag{5.112}$$

with $n_\rho(\pi) < \infty$.

Recall that τ is a real finite-dimensional orthogonal representation of K on the real Euclidean space E_τ, and that $C^{\mathfrak{g},Z,\tau,\rho}$ is the Casimir element of G acting on $C^\infty(Z,\mathcal{F}_\tau\otimes_{\mathbf{C}} F)$. By (5.112), we have

$$\ker C^{\mathfrak{g},Z,\tau,\rho} = \bigoplus_{\pi\in\widehat{G}_u,\chi_\pi(C^{\mathfrak{g}})=0} n_\rho(\pi)\big(V_{\pi,K}\otimes_{\mathbf{R}} E_\tau\big)^K, \tag{5.113}$$

By properties of elliptic operators, the sum on right-hand side of (5.113) is finite. We will apply (5.113) in the case $\tau=\Lambda^j(\mathfrak{p}^*)$ and also $\tau=\widehat{\eta}_j$.

The case $\tau=\Lambda^j(\mathfrak{p}^*)$. By (5.52), (5.113), and by the Hodge theory (1.5), we have

$$H^\cdot(Z,F) = \bigoplus_{\pi\in\widehat{G}_u,\chi_\pi(C^{\mathfrak{g}})=0} n_\rho(\pi)\big(V_{\pi,K}\otimes_{\mathbf{R}}\Lambda^\cdot(\mathfrak{p}^*)\big)^K. \tag{5.114}$$

The Hodge theory for Lie algebras [BorW00, Proposition II.3.1] tells us that the right-hand side of (5.114) has a cohomological interpretation,

$$H^\cdot(Z,F) = \bigoplus_{\pi\in\widehat{G}_u,\chi_\pi(C^{\mathfrak{g}})=0} n_\rho(\pi)H^\cdot(\mathfrak{g},K;V_{\pi,K}), \tag{5.115}$$

where $H^\cdot(\mathfrak{g},K;V_{\pi,K})$ is the $(\mathfrak{g},K)$-cohomology of the Harish-Chandra $(\mathfrak{g},K)$-module $V_{\pi,K}$. By the following property of $(\mathfrak{g},K)$-cohomology [VZu84, V84, SR99] (see also [Sh18, Theorem 8.9]), for $\pi\in\widehat{G}_u$,

$$\chi_\pi \text{ is trivial} \iff H^\cdot(\mathfrak{g},K;V_{\pi,K})\neq 0, \tag{5.116}$$

we see that the sum in (5.115) can be reduced to $\pi\in\widehat{G}_u$ with trivial infinitesimal character,

$$H^\cdot(Z,F) = \bigoplus_{\pi\in\widehat{G}_u,\chi_\pi \text{ trivial}} n_\rho(\pi)H^\cdot(\mathfrak{g},K;V_{\pi,K}), \tag{5.117}$$

and each summand does not vanish except for $n_\rho(\pi)=0$.

The case $\tau=\widehat{\eta}_j$. By (5.113), we have

$$r_j = \sum_{\pi\in\widehat{G}_u,\chi_\pi(C^{\mathfrak{g}})=0} n_\rho(\pi)\left(\dim\big(V_{\pi,K}\otimes_{\mathbf{R}}\widehat{\eta}_j^+\big)^K - \dim\big(V_{\pi,K}\otimes_{\mathbf{R}}\widehat{\eta}_j^-\big)^K\right). \tag{5.118}$$

As (5.115), in [Sh18, Theorem 8.14, Corollary 8.15], we give a cohomology interpretation of the right-hand side of (5.118),

$$\begin{aligned} r_j = \frac{1}{\chi(K/K\cap\mathcal{M})} &\sum_{\pi\in\widehat{G}_u,\chi_\pi(C^{\mathfrak{g}})=0} n_\rho(\pi) \\ &\times\sum_{i=0}^{\dim\mathfrak{p}_{\mathfrak{m}}}\sum_{j=0}^{\dim\mathfrak{n}}(-1)^{i+j}\dim H^i\big(\mathfrak{m},K\cap\mathcal{M};H_j(\mathfrak{n},V_{\pi,K})\otimes_{\mathbf{R}}E_\eta\big), \end{aligned} \tag{5.119}$$

where $H_j(\mathfrak{n}, V_{\pi,K})$ is the $\mathfrak{n}$-homology of $V_{\pi,K}$ and is a $(\mathfrak{m}, K\cap \mathcal{M})$-module. Moreover, as (5.117), [Sh18, Proposition 8.17, Corollary 8.18] implies that the first sum in (5.119) can be reduced to $\pi \in \widehat{G}_u$ with trivial infinitesimal character, i.e.,

$$\begin{aligned} r_j = \frac{1}{\chi(K/K\cap \mathcal{M})} \sum_{\pi\in\widehat{G}_u, \chi_\pi \text{ trivial}} n_\rho(\pi) \\ \times \sum_{i=0}^{\dim \mathfrak{p}_\mathfrak{m}} \sum_{j=0}^{\dim \mathfrak{n}} (-1)^{i+j} \dim H^i\big(\mathfrak{m}, K\cap \mathcal{M}; H_j(\mathfrak{n}, V_{\pi,K}) \otimes_{\mathbf{R}} E_\eta\big). \end{aligned} \tag{5.120}$$

Equations (5.117) and (5.120) can be considered as an analogue of (5.98). Now we prove (5.44).

The proof of (5.44). If $H^\cdot(Z,F) = 0$, by (5.117), we see that if χ_π is trivial, then $n_\rho(\pi) = 0$. By (5.120), we see that $r_j = 0$ for all j. By (5.99), we complete the proof of (5.44). □

6. Anosov flow

The purpose of this section is to study the Fried conjecture for the Anosov flow. This section is organized as follows. In Section 6.1, we introduce the Anosov flow.

In Section 6.2, we explain the meromorphic extension of the Ruelle dynamical zeta function [GiLiPo13, DyZ16].

In Section 6.3, we explain a proof that under certain resonance conditions the value at zero of the Ruelle dynamical zeta function does not depend on a small perturbation of the Anosov flow.

In Section 6.4, we study the Anosov flow on 3-manifolds.

6.1. Closed orbits of the Anosov flow

Let M be a closed manifold with a smooth vector field $V \in C^\infty(M, TM)$. Let $(\phi_t)_{t\in\mathbf{R}}$ be the flow on M generated by V.

Definition 6.1. A flow $\phi.$ is called Anosov if there is a ϕ_t-invariant continuous splitting[12]

$$TM = \mathbf{R}V \oplus E^u \oplus E^s \tag{6.1}$$

of C^0-vector bundles on M and there exist $C > 0, \theta > 0$ and a Riemannian metric on M such that for $v \in E^u_x$, $v' \in E^s_x$, and $t > 0$, we have

$$|\phi_{-t,*}v| \leqslant Ce^{-\theta|t|}\,|v|, \qquad |\phi_{t,*}v'| \leqslant Ce^{-\theta|t|}\,|v'|. \tag{6.2}$$

[12]This requires that $\mathbf{R}V$ is a line bundle on M. It implies $V(x) \neq 0$ for all $x \in M$, and so the Euler characteristic number $\chi(M)$ vanishes.

In this section, we assume $\phi_\cdot$ is Anosov. It is well known that the set of closed orbits $\overline{\wp}(\phi_\cdot)$ defined in (2.12) is discrete (see [Mar04] or [DyZ16, Appendix A]), so that Assumption 2.4 holds. More precisely, for $\gamma \in \overline{\wp}(\phi_\cdot)$, let $\ell_\gamma \in \mathbf{R}_+^*, m_\gamma \in \mathbf{N}^*$ be the period and the multiplicity of γ. Then (2.14) becomes

$$\wp(\phi_\cdot) = \coprod_{\gamma \in \overline{\wp}(\phi_\cdot)} \mathbb{S}^1 \times \{\ell_\gamma\}. \tag{6.3}$$

Moreover, by [Mar04, Theorem 1.1, p.78] or [DyZ16, Lemma 2.2] there is $C > 0$ such that for $r \geqslant 0$, we have

$$|\{\gamma \in \overline{\wp}(\phi_\cdot) : \ell_\gamma \leqslant r\}| \leqslant Ce^{Cr}. \tag{6.4}$$

Thanks to (6.4), the Ruelle dynamical zeta function is well defined for $\sigma \in \mathbf{C}$ with $\mathrm{Re}(\sigma) \gg 1$. Recall that for a prime closed orbit γ, $\Delta(\gamma) \in \{\pm 1\}$ is defined in (4.9). Proceeding as in the proof of Proposition 4.5, for $\sigma \in \mathbf{C}$ with $\mathrm{Re}(\sigma) \gg 1$, we have

$$R_{\phi,\rho}(\sigma) = \left(\prod_{\gamma:\text{prime}} \det\left(1 - \Delta(\gamma)\rho(\gamma)e^{-\sigma\ell_\gamma}\right) \right)^{(-1)^{\mathrm{rk}[E^u]+1}}, \tag{6.5}$$

where $\mathrm{rk}[E^u]$ denotes the rank of the C^0-vector bundle E^u. We note the similarity between (0.2) and (6.5).

Example 6.2. If (Z, g^{TZ}) is a negatively curved manifold, then the geodesic flow on the unit tangent bundle $M = SZ$ is Anosov [A67]. By (6.5), for $\sigma \in \mathbf{C}$ with $\mathrm{Re}(\sigma) \gg 1$,

$$R_{\phi,\rho}(\sigma) = \left(\prod_{\gamma:\text{prime}} \det\left(1 - \Delta(\gamma)\rho(\gamma)e^{-\sigma\ell_\gamma}\right) \right)^{(-1)^{\dim Z}}. \tag{6.6}$$

In addition, if Z is orientable, for all prime closed orbits γ, we have $\Delta(\gamma) = 1$ (see [GiLiPo13, Lemma B.1]).

6.2. The meromorphic extension

The proofs of the meromorphic extension of the Ruelle dynamical zeta function for the Anosov flow given in [GiLiPo13] and [DyZ16][13] are based on a spectral interpretation of $R_{\phi,\rho}$.

Let us begin with establishing a relation between $R_{\phi,\rho}$ and the Lie derivation L_V along V acting on $\Omega^\cdot(M,F)$. For $t > 0$, write

$$e^{tL_V} : u \in \Omega^\cdot(M,F) \to \phi_t^* u \in \Omega^\cdot(M,F). \tag{6.7}$$

[13] Unfortunately, there is a sign conflict between the convention used in these two papers and Fried's paper [F87]. In [GiLiPo13, (2.2)] and [DyZ16, (1.1),(B.1)], the sign in the Fuller index (2.18) is defined using $D\phi_{-\ell_\gamma}$. As we adopt Fried's convention, the statements in this section are slightly different with [GiLiPo13, DyZ16].

The Schwartz kernel $e^{tL_V}(x,y)$ of e^{tL_V} is a current on $\mathbf{R}_+^* \times M \times M$ with coefficients in

$$\mathbf{C} \boxtimes (\Lambda^\cdot(T^*M) \otimes_{\mathbf{R}} F) \boxtimes \Big((\Lambda^\cdot(T^*M) \otimes_{\mathbf{R}} F)^* \otimes_{\mathbf{R}} |\det(T^*M)| \Big), \tag{6.8}$$

where $|\det(T^*M)| = \Lambda^{\dim M}(T^*M) \otimes o(TM)$ is the density bundle on M. By (6.1), its wave front set is disjoint from the conormal bundle of the submanifold $\{(t,x,x) \in \mathbf{R}_+^* \times M \times M\} \subset \mathbf{R}_+^* \times M \times M$. So, the restriction on the diagonal $e^{tL_V}(x,x)$ is a well-defined current on $\mathbf{R}_+^* \times M$ with coefficients in

$$\mathbf{C} \boxtimes \Big(\mathrm{End}\left(F \otimes_{\mathbf{R}} \Lambda^\cdot(T^*M)\right) \otimes_{\mathbf{R}} |\det(T^*M)| \Big). \tag{6.9}$$

Definition 6.3. The flat trace of e^{tL_V} is a distribution on $\mathbf{R}_+^*$ defined by

$$\mathrm{Tr}^\flat\left[e^{tL_V}\right] = \int_M \mathrm{Tr}\left[e^{tL_V}(x,x)\right]. \tag{6.10}$$

Note that the flat trace is not a classical trace in the sense of the trace of a trace class operator. However, we can still show that the flat trace of the commutator of e^{tL_V} with a differential operator vanishes.

Let us give an explicit formula for $\mathrm{Tr}^\flat\left[e^{tL_V}\right]$, which is known as the Atiyah–Bott–Guillemin trace formula [Gui77]. Let $\gamma \in \overline{\wp}(\phi_\cdot)$ be a closed orbit. For $x \in \gamma$, we have a morphism

$$D\phi_{\ell_\gamma}(x) : T_xM \to T_xM. \tag{6.11}$$

Up to conjugation it does not depend on the choice of $x \in \gamma$, and is denoted by $D\phi_{\ell_\gamma}|_{T_\gamma M}$. It acts naturally on any tensor of $T_\gamma M$. Acting on $T_\gamma M/\mathbf{R}V$, it is just the linearized Poincaré return map.

For $0 \leqslant q \leqslant \dim M$, denote by $L_{V,q}$ the restriction of L_V on $\Omega^q(M,F)$. Let $N^{\Lambda^\cdot(T^*M)}$ be the number operator on $\Lambda^\cdot(T^*M)$, which sends $s \in \Omega^q(M,F)$ to $qs \in \Omega^q(M,F)$.

Proposition 6.4. *For $0 \leqslant q \leqslant \dim M$, the following identity holds,*

$$\mathrm{Tr}^\flat\left[e^{tL_{V,q}}\right] = \sum_{\gamma \in \overline{\wp}(\phi_\cdot)} \frac{\mathrm{Tr}^{\Lambda^q(T_\gamma^*M)}\left[\left(D\phi_{\ell_\gamma}\right)^{\mathrm{tr}}|_{\Lambda^q(T_\gamma^*M)}\right]}{|\det(1 - D\phi_{\ell_\gamma})|_{T_\gamma M/\mathbf{R}V}|} \mathrm{Tr}\left[\rho(\gamma)\right] \frac{\ell_\gamma}{m_\gamma} \delta_{\ell_\gamma}(t). \tag{6.12}$$

In particular, we have

$$\mathrm{Tr}_s^\flat\left[N^{\Lambda^\cdot(T^*M)} e^{tL_V}\right] = -\sum_{\gamma \in \overline{\wp}(\phi_\cdot)} \mathrm{sgn}\big(\det(1 - D\phi_{\ell_\gamma})|_{T_\gamma M/\mathbf{R}V}\big) \mathrm{Tr}\left[\rho(\gamma)\right] \frac{\ell_\gamma}{m_\gamma} \delta_{\ell_\gamma}(t). \tag{6.13}$$

Proof. The Schwartz kernel of $e^{tL_{V,q}}$ is given by

$$e^{tL_{V,q}}(x,y) = \left\{ \tau_0^t \otimes (D\phi_t)^{\mathrm{tr}}|_{\Lambda^q\left(T^*_{\phi_t(x)}M\right)} \right\} \cdot \delta_{\phi_t(x)}(y), \tag{6.14}$$

where $\tau_0^t \in \mathrm{Hom}(F_{\phi_t(x)}, F_x)$ is the parallel transport with respect to ∇^F along the curve $(\phi_s(x))_{0\leqslant s\leqslant t}$ from t to 0. So the restriction of the distribution to the diagonal $e^{tL_V}(x,x)$ is supported on the periodic set $\wp(\phi) \subset \mathbf{R}_+^* \times M$ and is given by[14]

$$\mathrm{Tr}\left[e^{tL_{V,q}}(x,x)\right] = \sum_{\gamma\in\overline{\wp}(\phi.)} \mathrm{Tr}\left[\rho(\gamma)\right] \frac{\mathrm{Tr}^{\Lambda^q(T_\gamma^*M)}\left[\left(D\phi_{\ell_\gamma}\right)^{\mathrm{tr}}|_{\Lambda^q(T_\gamma^*M)}\right]}{|\det(1-D\phi_{\ell_\gamma})|_{T_\gamma M/\mathbf{R}V}|}\delta_{\ell_\gamma}(t)\otimes\delta_{\gamma^\sharp}(x). \tag{6.15}$$

where $\delta_{\gamma^\sharp}(x)$ is the current of integration on the prime closed orbit $\gamma^\sharp$ associated to γ defined by

$$s \in C_c^\infty(M) \to \int_0^{\ell_\gamma/m_\gamma} s(\phi_t(x))dt, \tag{6.16}$$

where x is any point on γ. By (6.15), we get (6.12). □

By (2.21), (6.4), and (6.13), for $\mathrm{Re}\,(\sigma) \gg 1$, we have

$$\log R_{\phi,\rho}(\sigma) = -\int_0^\infty \mathrm{Tr}_{\mathrm{s}}^\flat\left[N^{\Lambda^\cdot(T^*M)}\exp\left(-t\left(\sigma - L_V\right)\right)\right]\frac{dt}{t}. \tag{6.17}$$

So formally, $R_{\phi,\rho}(\sigma)$ is a certain flat regularized determinant

$$\prod_{q=1}^{\dim M} \det{}^\flat\left(\sigma - L_{V,q}\right)^{(-1)^q q}. \tag{6.18}$$

Note the similarity between (1.8) and (6.18).

For $\mathrm{Re}\,(\sigma) \gg 1$, we write

$$\begin{aligned}\frac{\partial}{\partial\sigma}\log R_{\phi,\rho}(\sigma) &= \int_0^\infty \mathrm{Tr}_{\mathrm{s}}^\flat\left[N^{\Lambda^\cdot(T^*M)}\exp\left(-t\left(\sigma - L_V\right)\right)\right]dt \\ &= \int_\delta^\infty \mathrm{Tr}_{\mathrm{s}}^\flat\left[N^{\Lambda^\cdot(T^*M)}\exp\left(-t\left(\sigma - L_V\right)\right)\right]dt,\end{aligned} \tag{6.19}$$

where $\delta > 0$ is some positive number smaller than the minimum of the length spectrum. In the second identity of (6.19), we use the fact that the support of the distribution $\mathrm{Tr}_{\mathrm{s}}^\flat\left[N^{\Lambda^\cdot(T^*M)}\exp(-t\left(\sigma - L_V\right))\right]$ is away from 0.

To show the meromorphic extension of $R_{\phi,\rho}$, it is enough to show that $\frac{\partial}{\partial\sigma}\log R_{\phi,\rho}$ has a meromorphic extension to $\mathbf{C}$ with simple poles and integer residues. By (6.19), we write formally

$$\frac{\partial}{\partial\sigma}\log R_{\phi,\rho}(\sigma) = \mathrm{Tr}_{\mathrm{s}}^\flat\left[N^{\Lambda^\cdot(T^*M)}e^{-\delta(\sigma-L_V)}\left(\sigma - L_V\right)^{-1}\right]. \tag{6.20}$$

An important step is to give a proper sense of the operators on the right-hand side of (6.20) and to show its flat trace exists and has a meromorphic extension. We refer the reader to [DyZ16] for more details. Here we just state a weak version

[14] This can be obtained by proceeding as in the proof of Lefschetz fixed point formula. We refer the reader to [DyZ16, Appendix B] for more details.

of [DyZ16, Propositions 3.1–3.3], and explain the reason for which we need to introduce the small $\delta > 0$.

Let

$$T^*M = (\mathbf{R}V)^* \oplus E_u^* \oplus E_s^* \tag{6.21}$$

be the dual of the splitting (6.1). Let $\Delta \subset T^*(M \times M)$ be the diagonal of $T^*(M \times M)$. Set

$$\Omega^- = \left\{ \left(\phi_t(x), \left((D\phi_t)_x^{\mathrm{tr}} \right)^{-1} \cdot \xi, x, \xi \right) \in T^*(M \times M) : \langle V(x), \xi \rangle = 0, t \leqslant 0 \right\}, \tag{6.22}$$

where $\cdot^{\mathrm{tr}}$ denotes the transpose of a matrix. Denote by WF$'$ the wave front set of an operator (see [DyZ16, Appendix C.2]).

Theorem 6.5. *The operator*

$$(\sigma - L_V)^{-1} : \Omega^\cdot(M, F) \to \mathcal{D}'(M, \Lambda^\cdot(T^*M) \otimes_{\mathbf{R}} F) \tag{6.23}$$

defines a meromorphic family on $\mathbf{C}$. *If it is holomorphic at* σ_0, *then*

$$\mathrm{WF}' (\sigma_0 - L_V)^{-1} \subset \Delta \cup \Omega^- \cup (E_s^* \times E_u^*). \tag{6.24}$$

From (6.24), we see that $(\sigma_0 - L_V)^{-1}$ does not necessarily have a well-defined flat trace. But $e^{-\delta(\sigma - L_V)} (\sigma - L_V)^{-1}$ does.

6.3. The $R_{\phi,\rho}(0)$ as a topological invariant

The poles of the meromorphic family in (6.23) are called Ruelle–Pollicott resonances. The set of Ruelle–Pollicott resonances is denoted by $\mathrm{Res}_\rho(V)$. If $0 \notin \mathrm{Res}_\rho(V)$, then $R_{\phi,\rho}(\sigma)$ is regular at $\sigma = 0$.[15]

Set

$$\mathscr{V}_\rho(M) = \{V \in C^\infty(M, TM) : V \text{ is Anosov such that } 0 \notin \mathrm{Res}_\rho V\}. \tag{6.25}$$

Thanks to the stability of the Anosov flow [A67] and of its resonance [ButLi07, ButLi13], $\mathscr{V}_\rho(M)$ forms an open subset in $C^\infty(M, TM)$. The following theorem [DaGuRiSh20] tells us the value at zero of the Ruelle dynamical zeta function does not depend on a small perturbation of the flow.

For $V \in \mathscr{V}_\rho(M)$, denote by $\phi_\cdot^V$ the corresponding flow.

Theorem 6.6. *For any flat vector bundle* F, *the map*

$$V \in \mathscr{V}_\rho(M) \to R_{\phi^V,\rho}(0) \in \mathbf{C}^* \tag{6.26}$$

is locally constant.

---

[15] Due to the cancellation from the supertrace, the converse is not correct.

Proof. Let $(V_b)_{b\in\mathbf{R}}$ be a smooth family of vector fields in $\mathscr{V}_\rho(M)$. Let $R_{b,\rho}$ be the corresponding family of the Ruelle dynamical zeta functions. Take a smooth family $\alpha_b \in \Omega^1(M)$ such that $\alpha_b(V_b) = 1$. Write $\dot{V}_b = \frac{\partial}{\partial b} V_b$.

We claim that for $\sigma \in \mathbf{C}$ with $\mathrm{Re}(\sigma) \gg 1$, we have

$$\frac{\partial}{\partial b} \log R_{b,\rho}(\sigma) = -\sigma \int_\delta^\infty \mathrm{Tr}_{\mathrm{s}}^\flat \left[\alpha_b i_{\dot{V}_b} \exp\left(-t\left(\sigma - L_{V_b}\right)\right)\right] dt. \tag{6.27}$$

In [DaGuRiSh20], the proof of (6.27) is obtained by variation of the periods of closed orbits. Here, we give a proof via supersymmetry (cf. [RS71, Theorem 2.1]). We argue formally. The argument can be made rigorous easily. By (6.17), for $\mathrm{Re}(\sigma) \gg 1$, we have

$$\frac{\partial}{\partial b} \log R_{b,\rho}(\sigma) = -\int_\delta^\infty \mathrm{Tr}_{\mathrm{s}}^\flat \left[N^{\Lambda^\cdot(T^*M)} L_{\dot{V}_b} \exp\left(-t\left(\sigma - L_{V_b}\right)\right)\right] dt. \tag{6.28}$$

Using the Cartan identity[16] $L_{\dot{V}_b} = \left[d, i_{\dot{V}_b}\right]$, the fact that d commutes with L_{V_b}, and fact that the supertrace vanishes on the supercommutator, we have identities of distributions on $\mathbf{R}_+^*$,

$$\begin{aligned}\mathrm{Tr}_{\mathrm{s}}^\flat \left[N^{\Lambda^\cdot(T^*M)} L_{\dot{V}_b} \exp\left(tL_{V_b}\right)\right] &= \mathrm{Tr}_{\mathrm{s}}^\flat \left[N^{\Lambda^\cdot(T^*M)} \left[d, i_{\dot{V}_b}\right] \exp\left(tL_{V_b}\right)\right] \\ &= \mathrm{Tr}_{\mathrm{s}}^\flat \left[\left[N^{\Lambda^\cdot(T^*M)}, d\right] i_{\dot{V}_b} \exp\left(tL_{V_b}\right)\right] \\ &= \mathrm{Tr}_{\mathrm{s}}^\flat \left[d i_{\dot{V}_b} \exp\left(tL_{V_b}\right)\right].\end{aligned} \tag{6.29}$$

Since $\alpha_b(V_b) = 1$, we have $i_{\dot{V}_b} = \left[i_{V_b}, \alpha_b i_{\dot{V}_b}\right]$. By (6.29), and proceeding as before we have

$$\begin{aligned}\mathrm{Tr}_{\mathrm{s}}^\flat \left[N^{\Lambda^\cdot(T^*M)} L_{\dot{V}_b} \exp\left(tL_{V_b}\right)\right] &= \mathrm{Tr}_{\mathrm{s}}^\flat \left[d\left[i_{V_b}, \alpha_b i_{\dot{V}_b}\right] \exp\left(tL_{V_b}\right)\right] \\ &= \mathrm{Tr}_{\mathrm{s}}^\flat \left[\left[d, i_{V_b}\right] \alpha_b i_{\dot{V}_b} \exp\left(tL_{V_b}\right)\right] \\ &= \mathrm{Tr}_{\mathrm{s}}^\flat \left[L_{V_b} \alpha_b i_{\dot{V}_b} \exp\left(tL_{V_b}\right)\right] \\ &= \frac{\partial}{\partial t} \mathrm{Tr}_{\mathrm{s}}^\flat \left[\alpha_b i_{\dot{V}_b} \exp\left(tL_{V_b}\right)\right].\end{aligned} \tag{6.30}$$

By (6.28) and (6.30), we get (6.27).

By (6.27), using the method as in [DyZ16], we can show that for $\sigma \in \mathbf{C}$, $\mathrm{Re}(\sigma) \gg 1$,

$$\frac{\partial}{\partial b} \log R_{b,\rho}(\sigma) = -\sigma \mathrm{Tr}_{\mathrm{s}}^\flat \left[\alpha_b i_{\dot{V}_b} e^{-\delta(\sigma - L_{V_b})} \left(\sigma - L_{V_b}\right)^{-1}\right]. \tag{6.31}$$

Note that thanks to (6.24), the above flat trace is well defined. Moreover, the function $\sigma \to \mathrm{Tr}_{\mathrm{s}}^\flat \left[\alpha_b i_{\dot{V}_b} e^{-\delta(\sigma - L_{V_b})} \left(\sigma - L_{V_b}\right)^{-1}\right]$ has a meromorphic extension to $\mathbf{C}$, and is regular at $\sigma = 0$ since $V_b \in \mathscr{V}_\rho(M)$.

---

[16] Here $[a, b] = ab - (-1)^{\deg a \deg b} ba$ denotes the supercommutator of a and b (see [BeGeVe04, Section 1.3]).

To complete our proof, it remains to show that for a fixed σ near 0, the function $b \to R_{b,\rho}(\sigma)$ is C^1 and (6.31) holds near 0. This is somehow technical and we refer the reader to [DaGuRiSh20] for a detailed proof. □

6.4. Anosov flow on 3-manifold

Let us restrict ourself to an orientable 3-manifold, where we have a partial solution for the Fried conjecture.

The following proposition [DaGuRiSh20, Proposition 7.3, Lemma 7.4] gives a characterization of the acyclicity of a unitarily flat vector bundle via resonance. Its proof uses [DyZ17, Lemma 2.3] in an essential way.

Proposition 6.7. *Let F be a unitarily flat vector bundle on a closed orientable 3-manifold. For any volume preserving Anosov flow, F is acyclic if and only if 0 is not a resonance.*

Recall the following theorem due to Sánchez-Morgado [SM96a], whose proof is based on the Markov partition [Rat69] and Rugh's technique [Rug96].

Theorem 6.8. *Let F be an acyclic unitarily flat vector bundle with holonomy ρ on a closed orientable analytic 3-manifold M. If $\phi.$ is a transitive analytic Anosov flow, and if there is a prime closed orbit γ such that 1 and $\Delta(\gamma)$ are not eigenvalues of $\rho(\gamma)$, then $R_{\phi,\rho}(\sigma)$ is regular at 0 and*

$$|R_{\phi,\rho}(0)| = T_F(M). \tag{6.32}$$

Note that any smooth manifold has a unique compatible analytic structure, and that any volume preserving Anosov flow is transitive. Since we can always approximate a smooth Anosov flow by an analytic one, and since we can approximate a flat vector bundle by the one with specified holonomy condition in the above theorem provided $H^1(M) \neq 0$, using Theorem 6.6, in [DaGuRiSh20, Section 7.2], we deduce the following Theorem [DaGuRiSh20, Theorem 1].

Theorem 6.9. *Let F be an acyclic unitarily flat vector bundle with holonomy ρ on a closed orientable 3-manifold with $H^1(M) \neq 0$. For any flow $\phi.$ which is a volume preserving Anosov flow or a flow nearby*[17]*, we have*

$$|R_{\phi,\rho}(0)| = T_F(M). \tag{6.33}$$

Let us return to the case of the geodesic flow on the unit tangent bundle $M = SZ$ of a negatively curved orientable surface (Z, g^{TZ}). Recall that $a_0 \in \pi_1(M)$ is defined after (5.8). By Corollary 5.4 and Theorem 6.9, we get:

Corollary 6.10. *Let F be an acyclic unitarily flat vector bundle on the unit tangent bundle of a negatively curved orientable surface (Z, g^{TZ}). Then,*

$$|R_{\phi,\rho}(0)| = T_F(M) = \left|\det\left(1 - \rho(a_0)\right)\right|^{-\chi(Z)}. \tag{6.34}$$

The above corollary can be considered as a complementary of Dyatlov–Zworski's result [DyZ17], where ρ is assumed to be trivial.

---

[17] It is still an Anosov flow by the stability of Anosov flows [A67].

Theorem 6.11. *Assume that* (Z, g^{TZ}) *is a negatively curved orientable surface. There is* $C \in \mathbf{R}^*$ *such that as* $\sigma \to 0$, *we have*

$$R_{\phi,\mathrm{trivial}}(\sigma) = C\sigma^{-\chi(Z)}\big(1 + \mathcal{O}(\sigma)\big). \tag{6.35}$$

The above two results are generalizations of Fried's results [F86b, Corollaries 1 and 2] for hyperbolic surfaces.

References

[A67] D.V. Anosov, *Geodesic flows on closed Riemannian manifolds of negative curvature*, Trudy Mat. Inst. Steklov. **90** (1967), 209. MR 0224110

[ArMaz65] M. Artin and B. Mazur, *On periodic points*, Ann. of Math. (2) **81** (1965), 82–99. MR 0176482

[Ba18] V. Baladi, *Dynamical zeta functions and dynamical determinants for hyperbolic maps*, Ergebnisse der Mathematik und ihrer Grenzgebiete. 3. Folge, vol. 68, Springer International Publishing, 2018.

[BaGrSch85] W. Ballmann, M. Gromov, and V. Schroeder, *Manifolds of nonpositive curvature*, Progress in Mathematics, vol. 61, Birkhäuser Boston, Inc., Boston, MA, 1985. MR 823981

[BeGeVe04] N. Berline, E. Getzler, and M. Vergne, *Heat kernels and Dirac operators*, Grundlehren Text Editions, Springer-Verlag, Berlin, 2004, Corrected reprint of the 1992 original. MR 2273508 (2007m:58033)

[B05] J.-M. Bismut, *The hypoelliptic Laplacian on the cotangent bundle*, J. Amer. Math. Soc. **18** (2005), no. 2, 379–476 (electronic). MR 2137981 (2006f:35036)

[B11] J.-M. Bismut, *Hypoelliptic Laplacian and orbital integrals*, Annals of Mathematics Studies, vol. 177, Princeton University Press, Princeton, NJ, 2011. MR 2828080

[B19] J.-M. Bismut, *Eta invariants and the hypoelliptic Laplacian*, J. Eur. Math. Soc. 21 (2019), 2355–2515.

[BG01] J.-M. Bismut and S. Goette, *Families torsion and Morse functions*, Astérisque (2001), no. 275, x+293. MR 1867006 (2002h:58059)

[BG04] J.-M. Bismut and S. Goette, *Equivariant de Rham torsions*, Ann. of Math. (2) **159** (2004), no. 1, 53–216. MR 2051391 (2005f:58059)

[BLo95] J.-M. Bismut and J. Lott, *Flat vector bundles, direct images and higher real analytic torsion*, J. Amer. Math. Soc. **8** (1995), no. 2, 291–363. MR 1303026 (96g:58202)

[BZ92] J.-M. Bismut and W. Zhang, *An extension of a theorem by Cheeger and Müller*, Astérisque (1992), no. 205, 235, With an appendix by François Laudenbach. MR 1185803 (93j:58138)

[BZ94] J.-M. Bismut and W. Zhang, *Milnor and Ray-Singer metrics on the equivariant determinant of a flat vector bundle*, Geom. Funct. Anal. **4** (1994), no. 2, 136–212. MR 1262703 (96f:58179)

[BrMa13] J. Brüning and X. Ma, *On the gluing formula for the analytic torsion*, Math. Z. **273** (2013), no. 3-4, 1085–1117. MR 3030691

[Bo65] R. Bott, *The index theorem for homogeneous differential operators*, Differential and Combinatorial Topology (A Symposium in Honor of Marston Morse), Princeton Univ. Press, Princeton, N.J., 1965, pp. 167–186. MR 0182022 (31 #6246)

[BoTu82] R. Bott and L.W. Tu, *Differential forms in algebraic topology*, Graduate Texts in Mathematics, vol. 82, Springer-Verlag, New York-Berlin, 1982. MR 658304

[BorW00] A. Borel and N. Wallach, *Continuous cohomology, discrete subgroups, and representations of reductive groups*, second ed., Mathematical Surveys and Monographs, vol. 67, American Mathematical Society, Providence, RI, 2000. MR 1721403 (2000j:22015)

[BuO95] U. Bunke and M. Olbrich, *Selberg zeta and theta functions*, Mathematical Research, vol. 83, Akademie-Verlag, Berlin, 1995, A differential operator approach. MR 1353651 (97c:11088)

[ButLi07] O. Butterley and C. Liverani, *Smooth Anosov flows: correlation spectra and stability*, J. Mod. Dyn. **1** (2007), no. 2, 301–322. MR 2285731

[ButLi13] O. Butterley and C. Liverani, *Robustly invariant sets in fiber contracting bundle flows*, J. Mod. Dyn. **7** (2013), no. 2, 255–267. MR 3106713

[BWSh20] Y. Borns-Weil and S. Shen, *Dynamical zeta functions in the nonorientable case*, arXiv:2007.08043 (2020).

[C79] J. Cheeger, *Analytic torsion and the heat equation*, Ann. of Math. (2) **109** (1979), no. 2, 259–322. MR 528965 (80j:58065a)

[DMe12] X. Dai and R.B. Melrose, *Adiabatic limit, heat kernel and analytic torsion*, Metric and differential geometry, Progr. Math., vol. 297, Birkhäuser/Springer, Basel, 2012, pp. 233–298. MR 3220445

[DaGuRiSh20] N. V. Dang, C. Guillarmou, G. Rivière, and S. Shen, *The Fried conjecture in small dimensions*, Invent. Math. **220** (2020), no. 2, 525–579. MR 4081137

[DaRi19] N. V. Dang and G. Rivière, *Spectral analysis of Morse-Smale gradient flows*, Ann. Sci. Éc. Norm. Supér. (4) **52** (2019), no. 6, 1403–1458. MR 4061023

[DaRi20a] N. V. V. Dang and G. Rivière, *Spectral analysis of Morse-Smale flows, I: construction of the anisotropic spaces*, J. Inst. Math. Jussieu **19** (2020), no. 5, 1409–1465. MR 4138948

[DaRi20b] N.V. Dang and G. Rivière, *Spectral analysis of Morse-Smale flows, II: Resonances and resonant states*, Amer. J. Math. **142** (2020), no. 2, 547–593. MR 4084163

[DaRi20c] N.V. Dang and G. Rivière, *Topology of Pollicott-Ruelle resonant states*, to appear in Ann. Sc. Norm. Super. Pisa Cl. Sci, doi: 10.2422/2036-2145.201804_010

[DaRi21] N.V. Dang and G. Rivière, *Pollicott-Ruelle spectrum and Witten L aplacians*, to appear in J. Eur. Math. Soc. doi: 10.4171/JEMS/1044

[dR50] G. de Rham, *Complexes à automorphismes et homéomorphie différentiable*, Ann. Inst. Fourier Grenoble **2** (1950), 51–67 (1951). MR 0043468 (13,268c)

[DuKV79] J.J. Duistermaat, J.A.C. Kolk, and V.S. Varadarajan, *Spectra of compact locally symmetric manifolds of negative curvature*, Invent. Math. **52** (1979), no. 1, 27–93. MR 532745 (82a:58050a)

[DyGu16] S. Dyatlov and C. Guillarmou, *Pollicott–Ruelle resonances for open systems*, Ann. Henri Poincaré **17** (2016), no. 11, 3089–3146. MR 3556517

[DyGu18] S. Dyatlov and C. Guillarmou, *Afterword: dynamical zeta functions for Axiom A flows*, Bull. Amer. Math. Soc. (N.S.) **55** (2018), no. 3, 337–342. MR 3803156

[DyZ16] S. Dyatlov and M. Zworski, *Dynamical zeta functions for Anosov flows via microlocal analysis*, Ann. Sci. Éc. Norm. Supér. (4) **49** (2016), no. 3, 543–577. MR 3503826

[DyZ17] S. Dyatlov and M. Zworski, *Ruelle zeta function at zero for surfaces*, Invent. Math. **210** (2017), no. 1, 211–229. MR 3698342

[FaT17] F. Faure and M. Tsujii, *The semiclassical zeta function for geodesic flows on negatively curved manifolds*, Invent. Math. **208** (2017), no. 3, 851–998. MR 3648976

[Fra82] J.M. Franks, *Homology and dynamical systems*, CBMS Regional Conference Series in Mathematics, vol. 49, Published for the Conference Board of the Mathematical Sciences, Washington, D.C.; by the American Mathematical Society, Providence, R. I., 1982. MR 669378

[Fr35] W. Franz, *Über die Torsion einer Überdeckung*, J. Reine Angew. Math. **17** (1935), 245–254 (German).

[F86a] D. Fried, *Analytic torsion and closed geodesics on hyperbolic manifolds*, Invent. Math. **84** (1986), no. 3, 523–540. MR 837526 (87g:58118)

[F86b] D. Fried, *Fuchsian groups and Reidemeister torsion*, The Selberg trace formula and related topics (Brunswick, Maine, 1984), Contemp. Math., vol. 53, Amer. Math. Soc., Providence, RI, 1986, pp. 141–163. MR 853556 (88e:58098)

[F86c] D. Fried, *The zeta functions of Ruelle and Selberg. I*, Ann. Sci. École Norm. Sup. (4) **19** (1986), no. 4, 491–517. MR 875085

[F87] D. Fried, *Lefschetz formulas for flows*, The Lefschetz centennial conference, Part III (Mexico City, 1984), Contemp. Math., vol. 58, Amer. Math. Soc., Providence, RI, 1987, pp. 19–69. MR 893856 (88k:58138)

[F88] D. Fried, *Torsion and closed geodesics on complex hyperbolic manifolds*, Invent. Math. **91** (1988), no. 1, 31–51. MR 918235

[F95] D. Fried, *Meromorphic zeta functions for analytic flows*, Comm. Math. Phys. **174** (1995), no. 1, 161–190. MR 1372805

[Fu67] F.B. Fuller, *An index of fixed point type for periodic orbits*, Amer. J. Math. **89** (1967), 133–148. MR 0209600

[GeGraPS69] I.M. Gel′fand, M.I. Graev, and I.I. Pyatetskii-Shapiro, *Representation theory and automorphic functions*, Translated from the Russian by K.A. Hirsch, W.B. Saunders Co., Philadelphia, Pa.-London-Toronto, Ont., 1969. MR 0233772 (38 #2093)

[GiLiPo13] P. Giulietti, C. Liverani, and M. Pollicott, *Anosov flows and dynamical zeta functions*, Ann. of Math. (2) **178** (2013), no. 2, 687–773. MR 3071508

[Gui77] V. Guillemin, *Lectures on spectral theory of elliptic operators*, Duke Math. J. **44** (1977), no. 3, 485–517. MR 0448452

[H66] F. Hirzebruch, *Topological methods in algebraic geometry*, Third enlarged edition. New appendix and translation from the second German edition by R.L.E.

Schwarzenberger, with an additional section by A. Borel. Die Grundlehren der Mathematischen Wissenschaften, Band 131, Springer-Verlag New York, Inc., New York, 1966. MR 0202713

[K86] A.W. Knapp, *Representation theory of semisimple groups*, Princeton Mathematical Series, vol. 36, Princeton University Press, Princeton, NJ, 1986, An overview based on examples. MR 855239 (87j:22022)

[K02] A.W. Knapp, *Lie groups beyond an introduction*, second ed., Progress in Mathematics, vol. 140, Birkhäuser Boston, Inc., Boston, MA, 2002. MR 1920389 (2003c:22001)

[LoRo91] J. Lott and M. Rothenberg, *Analytic torsion for group actions*, J. Differential Geom. **34** (1991), no. 2, 431–481. MR 1131439

[Lü93] W. Lück, *Analytic and topological torsion for manifolds with boundary and symmetry*, J. Differential Geom. **37** (1993), no. 2, 263–322. MR 1205447

[LüScTh98] W. Lück, T. Schick, and T. Thielmann, *Torsion and fibrations*, J. Reine Angew. Math. **498** (1998), 1–33. MR 1629917

[Ma02] X. Ma, *Functoriality of real analytic torsion forms*, Israel J. Math. **131** (2002), 1–50. MR 1942300

[Ma17] X. Ma, *Geometric hypoelliptic Laplacian and orbital integral, after Bismut, Lebeau and Shen*, Séminaire Bourbaki, No. 1130, March 11 2017. available at http://www.bourbaki.ens.fr/TEXTES/1130.pdf. Video available at https://www.youtube.com/watch?v=dCDwN-HqcJw.

[MaMari15] X. Ma and G. Marinescu, *Exponential estimate for the asymptotics of Bergman kernels*, Math. Ann. **362** (2015), no. 3-4, 1327–1347. MR 3368102

[Mar04] G.A. Margulis, *On some aspects of the theory of Anosov systems*, Springer Monographs in Mathematics, Springer-Verlag, Berlin, 2004, With a survey by Richard Sharp: Periodic orbits of hyperbolic flows, Translated from the Russian by Valentina Vladimirovna Szulikowska. MR 2035655

[Mi68a] J. Milnor, *Infinite cyclic coverings*, Conference on the Topology of Manifolds (Michigan State Univ., E. Lansing, Mich., 1967), Prindle, Weber & Schmidt, Boston, Mass., 1968, pp. 115–133. MR 0242163 (39 #3497)

[Mi68b] J. Milnor, *A note on curvature and fundamental group*, J. Differential Geometry **2** (1968), 1–7. MR 0232311

[Mil78] J.J. Millson, *Closed geodesics and the η-invariant*, Ann. of Math. (2) **108** (1978), no. 1, 1–39. MR 0501204

[MoSt89] H. Moscovici and R.J. Stanton, *Eta invariants of Dirac operators on locally symmetric manifolds*, Invent. Math. **95** (1989), no. 3, 629–666. MR 979370

[MoSt91] H. Moscovici and R.J. Stanton, *R-torsion and zeta functions for locally symmetric manifolds*, Invent. Math. **105** (1991), no. 1, 185–216. MR 1109626 (92i:58199)

[MoSt18] H. Moscovici and R.J. Stanton, *Holomorphic torsion with coefficients and geometric zeta functions for certain hermitian locally symmetric manifolds*, appendix by J. Frahm, arXiv:1802.08886(2018)

[M78] W. Müller, *Analytic torsion and R-torsion of Riemannian manifolds*, Adv. in Math. **28** (1978), no. 3, 233–305. MR 498252 (80j:58065b)

[M93] W. Müller, *Analytic torsion and R-torsion for unimodular representations*, J. Amer. Math. Soc. **6** (1993), no. 3, 721–753. MR 1189689 (93m:58119)

[M12] W. Müller, *The asymptotics of the Ray–Singer analytic torsion of hyperbolic 3-manifolds*, Metric and differential geometry, Progr. Math., vol. 297, Birkhäuser/Springer, Basel, 2012, pp. 317–352. MR 3220447

[M20] W. Müller, *On Fried's conjecture for compact hyperbolic manifolds*, arXiv: 2005.01450 (2020).

[Pa09] J. Park, *Analytic torsion and Ruelle zeta functions for hyperbolic manifolds with cusps*, J. Funct. Anal. **257** (2009), no. 6, 1713–1758. MR 2540990

[PalMe82] J. Palis, Jr. and W. de Melo, *Geometric theory of dynamical systems*, Springer-Verlag, New York-Berlin, 1982, An introduction, Translated from the Portuguese by A.K. Manning. MR 669541

[Rat69] M.E. Ratner, *Markov decomposition for an $\mathcal{Y}$-flow on a three-dimensional manifold*, Mat. Zametki **6** (1969), 693–704. MR 0260977

[RS71] D.B. Ray and I.M. Singer, *R-torsion and the Laplacian on Riemannian manifolds*, Advances in Math. **7** (1971), 145–210. MR 0295381 (45 #4447)

[Re35] K. Reidemeister, *Homotopieringe und Linsenräume*, Abh. Math. Sem. Univ. Hamburg **11** (1935), no. 1, 102–109. MR 3069647

[Ru76a] D. Ruelle, *Generalized zeta-functions for Axiom A basic sets*, Bull. Amer. Math. Soc. **82** (1976), no. 1, 153–156. MR 0400311

[Ru76b] D. Ruelle, *Zeta-functions for expanding maps and Anosov flows*, Invent. Math. **34** (1976), no. 3, 231–242. MR 0420720

[Rug96] H.H. Rugh, *Generalized Fredholm determinants and Selberg zeta functions for Axiom A dynamical systems*, Ergodic Theory Dynam. Systems **16** (1996), no. 4, 805–819. MR 1406435

[SM93] H. Sánchez-Morgado, *Lefschetz formulae for Anosov flows on 3-manifolds*, Ergodic Theory Dynam. Systems **13** (1993), no. 2, 335–347. MR 1235476

[SM96a] H. Sánchez-Morgado, *R-torsion and zeta functions for analytic Anosov flows on 3-manifolds*, Trans. Amer. Math. Soc. **348** (1996), no. 3, 963–973. MR 1348868

[SM96b] H. Sánchez-Morgado, *Reidemeister torsion and Morse–Smale flows*, Ergodic Theory Dynam. Systems **16** (1996), no. 2, 405–414. MR 1389631

[SR99] S.A. Salamanca-Riba, *On the unitary dual of real reductive Lie groups and the $A_g(\lambda)$ modules: the strongly regular case*, Duke Math. J. **96** (1999), no. 3, 521–546. MR 1671213 (2000a:22023)

[Sa57] I. Satake, *The Gauss–Bonnet theorem for V-manifolds*, J. Math. Soc. Japan **9** (1957), 464–492. MR 0095520 (20 #2022)

[Sch74] P.A. Schweitzer, *Counterexamples to the Seifert conjecture and opening closed leaves of foliations*, Ann. of Math. (2) **100** (1974), 386–400. MR 0356086 (50 #8557)

[Se67] R.T. Seeley, *Complex powers of an elliptic operator*, Singular Integrals (Proc. Sympos. Pure Math., Chicago, Ill., 1966), Amer. Math. Soc., Providence, R.I., 1967, pp. 288–307. MR 0237943 (38 #6220)

[Sel56] A. Selberg, *Harmonic analysis and discontinuous groups in weakly symmetric Riemannian spaces with applications to Dirichlet series*, J. Indian Math. Soc. (N.S.) **20** (1956), 47–87. MR 0088511

[Sel60] A. Selberg, *On discontinuous groups in higher-dimensional symmetric spaces*, Contributions to function theory (internat. Colloq. Function Theory, Bombay, 1960), Tata Institute of Fundamental Research, Bombay, 1960, pp. 147–164. MR 0130324

[Sh18] S. Shen, *Analytic torsion, dynamical zeta functions, and the Fried conjecture*, Anal. PDE **11** (2018), no. 1, 1–74. MR 3707290

[Sh20a] S. Shen, *Complex valued analytic torsion and dynamical zeta function on locally symmetric spaces*, arXiv:2009.03427 (2020).

[Sh20b] S. Shen, *Analytic torsion, dynamical zeta function, and the Fried conjecture for admissible twists*, arXiv: 2009.04897 (2020).

[ShY17] S. Shen and J. Yu, *Flat vector bundles and analytic torsion on orbifolds*, arXiv: 1704.08369 (2017), to appear in Comm. Anal. Geom.

[ShY18] S. Shen and J. Yu, *Morse–Smale flow, Milnor metric, and dynamical zeta function*, arXiv: 1806.00662 (2018), to appear in J. Éc. polytech. Math.

[Sm67] S. Smale, *Differentiable dynamical systems*, Bull. Amer. Math. Soc. **73** (1967), 747–817. MR 0228014 (37 #3598)

[Sp20] P. Spilioti, *Twisted Ruelle zeta function and complex-valued analytic torsion*, arXiv:2004.13474 (2020).

[V84] D.A. Vogan, Jr., *Unitarizability of certain series of representations*, Ann. of Math. (2) **120** (1984), no. 1, 141–187. MR 750719 (86h:22028)

[VZu84] D.A. Vogan, Jr. and G.J. Zuckerman, *Unitary representations with nonzero cohomology*, Compositio Math. **53** (1984), no. 1, 51–90. MR 762307 (86k:22040)

[Vo87] A. Voros, *Spectral functions, special functions and the Selberg zeta function*, Comm. Math. Phys. **110** (1987), no. 3, 439–465. MR 891947

[W49] A. Weil, *Numbers of solutions of equations in finite fields*, Bull. Amer. Math. Soc. **55** (1949), 497–508. MR 0029393

[Wi66] F.W. Wilson, Jr., *On the minimal sets of non-singular vector fields*, Ann. of Math. (2) **84** (1966), 529–536. MR 0202155 (34 #2028)

[Wo08] A. Wotzke, *Die Ruellesche Zetafunktion und die analytische Torsion hyperbolischer Mannigfaltigkeiten*, Ph.D. thesis, Bonn, Bonner Mathematische Schriften (2008), no. Nr. 389.

Shu Shen
Institut de Mathématiques de Jussieu-Paris Rive Gauche
Sorbonne Université, 4 place Jussieu
F-75252 Paris Cedex 5, France
e-mail: shu.shen@imj-prg.fr

Progress in Mathematics, Vol. 338, 301–324

A Survey of the Additive Dilogarithm

Sinan Ünver

Abstract. Borel's construction of the regulator gives an injective map from the algebraic K-groups of a number field to its Deligne–Beilinson cohomology groups. This has many interesting arithmetic and geometric consequences. The formula for the regulator is expressed in terms of the classical polyogarithm functions. In this paper, we give a survey of the additive dilogarithm and the several different versions of the weight two regulator in the infinitesimal setting. We follow a historical approach which we hope will provide motivation for the definitions and the constructions.

Mathematics Subject Classification (2010). 19E15, 14C25.

Keywords. Additive dilogarithm, infinitesimal dilogarithm, Bloch group, regulators.

1. Introduction

The dilogarithm function, even though it has been known for a very long time, has become more prevalent in the past few decades because of its relation to regulators in algebraic K-theory, as was first observed in the pioneering work of Bloch [6]. Among others, this point of view was furthered through the far reaching conjectures of Beilinson on motivic cohomology [1], by the work of Zagier on his conjecture relating special values of Dedekind zeta functions of number fields to values of regulators [33] and in many works of Goncharov ([15], [16], [18] to name a few). The dilogarithm function also appears in hyperbolic geometry, conformal field theory and the theory of cluster algebras. The survey [34] is an excellent introduction to some aspects of this function.

In this note, we give a survey of the infinitesimal version of the above theory. Since the generalizations of the results in this survey to higher weights is still in progress, we restrict to the case of the dilogarithm. In §5.1, we will only briefly mention the construction of additive polylogarithms of higher weight on certain special linear configurations. The existence of this theory itself is quite surprising and is based on ideas of Cathelineau ([9], [10]), Bloch and Esnault [7] and

Goncharov [16], which we will describe in detail below. We emphasize that these functions cannot be deduced from their classical counterparts through a limiting process. We illustrate this point in the somewhat deceptively simple case of weight 1 as follows. The regulator over the complex numbers is given essentially by the real analytic map $\log|\cdot|:\mathbb{C}^{\times}\to\mathbb{R}$. On the other hand, in the infinitesimal case for k a field of characteristic 0, and $k_n := k[[t]]/(t^n)$, one has the algebraic map $\log^{\circ}: k_n^{\times}\to k_n$ defined by $\log^{\circ}(a) := \log(\frac{a}{a(0)})$. The use of the absolute value makes the first function non-algebraic, single-valued and dependent, in an essential way, on the local field in question. In the second case, the map $k\to k_n$, which is a section of the canonical projection from k_n to its quotient by its nilradical achieves the purpose of choosing a branch in an appropriate sense. We will see below that over a scheme with non-reduced structure such local splittings, which correspond to retractions of the scheme with the reduced induced structure, will play a role analogous to choosing branches.

In the second section, we briefly recall the definitions of the Bloch–Wigner dilogarithm, the Chow dilogarithm of Goncharov and Bloch's regulator function from $K_2(C)^{(2)}_{\mathbb{Q}}$ of a curve C. We emphasize the point of view of the Aomoto dilogarithms and scissors congruence class groups whose analogs will be the main motivation for the infinitesimal versions of the above functions.

In the third section, we give the infinitesimal analogs of these functions starting with the ideas of Cathelineau, Goncharov and Bloch–Esnault. We also recall the additive dilogarithm construction of Bloch–Esnault.

In the fourth section, we discuss the construction of the infinitesimal Chow dilogarithm, together with its application to algebraic cycles and Goncharov's strong reciprocity conjecture. We also describe the infinitesimal version of Bloch's regulator on curves.

In the last section, we discuss some partial results in higher weights and in characteristic p and some open problems.

Conventions. Except in §5.2, we will consider motivic cohomology always with $\mathbb{Q}$-coefficients. Therefore all the Bloch complexes, Aomoto complexes etc. are tensored with $\mathbb{Q}$. For example, the notation $\Lambda^2 k^{\times}$ means that the group $\Lambda^2_{\mathbb{Z}} k^{\times}$ is tensored with $\mathbb{Q}$. The cyclic homology and André–Quillen homology groups are always considered relative to $\mathbb{Q}$. The notation Ω^1_A for an algebra A over a field always means the Kähler differentials relative to the prime field. For an A-module I, $S^{\cdot}_A I$ denotes the symmetric algebra of I over A. For a ring A, $A^{\flat}$ denotes the set of all units a in A such that $1-a$ is also a unit. For a functor F from the category of pairs (R, I) of rings R and nilpotent ideals I to an abelian category, we let $F^{\circ}(R, I)$ denote the kernel of the map from $F(R, I)$ to $F(R/I, 0)$. We informally refer to this object as the *infinitesimal part of* F. We have the corresponding notion for the category of artin local algebras over a field, since their maximal ideals are nilpotent.

2. Bloch–Wigner dilogarithm and the scissors congruence class group

2.1. Aomoto dilogarithm

The general conjectures on motives expect that for any field k one has a tannakian category MTM_k over $\mathbb{Q}$ of mixed Tate motives over k. This gives a graded Hopf algebra $\mathscr{A}.(k)$ such that a mixed Tate motive over k is the same as a graded $\mathbb{Q}$-space with a co-module structure over $\mathscr{A}.(k)$.

Since the objects in MTM_k should be constructed from Tate objects by means of extensions, one expects MTM_k to have a *linear* algebraic description. In [2] a graded Hopf algebra $A.(k)$ was defined, using linear algebraic objects, such that one expects a natural map $A.(k) \to \mathscr{A}.(k)$.

This $A.(k)$ is the graded Hopf algebra of Aomoto polylogarithms over k defined in [2, §2]. An n-simplex L in $\mathbb{P}^n_k$ is an $(n+1)$-tuple $(L_0, \ldots, L_n)$ of hyperplanes. It is said to be non-degenerate if the hyperplanes are in general position. A pair of simplices (L, M) is said to be admissible if they do not have a common face. $A_n(k)$ is the $\mathbb{Q}$-space generated by pairs of admissible simplices $(L; M)$ in $\mathbb{P}^n_k$ subject to the following relations:

(i) $(L, M) = 0$, if one of the simplices is degenerate;
(ii) (L, M) is anti-symmetric with respect to the ordering of the hyperplanes in both of the n-simplices.
(iii) If L is an $n+2$-tuple of hyperplanes $(L_0, \ldots, L_{n+1})$ and $\hat{L}^j$ is the n-simplex obtained by omitting L_j, then $\sum_{0 \le j \le n+1} (\hat{L}^j, M) = 0$, and the corresponding relation for the second component.
(iv) For $\alpha \in \mathrm{GL}_{n+1}(k)$, $(\alpha(L), \alpha(M)) = (L, M)$.

There are certain configurations, called polylogarithmic configurations, in $A_n(k)$ that play an important role in understanding the motivic cohomology of k, since they act as building blocks for all configurations [2, §1.16]. Let $P_n(k)$ denote the subgroup of *prisms* in $A_n(k)$. This is the subgroup generated by configurations which come from products of configurations from lower dimensions. For every $a \in k^\flat := k^\times \setminus \{1\}$, there is a special configuration $(L, M_a) \in A_n(k)$ [15, Fig. 1.14], which corresponds to the value of the abstract polylogarithm at a. If z_i, $0 \le i \le n$ are the homogenous coordinates on $\mathbb{P}^n_k$, then L_i is defined by $z_i = 0$. The simplex M_a is defined by the following formulas. $M_0 : z_0 = z_1$; $M_1 : z_0 = z_1 + z_2$; $M_i : z_i = z_{i+1}$, for $2 \le i < n$; and $M_n : az_0 = z_n$.

This defines a map $l_n : \mathbb{Q}[k^\flat] \to A_n(k)/P_n(k)$, which sends the generator $[a]$ to the class of (L, M_a). Denoting the image of l_n by $B'_n(k)$, one expects the co-multiplication on $A.(k)$ to induce a complex $\Gamma'_k(n)$:

$$B'_n(k) \to B'_{n-1}(k) \otimes k^\times \to \cdots \to B'_2(k) \otimes \Lambda^{n-2} k^\times \to \Lambda^n k^\times,$$

which would compute the motivic cohomology of k of weight n.

For $n = 2$, there is a simpler complex, namely the Bloch complex $\Gamma_k(2)$ of weight two, which computes the motivic cohomology. Let $B_2(k)$ be the quotient of $\mathbb{Q}[k^\flat]$, the vector space with basis $[x]$ for $x \in k^\flat$, by the subspace generated by elements of the form

$$[x] - [y] + [y/x] - [(1-x^{-1})/(1-y^{-1})] + [(1-x)/(1-y)], \qquad (2.1.1)$$

for all $x, y \in k^\times$ such that $(1-x)(1-y)(1-x/y) \in k^\times$. The last equation is the 5-term functional equation of the dilogarithm. Let δ be the map that sends $[x]$ to $(1-x) \wedge x \in \Lambda^2 k^\times$. This map factors through $B_2(k)$ and we obtain a complex:

$$B_2(k) \xrightarrow{\delta} \Lambda^2 k^\times,$$

concentrated in degrees 1 and 2. We denote this complex by $\Gamma_k(2)$. This complex indeed computes the motivic cohomology of k with coefficients $\mathbb{Q}(2)$, assuming the Beilinson–Soule vanishing conjecture, by a theorem of Bloch. In other words, the sequence

$$0 \to K_3(k)^{(2)}_{\mathbb{Q}} \to B_2(k) \to \Lambda^2 k^\times \to K_2^M(k)_{\mathbb{Q}} \to 0$$

is exact ([6], [27]).

The map l_2 factors through the quotient $\mathbb{Q}[k^\flat] \to B_2(k)$ to induce an isomorphism:

$$l_2 : B_2(k) \to B_2'(k) = A_2(k)/P_2(k)$$

which we continue to denote with the same symbol [2, Proposition 3.7]. This can be thought of as the abstract motivic dilogarithm function.

2.2. Bloch–Wigner dilogarithm

The nth polylogarithm function is defined inductively by $\ell i_1(z) = -\log(1-z)$ and

$$d\ell i_k(z) = \ell i_{k-1}(z)\frac{dz}{z},$$

with $\ell i_k(0) = 0$.

These functions have the power series expansion $\ell i_k(z) = \sum_{1 \leq n} \frac{z^n}{n^k}$, in the unit disc around 0, and have multi-valued analytic continuations to $\mathbb{C}^\times \setminus \{1\}$. They appear as coordinates of a matrix which describe a canonical quotient of the fundamental groupoid associated to the Hodge realization of the unipotent fundamental group of $\mathbb{P}^1 \setminus \{0, 1, \infty\}$ [3]. The specialization of this construction at a point $a \in \mathbb{C}^\flat$ gives a motive which coincides with the motive associated to the configuration $l_2(a)$ in §2.1.

The Hodge realization of this motive (specialized at a point) as well as of the motive above defined by the configurations in §2.1 above are Hodge–Tate structures. An $\mathbb{R}$-Hodge–Tate structure is a mixed $\mathbb{R}$-Hodge structure such that for every $r \in \mathbb{Z}$, its graded piece of degree $-2r$ with respect to the weight filtration are direct sums of the Tate structures $\mathbb{R}(r)$, of weight $-2r$; and its graded pieces of odd degree are equal to 0. Let $\mathscr{H}_\cdot$ denote the graded Hopf algebra associated to the tannakian category of $\mathbb{R}$-Hodge–Tate structures. The Hodge realization functor should give a morphism $\mathscr{A}_\cdot(\mathbb{C}) \to \mathscr{H}_\cdot$ of graded Hopf algebras.

A construction of Beilinson and Deligne (§2.5, [3]; pp. 248–249, [15]) associates to each framed $\mathbb{R}$-Hodge–Tate structure a number. Associated to the variation of Hodge structures on $\mathbb{G}_m$ that gives the function $\log(z)$ one gets the corresponding single-valued function $\log|z|$. This construction gives a map $p_{\mathscr{H},n} : \mathscr{H}_n \to \mathbb{R}$. It turns out that this map vanishes on the products [15]. Hence composing with the Hodge realization map associated to the Aomoto configurations, the corresponding map vanishes on prisms and one gets a map $B_2'(\mathbb{C}) = A_2(\mathbb{C})/P_2(\mathbb{C}) \to \mathbb{R}$. The composition of this map with $l_2 : B_2(\mathbb{C}) \to B_2'(\mathbb{C})$ turns out to be, up to scaling, the Bloch–Wigner dilogarithm D defined by

$$D(z) = \mathrm{Im}(\ell i_2(z)) + \arg(1-z)\log|z|.$$

The main importance of the Bloch–Wigner dilogarithm comes from the fact that they are regulators.

Composing $p_{\mathscr{H},n}$ with the Hodge realization would give a map

$$\mathrm{vol}_n : \mathscr{A}_n(\mathbb{C}) \to \mathbb{R},$$

which is an analog of the volume map on the scissors congruence class groups below and its infinitesimal version is the main concern of this survey.

2.3. Chow dilogarithm

If $X/\mathbb{C}$ is a smooth and projective curve over $\mathbb{C}$, there is a version of the dilogarithm above which gives certain regulators of X. Namely $\mathrm{H}^3_{\mathcal{M}}(X, \mathbb{Q}(3)) \simeq K_3(X)^{(3)}_{\mathbb{Q}}$ and applying the Leray–Serre spectral sequence to the map $X \to \mathbb{C}$, there would be a map $K_3(X)^{(3)}_{\mathbb{Q}} \simeq \mathrm{H}^3_{\mathcal{M}}(X, \mathbb{Q}(3)) \to \mathrm{H}^1_{\mathcal{M}}(\mathbb{C}, H^2(X/\mathbb{C})(3)) = \mathrm{H}^1_{\mathcal{M}}(\mathbb{C}, \mathbb{Q}(2)) \simeq K_3(\mathbb{C})^{(2)}_{\mathbb{Q}}$. Combining with the regulator $K_3(\mathbb{C})^{(2)}_{\mathbb{Q}} \to \mathbb{R}$, given by the Bloch–Wigner dilogarithm above, one would get a map

$$K_3(X)^{(3)} \to \mathbb{R}.$$

This map is given by the following Chow dilogarithm of Goncharov.

If f_1, f_2, and f_3 are rational functions on X. Let

$$r_2(f_1,f_2,f_3) := \mathrm{Alt}_3\left(\frac{1}{6}\log|f_1|\cdot d\log|f_2|\wedge d\log|f_3| - \frac{1}{2}\log|f_1|\cdot d\arg f_2 \wedge d\arg f_3\right),$$

which has the formal property that $d(r_2(f_1, f_2, f_3)) = \mathrm{Re}(d\log(f_1) \wedge d\log(f_3) \wedge d\log(f_3))$. The map $\rho_{\mathbb{R}} : \Lambda^3\mathbb{C}(X)^{\times} \to \mathbb{R}$, given by

$$\rho_{\mathbb{R}}(f_1 \wedge f_2 \wedge f_3) := \int_{X(\mathbb{C})} r_2(f_1, f_2, f_3),$$

is, up to a constant multiple, the Chow dilogarithm [18, p. 4]. The middle cohomology of the complex

$$\cdots \to B_2(\mathbb{C}(X)) \to (\oplus_{x\in X} B_2(\mathbb{C})) \oplus \Lambda^3\mathbb{C}(X)^{\times} \to \oplus_{x\in X}\Lambda^2\mathbb{C}^{\times} \to \cdots$$

is $K_3(X)^{(3)}_{\mathbb{Q}}$ and the map $(\oplus_{x\in X} D) \oplus \rho_{\mathbb{R}}$ obtained by using the Bloch–Wigner and the Chow dilogarithm, gives the regulator.

2.4. Bloch's regulator on curves and the tame symbol

There is another regulator which is based on a version of the dilogarithm. Again assume that $X/\mathbb{C}$ is a smooth and projective curve. This regulator is essentially the map from $K_2(X)_{\mathbb{Q}}^{(2)}$ to the corresponding Deligne cohomology group: $K_2(X)_{\mathbb{Q}}^{(2)} \to \mathrm{H}_{D}^{2}(X_{an}, \mathbb{Q}(2)) \simeq H^1(X_{an}, \mathbb{C}/\mathbb{Q}(2))$. Dividing by $2\pi i$ and using the exponential map on the coefficients, the last cohomology group is identified with $H^1(X_{an}, \mathbb{C}_{\mathbb{Q}}^{\times})$. Since $H^1(X_{an}, \mathbb{C}^{\times})$ coincides with local systems of complex vector spaces of rank 1 and hence with analytic line bundles with connection, the above map can also be deduced from the local and analytic construction of Deligne, which associates to each pair f, g of meromorphic functions on X, a line bundle with connection on X_{an}, such that the monodromy at each point is given by the tame symbol of f and g at that point [13]. Explicitly, if $\log(f)$ is a choice of a branch of f, locally analytically, then the line bundle in question is the trivial line bundle with the connection ∇ given by $\nabla(1) = \frac{1}{2\pi i} \log(f) \frac{dg}{g}$. For a different choice $\log(f) + n2\pi i$ of a logarithm of f, the isomorphism between the line bundles with connection is given as multiplication by g^{-n} [13, §2.3].

When X is defined over a number field, the Bloch regulator is fundamental in the study of certain special values of the L-function of X [24]. It also appears, for example, in the geometric study of cycles on $X/\mathbb{C}$ [20].

3. Additive dilogarithm and the infinitesimal scissors congruence class group

In this section, we start with the 4-term functional equation for the entropy function which is also satisfied by an infinitesimal version of the Dehn invariant for scissors congruence class groups. This 4-term functional equation of Cathelineau can be thought of as a deformation of the 5-term functional equation that is restricted to certain special elements. The precise relation is explained in §3.4.2. Next we describe Goncharov's idea that the hyperbolic scissors congruence class group can be thought of degenerating to the Euclidean one as the model for the hyperbolic space blows up. We continue the section with describing the construction of the additive dilogarithm by Bloch and Esnault based on the localization sequence in K-theory and end the section on our construction of the additive dilogarithm on the Bloch group.

3.1. The 4-term functional equation

In information theory, Shannon's binary entropy function H is defined as

$$H(p) := -p\log(p) - (1-p)\log(1-p),$$

for the probability p. This function satisfies the following fundamental functional equation of information theory:

$$H(p) + (1-p)H\left(\frac{q}{1-p}\right) = H(q) + (1-q)H\left(\frac{p}{1-q}\right). \tag{3.1.1}$$

The same functional equation reappeared in ([9], [12]) as follows. For a field k of characteristic 0, let $\beta_2(k)$ is the vector space over k generated by the symbols $\langle a \rangle$, for $a \in k^\flat$ with relations generated by

$$\langle p \rangle - \langle q \rangle + p \left\langle \frac{q}{p} \right\rangle + (1-p) \left\langle \frac{1-q}{1-p} \right\rangle = 0 \tag{3.1.2}$$

when $p \neq q$. These relations already imply that $\langle p \rangle = \langle 1-p \rangle$ and $\langle \frac{1}{p} \rangle = -\frac{1}{p} \langle p \rangle$ and using these the two relations (3.1.1) and (3.1.2) are equivalent. In [9, Theorème 1], Cathelineau proves that the following sequence

$$0 \longrightarrow \beta_2(k) \xrightarrow{D} k \otimes k^\times \xrightarrow{L} \Omega^1_k \longrightarrow 0$$

is exact, where D is defined on the generators by $D(\langle a \rangle) := a \otimes a + (1-a) \otimes (1-a)$ and L sends $a \otimes b$ to $a\frac{db}{b}$. This was used in [9] in order to show that for an algebraically closed field k of characteristic 0, the homology groups of $\mathrm{SL}(2,k)$ with adjoint action on its Lie algebra $\mathfrak{sl}(2,k)$ are given by:

$$\mathrm{H}_1(\mathrm{SL}(2,k), \mathfrak{sl}(2,k)) \simeq \Omega^1_k$$
$$\mathrm{H}_2(\mathrm{SL}(2,k), \mathfrak{sl}(2,k)) = 0\,.$$

This is in analogy with the computation of the homology of the discrete special orthogonal group $\mathrm{SO}^\delta(3,\mathbb{R})$ with the standard action on $\mathbb{R}^3$:

$$\mathrm{H}_1(\mathrm{SO}^\delta(3,\mathbb{R}), \mathbb{R}^3) \simeq \Omega^1_{\mathbb{R}}$$
$$\mathrm{H}_2(\mathrm{SO}^\delta(3,\mathbb{R}), \mathbb{R}^3) = 0\,.$$

This is a restatement of Sydler's theorem that the Dehn invariant and the volume completely determine the scissors congruence class. In this Euclidean case, the analog of $k \otimes k^\times$ is the group $\mathbb{R} \otimes \mathbb{R}/\pi\mathbb{Z}$ and the analog of the map L above is the map

$$\mathbb{R} \otimes \mathbb{R}/\pi\mathbb{Z} \to \Omega^1_{\mathbb{R}}$$

that sends $l \otimes \theta$ to $l\frac{d(\cos\theta)}{\sin\theta}$.

The above D can be thought of as the infinitesimal version of the Dehn invariant and the functional equation above can be thought of as the infinitesimal version of the functional equation of the dilogarithm in the following sense.

3.2. Hyperbolic space degenerating to Euclidean space

In this section, we describe how Goncharov's idea on the degeneration of hyperbolic space to Euclidean space and the analogy between the scissors congruence class groups and mixed Tate motives leads one to expect a volume map on mixed Tate motives over dual numbers which is reminiscent of the polylogarithm functions.

3.2.1. If $\mathscr{G}^n$ is one of the three n-dimensional classical geometries: $\mathscr{E}^n$, the Euclidean; $\mathscr{H}^n$, the hyperbolic; or $\mathscr{S}^n$, the spherical, then let $\mathscr{P}(\mathscr{G}^n)$ denote the scissors congruence class group corresponding to $\mathscr{G}^n$. The Dehn invariant map:

$$D_n^{\mathscr{G}} : \mathscr{P}(\mathscr{G}^n) \to \oplus_{i=1}^{n-2} \mathscr{P}(\mathscr{G}^i) \otimes \mathscr{P}(\mathscr{S}^{n-i-1})$$

endows $\oplus \mathscr{P}(\mathscr{S}^{\cdot})$ with the structure of a co-algebra and, $\oplus \mathscr{P}(\mathscr{H}^{\cdot})$ and $\oplus \mathscr{P}(\mathscr{E}^{\cdot})$ with structures of co-modules over this co-algebra [16].

There exists a map from $\mathscr{P}(\mathscr{H}^{2n-1})$ to $\mathscr{A}_n(\mathbb{C})$, defined by Goncharov, which attaches a framed mixed Tate motive to an element in the hyperbolic scissors congruence class group [16]. If one considers the Cayley spherical model for the hyperbolic geometry then as the sphere gets bigger the hyperbolic geometry approaches the Euclidean geometry [16]. Therefore, in the limit case one would expect to have a map $\mathscr{P}(\mathscr{E}^{2n-1}) \to \mathscr{A}_n^{\circ}(\mathbb{C}_2)$.

These suggest a close similarity between the structures of $\mathscr{A}_n^{\circ}(k_2)$ and $\mathscr{P}(\mathscr{E}_k^{2n-1})$ [16], [17]. The Euclidean scissors congruence class group has a volume map

$$\mathscr{P}(\mathscr{E}_k^{2n-1}) \to k,$$

which is conjectured to induce an isomorphism from $\mathrm{H}^1(\oplus_{2n-1}\mathscr{P}(\mathscr{E}_k^{\cdot}))$, the kernel in $\mathscr{P}(\mathscr{E}_k^{2n-1})$ of the Dehn invariant map, to k. For $n = 2$ and $k = \mathbb{R}$, this is Sydler's theorem. In analogy, we expect a map:

$$\mathrm{vol}_n^{\circ} : \mathscr{A}_n^{\circ}(k_2) \to k,$$

which induces an isomorphism from $\mathrm{H}^1(\mathscr{A}_{\cdot}^{\circ}(k_2)(n))$ to k. Moreover, we should have the identity $\mathrm{vol}_n^{\circ} \circ \rho_\lambda = \lambda^{2n-1}\mathrm{vol}_n^{\circ}$, for $\lambda \in k^{\times}$. This map would be an analog of both the map $\mathscr{A}_n(\mathbb{C}) \to \mathbb{R}$ that is constructed using the Beilinson–Deligne construction and of the volume map on Euclidean scissors congruence class groups.

3.2.2. Given an element (L, M) in $A_n(\mathbb{C})$, this defines a framed mixed Tate motive in $\mathscr{A}_n(\mathbb{C})$ whose associated mixed Hodge structure $\mathrm{H}^n(\mathbb{P}_{\mathbb{C}}^n \setminus L, M \setminus L)$, is Hodge–Tate. Therefore, using the construction of Beilinson and Deligne described above which attaches a real number to $\mathbb{R}$-Hodge–Tate structures, we get $\mathrm{vol}_n(L, M) \in \mathbb{R}$. This vanishes on the products [15] to give: $\mathrm{vol}_n : A_n(\mathbb{C})/P_n(\mathbb{C}) \to \mathbb{R}$. Composing with the abstract polylogarithm map induces

$$\mathrm{vol}_n \circ l_n : B_n'(\mathbb{C}) \to \mathbb{R}.$$

This has the following description. Let $\mathscr{L}_n$ be the real single-valued version of the n-polylogarithm:

$$\mathscr{L}_n(z) := \mathscr{R}_n\Big(\sum_{j=0}^{n} \frac{2^j B_j}{j!}(\log|z|)^j \ell i_{n-j}(z)\Big),$$

where B_n is the nth Bernoulli number; $\mathscr{R}_n$ is the real part if n is odd and the imaginary part if n is even; and $\ell i_0(z) := -1/2$. Then for $z \in \mathbb{C}^{\flat}$, $\mathrm{vol}_n \circ l_n(z) = \mathscr{L}_n(z)$ [15].

3.2.3. Let k be any field of characteristic 0. The definitions of $A_n(k)$, $P_n(k)$, l_n and $B'_n(k)$ exactly carry over to the case of k_2 to define the groups $A_n(k_2)$, $P_n(k_2)$, and $B'_n(k_2)$, and a map, $l_n : \mathbb{Q}[k_2^\flat] \to A_n(k_2)$. One would like to define a map

$$\mathrm{vol}_n^\circ : A_n(k_2)/P_n(k_2) \to k,$$

which would be an analog of the map defined above over the complex numbers using the Beilinson–Deligne construction. This map would be the composition of the natural map from $A_n(k_2)$. In this context the analog of the single-valued polylogarithm $\mathscr{L}_n$ would be the composition $\mathrm{vol}_n^\circ \circ l_n$.

3.2.4. As in §2.1, one has a complex $\Gamma'_{k_2}(n)$, concentrated in degrees $[1, n]$:

$$B'_n(k_2) \to B'_{n-1}(k_2) \otimes k_2^\times \to \cdots \to B'_2(k_2) \otimes \Lambda^{n-2} k_2^\times \to \Lambda^n k_2^\times$$

induced by the co-multiplication map on $A.(k_2)$ and such that $\{x\}_i \otimes y \in B'_i(k_2) \otimes \Lambda^{n-i} k_2^\times$ is mapped to:

$$\{x\}_{i-1} \otimes x \wedge y \in B'_{i-1}(k_2) \otimes \Lambda^{n-i+1} k_2^\times$$

if $i \geq 3$, and to

$$(1-x) \wedge x \wedge y \in \Lambda^n k_2^\times$$

if $i = 2$.

One expects the cohomology groups to be given by $\mathrm{H}^i(\Gamma'_{k_2}(n)) \simeq K_{2n-i}(k_2)^{(n)}_{\mathbb{Q}}$. By Goodwillie's theorem [19], we have, $K^\circ_{2n-i}(k_2)^{(n)}_{\mathbb{Q}} \simeq \mathrm{HC}^\circ_{2n-i-1}(k_2)^{(n-1)}$. The infinitesimal part of the cyclic homology of k_2 is computed as $\mathrm{HC}^\circ_n(k_2)^{(m)} \simeq \Omega_k^{2m-n}$, for $[\frac{n+1}{2}] \leq m \leq n$, and is 0 otherwise [11]. Moreover, for $\lambda \in k^\times$, the automorphism ρ_λ of k_2 that sends t to λt induces multiplication by $\lambda^{2(n-m)+1}$ on Ω_k^{2m-n} [11]. Combining these, one expects the infinitesimal part of the cohomology of $\Gamma'_{k_2}(n)$ to be:

$$\mathrm{H}^i(\Gamma'^\circ_{k_2}(n)) \simeq \Omega_k^{i-1},$$

for $1 \leq i \leq n$, and that ρ_λ induces multiplication by $\lambda^{2(n-i)+1}$ on Ω_k^{i-1}. Note that when $i = 1$, this map scales by λ^{2n-1}, exactly like the volume map in §3.2.1.

3.3. Bloch and Esnault's construction of the additive dilogarithm on the localization sequence

The work of Bloch and Esnault was the principal motivation for the various generalizations of the additive dilogarithm. Here we briefly describe their work, generalized to the case of higher moduli. The proofs of the statements can be found in [7] and in [28, §6.2]. In this section, we assume that k is algebraically closed in addition to being of characteristic 0.

Let $\mathcal{O}$ be the local ring of $\mathbb{A}_k^1$ at 0. The localization sequence of the pair $(k[t], (t^m))$ gives the following two exact sequences:

$$K_2(k[t], (t^m)) \to K_2(\mathcal{O}, (t^m)) \xrightarrow{\partial} \oplus_{x \in k^\times} K_1(k) \to K_1(k[t], (t^m)) \to 0$$

and

$$0 \to K_1(\mathcal{O},(t^m)) \xrightarrow{\ \partial\ } \oplus_{x\in k^\times} K_0(k) \to K_0(k[t],(t^m)) \to 0.$$

The group

$$T_mB_2(k) := (K_2(\mathcal{O},(t^m))/\operatorname{im}(K_1(k)\cdot K_1(\mathcal{O},(t^m)))_{\mathbb{Q}},$$

is the infinitesimal analog of the Bloch group. Since $K_0(k[t],(t^m)) \simeq 1+(t) = (k_m^\times)^\circ \subseteq k_m^\times$, the quotient $\oplus_{x\in k^\times} K_1(k)/\partial(K_1(k)\cdot K_1(\mathcal{O},(t^m))) \simeq k^\times \otimes (k_m^\times)^\circ$. This gives the complex:

$$T_mB_2(k) \to k^\times \otimes (k_m^\times)^\circ,$$

which is the analog of the Bloch complex and is denoted by $T_m\mathbb{Q}(2)(k)$. The cohomology groups of this complex in degrees 1 and 2 are respectively, $K_3^\circ(k_m)_{\mathbb{Q}}^{(2)}$ and $K_2^M(k_m)_{\mathbb{Q}}^\circ$, and the natural map from $K_2(k[t],(t^m))_{\mathbb{Q}}$ to $T_mB_2(k)$ obtained from the localization sequence surjects to this $K_3^\circ(k_m)_{\mathbb{Q}}^{(2)}$ as one can see by considering the reduction modulo (t^{2m-1}) map below.

The reduction modulo (t^{2m-1}) map:

$$\begin{aligned}&(K_2(\mathcal{O},(t^m))/\operatorname{im}(K_1(k)\cdot K_1(\mathcal{O},(t^m)))_{\mathbb{Q}}\\ &\quad\to (K_2(k_{2m-1},(t^m))/\operatorname{im}(K_1(k)\cdot K_1(k_{2m-1},(t^m)))_{\mathbb{Q}},\end{aligned}$$

from $T_mB_2(k)$ to $(K_2(k_{2m-1},(t^m))/\operatorname{im}(K_1(k)\cdot K_1(k_{2m-1},(t^m)))_{\mathbb{Q}} \simeq K_3^\circ(k_m)_{\mathbb{Q}}^{(2)} \simeq \oplus_{m<w<2m} t^w k$ is the additive dilogarithm map in this context.

If one starts with the localization sequence for the ideal $(t(1-t))$ instead of the one for (t^m) above, one obtains a similar complex which computes the ordinary weight two motivic cohomology of k. This was carried out in the fundamental work [6].

3.4. The additive dilogarithm as an infinitesimal dilogarithm

In the first part, we describe the infinitesimal analog of the Bloch–Wigner dilogarithm. In the second part, we explain the relation of our complex to that of Cathelineau's and that of Bloch–Esnault's. We also describe how the 4-term functional equation is related to the 5-term functional equation.

3.4.1. Construction of $\ell i_{m,w}$. For any local $\mathbb{Q}$-algebra A, we let $B_2(A)$ denote the $\mathbb{Q}$-space generated by $[x]$ with $x \in A^\flat := \{x|x(1-x)\in A^\times\}$ subject to the relations (2.1.1), for all $x,\ y \in A^\times$ such that $(1-x)(1-y)(1-x/y) \in A^\times$. We then have a complex $\Gamma_A(2)$ as in §2.1.

Let k be a field of characteristic 0, $k_\infty := k[[t]]$, the formal power series over k, and for $1 \le m$, $k_m := k_\infty/(t^m)$. Recall that the Bloch–Wigner dilogarithm D defines a map $B_2(\mathbb{C}) \to \mathbb{R}$ and is the unique measurable function, up to multiplication, with this property [6]. Its restriction to $K_3(\mathbb{C})_{\mathbb{Q}}^{(2)}$ is, up to a rational multiple, the Borel regulator. We have the corresponding theorem for the infinitesimal part of $B_2(k_m)$. In order to describe the infinitesimal analogs of D, first note that the corresponding cohomology group $\mathrm{H}^1(\Gamma_{k_m}^\circ(2))$ should be $K_3^\circ(k_m)_{\mathbb{Q}}^{(2)}$. This last group, by Goodwillie's theorem [19], can be expressed in terms of cyclic homology, relative to $\mathbb{Q}$, as $\mathrm{HC}_2^\circ(k_m)^{(1)}$. There is an action, which we denote by $\star$, of $k^\times$ on k_m such that $\lambda \in k^\times$ acts by sending t to $\lambda \star t := \lambda t$. The induced action on $\mathrm{HC}_2^\circ(k_m)^{(1)}$ decomposes this group into a direct sum $\mathrm{HC}_2^\circ(k_m)^{(1)} = \oplus_{m<w<2m} k$, with respect to the weights of the $\star$ action. The action of $\lambda \in k^\times$ on k, in the component of $\star$-weight w, is the one which sends $a \in k$ to $\lambda^w a \in k$ ([28], [11]). This suggests that corresponding to each $\star$-weight w between m and $2m$, there is a dilogarithm $\ell i_{m,w} : B_2(k_m) \to k$, which vanishes on the image of $B_2(k)$ in $B_2(k_m)$ and induces an isomorphism between the $\star$-weight w component in $\mathrm{HC}_2^\circ(k_m)^{(1)}$ and the target.

We describe this dilogarithm as follows. Let $\log^\circ : k_\infty^\times \to k_\infty$, be defined as $\log^\circ(\alpha) := \log(\frac{\alpha}{\alpha(0)})$. If $q = \sum_{0 \le i} q_i t^i \in k_\infty$ and $1 \le a$ then $q|_a := \sum_{0 \le i < a} q_i t^i$, and $t_a(q) := q_a$. If $u \in t k_\infty$ and $s(1-s) \in k^\times$, we let

$$\ell i_{m,w}(se^u) := t_{w-1}\left(\log^\circ(1-se^{u|_m}) \cdot \frac{\partial u}{\partial t}\Big|_{w-m}\right), \tag{3.4.1}$$

for $m < w < 2m$.

The Bloch complex $\Gamma_{k_m}(2)$ computes the motivic cohomology of weight two over the truncated polynomial ring k_m. Namely the sequence:

$$0 \to K_3(k_m)_{\mathbb{Q}}^{(2)} \longrightarrow B_2(k_m) \xrightarrow{\delta_m} \Lambda^2 k_m^\times \longrightarrow K_2^M(k_m)_{\mathbb{Q}} \to 0$$

is exact. We can state the combination of these as [28]:

Theorem. *The complex $B_2^\circ(k_m) \xrightarrow{\delta^\circ} (\Lambda^2 k_m^\times)^\circ$ computes the infinitesimal part of the weight two motivic cohomology of k_m, the maps $\ell i_{m,w}$ satisfy the functional equation for the dilogarithm and descend to give maps from $B_2(k_m)$ to k, such that $\oplus_{m<w<2m} \ell i_{m,w}$ induces an isomorphism*

$$\mathrm{HC}_2^\circ(k_m)^{(1)} \simeq K_3^\circ(k_m)_{\mathbb{Q}}^{(2)} \simeq ker(\delta^\circ) \xrightarrow{\sim} k^{\oplus(m-1)}.$$

We sketch the main points of the proof in [28]. First we describe the map $\ell i_{m,w}$ in terms of the map δ. In order to specify the range and domain of δ, we denote the δ from $B_2(k_m)$ to $\Lambda^2 k_m^\times$ by the symbol δ_m. For $i < w$, let $\ell_i : k_w^\times \to k$ be defined by $\ell_i(\alpha) := t_i(\log^\circ \alpha)$, and $\ell_i \wedge \ell_j : \Lambda^2 k_w^\times \to k$, as $(\ell_i \wedge \ell_j)(a \wedge b) := \ell_i(a) \cdot \ell_j(b) - \ell_i(b) \cdot \ell_j(a)$.

The following diagram

$$\begin{array}{ccc} B_2(k_w) & \xrightarrow{\delta_w} & \Lambda^2 k_w^\times \\ \downarrow{\scriptstyle \pi_{w,m}} & & \downarrow{\scriptstyle \sum_{1 \le i \le w-m} i \cdot \ell_{w-i} \wedge \ell_i} \\ B_2(k_m) & \overset{\ell i_{m,w}}{\dashrightarrow} & k, \end{array}$$

where $\pi_{w,m} : B_2(k_w) \to B_2(k_m)$ is the natural projection, commutes. This shows that $\ell i_{m,w}$ satisfies the same five term functional equation as the Bloch–Wigner dilogarithm.

Next by the stabilization theorem of Suslin

$$\mathrm{H}_3(\mathrm{GL}(k_m),\mathbb{Q}) = \mathrm{H}_3(\mathrm{GL}_3(k_m),\mathbb{Q}),$$

and by an argument of Goncharov, we have a map

$$\mathrm{H}_3(\mathrm{GL}_3(k_m),\mathbb{Q}) \to \mathrm{H}_3(\mathrm{GL}_2(k_m),\mathbb{Q}).$$

Studying the action of GL_2 on configurations of points on $\mathbb{P}^1$, it is easy to construct a map from $\mathrm{H}_3(\mathrm{GL}_2(k_m),\mathbb{Q})$ to $\ker(\delta_m)$. Combining these, we obtain a map from $\mathrm{H}_3(\mathrm{GL}(k_m),\mathbb{Q})$ to $\ker(\delta_m)$.

Using Volodin's construction of K-theory, we can then make Goodwillie's theorem explicit by constructing a map from $\mathrm{HC}_2^\circ(k_m)$ to $H_3(\mathrm{GL}(k_m),\mathbb{Q})$. Combining with the above, this gives a surjection from $\mathrm{HC}_2^\circ(k_m)$ to $\ker(\delta_m)$, Finally, by an explicit computation of $\ell i_{m,w}$ on the image of a basis of $\mathrm{HC}_2^\circ(k_m)$ in $\ker(\delta_m)$, we see that $\oplus_{m<w<2m}\ell i_{m,w}$ is injective on $\mathrm{HC}_2^\circ(k_m)$. This implies that the above surjection is an isomorphism and that $\oplus_{m<w<2m}\ell i_{m,w}$ is injective on it.

Example. Using the formula (3.4.1) above one can explicitly compute the additive dilogarithms. For example, $\ell i_{2,3} : B_2(k_2) \to k$ is given by

$$\ell i_{2,3}([s+at]) = -\frac{a^3}{2s^2(1-s)^2}.$$

The above theorem is an exact analog of Sydler's theorem which provides a solution to Hilbert's 3rd problem. This states that the scissors congruence class of a three-dimensional polyhedron is completely determined by its Dehn invariant and volume. In this context δ_m corresponds to the Dehn invariant map and $\oplus_{m<w<2m}\ell i_{m,w}$ is the sum of volumes of different $\star$-weights. When $m=2$, this analogy gets even more precise. In this case, there is only one dilogarithm of $\star$-weight 3, and the corresponding complex, which can be thought of as the deformation of the hyperbolic scissors complex, is analogous to the Euclidean scissors congruence complex and on this complex the volume map, which is the analog of the dilogarithm, scales by the cube of the dilation factor.

3.4.2. Comparison of $\Gamma_{k_m}(2)$ to $T_m\mathbb{Q}(2)(k)$ and $\beta_2(k)$. We first describe a subcomplex $\overline{\Gamma}^\circ_{k_m}(2)$ of $\Gamma^\circ_{k_m}(2)$. This is the complex whose degree 2 term is $k^\times \otimes (1+(t)) = k^\times \otimes (k_m^\times)^\circ \subseteq \Lambda^2 k_m^\times$, and degree 1 term is $\delta_m^{-1}(k^\times \otimes (k_m^\times)^\circ) \subseteq B_2^\circ(k_m)$. We denote this last group by $\overline{B}_2^\circ(k_m)$. Then the inclusion is a quasi-isomorphism from $\overline{\Gamma}^\circ_{k_m}(2)$ to $\Gamma^\circ_{k_m}(2)$ [28, Proposition 6.1.2]. In [28, Corollary 1.4.1], noticing that the terms in degree 2 are the same in both of the complexes and using the dilogarithm in degree 1 we deduce that the complexes $\overline{\Gamma}^\circ_{k_m}(2)$ and $T_m\mathbb{Q}(2)(k)$ are isomorphic.

Let $\tilde{\beta}_2(k)$ denote the $\mathbb{Q}[k^\times]$-module generated by $\langle a\rangle$, for $a \in k^\flat$, with the action of $\lambda \in k^\times$ on α by $\lambda \star \alpha$, subject to the relations generated by

$$\langle a\rangle - \langle b\rangle + a \star \left\langle \frac{b}{a}\right\rangle - (a-1)\star\left\langle \frac{1-b}{1-a}\right\rangle = 0,$$

$$(-1)\star\langle 1-a\rangle = -\langle a\rangle \qquad \text{and} \qquad \star\,\langle a^{-1}\rangle = -\langle a\rangle.$$

For $a \in k^\flat$, let $\langle a \rangle := a + a(1-a)t \in k_2$. Then we have the following relations,

$$\frac{\langle b \rangle}{\langle a \rangle} = a \star \left\langle \frac{b}{a} \right\rangle, \quad \frac{1-\langle a \rangle}{1-\langle b \rangle} = (b-1) \star \left\langle \frac{1-a}{1-b} \right\rangle, \quad \text{and} \quad 1-\langle a \rangle^{-1} = (1-a^{-1})(1-t),$$

and hence

$$\frac{1-\langle a \rangle^{-1}}{1-\langle b \rangle^{-1}} = \frac{1-a^{-1}}{1-b^{-1}}.$$

These imply, by the 5-term relation, that

$$[\langle a \rangle] - [\langle b \rangle] + a \star \left[\left\langle \frac{b}{a} \right\rangle\right] + (b-1) \star \left[\left\langle \frac{1-a}{1-b} \right\rangle\right] = 0$$

in $B_2^\circ(k_2)$. Since for any $x \in k_2^\flat$, $[1-x] = -[x]$ and $[x^{-1}] = -[x]$ in $B_2(k_2)$, we have $(-1) \star [\langle 1-a \rangle] = -[\langle a \rangle]$ and $a \star [\langle a^{-1} \rangle] = -[\langle a \rangle]$. These relations imply that the map that sends $\langle a \rangle$ to $[\langle a \rangle] \in B_2^\circ(k_2)$ factors through $\tilde{\beta}_2(k)$. There is also a natural surjection from $\tilde{\beta}_2(k)$ to $\beta_2(k)$, with $\beta_2(k)$ defined as in §3.1. These maps describe the 4-term functional equation of Cathelineau as a deformation of the standard 5-term functional equation computed on special elements, where one of the terms vanish since it has no infinitesimal part.

4. Infinitesimal Chow dilogarithm and the infinitesimal Bloch regulator

In this section, we will define variants of the additive dilogarithm in order to be able to construct regulators in different settings. The first section could be thought of as removing the restriction of considering only linear configurations when defining additive dilogarithms and is the essential step in being able to apply additive dilogarithms in an algebro geometric setting. In the second part, we describe the infinitesimal version of the Bloch regulator on curves, removing the restriction of being a curve. This is the infinitesimal version of the tame symbol construction of Deligne [13].

4.1. Infinitesimal Chow dilogarithm

In this section, we construct the infinitesimal analog of the Chow dilogarithm described in §2.3. The details of the construction are in [31]. We will only consider the case of k_2; the generalization of this construction to the higher modulus case is current work in progress. The specialization of this construction to the curve $\mathbb{P}^1_{k_2}$ and to the three linear fractional functions $1-z$, z and $1-\frac{a}{z}$ gives the additive dilogarithm $\ell i_{2,3}$ constructed in §3.4.1, [31, Lemma 3.5.1].

4.1.1. Construction of the infinitesimal Chow dilogarithm. In this section, we continue to assume that k is a field of characteristic 0. Let C_2 be a smooth and projective curve over k_2. We do *not* assume that C_2 comes as a product of a curve over k and k_2. Let $\underline{C}_2$ denote the fiber of C_2 over the closed point of $\mathrm{Spec}(k_2)$. Given c a (closed) point in $\underline{C}_2$, we call an element $\pi_{2,c} \in \mathcal{O}_{C_2,c}$ in the local ring of

C_2 at c a *uniformizer*, if its reduction is a uniformizer in the local ring of $\underline{C}_2$ at c. We call an element y in the local ring $k(C_2)$ of C_2 at its generic point, $\pi_{2,c}$-*good* if there exists $a \in \mathbb{Z}$ such that $y = \pi_{2,c}^a u$, for some unit u in $\mathcal{O}_{C_2,c}$. Note that $k(C_2)$ is an artin ring with residue field equal to the function field of $\underline{C}_2$. Fix a set $\mathscr{P}_2 := \{\pi_{2,c} | c \in \underline{C}_2\}$ of uniformizers. We say that y is $\mathscr{P}_2$-*good*, if it is $\pi_{2,c}$-good, for all $c \in \underline{C}_2$. Letting $k(C_2, \mathscr{P}_2)^\times \subseteq k(C_2)^\times$ denote the group of functions which are $\mathscr{P}_2$-good, the infinitesimal Chow dilogarithm ρ is a map $\rho : \Lambda^3 k(C_2, \mathscr{P}_2)^\times \to k$.

In the previous section, we defined the additive dilogarithm $\ell i_{2,3} : B_2(k_2) \to k$, by

$$\ell i_{2,3}([s + at]) = -\frac{a^3}{2s^2(1-s)^2},$$

and interpreted this function as the function induced by the composition

$$(\ell_2 \wedge \ell_1) \circ \delta_\infty : B_2(k_\infty) \to \Lambda^2 k_\infty^\times \to k$$

via the canonical map $B_2(k_\infty) \to B_2(k_2)$. Let us denote the map $\ell_2 \wedge \ell_1 : \Lambda^2 k_\infty^\times \to k$ by ℓ.

If $\mathcal{A}/k_\infty$ is a smooth algebra over k_∞ of relative dimension 1, c is a closed point of the spectrum of its reduction $\underline{\mathcal{A}}$ modulo (t), then we call an element $\tilde{\pi}_c$ of the local ring $\mathcal{A}$ at c, a uniformizer, if its reduction is a uniformizer in the corresponding local ring at $\underline{\mathcal{A}}$. we have similar notions of *goodness* with respect to $\tilde{\pi}_c$. If $\tilde{f}$, $\tilde{g}$ and $\tilde{h}$ are three functions in the local ring of $\mathcal{A}$ at the generic point of $\mathcal{A}$, which are $\tilde{\pi}_c$-good, then one can define their residue along $\tilde{\pi}_c$:

$$res_{\tilde{\pi}_c}(\tilde{f} \wedge \tilde{g} \wedge \tilde{h}) \in \Lambda^2(\mathcal{A}/(\tilde{\pi}_c))^\times.$$

As k_∞-algebras $\mathcal{A}/(\tilde{\pi}_c)$ is canonically isomorphic to k'_∞ for the finite extension k' of k which is the residue field of c. Therefore, we have a well-defined element $\ell(res_{\tilde{\pi}_c}(\tilde{f} \wedge \tilde{g} \wedge \tilde{h})) \in k'$ whose trace $\mathrm{Tr}_k(\ell(res_{\tilde{\pi}_c}(\tilde{f} \wedge \tilde{g} \wedge \tilde{h})))$ from k' to k will be essential in defining the local contribution to the Chow dilogarithm.

In case C_2/k_2 has a global lifting $\tilde{C}/k_\infty$ to a smooth and projective curve and f, g, and h have global liftings $\tilde{f}$, $\tilde{g}$, and $\tilde{h}$ to functions on $\tilde{C}$ which are good with respect to a system of uniformizers $\tilde{\mathscr{P}} := \{\tilde{\pi}_c | c \in |\tilde{C}_s|\}$ on $\tilde{C}$ that lift $\mathscr{P}_2$ then

$$\rho(f, g, h) = \sum_{c \in |\underline{C}_2|} \mathrm{Tr}_k(\ell(res_{\tilde{\pi}_c}(\tilde{f} \wedge \tilde{g} \wedge \tilde{h}))).$$

In general, we cannot expect such global liftings to exist. The method of defining ρ is then to choose a generic lifting of the curve and arbitrary liftings of the functions and for each point of the curve to choose also local liftings of the curve together with good local liftings of the functions and then to use the residues of a 1-form which measures the defect between choosing different models. We next describe this in detail.

The 1-form in question is defined as follows. We attach an element $\omega(\tilde{p}, \hat{p}, \chi) \in \Omega^1_{\underline{\hat{A}}/k}$ to the following data: smooth affine schemes $\tilde{\mathcal{A}}$, $\hat{\mathcal{A}}/k_\infty$ of relative dimension one, an isomorphism $\chi : \tilde{\mathcal{A}}/(t^2) \xrightarrow{\sim} \hat{\mathcal{A}}/(t^2)$ and triples of functions $\tilde{p} := (\tilde{f}, \tilde{g}, \tilde{h})$

in $\tilde{\mathcal{A}}^{\times}$ and $\hat{p} := (\hat{f}, \hat{g}, h)$ in $\hat{\mathcal{A}}^{\times}$, whose reductions modulo t^2 map to each other via χ. Let $\overline{\chi} : \tilde{\mathcal{A}} \xrightarrow{\sim} \hat{\mathcal{A}}$ be any lifting of χ, and $\varphi : \underline{\hat{\mathcal{A}}} \to \hat{\mathcal{A}}$ be any splitting of the canonical projection, which exist because of the smoothness assumptions. Denote by $\overline{\varphi}$ the corresponding isomorphism $\underline{\hat{\mathcal{A}}}[[t]] \xrightarrow{\sim} \hat{\mathcal{A}}$. Then we let:

$$\omega(\tilde{p}, \hat{p}, \chi) := \Omega(\overline{\varphi}^{-1}(\overline{\chi}(\tilde{p})), \overline{\varphi}^{-1}(\hat{p})),$$

with Ω as below.

Let $\tilde{q} = (\tilde{y}_1, \tilde{y}_2, \tilde{y}_3)$ and $\hat{q} = (\hat{y}_1, \hat{y}_2, \hat{y}_3)$, with $\tilde{y}_i, \hat{y}_i \in \underline{\hat{\mathcal{A}}}[[t]]^{\times}$, and $\hat{y}_i - \tilde{y}_i \in (t^2)$, for all $1 \leq i \leq 3$. Then we can write uniquely, $\hat{y}_i = \alpha_{0i} e^{t\alpha_{1i} + t^2 \hat{\alpha}_{2i} + \cdots}$ and $\tilde{y}_i = \alpha_{0i} e^{t\alpha_{1i} + t^2 \tilde{\alpha}_{2i} + \cdots}$, with $\alpha_{ji}, \hat{\alpha}_{ki}, \tilde{\alpha}_{ki} \in \underline{\hat{\mathcal{A}}}$, for $0 \leq j \leq 1$, $2 \leq k$, and $1 \leq i \leq 3$. We then define

$$\Omega(\tilde{q}, \hat{q}) := \sum_{\sigma \in S_3} (-1)^{\sigma} \alpha_{1\sigma(1)} (\tilde{\alpha}_{2\sigma(3)} - \hat{\alpha}_{2\sigma(3)}) \cdot d\log(\alpha_{0\sigma(2)}) \in \Omega^1_{\underline{\hat{\mathcal{A}}}/k}.$$

The definition of $\omega(\tilde{p}, \hat{p}, \chi)$ is then independent of all the choices involved.

Suppose that p is a triple (f, g, h) of functions on C_2 which are $\mathscr{P}_2$-good, i.e., in $k(C_2, \mathscr{P}_2)^{\times}$. In order to define $\rho(p)$, we first choose generic and local liftings of C_2 as follows. Let $\tilde{\mathcal{A}}$ be a generic lifting of C_2. More precisely, $\tilde{\mathcal{A}}/k_{\infty}$ is a smooth algebra together with an isomorphism $\alpha : \tilde{\mathcal{A}}/(t^2) \xrightarrow{\sim} \mathcal{O}_{C_2,\eta}$. Let $\tilde{p}_{\eta}$ be a triple of functions in $\tilde{\mathcal{A}}$, whose reductions modulo (t^2) map to the germs p_{η} of the functions p at η. For each $c \in |C_2|$, let $\widetilde{\mathcal{B}^{\circ}_c}$ be a local lifting of C_2 at c. In other words, $\widetilde{\mathcal{B}^{\circ}_c}/k_{\infty}$ is a smooth algebra together with an isomorphism $\tilde{\gamma}_c : \widetilde{\mathcal{B}^{\circ}_c}/(t^2) \xrightarrow{\sim} \hat{\mathcal{O}}_{C_2,c}$, from the reduction of $\widetilde{\mathcal{B}^{\circ}_c}$ modulo (t^2) to the completion of the local ring of C_2 at c. Let $\tilde{q}_c$ be a triple of functions on the localization of $\widetilde{\mathcal{B}^{\circ}_c}$ at the prime ideal (t), which map to the image of p via the map $\tilde{\gamma}_c$ and which are good with respect to a lift of the uniformizer on $\widetilde{\mathcal{B}^{\circ}_c}$. Because of this goodness assumption on $\tilde{q}_c$, its residue is well defined. We can add a term which measures the defect between the choices of the local liftings and the generic lifting and define the value of ρ on p as:

$$\rho(p) := \sum_{c \in |C|} \mathrm{Tr}_k(\ell(res_c(\tilde{q}_c)) + res_c \omega(\tilde{p}_{\eta}, \tilde{q}_c, \tilde{\gamma}_{c,\eta}^{-1} \circ \alpha_c)).$$

It turns out that this definition is independent of all the choices involved and define a map from $\Lambda^3 k(C_2, \mathscr{P}_2)^{\times}$ to k.

One can define a version of the Bloch group $B_2(k(C_2, \mathscr{P}_2))$ consisting of functions which are $\mathscr{P}_2$-good on C_2 as in [31, §3.3] and define a map

$$\Delta : B_2(k(C_2, \mathscr{P}_2)) \otimes k(C_2, \mathscr{P}_2)^{\times} \to \Lambda^3 k(C_2, \mathscr{P}_2)^{\times},$$

sending $[f] \otimes g$ to $(1 - f) \wedge f \wedge g$. This can then be sheafified, and using the residue map, made into a complex which computes the motivic cohomology group $K^{\circ}_3(C_2)^{(3)}_{\mathbb{Q}}$ as we described in §2.3 above, in the complex case. The infinitesimal Chow dilogarithm ρ and the additive dilogarithm in the previous section can then be joined together to define a regulator from $K^{\circ}_3(C_2)^{(3)}_{\mathbb{Q}}$ to k. We will construct and analyze this map in a future paper.

4.1.2. Goncharov's strong reciprocity conjecture in the infinitesimal case. The infinitesimal Chow dilogarithm allows us to state and prove an infinitesimal version of the strong reciprocity conjecture of Goncharov [18] with an explicit formula for the homotopy map. Let us first state the original version of the conjecture over a field which was proved recently by Rudenko [25].

Let C/k be a smooth and projective curve over an algebraically closed field k of characteristic 0. Taking the sum of the residue maps for all $c \in |C|$, we obtain a commutative diagram

$$\begin{array}{ccccc} B_3(k(C)) & \longrightarrow & B_2(k(C)) \otimes k(C)^\times & \xrightarrow{\Delta} & \Lambda^3 k(C)^\times \\ & & \Big\downarrow {\scriptstyle res_{|C|}} & & \Big\downarrow {\scriptstyle res_{|C|}} \\ & & B_2(k) & \xrightarrow{\quad\delta\quad} & \Lambda^2 k^\times . \end{array}$$

Suslin's reciprocity theorem implies that the image of the residue map from $\Lambda^3 k(C)^\times$ to $\Lambda^2 k^\times$ is in the image of δ. Goncharov's strong reciprocity conjecture states that the residue map between the complexes above is in fact homotopic to 0 with an explicit homotopy.

In the infinitesimal setting, we start with a smooth and projective curve C_2/k_2, where k is algebraically closed and of characteristic 0. We have the following commutative diagram:

$$\begin{array}{ccc} B_2(k(C_2, \mathscr{P}_2)) \otimes k(C_2, \mathscr{P}_2)^\times & \xrightarrow{\Delta} & \Lambda^3 k(C_2, \mathscr{P}_2)^\times \\ {\scriptstyle \oplus res_c} \Big\downarrow & & {\scriptstyle \oplus res_c} \Big\downarrow \\ \oplus_{c \in |\underline{C}_2|} B_2(k_2) & \xrightarrow{\delta} & \oplus_{c \in |C|} \Lambda^2 k_2^\times . \end{array}$$

By [31, Proposition 3.3.3], the composition $(\oplus \ell i_{2,3}) \circ (\oplus res_c)$ from $B_2(k(C_2, \mathscr{P}_2)) \otimes k(C_2, \mathscr{P}_2)^\times$ to k is equal to $\rho \circ \Delta$. Then the analog of Hilbert's third problem which determines structure of $B_2^\circ(k_2)$ in the previous section implies the following infinitesimal analog of Goncharov's strong reciprocity conjecture [31, Theorem 3.4.4].

Theorem. There is an explicit map $h : \Lambda^3 k(C_2, \mathscr{P}_2)^\times \to B_2(k_2)^\circ$ which makes the diagram

$$\begin{array}{ccc} B_2(k(C_2, \mathscr{P}_2)) \otimes k(C_2, \mathscr{P}_2)^\times & \xrightarrow{\Delta} & \Lambda^3 k(C_2, \mathscr{P}_2)^\times \\ \Big\downarrow {\scriptstyle res_{|\underline{C}_2|}} & \overset{h}{\swarrow} & \Big\downarrow {\scriptstyle res_{|\underline{C}_2|}} \\ B_2^\circ(k_2) & \xrightarrow{\quad\delta^\circ\quad} & (\Lambda^2 k_2^\times)^\circ \end{array}$$

commute and has the property that $h(k_2^\times \wedge \Lambda^2 k(C_2, \mathscr{P}_2)^\times) = 0$.

4.1.3. An infinitesimal invariant of cycles. The above construction gives an infinitesimal invariant of cycles of codimension 2 in the 3-dimensional space. We briefly describe this invariant in this section and refer to [31, §4] for the details. This invariant is a generalization of the invariant defined in [22]. The approach taken in

[22] for considering the infinitesimal part of the motivic cohomology of k_2 is to use the additive chow groups defined in [8], where one considers cycles on $\mathbb{A}^1_k$ which are close to the zero cycle with multiplicity 2 near the origin. In the approach taken here, we consider all cycles on $\mathbb{A}^1_k$, but identify them if they have the same reduction modulo (t^2). For the Milnor range, the additive cycle approach is the one taken in [26], whereas the analog of the approach of this section is the one in [23].

Let S denote $\mathrm{Spec}(k_\infty)$, with s being the closed and η the generic point, $\square_k := \mathbb{P}^1_k \setminus \{1\}$ and $\square^n_k$ the n-fold product of $\square_k$ with itself over k, with the coordinate functions $y_1, \ldots, y_n$. For a smooth k-scheme X, we let $\square^n_X := X \times_k \square^n_k$. Considering the free abelian group of *admissable* cycles, the cycles which intersect each of the faces properly, of codimension q on $\square^n_X$ for varying n, one gets a complex $(\underline{z}^q(X, \cdot), \partial)$. This complex considered modulo the complex of degenerate cycles is the Bloch's cubical higher Chow complex and its cohomology groups are Bloch's higher Chow groups which compute the motivic cohomology of X [5].

Let $\overline{\square}_k := \mathbb{P}^1_k$, $\overline{\square}^n_k$, the n-fold product of $\overline{\square}_k$ with itself over k, and $\overline{\square}^n_S := \overline{\square}^n_k \times_k S$. Let $\underline{z}^q_f(S, \cdot) \subseteq \underline{z}^q(S, \cdot)$ be the subgroup generated by integral, closed subschemes $Z \subseteq \square^n_S$ which are admissible and have *finite reduction*, i.e., $\overline{Z}$ intersects each $s \times \overline{F}$ properly on $\overline{\square}^n_S$, for every face F of $\square^n_k$. Modding out by degenerate cycles, we have the complex $z^q_f(S, \cdot)$.

An irreducible cycle p in $\underline{z}^2_f(S, 2)$ is given by a closed point p_η of $\square^2_\eta$ whose closure $\overline{p}$ in $\overline{\square}^2_S$ does not meet $(\{0, \infty\} \times \overline{\square}_S) \cup (\overline{\square}_S \times \{0, \infty\})$. Let $\tilde{p}$ denote the normalisation of $\overline{p}$ and $\{s_1, \ldots, s_m\}$ the closed fiber of $\tilde{p}$. We have surjections $\hat{\mathcal{O}}_{\tilde{p}, s_i} \to k(s_i)$. Since $k(s_i)/k$ is finite étale there is a unique splitting $\sigma_{\tilde{p}, s_i} : k(s_i) \to \hat{\mathcal{O}}_{\tilde{p}, s_i}$. We define $\log^\circ_{\tilde{p}, s_i} : \hat{\mathcal{O}}^\times_{\tilde{p}, s_i} \to \hat{\mathcal{O}}_{\tilde{p}, s_i}$, by

$$\log^\circ_{\tilde{p}, s_i}(y) = \log\Big(\frac{y}{\sigma_{\tilde{p}, s_i}(y(s_i))}\Big).$$

Let

$$l(p) := \sum_{1 \leq i \leq m} \mathrm{Tr}_k\Big(res_{\tilde{p}, s_i} \Big(\frac{1}{t^3} \big(\log^\circ_{\tilde{p}, s_i}(y_1) \cdot d\log(y_2) - \log^\circ_{\tilde{p}, s_i}(y_2) \cdot d\log(y_1) \big) \Big) \Big).$$

The infinitesimal invariant $\rho_f : \underline{z}^2_f(S, 3) \to k$ is then defined as the composition $l \circ \partial$. Since $\partial^2 = 0$, it is immediate that it vanishes on boundaries. The following property is the most essential property of ρ_f, which roughly states that ρ_f depends only on the reduction of Z modulo (t^2).

Suppose that Z_i for $i = 1, 2$ are two irreducible cycles in $\underline{z}^2_f(S, 3)$. We say that Z_1 and Z_2 are equivalent modulo t^m if the following condition (M_m) holds:

(i) $\overline{Z}_i/S$ are smooth with $(\overline{Z}_i)_s \cup (\cup_{j,a} |\partial^a_j Z_i|)$ a strict normal crossings divisor on $\overline{Z}_i$,

and more importantly

(ii) $\overline{Z}_1|_{t^m} = \overline{Z}_2|_{t^m}$.

Then we have the following theorem [31]:

Theorem. *If $Z_i \in \underline{z}^2_f(S,3)$, for $i=1,2$, satisfy the condition (M_2) then they have the same infinitesimal regulator value:*

$$\rho_f(Z_1) = \rho_f(Z_2).$$

Another essential property, which would justify calling ρ_f a regulator, is that it vanishes on products. More precisely, if $Z \in \underline{z}^2_f(S,3)$ and there is $1 \leq i \leq 3$ such that y_i restricted to $Z|_{t^2}$ is in $k_2^\times$ then $\rho_f(Z)=0$. After we take the quotient with degenerate cycles, mod (t^2) equivalence and boundaries, we expect ρ_f to be injective, but we are very far away from proving such a result.

4.2. Infinitesimal Bloch regulator

In this section, we briefly describe the infinitesimal version of the classical Bloch regulator described above in §2.4. Details of the construction will appear in [32]. Unlike the classical case, we do not need to restrict ourselves to the case of curves. Moreover, we do not need to assume that our schemes are smooth over a truncated polynomial ring.

We assume that X/k is a scheme over a field k of characteristic 0, and if $\underline{X} \hookrightarrow X$ is the scheme X together with the reduced induced structure then $\underline{X}/k$ is smooth and connected, the ideal sheaf $\mathcal{I}$ of $\underline{X}$ in X is square-zero, and is locally free, as a sheaf on $\underline{X}$.

There is a complex which computes the infinitesimal motivic cohomology of X of weight two. Namely, for a ring A, let $\Gamma_A(2)$ denote the complex $B_2(A) \to \Lambda^2 A^\times$, and if A comes together with a square-zero ideal I, we let $\Gamma^\circ_A(2)$ denote the cone of the map $\Gamma_A(2) \to \Gamma_{\underline{A}}(2)$, with $\underline{A} := A/I$. Even though $\Gamma^\circ_A(2)$ depends on I, we suppress this dependence in the notation since I will be fixed in what follows. This complex is quasi-isomorphic to the complex $B^\circ_2(A) \to (\Lambda^2 A^\times)^\circ$. Sheafifying this we obtain the complex of sheaves $\Gamma^\circ_X(2)$ on X. Let $F\Gamma^\circ_X(2) \subseteq \Gamma^\circ_X(2)$ be the subcomplex which agrees with $\Gamma^\circ_X(2)$ in degree 1 and is the image of δ, in degree 2. In other words, it is the subcomplex

$$\begin{array}{ccc} B^\circ_2(\mathcal{O}_X) & \xrightarrow{\delta} & \delta(B^\circ_2(\mathcal{O}_X)) \\ \downarrow & & \downarrow \\ B^\circ_2(\mathcal{O}_X) & \xrightarrow{\delta} & (\Lambda^2\mathcal{O}_X^\times)^\circ. \end{array}$$

The analog of the Bloch regulator in this case is the following construction. We have regulators:

$$\rho_1 : \mathrm{H}^2(X, \Gamma^\circ_X(2)) \to \mathrm{H}^0(X, \Omega^1_X/d\mathcal{O}_X)^\circ$$

and

$$\rho_2 : \mathrm{H}^2(X, F\Gamma^\circ_X(2)) = \ker(\rho_1) \to \mathrm{H}^1(X, D_1(\mathcal{O}_X)).$$

The first regulator ρ_1 is defined as follows. On a ring A, with a square-zero ideal I as above, we define

$$\log d\log : (\Lambda^2 A^\times)^\circ \to (\Omega^1_A/dA)^\circ$$

by sending $a \wedge b$ with $a \in (A^\times)^\circ = 1 + I$, and $b \in A^\times$ to $\log(a) d\log(b)$.

The map is well defined and vanishes on the boundaries coming from $B_2^\circ(A)$. Therefore it can be sheafified to obtain the map ρ_1.

The more interesting part of the regulator is ρ_2. First let us give the local construction, then we will show how to globalize this construction using a homotopy map. For the local version, we give two equivalent constructions in [32], one of them computational, the other one conceptual. In this survey, we only describe the computational one since it is shorter.

Suppose that A is a k-algebra with a square-zero ideal I as above, such that $\underline{A}/k$ is smooth. The smoothness assumption implies that there is a splitting $\tau : \underline{A} \to A$ of the canonical projection. We will define a branch of the dilogarithm corresponding to this splitting. First, using the splitting τ we regard A as an $\underline{A}$-algebra. Then we express A as a quotient $B \twoheadrightarrow A$ of a smooth $\underline{A}$-algebra B. Let $\hat{B}$ denote the completion of B along the kernel of this map, $\hat{\tau}$ denote the structure map from $\underline{A}$ to $\hat{B}$, $\hat{J}$ be the kernel of the projection $\hat{B} \twoheadrightarrow A$, and $\hat{I}$ be the inverse image of I in $\hat{B}$. Since $I^2 = 0$, we have $\hat{I}^2 \subseteq \hat{J}$. Given this presentation, the first André–Quillen homology $D_1(A)$ of A relative to $\mathbb{Q}$ is given by $D_1(A) = \ker(\hat{J}/\hat{J}^2 \xrightarrow{d} \Omega^1_{\hat{B}}/\hat{J}\Omega^1_{\hat{B}})$.

We define a map $\ell i_{2,\tau} : \mathbb{Q}[A^\flat] \to D_1(A)$, by sending $[a]$ to

$$-\frac{1}{2}\frac{(\tilde{a}-\hat{\tau}(\underline{a}))^3}{\hat{\tau}(\underline{a})^2(\hat{\tau}(\underline{a})-1)^2} \in \ker(\hat{J}/\hat{J}^2 \xrightarrow{d} \Omega^1_{\hat{B}}/\hat{J}\Omega^1_{\hat{B}}),$$

where $\hat{\underline{a}} := \hat{\tau}(\underline{a})$, with $\underline{a}$ is the image of a under the map $A \twoheadrightarrow \underline{A}$, and $\tilde{a}$ is any lifting of $a \in A$ to an element in $\hat{B}$. It turns out that the definition is independent of the lifting of a to an element of $\hat{B}$ and that there is a natural commutative diagram corresponding to different presentations of A as quotients of smooth $\underline{A}$-algebras. From the definition it is immediate that $\ell i_{2,\tau}$ vanishes on the image of $\tau(\underline{A}^\flat)$. Moreover, it satisfies the 5-term relation and hence descends to a map $\ell i_{2,\tau} : B_2(A) \to D_1(A)$.

In order to compare dilogarithms corresponding to different splittings, it is necessary to restrict to the subgroup $B_2^\circ(A)$ of $B_2(A)$. Again there is a more conceptual description of this homotopy map, and again we are going to take the explicit approach.

Suppose that A_i, for $i = 1, 2$ are k-algebras, with square-zero ideals I_i, which are locally free $\underline{A}_i$-modules. Suppose that $f : A_1 \to A_2$ is a k-algebra homomorphism and that

$$\tau_1 : \underline{A}_1 \to A_1 \qquad \text{and} \qquad \tau_2 : \underline{A}_2 \to A_2,$$

are splittings which are *not* necessarily compatible with f. The homotopy, that we mentioned above, in this context is a map

$$h_f(\tau_1, \tau_2) : F(\Lambda^2 A_1^\times)^\circ \to D_1(A_2),$$

with the property that, for every $\alpha \in B_2^\circ(A_1)$,

$$\ell i_{2,\tau_2}(f(\alpha)) - f_*(\ell i_{2,\tau_1}(\alpha)) = h_f(\tau_1, \tau_2)(\delta(\alpha)).$$

This map is a measure of the difference between $f \circ \tau_1$ and $\tau_2 \circ \underline{f}$, where $\underline{f} : \underline{A}_1 \to \underline{A}_2$ is the induced map. We give the definition of $h_f(\tau_1, \tau_2)$ below. First, note that the map $\theta : \underline{A}_1 \to I_2$, given by $\theta(a) := f(\tau_1(a)) - \tau_2(\underline{f}(a))$ is an $\underline{f}$-derivation. Let $\varphi : I_1 \to S^2_{\underline{A}_2} I_2$ be any additive map such that for all $a \in \underline{A}_1$ and $\alpha \in I_1$:

$$\varphi(a\alpha) = \theta(a) \otimes f(\alpha) + \underline{f}(a)\varphi(\alpha).$$

Since I_1 is by assumption a locally free $\underline{A}_1$-module, such a map exists locally. Let

$$H_\varphi : (\Lambda^2 A_1^\times)^\circ \to S^3_{\underline{A}_2} I_2$$

be the map that sends $(1+\alpha)\wedge(1+\beta)$ with $\alpha, \beta \in I_1$ to $f(\alpha)\otimes\varphi(\beta) - \varphi(\alpha)\otimes f(\beta)$, and sends $(1+\alpha)\wedge\tau_1(a)$, with $\alpha \in I_1$ and $a \in \underline{A}_1^\flat$, to $-\varphi(\alpha)\otimes \frac{\theta(a)}{\underline{f}(a)}$. The restriction of $F(\Lambda^2 A_1^\times)^\circ$ does not depend on φ and the image lands in $D_1(A_2)$, if we use the presentation $S^\cdot_{\underline{A}_2} I_2 \to A_2$ to compute $D_1(A_2)$. With these identifications, $h_f(\tau_1, \tau_2)$ is the restriction of $-\frac{3}{2}H_\varphi$ to $F(\Lambda^2 A_1^\times)^\circ$.

This can now be used to define ρ_2. Let $\{U_i\}_{i\in I}$ be an open affine cover of X and τ_i be splittings of $\underline{U}_i \hookrightarrow U_i$. Let $\{a_{ij}\}_{i,j\in I}$ be local sections of $\underline{B}_2^\circ$ on U_{ij} and $\{b_i\}_{i\in I}$ be local sections of $F(\Lambda^2 \mathcal{O}_X^\times)^\circ$ on U_i such that $\delta(a_{ij}) = b_j|_{U_{ij}} - b_i|_{U_{ij}}$, and $a_{jk}|_{U_{ijk}} - a_{ik}|_{U_{ijk}} + a_{ij}|_{U_{ijk}} = 0$. This defines an element of $\mathrm{H}^2(X, F\Gamma^\circ_X(2))$.

Consider the elements

$$\gamma_{ij} := \ell i_{2,\tau_i}(a_{ij}) + h(\tau_i, \tau_j)(b_j) \in D_1(\mathcal{O}_X(U_{ij}))),$$

for each i, $j \in I$. These define a cocycle which gives the element in $\mathrm{H}^1(X, D_1(\mathcal{O}_X))$ which is the image of $(\{a_{ij}\}_{i,j\in I}, \{b_i\}_{i\in I})$ under ρ_2. This element does not depend on any of the choices made. Using Goodwillie's theorem, we also prove in [32] that the map ρ_2 is injective. Therefore together with ρ_1, they describe the motivic cohomology group $\mathrm{H}^2(X, \Gamma^\circ_X(2))$ completely.

5. Complements

In this last section, we describe some results which are very much incomplete: first the case of higher weights, then the case of characteristic p. In the last section, we discuss some open problems.

5.1. Additive polylogarithms of higher weight

In [29], we constructed an analog of the single-valued n-polylogarithms $\mathscr{L}_n$ of [3] described in §3.2.2 above. These functions, which we denote by li_n are the higher weight analogs of the functions defined in §3.4. In this higher weight case, so far we can define these functions only in modulus (t^2), i.e., for k_2. They should, of course, exist for all k_m.

Theorem. *For $s + at \in k_2^\flat$, let us define*

$$li_n(s+at) = \frac{(-1)^n}{n!}(a/s)^{2n-1}\log\left(\frac{1-se^u}{1-s}\right)^{(n)}(0),$$

where the derivative is with respect to u. If $A.(k_2)$ has a comultiplication Δ such that $\Delta_{n-1,1}(\{x\}_n) = \{x\}_{n-1} \otimes x$ for $n \geq 3$ and $\Delta_{1,1}(\{x\}_n) = (1-x) \wedge x$ then the above function descends to give a map $B'_n(k_2) \to k$.

We also proved that these infinitesimal polylogarithms satisfy the functional equations that the ordinary polylogarithms satisfy in [29]. One should in principle be able to define such functions on all of $A_n(k_2)$, rather than only on the Bloch group part.

5.2. Partial results in characteristic p

In the previous sections we assumed the base field k to be of characteristic 0. We would expect similar constructions in characteristic p. We first note that in this section we do not assume our complexes to be tensored with $\mathbb{Q}$. Otherwise, most of the objects in question will be equal to 0. Therefore, for any local ring with infinite residue field A, we let $B_2(A)_{\mathbb{Z}}$ denote the free abelian group generated by $[x]$, with $x(1-x) \in A^\times$ modulo the subgroup generated by (2.1.1) with $xy(1-x)(1-y)(1-\frac{x}{y}) \in A^\times$. This gives a complex $B_2(A)_{\mathbb{Z}} \to \Lambda^2_{\mathbb{Z}} A^\times$ of abelian groups. We will explore this complex for k_2, when k is of characteristic p.

More specifically, fix $p \geq 5$ and let $\mathbb{F}$ denote an algebraic closure of the field with p elements. In the following, we let $\mathbb{F}_m := \mathbb{F}[t]/(t^m)$. In particular, $\mathbb{F}_p = \mathbb{F}[t]/(t^p)$, and *not* the field with p elements. The additive dilogarithm $\ell i_{2,3}$, with the same formula, defines a map $\ell i_{2,3} : B_2(\mathbb{F}_2) \to \mathbb{F}$ of $\star$-weight 3. In characteristic p, there is another additive dilogarithm of $\star$-weight 1 which does *not* come from characteristic 0. Recall that a finite version of the logarithm, called the $1\frac{1}{2}$-logarithm was defined by Kontsevich [21] as:

$$\pounds_1(s) = \sum_{1 \leq k \leq p-1} \frac{s^k}{k},$$

for $s \in \mathbb{F}$. This functions satisfies: $\pounds_1(x) = -x^p \pounds_1\left(\frac{1}{x}\right)$, $\pounds_1(x) = \pounds_1(1-x)$, and

$$\pounds_1(x) - \pounds_1(y) + x^p \pounds_1\left(\frac{y}{x}\right) + (1-x)^p \pounds_1\left(\frac{1-y}{1-x}\right) = 0.$$

Therefore $\pounds_1^{1/p}$ satisfies the 4-term functional equation of the entropy function. If we let

$$\mathfrak{Li}_2([s+\alpha t]) := \frac{\alpha}{s(1-s)} \sum_{1 \leq k \leq p-1} \frac{s^{k/p}}{k},$$

then we have [30]:

Theorem. *$\mathfrak{Li}_2$ descends to give a map*

$$\mathfrak{Li}_2 : B_2^\circ(\mathbb{F}_2) \to \mathbb{F}.$$

and together with $\ell i_{2,3}$ they give the regulator from $B_2^\circ(\mathbb{F}_2)$ to $\mathbb{F} \oplus \mathbb{F}$ which gives an isomorphism $K_3^\circ(\mathbb{F}_2) \to B_2^\circ(\mathbb{F}_2) \to \mathbb{F} \oplus \mathbb{F}$.

Surprisingly Kontsevich's logarithm can be obtained using δ over a truncated polynomial ring of higher modulus in an analogous manner that $\ell i_{2,3}$ can be obtained by using δ on the Bloch complex over $\mathbb{F}_3$. However, in order to obtain $\mathfrak{Li}_2$, one needs to use a much higher modulus, namely one needs to lift the elements to $\mathbb{F}_p$.

Using the notation in §3.4.1, for each $1 \le i < p$, we have a map $\ell_i : \mathbb{F}_p^\times \to \mathbb{F}$ and a commutative diagram:

$$\begin{array}{ccc} B_2(\mathbb{F}_p)_{\mathbb{Z}} & \xrightarrow{\delta_p} & \Lambda^2_{\mathbb{Z}}\mathbb{F}_p^\times \\ \downarrow & & \downarrow {\scriptstyle -\frac{1}{2}(\sum_{1\le i<p} a\cdot\ell_a\wedge\ell_{p-a})^{1/p}} \\ B_2(\mathbb{F}_2)_{\mathbb{Z}} & \xrightarrow{\mathfrak{Li}_2} & \mathbb{F}, \end{array}$$

which expresses $\mathfrak{Li}_2$ in the manner we were looking for.

We would like to mention [4] for an approach to finite polylogarithms that relates them to p-adic polylogarithms and [14] for the relation of functional equations of finite polylogarithms to those of the classical polylogarithms.

5.3. Further problems

As can be seen from the discussion above, the above theory is only the starting point of a general theory of infinitesimal regulators. There are many open questions, some of which will be considered in future papers.

In the linear part of the question, the most fundamental one is that of defining the maps

$$vol^\circ_{n,w} : A_n(k_m)/P_n(k_m) \to k$$

of $\star$-weight w, for each $(n-1)m < w < nm$. These maps are defined above for $n = 2$. They are also defined on the subspace $B'_n(k_2)$ when $m = 2$. These are the analogs of the volume maps. Using these maps, one would then try to construct maps from each cohomology group of the complex $A.(k_m)$ to various Ω^i_k. One expects, by Goodwillie's theorem and by the computation of the cyclic homology of truncated polynomial algebras that the combination of these regulators mapping to the direct sum of $(m-1)$-copies of Ω^i_k gives an isomorphism from the infinitesimal part of the corresponding cohomology group.

Solving the linear part of the above problem, we expect that one could use these maps to define regulators for smooth projective schemes X/k_m. This would be the generalization of the construction of the infinitesimal Chow dilogarithm. One would have infinitesimal invariants of higher Chow groups which are expected to give all the infinitesimal invariants of the Chow groups. This last part would require significantly new ideas.

Another main problem is to do all of the above constructions in characteristic p. As we saw above in characteristic p there are significantly more regulators. An essential computation in the Milnor case is done by Rülling in [26] in the context of the additive Chow groups. In this theory, one would have to use the residue construction in the de Rham–Witt complex rather than the ordinary de Rham complex.

Finally, some the aspects of the construction can be done for any artin algebra over a field. This was done in the section on infinitesimal Bloch regulator in weight two. The aim would be to generalize the above to all artin algebras over a field. For the mixed characteristic case, let us say for $W_m(k)$, the case of truncated Witt vectors over a perfect field k, some of the regulators are of the above form. One could aim to study them using the methods above.

References

[1] A. Beilinson. *Higher regulators and values of L-functions.* Current problems in mathematics, Vol. 24, 181–238, Itogi Nauki i Tekhniki, Akad. Nauk SSSR, Vsesoyuz. Inst. Nauchn. i Tekhn. Inform. (1984).

[2] A. Beilinson, A. Goncharov, V. Schechtman, A. Varchenko. *Aomoto dilogarithms, mixed Hodge structures and motivic cohomology of pairs of triangles on the plane.* Grothendieck Festschrift vol. 1, Progr. Math. 86 (1990), 135–172.

[3] A. Belinson, P. Deligne. *Interprétation motivique de la conjecture de Zagier reliant polylogarithmes et régulateurs.* Motives (Seattle, WA, 1991), 97–121, Proc. Sympos. Pure Math., 55, Part 2, Amer. Math. Soc. (1994).

[4] A. Besser. *Finite and p-adic polylogarithms.* Compositio Math. 130 (2002) 215–223.

[5] S. Bloch. *Algebraic cycles and higher K-theory*, Adv. Math., 61, (1986), no. 3, 267–304.

[6] S. Bloch. *Higher regulators, algebraic K-theory, and zeta functions of elliptic curves.* CRM Monograph Series, 11, (2000).

[7] S. Bloch, H. Esnault. *The additive dilogarithm*, Doc. Math., Extra Vol., (2003), 131–155.

[8] S. Bloch, H. Esnault. *An additive version of higher Chow groups.* Ann. Sci. École Norm. Sup. (4) 36 (2003), no. 3, 463–477.

[9] J.-L. Cathelineau. *Sur l'homologie de SL_2 à coefficients dans l'action adjointe.* Math. Scand. 63 (1988), 51–86.

[10] J.-L. Cathelineau. *Remarques sur les différentielles des polylogarithmes uniformes.* Ann. Inst. Fourier 46 (1996), no. 5, 1327–1347.

[11] J.-L. Cathelineau. *λ-structures in algebraic K-theory and cyclic homology.* K-Theory 4 (1991), no. 6, 591–606.

[12] J.-L. Cathelineau. *Infinitesimal dilogarithms, extensions and cohomology.* J. of Algebra 332 (2011), 87–113.

[13] P. Deligne. *Le symbole modéré.* Inst. Hautes Études Sci. Publ. Math. No. 73 (1991), 147–181.

[14] P. Elbaz-Vincent, H. Gangl. *On poly(ana)logs. I.* Compos. Math. 130 (2) (2002) 161–210.

[15] A. Goncharov. *Geometry of configurations, polylogarithms, and motivic cohomology.* Adv. Math. 114 (1995), 197–318.

[16] A. Goncharov. *Volumes of hyperbolic manifolds and mixed Tate motives.* J. Amer. Math. Soc. 12 (1999), no. 2, 569–618.

[17] A. Goncharov. *Euclidean scissor congruence groups and mixed Tate motives over dual numbers.* Math. Res. Letters 11 (2004), 771–784.

[18] A. Goncharov. *Polylogarithms, regulators, and Arakelov motivic complexes.* J. Amer. Math. Soc., 18 (2005), no. 1, 1–60.

[19] T. Goodwillie. *Relative algebraic K-theory and cyclic homology.* Ann. of Math. (2) 124 (1986), no. 2, 347–402.

[20] M. Green, P. Griffiths. *The regulator map for a general curve.* Symposium in Honor of C.H. Clemens (Salt Lake City, UT, 2000), Contemporary Mathematics 312 (2002), 117–127.

[21] M. Kontsevich. *The* $1\frac{1}{2}$ *-logarithm.* Appendix to On poly(ana)logs. I by P. Elbaz-Vincent and H. Gangl, see [14].

[22] J. Park. *Regulators on additive higher Chow groups*, Amer. J. Math. 131 (2009), no. 1, 257–276.

[23] J. Park, S. Ünver. *Motivic cohomology of fat points in Milnor range.* Doc. Math. 23 (2018), 759–798.

[24] D. Ramakrishnan. *Regulators, algebraic cycles, and values of L-functions.* Algebraic K-theory and algebraic number theory (Honolulu, HI, 1987), Contemporary Mathematics 83 (1989), 183–310.

[25] D. Rudenko. *Scissor congruence and Suslin reciprocity law.* Preprint. arXiv:1511.00520

[26] K. Rülling. *The generalized de Rham–Witt complex over a field is a complex of zero cycles.* J. Algebraic Geom. 16 (2007) 109–169.

[27] A. Suslin, *K_3 of a field, and the Bloch group.* Trudy Mat. Inst. Steklov. 183 (1990), 180–199, 229. In Russian; translated in Proc. Steklov Inst. Math. 4 (1991), 217–239.

[28] S. Ünver. *On the additive dilogarithm.* Algebra Number Theory 3:1 (2009), 1–34.

[29] S. Ünver. *Additive polylogarithms and their functional equations.* Math. Ann. 348 (2010), no. 4, 833–858.

[30] S. Ünver. *Deformations of Bloch groups and Aomoto dilogarithms in characteristic p.* J. Number Theory 131 (2011), no. 8, 1530–1546.

[31] S. Ünver. *Infinitesimal Chow dilogarithm.* J. Algebraic Geom. (to appear). DOI: 10.1090/jag/746

[32] S. Ünver. *Infinitesimal Bloch regulator.* J. Algebra 559 (2020), 203–225.

[33] D. Zagier. *The Bloch–Wigner–Ramakrishnan polylogarithm function.* Math. Ann. 286 (1990), no. 1-3, 613–624.

[34] D. Zagier. *The dilogarithm function.* Frontiers in number theory, physics, and geometry. II, 3–65, Springer, (2007).

Sinan Ünver
Koç University, Mathematics Department
Rumelifeneri Yolu, 34450, Istanbul, Turkey
e-mail: sunver@ku.edu.tr

Printed by Printforce, the Netherlands